The Statisti

A Course in Methods of Data Analysis
Custom Edition for University of Connecticut

Third Edition

Fred L. Ramsey | Daniel W. Schafer

CENGAGE Learning®

Australia • Brazil • Japan • Korea • Mexico • Singapore • Spain • United Kingdom • United States

CENGAGE Learning

The Statistical Sleuth: A Course in Methods of Data Analysis, Custom Edition for University of Connecticut, Third Edition

Senior Project Development Manager:
Linda deStefano

Market Development Manager:
Heather Kramer

Senior Production/Manufacturing Manager:
Donna M. Brown

Production Editorial Manager:
Kim Fry

Sr. Rights Acquisition Account Manager:
Todd Osborne

Printed in the United States of America

The Statistical Sleuth, A Course in Methods of Data Analysis, Custom Edition for University of Connecticut, Third Edition
Ramsey | Schafer

STATISTICAL SLEUTH 2E
RAMSEY/SCHAFER

ISBN-13: 978-1-305-02430-4

ISBN-10 1-305-02430-3

Cengage Learning
5191 Natorp Boulevard
Mason, Ohio 45040
USA

Cengage Learning is a leading provider of customized learning solutions with office locations around the globe, including Singapore, the United Kingdom, Australia, Mexico, Brazil, and Japan. Locate your local office at:
international.cengage.com/region.
Cengage Learning products are represented in Canada by Nelson Education, Ltd.
For your lifelong learning solutions, visit **www.cengage.com/custom.**
Visit our corporate website at **www.cengage.com.**

Brief Contents

The Statistical Sleuth, A Course in Methods of Data Analysis, Custom Edition for University of Connecticut, Third Edition
Fred L. Ramsey/Daniel W. Schafer

Chapters 1 – 12, 18 - 22 excerpted from:
Statistical Sleuth, A Course In Methods of Data Analysis
Third Edition
Fred L. Ramsey/Daniel W. Schafer

Appendix A reprinted from:
Statistical Sleuth, A Course In Methods of Data Analysis
Second Edition
Fred L. Ramsey/Daniel W. Schafer

Contents

CHAPTER 1

Drawing Statistical Conclusions

Statistical sleuthing means carefully examining data to answer questions of interest. This book is about the process of statistical sleuthing, including strategies and tools for answering questions of interest and guidelines for interpreting and communicating results.

This chapter is about interpreting statistical results—a process crucially linked to study design. The setting for this and the next three chapters is the two-sample problem, where it is convenient to illustrate concepts and strategies employed in all statistical analysis.

An answer to a question is accompanied by a statistical measure of uncertainty, which is based on a probability model. When that probability model is based on a chance mechanism—like the flipping of coins to decide which subjects get which treatment or the drawing of lottery tickets to decide which members of a population get selected in a sample—measures of uncertainty and the inferences drawn from them are formally justified. Often, however, chance mechanisms are invented as conceptual frameworks for drawing statistical conclusions. For understanding and communicating statistical conclusions it is important to understand whether a chance mechanism was used and, if so, whether it was used for sample selection, group allocation, or both.

1.1 CASE STUDIES

1.1.1 Motivation and Creativity—A Randomized Experiment

Do grading systems promote creativity in students? Do ranking systems and incentive awards increase productivity among employees? Do rewards and praise stimulate children to learn? Although reward systems are deeply embedded in schools and in the workplace, a growing body of evidence suggests that rewards may operate in precisely the opposite way from what is intended.

A remarkable demonstration of this phenomenon was provided by psychologist Teresa Amabile in an experiment concerning the effects of intrinsic and extrinsic motivation on creativity. Subjects with considerable experience in creative writing were randomly assigned to one of two treatment groups: 24 of the subjects were placed in the "intrinsic" treatment group, and 23 in the "extrinsic" treatment group, as indicated in Display 1.1. The "intrinsic" group completed the questionnaire at the top of Display 1.2. The questionnaire, which involved ranking intrinsic reasons for writing, was intended as a device to establish a thought pattern concerning intrinsic motivation—doing something because doing it brings satisfaction. The "extrinsic" group completed the questionnaire at the bottom of Display 1.2, which was used as a device to get this group thinking about extrinsic motivation—doing something because a reward is associated with its completion.

DISPLAY 1.1 Creativity scores in two motivation groups, and their summary statistics

	Intrinsic group		Extrinsic group	
	12.0	20.5	5.0	17.4
	12.0	20.6	5.4	17.5
	12.9	21.3	6.1	18.5
	13.6	21.6	10.9	18.7
	16.6	22.1	11.8	18.7
	17.2	22.2	12.0	19.2
	17.5	22.6	12.3	19.5
	18.2	23.1	14.8	20.7
	19.1	24.0	15.0	21.2
	19.3	24.3	16.8	22.1
	19.8	26.7	17.2	24.0
	20.3	29.7	17.2	
Sample Size:	24		23	
Average:	19.88		15.74	
Sample Standard Deviation:	4.44		5.25	

After completing the questionnaire, all subjects were asked to write a poem in the Haiku style about "laughter." All poems were submitted to 12 poets, who evaluated them on a 40-point scale of creativity, based on their own subjective views. Judges were not told about the study's purpose. The average ratings given by the 12 judges are shown for each of the study subjects in Display 1.1. (Data based

DISPLAY 1.2 Questionnaires given creative writers, to rank intrinsic and extrinsic reasons for writing

INSTRUCTIONS: Please rank the following list of reasons for writing, in order of personal importance to you (1 = highest, 7 = lowest).

__ You get a lot of pleasure out of reading something good that you have written.
__ You enjoy the opportunity for self-expression.
__ You achieve new insights through your writing.
__ You derive satisfaction from expressing yourself clearly and eloquently.
__ You feel relaxed when writing.
__ You like to play with words.
__ You enjoy becoming involved with ideas, characters, events, and images in your writing.

List of ***extrinsic*** *reasons for writing.*

List of ***intrinsic*** *reasons for writing.*

INSTRUCTIONS: Please rank the following list of reasons for writing, in order of personal importance to you (1 = highest, 7 = lowest).

__ You realize that, with the introduction of dozens of magazines every year, the market for free-lance writing is constantly expanding.
__ You want your writing teachers to be favorably impressed with your writing talent.
__ You have heard of cases where one best-selling novel or collection of poems has made the author financially secure.
__ You enjoy public recognition of your work.
__ You know that many of the best jobs available require good writing skills.
__ You know that writing ability is one of the major criteria for acceptance into graduate school.
__ Your teachers and parents have encouraged you to go into writing.

on the study in T. Amabile, "Motivation and Creativity: Effects of Motivational Orientation on Creative Writers," *Journal of Personality and Social Psychology* 48(2) (1985): 393–99.) Is there any evidence that creativity scores tend to be affected by the type of motivation (intrinsic or extrinsic) induced by the questionnaires?

Statistical Conclusion

This experiment provides strong evidence that receiving the "intrinsic" rather than the "extrinsic" questionnaire caused students in this study to score higher on poem creativity (two-sided p-value $= 0.005$ from a two-sample t-test as an approximation to a randomization test). The estimated treatment effect—the increase in score attributed to the "intrinsic" questionnaire—is 4.1 points (95% confidence interval: 1.3 to 7.0 points) on a 0–40-point scale.

Note: The statistical conclusions associated with the case studies in each chapter use tools and, in some cases, terms that are introduced later. You should read the conclusions initially to appreciate the scientific statements that can be drawn from the chapter's tools, without worrying about the parts you don't understand, and then return to them again after reading the chapter for a more complete understanding.

Scope of Inference

Since this was a randomized experiment, one may infer that the difference in creativity scores was *caused* by the difference in motivational questionnaires. Because the subjects were not selected randomly from any population, extending this inference to any other group is speculative. This deficiency, however, is minor; the causal conclusion is strong even if it applies only to the recruited subjects.

1.1.2 Sex Discrimination in Employment—An Observational Study

Did a bank discriminatorily pay higher starting salaries to men than to women? The data in Display 1.3 are the beginning salaries for all 32 male and all 61 female skilled, entry-level clerical employees hired by the bank between 1969 and 1977. (Data from a file made public by the defense and described by H. V. Roberts, "Harris Trust and Savings Bank: An Analysis of Employee Compensation" (1979), Report 7946, Center for Mathematical Studies in Business and Economics, University of Chicago Graduate School of Business.)

DISPLAY 1.3 Starting salaries ($U.S.) for 32 male and 61 female clerical hires at a bank

Males			Females					
4,620	5,700	6,000	3,900	4,500	4,800	5,220	5,400	5,640
5,040	6,000	6,000	4,020	4,620	4,800	5,220	5,400	5,700
5,100	6,000	6,000	4,290	4,800	4,980	5,280	5,400	5,700
5,100	6,000	6,300	4,380	4,800	5,100	5,280	5,400	5,700
5,220	6,000	6,600	4,380	4,800	5,100	5,280	5,400	5,700
5,400	6,000	6,600	4,380	4,800	5,100	5,400	5,400	5,700
5,400	6,000	6,600	4,380	4,800	5,100	5,400	5,400	6,000
5,400	6,000	6,840	4,380	4,800	5,100	5,400	5,520	6,000
5,400	6,000	6,900	4,440	4,800	5,100	5,400	5,520	6,120
5,400	6,000	6,900	4,500	4,800	5,160	5,400	5,580	6,300
	6,000	8,100						6,300

Statistical Conclusion

As evident in Display 1.4, the data provide convincing evidence that the male mean is larger than the female mean (one-sided p-value < 0.00001 from a two-sample t-test). The male mean exceeds the female mean by an estimated $818 (95% confidence interval: $560 to $1,076).

DISPLAY 1.4 Histograms for male and female starting salaries

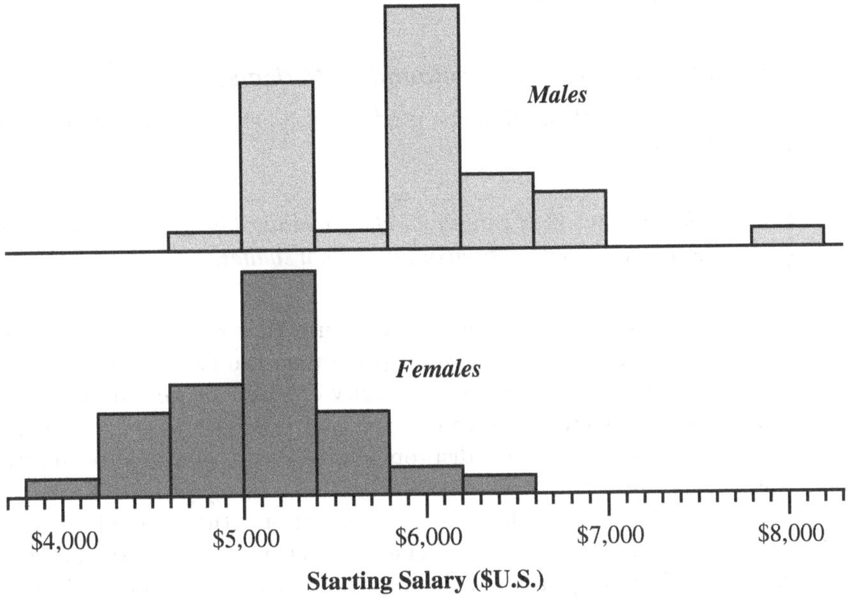

Scope of Inference

Although there is convincing evidence that the males, as a group, received larger starting salaries than the females, the statistics alone cannot address whether this difference is attributable to sex discrimination. The evidence is consistent with discrimination, but other possible explanations cannot be ruled out; for example, the males may have had more years of previous experience.

1.2 STATISTICAL INFERENCE AND STUDY DESIGN

The inferences one may draw from any study depend crucially on the study's design. Two distinct forms of inference—causal inference and inference to populations—can be justified by the proper use of random mechanisms.

1.2.1 Causal Inference

In a *randomized experiment*, the investigator controls the assignment of experimental units to groups and uses a chance mechanism (like the flip of a coin) to make the assignment. The motivation and creativity study is a randomized experiment, because a random mechanism was used to assign the study subjects to the two motivation groups. In an *observational study*, the group status of the subjects is established beyond the control of the investigator. The sex discrimination study is observational because the group status (sex of the employee) was, obviously, not decided by the investigator.

Causal Conclusions and Confounding Variables

Can statistical analysis alone be used to establish causal relationships? The answer is simple and concise:

> *Statistical inferences of cause-and-effect relationships can be drawn from randomized experiments, but not from observational studies.*

In a rough sense, randomization ensures that subjects with different, and possibly relevant, features are mixed up between the two groups. Surely some of the subjects in the motivation and creativity experiment were naturally more creative than others. In view of the randomization, however, there is no reason to expect that they would be placed disproportionately in the intrinsic motivation group, since every subject had the same chance of being placed in that group. There is, of course, a chance that the randomization—the particular result of the random assignment—turned out by chance in such a way that the intrinsic motivation group received many more of the naturally creative writers. *This chance, however, is incorporated into the statistical tools that are used to express uncertainty.*

In an observational study, it is impossible to draw a causal conclusion from the statistical analysis alone. One cannot rule out the possibility that confounding variables are responsible for group differences in the measured outcome.

> *A **confounding variable** is related both to group membership and to the outcome. Its presence makes it hard to establish the outcome as being a direct consequence of group membership.*

In fact, the males in the sex discrimination example generally did have more years of education than the females; and this, not sex, may have been responsible for the disparity in starting salaries. Thus, the effect of sex cannot be separated from—it is confounded with—the effect of education. Tools exist for comparing male and female salaries after accounting for the effect of education (see Chapter 9), and these help clarify matters to some extent. Other confounding variables, however, may not be recognized or measured, and these consequently cannot be accounted for in the analysis. Therefore, it may be possible to conclude that males tend to get larger starting salaries than females, even after accounting for years of education, and yet still not be possible to conclude, from the statistics alone, that this happens *because* they are males.

Do Observational Studies Have Value?

Is there any role at all for observational data in serious scientific inquiry? The following points indicate that the answer is yes.

1. *Establishing causation is not always the goal.* A study was conducted on 10 American men of Chinese descent and 10 American men of European descent to examine the effect of a blood-pressure-reducing drug. The result—that the men of Chinese ancestry tended to exhibit a different response to the drug—does not prove that being of Chinese descent is responsible for the difference. Diets, for example, may differ in the two groups and may be responsible for the different sensitivities. Nevertheless the study does provide important information for doctors prescribing the drug to people from these populations.
2. *Establishing causation may be done in other ways.* Although statistical methods cannot eliminate the possibility of confounding factors, there may be strong theoretical reasons for doing so. Radiation biologists counted chromosomal aberrations in a sample of Japanese atomic bomb survivors who received radiation from the blast, and compared these to counts on individuals who were far enough from the blast not to have received any radiation. Although the data are observational, the researchers are certain that higher counts in the radiation group can only be due to the radiation, and the data can therefore be used to estimate the dose–response relationship between radiation and chromosomal aberration.
3. *Analysis of observational data may lend evidence toward causal theories and suggest the direction of future research.* Many observational studies indicated an association between smoking and lung cancer, but causation was accepted only after decades of observational studies, experimental studies on laboratory animals, and a scientific theory for the carcinogenic mechanism of smoking.

Although observational studies are undoubtedly useful, the inappropriate practice of claiming or implying cause-and-effect relationships from them—sometimes in subtle ways—is widespread.

1.2.2 Inference to Populations

A second distinction between study designs relates to how units are selected. In a ***random sampling*** study, units are selected by the investigator from a well-defined population. All units in the population have a chance of being selected, and the investigator employs a chance mechanism (like a lottery) to determine actual selection. Neither of the case studies in Section 1.1 used random sampling. In contrast, the study units often are ***self-selected.*** The subjects of the creativity study volunteered their participation. The decision to volunteer can have a strong relationship with the outcome (creativity score), and this precludes the subjects from representing a broader population.

Again, the inferential situation is straightforward:

> *Inferences to **populations** can be drawn from random sampling studies, but not otherwise.*

Random sampling ensures that all subpopulations are represented in the sample in roughly the same mix as in the overall population. Again, random selection has a chance of producing nonrepresentative samples, but *the statistical inference procedures incorporate measures of uncertainty that describe that chance.*

Simple Random Sampling

The most basic form of random sampling is simple random sampling.

> *A* ***simple random sample*** *of size* n *from a population is a subset of the population consisting of* n *members selected in such a way that every subset of size* n *is afforded the same chance of being selected.*

A typical method assigns each member of the population a computer-generated random number. When both lists are re-ordered according to increasing values of the random numbers, the sample of population members appears at the top of the list. Most statistical programs can do this automatically with a "sample" command.

1.2.3 Statistical Inference and Chance Mechanisms

Statistical analysis is used to make statements from available data in answer to questions of interest about some broader context than the study at hand. No such statement about the broader context can be made with absolute certainty, so every statement includes some measure of uncertainty. In statistical analysis, uncertainty is explained by means of chance models, which relate the available data to the broader context. These ideas are expressed in the following definitions:

> *An* ***inference*** *is a conclusion that patterns in the data are present in some broader context.*
> *A* ***statistical inference*** *is an inference justified by a probability model linking the data to the broader context.*

The probability models are associated with the chance mechanisms used to select units from a population or to assign units to treatment groups. They enable the investigator to calculate measures of uncertainty to accompany inferential conclusions.

When a chance mechanism has been used in the study, uncertainty measures accompanying the researcher's inferences are backed by a *bona fide* probability structure that exactly describes the study. Often, however, units do not arise from random sampling of real populations; instead, the available units are self-selected or are the result of a haphazard selection procedure. For randomized experiments, this may be no problem, since cause-and-effect conclusions can be drawn regarding the effects on the particular units selected (as in the motivation and creativity study). These conclusions can be quite strong, even if the observed pattern cannot be inferred to hold in some general population. For observational studies, the lack of truly random samples is more worrisome, because making an inference about some larger population is usually the goal. It may be possible to *pretend* that the units are as representative as a random sample, but the potential for bias from haphazard or convenience selection remains a serious concern.

Example

Researchers measured the lead content in teeth and the intelligence quotient (IQ) test scores for all 3,229 children attending first and second grades in the period between 1975 and 1978 in Chelsea and Somerville, Massachusetts. The IQ scores for those with low lead concentrations were found to be significantly higher than for those with high lead concentrations. What can be inferred? There is no random sample. In a strict sense, the statistical results apply only to the children actually measured; any extrapolation of the pattern to other children comes from supposing that the relationship between lead and IQ is similar for others. This is not necessarily a bad assumption. The point is that extending the inference to other children is surely open to question.

Statistical Inferences Based on Chance Mechanisms

Four situations are exhibited in Display 1.5, along with the types of inference that may be drawn in each. In observational studies, obtaining random samples from the populations of interest is often impractical or impossible, and inference based on

DISPLAY 1.5 Statistical inferences permitted by study designs

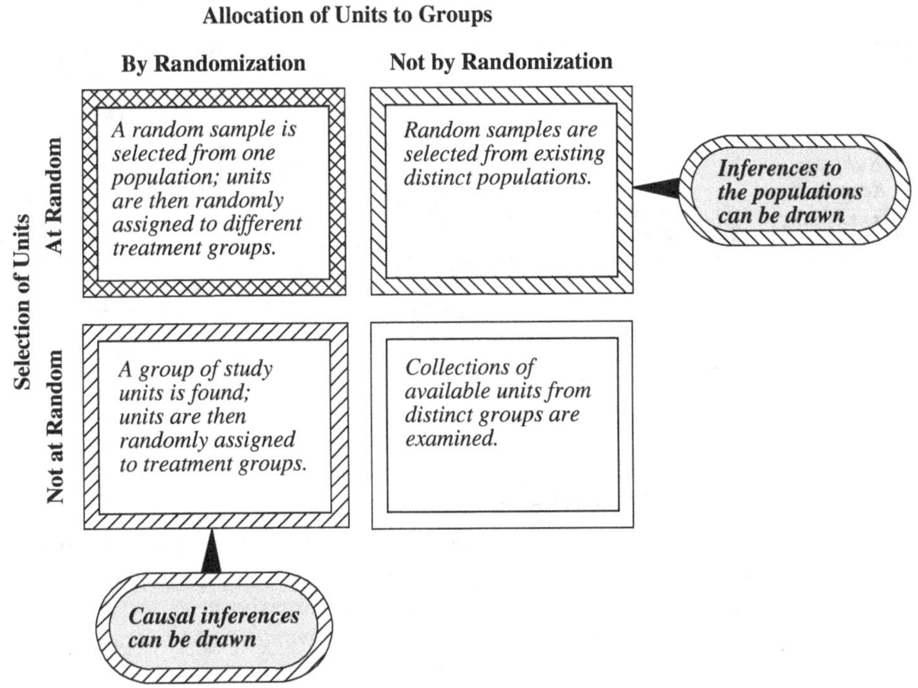

assumed models may be better than no inference at all. In controlled experiments, however, there is no excuse for avoiding randomization.

1.3 MEASURING UNCERTAINTY IN RANDOMIZED EXPERIMENTS

The critical element in understanding statistical measures of uncertainity is visualizing hypothetical replications of a study under different realizations of the chance mechanism used for sample selection or group assignment. This section describes the probability model for randomized experiments and illustrates how a measure of uncertainty is calculated to express the evidence for a treatment difference in the creativity study.

1.3.1 A Probability Model for Randomized Experiments

A schematic diagram for the creativity study (Display 1.6) is typical of a randomized experiment. The chance mechanism for randomizing units to treatment groups ensures that every subset of 24 subjects gets the same chance of becoming the intrinsic group. For example, 23 red and 24 black cards could be shuffled and dealt, one to each subject. Subjects are treated according to their assigned group, and the different treatments change their responses.

DISPLAY 1.6 Randomized experiment with two treatment groups

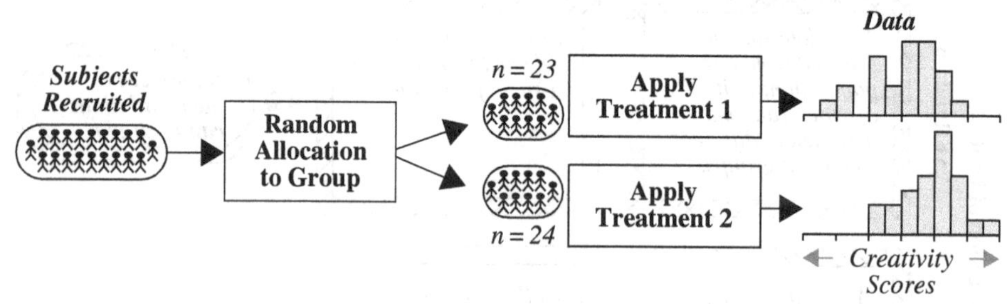

An Additive Treatment Effect Model

Let Y denote the creativity score that a subject would receive after exposure to the extrinsic questionnaire. A general model for the experiment would assert that this same subject would receive a different creativity score, Y^*, after exposure to the intrinsic questionnaire. An *additive treatment effect* model postulates that $Y^* = Y + \delta$. The treatment effect, δ, is a *parameter*—an unknown constant that describes a key feature in the model for answering questions of interest.

1.3.2 A Test for Treatment Effect in the Creativity Study

This section illustrates a statistical inference arising from the additive treatment effect model for the creativity study. The question concerns whether the treatment effect is real. The *p-value* will be introduced as a measure of uncertainty associated with the answer.

Null and Alternative Hypotheses

Questions of interest are translated into questions about parameters in probability models. This usually involves some simplification—such as the assumption that the intrinsic questionnaire adds the same amount to anyone's creativity score—but the expression of a question in terms of a single parameter crystallizes the situation into one where powerful statistical tools can be applied.

In the creativity study, the question "Is there a treatment effect?" becomes a question about whether the parameter δ has a nonzero value. The statement that $\delta = 0$ is called the ***null hypothesis***, while a proposition that $\delta \neq 0$ is an ***alternative hypothesis***. In general, the null hypothesis is the one that specifies a simpler state of affairs; typically—as in this case—an absence of an effect.

A Test Statistic

A ***statistic*** is a numerical quantity calculated from data, like the average creativity score in the intrinsic group. A ***test statistic*** is a statistic used to measure the plausibility of an alternative hypothesis relative to a null hypothesis. For the hypothesis that δ equals zero, the test statistic is the difference in averages, $\overline{Y}_2 - \overline{Y}_1$. Its value should be "close to" zero if the null hypothesis is true and "far from" zero if the alternative hypothesis is true. The value of this test statistic for the creativity scores is 4.14. Is this "close to" or "far from" zero? The answer to that question comes from what is known about how the test statistic might have turned out in other randomization outcomes if there were no effect of the treatments.

Randomization Distribution of the Test Statistic

If the questionnaires had no effect on creativity, the subjects would have received exactly the same creativity scores regardless of the group to which they were assigned. Therefore, it is possible to determine what test statistic values would have occurred had the randomization process turned out differently. Display 1.7 has an example. The first column lists the creativity scores of all 47 participants in the study. The second column lists the groups to which they were assigned. The third column lists another possible way that the group assignments could have turned out. With that grouping, the difference in averages is 2.07.

It is conceptually possible to calculate $\overline{Y}_2 - \overline{Y}_1$ for every possible outcome of the randomization process (if there is no treatment effect). A histogram of all these values describes the ***randomization distribution*** of $\overline{Y}_2 - \overline{Y}_1$ if the null hypothesis is true. The number of possible outcomes of the random assignment is prohibitively large in this example—there are 1.6×10^{13} different groupings, which would take a computer capable of one million assignments per second half a year to complete.

DISPLAY 1.7 A different group assignment for the creativity study, and a different result

Creativity score	Actual grouping	Another grouping	Creativity score	Actual grouping	Another grouping
12.0	Intrinsic(2)	1	5.0	Extrinsic(1)	2
12.0	Intrinsic	2	5.4	Extrinsic	2
12.9	Intrinsic	1	6.1	Extrinsic	1
13.6	Intrinsic	2	10.9	Extrinsic	2
16.6	Intrinsic	2	11.8	Extrinsic	1
17.2	Intrinsic	1	12.0	Extrinsic	1
17.5	Intrinsic	2	12.3	Extrinsic	1
18.2	Intrinsic	2	14.8	Extrinsic	2
19.1	Intrinsic	1	15.0	Extrinsic	2
19.3	Intrinsic	2	16.8	Extrinsic	2
19.8	Intrinsic	2	17.2	Extrinsic	2
20.3	Intrinsic	2	17.2	Extrinsic	1
20.5	Intrinsic	1	17.4	Extrinsic	2
20.6	Intrinsic	2	17.5	Extrinsic	2
21.3	Intrinsic	1	18.5	Extrinsic	2
21.6	Intrinsic	2	18.7	Extrinsic	1
22.1	Intrinsic	1	18.7	Extrinsic	1
22.2	Intrinsic	2	19.2	Extrinsic	1
22.6	Intrinsic	1	19.5	Extrinsic	1
23.1	Intrinsic	1	20.7	Extrinsic	1
24.0	Intrinsic	1	21.2	Extrinsic	1
24.3	Intrinsic	1	22.1	Extrinsic	2
26.7	Intrinsic	1	24.0	Extrinsic	2
29.7	Intrinsic	1			

Averages from actual grouping

Group	Average	Difference
Intrinsic (2)	19.88	4.14
Extrinsic (1)	15.74	

Averages from another grouping

Group	Average	Difference
Group 1	18.87	2.07
Group 2	16.80	

Display 1.8, however, shows an estimate of the randomization distribution, obtained as the histogram of the values of $\overline{Y}_2 - \overline{Y}_1$ from 500,000 random regroupings of the numbers.

Display 1.8 suggests several things. First, it appears just as likely that test statistics will be negative as positive. Second, the majority of values fall in the range from −3.0 to +3.0. Third, only 1,335 of the 500,000 (0.27%) random regroupings produced test statistics as large as 4.14. This last point indicates that 4.14 is a value corresponding to an unusually uneven randomization outcome, if the null hypothesis is correct.

DISPLAY 1.8 Histogram of differences between group averages, from 500,000 regroupings of the creativity study data

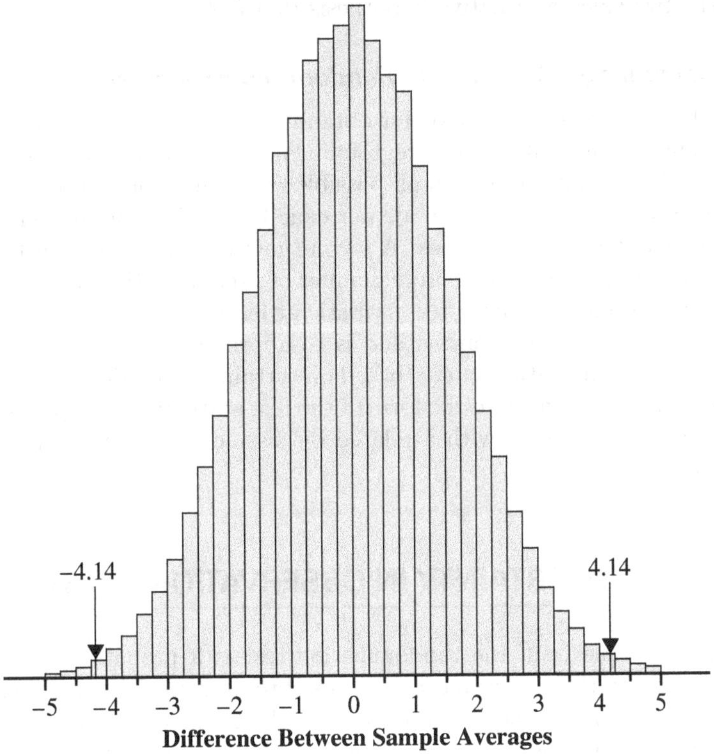

The p-Value of the Test

Because the observed test statistic is improbably large under the null hypothesis, one is led to believe that the null hypothesis is wrong, that the treatment—not the randomization—caused the big difference. That conclusion could be incorrect, however, because the randomization is capable of producing such an extreme. The chance of its doing so is called the *(observed) p-value. In a randomized experiment the p-value is the probability that randomization alone leads to a test statistic as extreme as or more extreme than the one observed.* The smaller the *p*-value, the more unlikely it is that chance assignment is responsible for the discrepancy between groups, and the greater the evidence that the null hypothesis is incorrect. A general definition appears in Chapter 2.

Since 1,335 of the 500,000 regroupings produced differences as large or larger than the observed difference, the *p*-value for this study is estimated from the simulation to be $1{,}335/500{,}000 = 0.00267$. This is a *one-sided p*-value (for the alternative hypothesis that $\delta > 0$) because it counted as extremes only those outcomes with test statistics as large as or larger than the observed one. Statistics that are smaller than -4.14 may provide equally strong evidence against the null hypothesis, favoring the alternative hypothesis that $\delta < 0$. If those (1,302 of them) are included, the result is a *two-sided* p-value of $2{,}637/500{,}000 = 0.005274$, which would be appropriate for the two-sided alternative hypothesis that $\delta \neq 0$.

Computing p-Values from Randomized Experiments

There are several methods for obtaining p-values in a randomized experiment. An enumeration of all possible regroupings of the data would represent all the ways that the data could turn out in all possible randomizations, absent any treatment effect. This would determine the answer exactly, but it is often overburdensome—as the creativity study illustrates. A second method, used previously, is to estimate the p-value by simulating a large number of randomizations and to find the proportion of these that produce a test statistic at least as extreme as the observed one.

The most common method is to approximate the randomization distribution with a mathematical curve, based on certain assumptions about the distribution of the measurements and the form of the test statistic. This approach is given formally in Chapter 2 along with a related discussion of confidence intervals for treatment effects.

1.4 MEASURING UNCERTAINTY IN OBSERVATIONAL STUDIES

The p-value from the randomization test is a probability statement that is tied directly to the chance mechanism used to assign subjects to treatment groups. For observational studies there is no such chance mechanism for group assignment, so the thinking about statistical inference has to be different. Two approaches for observational studies are discussed here.

Section 1.4.1 introduces a probability model that connects the distribution of a test statistic to a chance mechanism for *sample selection* (which, remember, is something quite different from group assignment). In the sex discrimination study, however, the subjects were not selected randomly from any populations and, in fact, constitute the *entire* populations of males and females in their particular job category. Section 1.4.2 applies the same "regrouping" idea that led to the randomization test in order to say whether the salaries in the two groups differed more than would be expected from a random assignment of salaries to employees. Since this approach is tied to a fictitious model and not a bona fide probability model stemming from an actual chance mechanism, the inference is substantially weaker than the one from the randomized experiment. This is such an important distinction that the test goes by a different name—the permutation test—even though the calculations are identical to those of the randomization test.

1.4.1 A Probability Model for Random Sampling

Display 1.9 depicts random sampling from two populations. A chance mechanism, such as a lottery, selects a subset of n_1 units from population 1 in such a way that all subsets of size n_1 have the same chance of selection. The mechanism is used

DISPLAY 1.9 Random sampling study with two populations

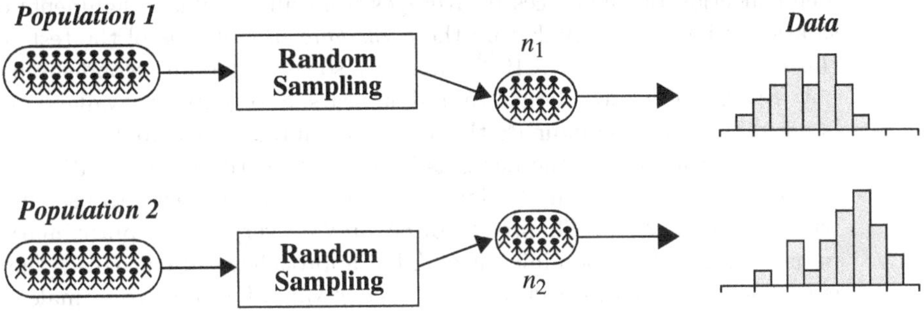

again to select a random sample of size n_2 from population 2. The samples are drawn independently; that is, the particular sample drawn from one population does not influence the process of selection from the other.

Questions of interest in sampling studies center on how features of the populations differ. In many cases, the difference between the populations can be described by the difference between their means. If the means of populations 1 and 2 are μ_1 and μ_2, respectively, then $\mu_2 - \mu_1$ becomes the single parameter for answering the questions of interest. Inferences and uncertainty measures are based on the difference, $\overline{Y}_2 - \overline{Y}_1$, between sample averages, which estimates the difference in population means.

As in Display 1.8, a *sampling distribution* for the statistic $\overline{Y}_2 - \overline{Y}_1$ is represented by a histogram of all values for the statistic from all possible samples that can be drawn from the two populations. The *p*-value for testing a hypothesis and the confidence intervals for estimating the parameter follow from an understanding of the sampling distribution. Full exposition appears in Chapter 2.

1.4.2 Testing for a Difference in the Sex Discrimination Study

In the sex discrimination study, there is no interest in the starting salaries of some larger population of individuals who were never hired, so a random sampling model is not relevant. Similarly, it makes no sense to view the sex of these individuals as randomly assigned. Neither the random sampling nor the randomized experiment model applies. Any interpretation that supports a statistical analysis must be based on a fictitious chance model.

One fictitious probability model for examining the difference between the average starting salary given to males and the average starting salary given to females assumes the employer assigned the set of starting salaries to the employees *at random.* That is, the employer shuffled a set of cards, each having one of the starting salaries written on it, and dealt them out. With this model, the investigator may ask whether the observed difference is a likely outcome.

Permutation Distribution of a Test Statistic

The collection of differences in averages from all possible assignments of starting salaries to individuals makes up the *permutation distribution* of the test statistic for this model. There are 8.7×10^{24} outcomes, so the million-a-second computer would now struggle along for about 275 million years. Shortcuts are available, however. It is not difficult to approximate the actual permutation distribution of a statistic by random regroupings of the data, as illustrated for the randomization distribution in the creativity study in Display 1.8. In addition, the easier-to-use t-tools, which will be introduced in Chapter 2, often provide very good approximations to the permutation distribution inferences. That approach was used to draw the statistical conclusion reported in Section 1.1.2. The p-value for comparing male and female distributions was less than 0.00001, indicating that the difference between male and female salaries was greater than what can reasonably be explained by chance. *It is calculated using the t-tools (see Chapter 2), which provide a close approximation.*

Scope of the Inferences

Inferences from fictitious models must be stated carefully. What the statistical argument shows is that the employer did not assign starting salaries at random—which was known at the outset. The strength of the argument comes from the demonstration that the actual assignment differs substantially from the expected outcome in *one model* where salary assignment is sex-blind.

1.5 RELATED ISSUES

1.5.1 Graphical Methods

This text relies heavily on the use of graphical methods for making decisions regarding the choice of statistical models. This section briefly reviews some important tools for displaying sets of numbers.

Relative Frequency Histograms

Histograms, like those in Display 1.4, are standard tools for displaying the general characteristics of a data set. Looking at the display, one should conclude that the central male starting salary is around \$6,000, that the central female starting salary is around \$5,200, that the two distributions have about the same spread—most salaries are within about \$1,000 of the centers—that both shapes appear reasonably symmetric, but that there is one male starting salary that appears very high in relation to the rest of the male salaries.

A histogram is a graph where the horizontal axis displays ranges for the measurement and the vertical axis displays the relative frequency per unit of measurement. Relative frequency is therefore depicted by area.

Other features are of less concern. Choices of the number of ranges (*bins*) and their boundaries can have some influence on the final picture. However, a histogram is ordinarily used to show broad features, not exquisite detail, and the broad features will be apparent with many choices.

Stem-and-Leaf Diagrams

A *stem-and-leaf diagram* is a cross between a graph and a table. It is used to get a quick idea of the distribution of a set of measurements with pencil and paper or to present a set of numbers in a report.

Display 1.10 shows stem-and-leaf diagrams for the two sets of creativity scores. The digits in each observation are separated into a stem and a leaf. Each number in a set is represented in the diagram by its leaf on the same line as its stem. All possible stem values are listed in increasing order from top to bottom, whether or not there are observations with those stems. At each stem, all corresponding leaves are listed, in increasing order. *Outliers* may require a break in the string of stems.

The stem-and-leaf diagrams show the centers, spreads, and shapes of distributions in the same way histograms do. Their advantages include exact depiction of each number, ease of determining the median and quartiles of each set, and ease of construction. Disadvantages include difficulty in comparing distributions when the numbers of observations in data sets are very different and severe clutter with large data sets.

DISPLAY 1.10 Back-to-back stem-and-leaf diagrams for the creativity study data

Creativity Scores

Extrinsic Group
$n = 23$
Average = 15.74
SD = 5.25

Intrinsic Group
$n = 24$
Average = 19.88
SD = 4.44

Extrinsic	Stem	Intrinsic
40	5	
1	6	
	7	
	8	
	9	
9	10	
8	11	
30	12	009
	13	6
8	14	
0	15	
8	16	6
5422	17	25
775	18	2
52	19	138
7	20	**356**
2	21	36
1	22	126
	23	1
0	24	03
	25	
	26	7
	27	
	28	
	29	7
	30	

The median is17.2, the 12th smallest observation.

The median is 20.4, halfway between the 12th and 13th smallest observations.

Leaves are ordered at each stem.

The ***stem*** *of 29.7.*

The ***leaf*** *of 29.7.*

Legend: | 29 | 7 *represents* 29.7

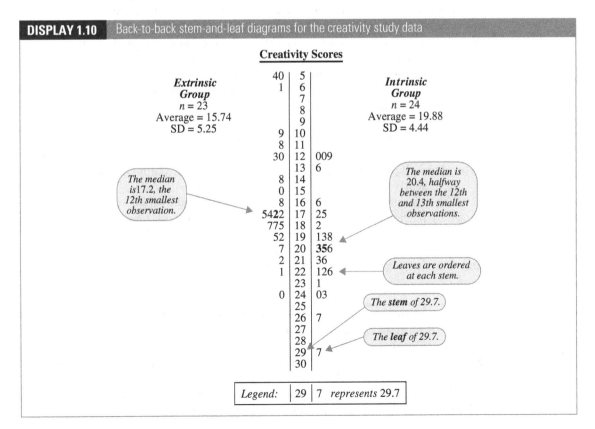

Finding the Median from a Stem-and-Leaf Diagram

Because stem-and-leaf diagrams show the order of numbers in a set, they offer a convenient presentation for locating the *median* of the data. At least half the values in the data are greater than or equal to the median; at least half are less than or equal to it. A median is found by the "$(n + 1)/2$" rule: calculate $k = (n + 1)/2$; if k is an integer, the median is the kth smallest observation; if k is halfway between two integers, the median is halfway between the corresponding observations (see Display 1.10).

Box Plots

A *box plot*, or box-and-whisker plot, is a graphical display that represents the middle 50% of a group of measurements by a box and highlights various features of the upper and lower 25% by other symbols. The box represents the body of the distribution of numbers, and the whiskers represent its tails. The graph gives an uncluttered view of the center, the spread, and the skewness of the distribution and indicates the presence of unusually small (or large) outlying values. Box plots are particularly useful for comparing several samples side by side. Unlike stem-and-leaf diagrams, box plots are not typically drawn by hand, but by a statistical computing program.

Most box plots use the *interquartile range* (IQR) to distinguish distant values from those close to the body of the distribution. The IQR is the difference between the upper and lower quartiles—the range of the middle 50% of the data.

Display 1.11 shows a box plot of gross domestic product (GDP) per capita in 228 countries. (See Exercise 16.) The lower quartile is $2,875 per year and the upper quartile is $22,250. The median income is marked by a line through the box at $9,300 per year. Whiskers extend from the box out through all values that are within 1.5 IQRs of the box.

Observations that are more than 1.5 IQRs away from the box are far enough from the main body of data that they are indicated separately by dots. Observations more than 3 IQRs from the box are quite distant from the main body of data and are prominently marked, as in Display 1.11, and often named.

The difference in the distributions of starting salaries for men and women is shown by placing box plots *side by side*, as in Display 1.12. One can see at a glance that the middle starting salaries for men were about $750 higher than for women, that the ranges of starting salaries were about the same for men and for women, and that the single starting salary of $8,100 is unusually large.

Notes

1. Box-plotting routines are widely available in statistical computing packages. The definitions of quartiles, extreme points, and very extreme points, however, may differ slightly among packages.
2. The choices of 1.5 and 3 IQRs are arbitrary but useful cutoffs for highlighting observations distant and quite distant from the main body.
3. The width of the box is chosen simply with a view toward making the box look nice; it does not represent any aspect of the data.

DISPLAY 1.11 Box plot of per capita GDP for 228 countries in 2010 ($U.S.)

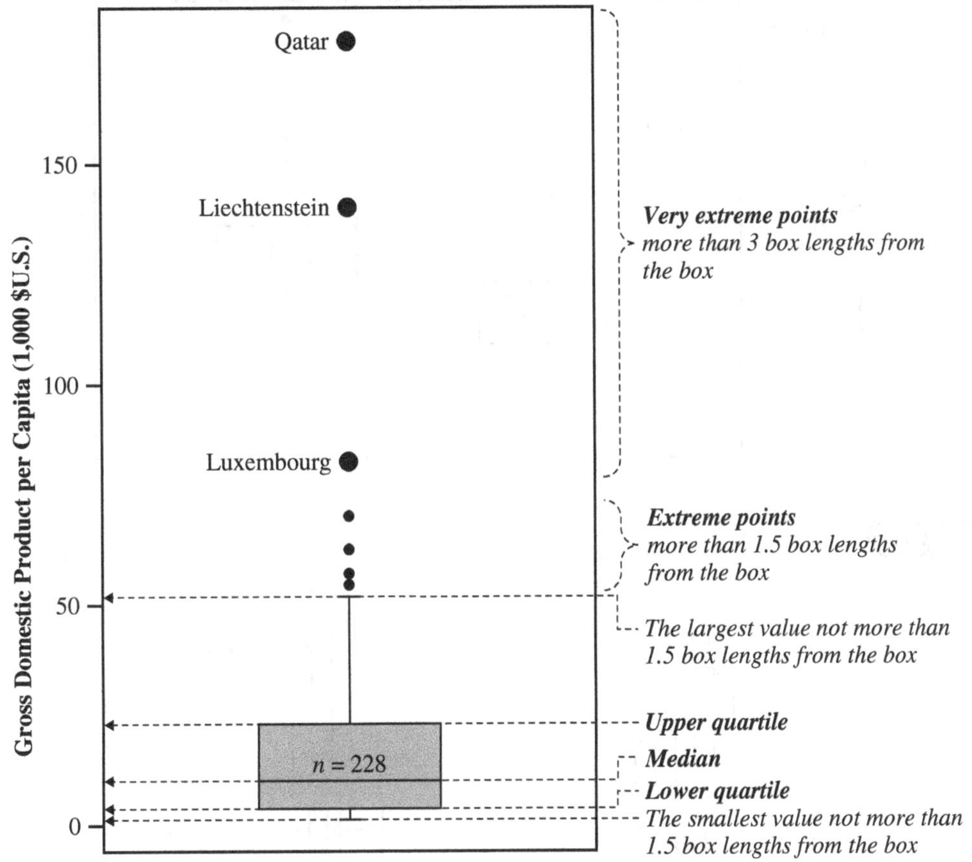

4. For presentation in a journal, it is common to define the box plot so that the whiskers extend to the 10th and 90th percentiles of the sample. This is easier to explain to readers who might be unfamiliar with box plots.
5. The data sets in Display 1.13 show a variety of different kinds of distributions. Box plots are matched with histograms of the same data sets.
6. Some statistical computer packages draw horizontal box plots.

1.5.2 Standard Statistical Terminology

A *parameter* is an unknown numerical value describing a feature of a probability model. Parameters are indicated by Greek letters. A *statistic* is any quantity that can be calculated from the observed data. Statistics are represented by Roman letter symbols. An *estimate* is a statistic used as a guess for the value of a parameter. The notation for the estimate of a parameter θ is the parameter symbol with a hat on

DISPLAY 1.12 Side-by-side box plots for the starting salary data

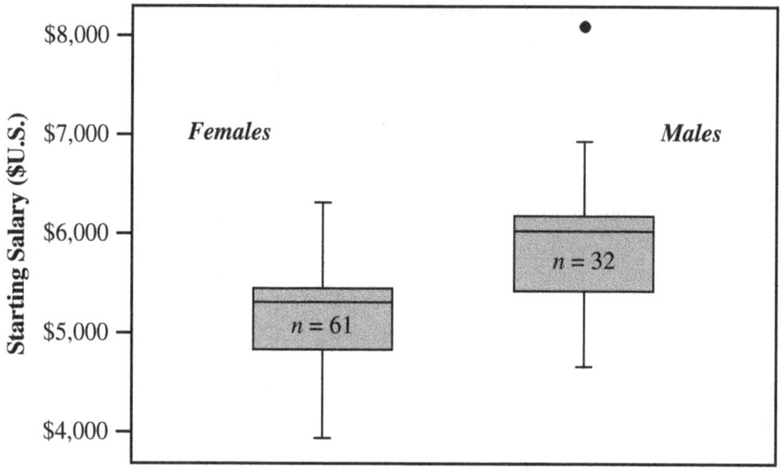

DISPLAY 1.13 Histograms and box plots for 100 observations from four distributions

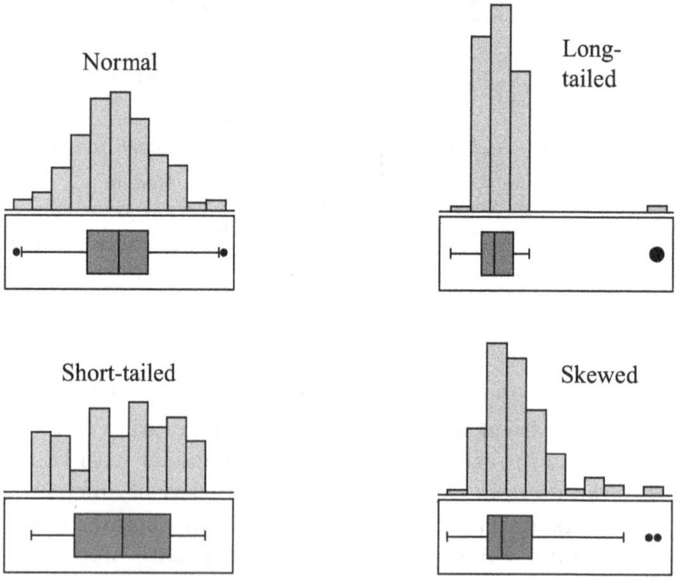

it, $\hat{\theta}$. Remember that the estimate can be calculated, but the parameter remains unknown.

The *Statistical Sleuth* uses the word *mean* when referring to an average calculated over an entire population. A mean is therefore a parameter. When referring to the average in a sample—which is both a statistic and an estimate of the population mean—the *Statistical Sleuth* uses the word *average*. Standard notation uses μ for the mean and $\overline{Y}$ for an average.

The *standard deviation* of a set of numbers $Y_1, \ldots, Y_n$ is defined as

$$\sqrt{\left(\sum_{i=1}^{n}(Y_i - \overline{Y})^2\right) \Big/ (n-1)}.$$

The symbol for a population standard deviation is σ; for a sample standard deviation, the symbol is s. The standard deviation is a measure of spread, interpreted as the typical distance between a single number and the set's average.

1.5.3 Randomization of Experimental Units to Treatments

Experimental units are the human or animal subjects, plots of land, samples of material, or other entities that are assigned to treatment groups in an experiment and to which a treatment is applied. In some cases, the experimental units are groups of individuals, such as classrooms of students that are assigned, as a class, to receive one of several teaching methods; the classrooms—not the students—are the experimental units in this case. In the creativity example, the experimental units were the 47 people. The treatments applied to them were the motivation questionnaires.

In a two-treatment randomized experiment, all available experimental units have the same chance of being placed in group 1. Drawing cards from a thoroughly shuffled deck of red and black cards is one method. Tables of random numbers or computer-generated random numbers may also be used. These have the advantage that the randomization can be documented and checked against certain patterns.

Most statistical computer programs have routines that generate random numbers. To use such a list for randomly assigning units to two experimental groups, first list the units (in any order). Then generate a series of random numbers, assigning them down the list to the units. Reorder the list so that the random numbers are listed by increasing order, keeping each subject matched with its random number. Then assign the subjects with the smallest random numbers (at the top of the reordered list) to treatment 1 and those with the largest random numbers (at the bottom) to treatment 2. Ties in the random numbers can be broken by generating additional random numbers.

1.5.4 Selecting a Simple Random Sample from a Population

A *simple random sample* (of size n) is a subset of a population obtained by a procedure giving all sets of n distinct items in the population an equal chance of being chosen. To accomplish this in practice requires a *frame*: a numbered list of all the subjects. If a population of 20,000 members is so listed, an integer from 1 to 20,000 is associated with each. If a random sample of 100 is desired, it is only necessary to generate 100 distinct random integers such that each of the integers 1 through 20,000 has an equal chance of being selected. This can be accomplished with a statistical computer program.

Other Random Sampling Procedures

A commonly used method for selecting a sample of 100 from the list of 20,000 is to pick a single random start from the positions 1 to 200 and then to select every 200th subject going down the frame. This is called *systematic random sampling*. Another method is to group the frame into 200 blocks of 100 consecutive subjects each and to select one of the blocks at random. This is called *random cluster sampling*. In sampling units such as lakes of different sizes, it is sometimes useful to allow larger units to have higher probabilities of being sampled than smaller units. This constitutes *variable probability sampling*. These and other random sampling schemes can be useful, but they differ from simple random sampling in fundamental ways.

1.5.5 On Being Representative

Random samples and randomized experiments are representative in the same sense that flipping a coin to see who takes out the garbage is fair. Flipping is always fair before the coin hits the table, but the outcome is always unfair to the loser. In the same way, close examination of the results of randomization or random sampling can usually expose ways in which the chosen sample is not representative. The key, however, is not to abandon the procedure when its result is suspect. Uncertainty about representativeness will be incorporated into the statistical analysis itself. If randomization were abandoned, there would be no way to express uncertainty accurately.

1.6 SUMMARY

Cause-and-effect relationships can be inferred from randomized experiments but not from observational studies. The problem with observational studies is that confounding variables—identifiable or not—may be responsible for observed differences. Randomized experiments eliminate this problem by ensuring that differences between groups (other than those of the assigned treatments) are due to chance alone. Statistical measures of uncertainty account for this chance.

Statistically, the statements that generalize sample results to more general contexts are based on a probability model. When the model corresponds to the planned use of randomization or random sampling, it provides a firm basis for drawing inferences. The probability model may also be a fiction, created for the purposes of assessing uncertainty. Results from fictitious probability models must be viewed with skepticism.

1.7 EXERCISES

Conceptual Exercises

1. **Creativity Study.** In the motivation and creativity experiment, the poems were given to the judges in random order. Why was that important?

2. Sex Discrimination Study. Explain why it is difficult to prove sex discrimination (that males in a company receive higher starting salaries because they are males) even if it has occurred.

3. A study found that individuals who lived in houses with more than two bathrooms tended to have higher blood pressure than individuals who lived in houses with two or fewer bathrooms. (a) Can cause and effect be inferred from this? (b) What confounding variables may be responsible for the difference?

4. A researcher performed a comparative experiment on laboratory rats. Rats were assigned to group 1 haphazardly by pulling them out of the cage without thinking about which one to select. Should others question the claim that this was as good as a randomized experiment?

5. In 1930 an experiment was conducted on 20,000 school children in England. Teachers were responsible for randomly assigning their students to a treatment group—to receive $\frac{3}{4}$ pint of milk each day—or to a control group—to receive no milk supplement. Weights and heights were measured before and after the four-month experiment. The study found that children receiving milk gained more weight during the study period. On further investigation, it was also found that the controls were heavier and taller than the treatment group *before* the milk treatment began (more so than could be attributed to chance). What is the likely explanation and the implication concerning the validity of the experiment?

6. Ten marijuana users, aged 14 to 16, were drawn from patients enrolled in a drug abuse program and compared to nine drug-free volunteers of the same age group. Neuropsychological tests for short-term memory were given, and the marijuana group average was found to be significantly lower than the control group average. The marijuana group was held drug-free for the next six weeks, at which time a similar test was given with essentially the same result. The researchers concluded that marijuana use caused adolescents to have short-term memory deficits that continue for at least six weeks after the last use of marijuana. (a) Can a genuine causal relationship be established from this study? (b) Can the results be generalized to other 14- to 16-year-olds? (c) What are some potential confounding factors?

7. Suppose that random samples of Caucasian-American and Chinese-American individuals are obtained from patient records of doctors participating in a study to compare blood pressures of the two populations. Suppose that the individuals selected are asked whether they want to participate and that some decline. The study is conducted only on those that volunteer to participate, and a comparison of the distributions of blood pressures is conducted. Where does this study fit in Display 1.5? What assumption would be necessary to allow inferences to be made to the sampled populations?

8. More people get colds during cold weather than during warm weather. Does that prove that cold temperatures cause people to get colds? What is a potential confounding factor?

9. A study showed that children who watch more than two hours of television each day tend to have higher cholesterol levels than children who watch less than two hours of television each day. Can you think of any use for the result of this study?

10. What is the difference between a randomized experiment and a random sample?

11. A number of volunteers were randomly assigned to one of two groups, one of which received daily doses of vitamin C and one of which received daily placebos (without any active ingredient). It was found that the rate of colds was lower in the vitamin C group than in the placebo group. It became evident, however, that many of the subjects in the vitamin C group correctly guessed that they were receiving vitamin C rather than placebo, because of the taste. Can it still be said that the difference in treatments is what caused the difference in cold rates?

12. Fish Oil and Blood Pressure. Researchers used 7 red and 7 black playing cards to randomly assign 14 volunteer males with high blood pressure to one of two diets for four weeks: a fish oil diet and a standard oil diet. The reductions in diastolic blood pressure are shown in Display 1.14.

DISPLAY 1.14 Reductions in diastolic blood pressure (mm of mercury) for 14 men after 4 weeks of a special diet containing fish oil or a regular, nonfish oil

Fish oil diet:	8	12	10	14	2	0	0
Regular oil diet:	−6	0	1	2	−3	−4	2

(Based on a study by H. R. Knapp and G. A. FitzGerald, "The Antihypertensive Effects of Fish Oil," *New England Journal of Medicine* 320 (1989): 1037–43.) Why might the results of this study be important, even though the volunteers do not constitute a random sample from any population?

13. Why does a stem-and-leaf diagram require less space than an ordinary table?

14. What governs the *width* of a box plot?

15. What general features are evident in a box plot of data from a normal distribution? from a skewed distribution? from a short-tailed distribution? from a long-tailed distribution?

Computational Exercises

16. Gross Domestic Product (GDP) per Capita. The data file ex0116 contains the gross domestic product per capita for 228 countries—the data used to construct the box plot in Display 1.11. (a) Make a box plot of the per capita GDPs with a statistical computer program. Include a y-axis label (for example, "Gross Domestic Product per Capita in \$U.S."). (b) In what ways, if any, is the display from your software different from Display 1.11? (c) Use a statistical computer program to draw a *histogram* of the per capita GDPs. Include an x-axis label. Use the program's default bin width. Report that bin width. (d) The program's default bin width for histograms is not necessarily the best choice. If it's possible with your statistical computer program, redraw the histogram of part (c) using bin widths of \$5,000. (Data from Central Intelligence Agency, "Country Comparison: GDP — per capita (PPP)," *The World Factbook*, June 24, 2011 https://www.cia.gov/library/publications/the-world-factbook/rankorder/2004rank.html (June 30, 2011).)

17. Seven students volunteered for a comparison of study guides for an advanced course in mathematics. They were randomly assigned, four to study guide A and three to study guide B. All were instructed to study independently. Following a two-day study period, all students were given an examination about the material covered by the guides, with the following results:

Study Guide A scores: 68, 77, 82, 85
Study Guide B scores: 53, 64, 71

Perform a randomization test by listing all possible ways that these students could have been randomized to two groups. There are 35 ways. For each outcome, calculate the difference between sample averages. Finally, calculate the two-sided p-value for the observed outcome.

18. Using the creativity study data (Section 1.1.1) and a computer, assign a set of random numbers to the 47 subjects. Order the entire data set by increasing values of the random numbers, and divide the subjects into group 1 with the 24 lowest random numbers and group 2 with the 23 highest. Compute the difference in averages. Repeat this process five times, using different sets of random numbers. Did you get any differences larger than the one actually observed (4.14)?

19. Write down the names and ages of 10 people. Using coin flips, divide them into two groups, as if for a randomized experiment. Did one group tend to get many of the older subjects? Was there any way to predict which group would have a higher average age in advance of the coin flips?

20. Repeat Exercise 19 using a randomization mechanism that ensures that each group will end up with exactly five people.

21. Read the methods and design sections of five published studies in your own field of specialization. (a) Categorize each according to Display 1.5. (b) Now read the conclusions sections. Determine whether inferential statements are limited to or go beyond the scope allowed in Display 1.5.

22. Draw back-to-back stem-and-leaf diagrams of the creativity scores for the two motivation groups in Display 1.1 (by hand).

23. Use the computer to draw side-by-side box plots for the creativity scores in Display 1.1.

24. Using the stem-and-leaf diagrams from Exercise 22, compute the median, lower quartile, and upper quartile for each of the motivation groups. Identify these on the box plots from Exercise 23. Using a ruler, measure the length of the box (the IQR) for each group, and make horizontal lines at 1.5 IQRs above and below each box and at 3 IQRs above and below the box. Are there any *extreme points* in either group? Are there any *very extreme points* in either group?

25. The following are zinc concentrations (in mg/ml) in the blood for two groups of rats. Group A received a dietary supplement of calcium, and group B did not. Researchers are interested in variations in zinc level as a side effect of dietary supplementation of calcium.

Group A: 1.31 1.45 1.12 1.16 1.30 1.50 1.20 1.22 1.42 1.14 1.23 1.59 1.11 1.10
1.53 1.52 1.17 1.49 1.62 1.29

Group B: 1.13 1.71 1.39 1.15 1.33 1.00 1.03 1.68 1.76 1.55 1.34 1.47 1.74 1.74
1.19 1.15 1.20 1.59 1.47

(a) Make a back-to-back stem-and-leaf diagram (by hand) for these measurements. (b) Use the computer to draw side-by-side box plots.

Data Problems

26. Environmental Voting of Democrats and Republicans in the U.S. House of Representatives. Each year, the League of Conservation Voters (LCV) identifies legislative votes taken in each house of the U.S. Congress—votes that are highly influential in establishing policy and action on environmental problems. The LCV then publishes whether each member of Congress cast a pro-environment or an anti-environment vote. Display 1.15 shows these votes during the years 2005, 2006, and 2007 for members of the House of Representatives. Evaluate the evidence supporting party differences in the percentage of pro-environment votes. Write a brief report of your conclusion, including a graphical display and summary statistics.

DISPLAY 1.15 Number of pro- and anti-environment votes in 2005, 2006, and 2007, according to the League of Conservation Voters, of Republican (R) and Democratic (D) members of the U.S. House of Representatives; and the total percentage of their votes that were deemed pro-environment; first 5 of 492 rows

State	Representative	Party	Pro05	Anti05	Pro06	Anti06	Pro07	Anti07	PctPro
Alabama	Bonner	R	2	16	3	9	2	18	14.0
Alabama	Everett	R	0	18	1	11	2	18	6.0
Alabama	Rogers	R	1	17	2	10	3	17	12.0
Alabama	Aderholt	R	0	18	0	12	2	18	4.0
Alabama	Cramer	D	5	13	4	7	14	6	46.9

27. Environmental Voting of Democrats and Republicans in the U.S. Senate. Display 1.16 shows the first five rows of a data set with pro- and anti-environment votes (according to the League of Conservation Voters; see Exercise 26) during 2005, 2006, and 2007, cast by U.S. senators. Evaluate the evidence supporting party differences in the percentage of pro-environment votes. Write a brief report of your conclusion, and include a graphical display and summary statistics.

DISPLAY 1.16 Number of pro- and anti-environment votes in 2005, 2006, and 2007, according to the League of Conservation Voters, of Republican (R) and Democratic (D) members of the U.S. Senate; and the total percentage of their votes that were deemed pro-environment; first 5 of 112 rows

State	Senator	Party	Pro05	Anti05	Pro06	Anti06	Pro07	Anti07	PctPro
Alabama	Session	R	1	18	0	7	2	13	7.3
Alabama	Shelby	R	1	19	0	7	1	14	4.8
Alaska	Murkowski	R	2	17	1	6	6	9	22.0
Alaska	Stevens	R	1	19	1	6	4	11	14.3
Arizona	Kyle	R	1	19	2	5	2	13	11.9

Answers to Conceptual Exercises

1. If one of the two treatment groups had their poems judged first, then the effect of motivation treatment would be confounded with time effects in the judges' marking of creativity scores. Judges are influenced by memories of previous cases. They may also discern a change in average quality, which could alter their expectations.

2. Statistically, the distributions of male and female starting salaries may be compared after adjusting for possible confounding variables. Since the data are necessarily observational, a difference in distribution cannot be linked to a specific cause. Once more, for emphasis: The best possible statistical analysis using the best possible data cannot establish causation in an observational study, but observational data are the only data likely to be available for discrimination cases. If, therefore, courts required plaintiffs to produce scientifically defensible proof of discrimination in order to prevail, defendants would win all discrimination cases *by definition*. As a result, some courts that wish to give weight to statistical information in discrimination cases adopt rules of evidence that allow proof to be established negatively—by the lack of an adequate rebuttal.

3. (a) No. (b) Wealth and richness of diet.

4. Yes. First of all, there is the possibility that the rats that were easier to get out of the cage are different from the others—bigger, less mobile, perhaps. Second, even if there is no obvious reason why the rats in the two groups might be different, that does not ensure they are not. Researchers must maintain skepticism about the conclusions in light of the uncertainty. With proper randomization, which is easy to carry out, there would be no doubt.

5. Teachers apparently gave the milk to the students they thought would most benefit from it. As a consequence, the results of the experiment were not valid.

6. (a) No. In this observational study, it is possible that the drug users were different from the nonusers in ways other than drug use. (b) No, because the samples are volunteers, not random samples. (c) Happiness of the child, stability of family, success of child in school.

7. It is an observational study. The population that is randomly sampled is the population of consenting subjects. To draw inference to all subjects requires the assumption that the response is unrelated to the reasons for consent.

8. No. The amount of time people spend indoors.

9. It may be possible for doctors to identify children who are likely to have high cholesterol by asking about their television watching habits. This requires no causative link. It is a minor use, however, and there is apparently no other.

10. In a randomized experiment a random mechanism is used to allocate the available subjects to treatment groups. In a random sample a random mechanism is used to select subjects from the populations of interest.

11. Yes, sort of. The treatment difference caused the different responses, but the actual "treatment" received in the vitamin C group was both a daily dose of vitamin C and knowledge that it was vitamin C. It's possible that the second aspect of this treatment is what was responsible for the difference. Researchers must make sure that the two groups are treated as similarly as possible in all respects, except for the specific agent under comparison.

12. The conclusion that the fish oil diet causes a reduction in blood pressure for these volunteers is a strong and useful one, even if it formally applies only to these particular individuals.

13. The leading digit or digits (the stems) are listed only once.

14. The width of the box is chosen to make the overall picture pleasing to the eye. It does not represent anything. For side-by-side box plots, the widths of the two boxes should be equal.

15. Normal: Median line in the middle of the box; equal whiskers; few if any extreme points; no very extreme points. Skewed: extreme points in one direction only; very extreme points possible, but few; long whisker on side of extreme points and short whisker (if any) on other side; median line closer to short whisker than to long one. Short-tailed: Like normal with no extreme points and very short whiskers. Long-tailed: Like normal (roughly symmetric) with extreme points strung out in both tails; some very extreme points possible.

CHAPTER 2

Inference Using *t*-Distributions

The first of the case studies in this chapter illustrates the structure of two independent samples and the second illustrates the structure of a single sample of pairs. This chapter has the dual purposes of detailing the inferential tools for these important structures and emphasizing the conceptual basis of confidence intervals and p-values based on the t-distribution more generally. These "t-tools" are useful in regression and analysis of variance structures that will be taken up in subsequent chapters, but their main features can be conveyed in the relatively simple setting of a two-group comparison. The tools are derived under random sampling models when populations are normally distributed. As will be seen, the resulting tools also find application as approximations to randomization tests. Furthermore, as described in Chapter 3, they often work quite well even when the populations are not normal.

2.1 CASE STUDIES

2.1.1 Evidence Supporting Darwin's Theory of Natural Selection—An Observational Study

In the search for evidence supporting Charles Darwin's theory of natural selection, Karl Pearson (1857–1936) made a simple yet profound contribution to human reasoning. Pearson was encouraged by Darwin's cousin, Francis Galton, to use mathematical analysis of numerical characteristics of animals to find evidence of species adaptation. Darwin had observed, for example, that beaks of various finch species in the Galápagos Islands differed in ways that seemed well suited to survival in their respective environments, but he had no direct evidence that the beak characteristics actually evolved to adapt to their environments. In thinking about the use of mathematics, Pearson puzzled over how to show that a numerical characteristic, such as beak size, differs in populations before and after an environmental disturbance if, in fact, the characteristic is *variable* among individuals within each population. His profound contribution was to articulate that the scientific questions could be addressed through a comparison of *population distributions* of the characteristic. Even if two populations of finches showed considerable overlap in beak sizes, he argued, a demonstration that the population distributions differed would suffice as evidence of a species difference. Although Pearson never found the evidence he was looking for, his insights and investigations into methodology for comparing population distributions had a major impact on the development of modern statistics.

In the 1980s, biologists Peter and Rosemary Grant and colleagues found what Pearson had been looking for. Over the course of 30 years, the Grants' research team caught and measured all the birds from more than 20 generations of finches on the Galápagos island of Daphne Major. In one of those years, 1977, a severe drought caused vegetation to wither, and the only remaining food source was a large, tough seed, which the finches ordinarily ignored. Were the birds with larger and stronger beaks for opening these tough seeds more likely to survive that year and did they tend to pass this characteristic to their offspring?

The Grants measured beak depths (height of the beak at its base) of all 751 Daphne Major finches the year before the drought (1976) and all 89 finches captured the year after the drought (1978). Display 2.1 shows side-by-side stem-and-leaf diagrams comparing the 89 post-drought finch bill depths with an equal-sized random sample of the pre-drought bill depths. (For the full set of 1976 finches, see Exercise 2.18.) Is there evidence of a difference between the population distributions of beak depths in 1976 and 1978? (The data were read from a histogram in P. Grant, 1986, *Ecology and Evolution of Darwin's Finches*, Princeton University Press, Princeton, N.J.)

Statistical Conclusion

These data provide overwhelming evidence that the mean beak depth increased from 1976 to 1978 (one-sided p-value < 0.00001 from a two-sample t-test). The

DISPLAY 2.1 Beak depths (mm) of Darwin finches on Daphne Major in 1976, pre-drought, and 1978, post-drought

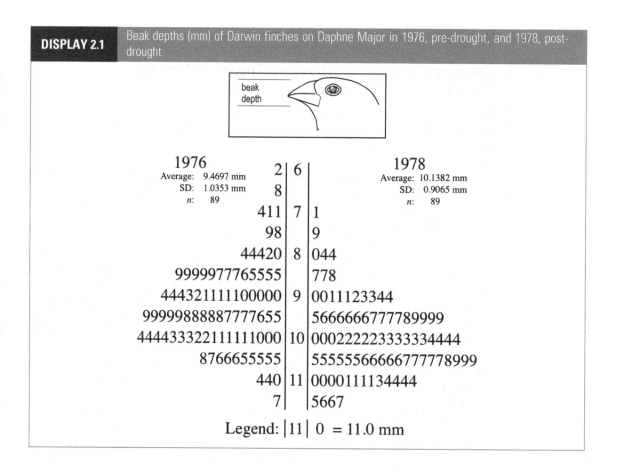

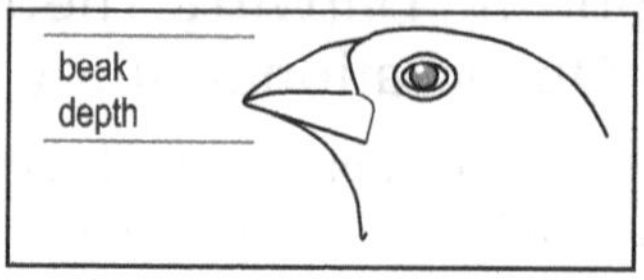

1976 — Average: 9.4697 mm; SD: 1.0353 mm; n: 89

1978 — Average: 10.1382 mm; SD: 0.9065 mm; n: 89

1976		1978
2	6	
8		
411	7	1
98		9
44420	8	044
9999977765555		778
444321111100000	9	0011123344
9999988888777655		5666666777789999
4444333221111111000	10	000222223333334444
8766655555		555555666667777778999
440	11	0000111134444
7		5667

Legend: |11| 0 = 11.0 mm

1978 (post-drought) mean was estimated to exceed the 1976 (pre-drought) mean by 0.67 mm (95% confidence interval: 0.38 mm to 0.96 mm).

Scope of Inference

Since this was an observational study, a causal conclusion—that the drought *caused* a change in the mean beak size—does not follow directly from the statistical evidence of a difference in the means. A lack of alternative explanations, though, might make biologists reasonably confident in the speculation that natural selection in response to the drought is responsible for the change. The Grants measured every finch in the 1976 and 1978 populations. Even though the entire populations were measured, a two-sample t-test is appropriate for showing that the difference in the two populations is greater than can be explained by chance. A more serious problem, though, is that the population in 1978 likely includes birds that were also in the 1976 population or offspring of birds in the 1976 population. If so, the assumption of independence of the two samples would be violated.

2.1.2 Anatomical Abnormalities Associated with Schizophrenia—An Observational Study

Are any physiological indicators associated with schizophrenia? Early studies, based largely on postmortem analysis, suggest that the sizes of certain areas of the brain may be different in persons afflicted with schizophrenia than in others. Confounding variables in these studies, however, clouded the issue considerably. In a 1990 article, researchers reported the results of a study that controlled for genetic and socioeconomic differences by examining 15 pairs of monozygotic twins, where one of the twins was schizophrenic and the other was not. The twins were located through an intensive search throughout Canada and the United States. (Data from R. L. Suddath et al., "Anatomical Abnormalities in the Brains of Monozygotic Twins Discordant for Schizophrenia," *New England Journal of Medicine* 322(12) (1990): 789–93.)

The researchers used magnetic resonance imaging to measure the volumes (in cm^3) of several regions and subregions inside the twins' brains. Display 2.2 presents data based on the reported summary statistics from one subregion, the left hippocampus. What is the magnitude of the difference in volumes of the left hippocampus between the unaffected and the affected individuals? Can the observed difference be attributed to chance?

Summary of Statistical Analysis

There is substantial evidence that the mean difference in left hippocampus volumes between schizophrenic individuals and their nonschizophrenic twins is nonzero (two-sided p-value $= 0.006$, from a paired t-test). It is estimated that the mean

DISPLAY 2.2 Differences in volumes (cm^3) of left hippocampus in 15 sets of monozygotic twins where one twin is affected by schizophrenia

Pair #	Unaffected	Affected	Difference
1	1.94	1.27	0.67
2	1.44	1.63	−0.19
3	1.56	1.47	0.09
4	1.58	1.39	0.19
5	2.06	1.93	0.13
6	1.66	1.26	0.40
7	1.75	1.71	0.04
8	1.77	1.67	0.10
9	1.78	1.28	0.50
10	1.92	1.85	0.07
11	1.25	1.02	0.23
12	1.93	1.34	0.59
13	2.04	2.02	0.02
14	1.62	1.59	0.03
15	2.08	1.97	0.11

Differences

Stem	Leaf
−2	
−1	9
−0	
0	23479
1	0139
2	3
3	
4	0
5	09
6	7
7	

Average: 0.199
Sample SD: 0.238
n: 15

Legend: 6 | 7 represents 0.67 cm^3

volume is $0.20\,\text{cm}^3$ smaller for those with schizophrenia (about 11% smaller). A 95% confidence interval for the difference is from 0.07 to $0.33\,\text{cm}^3$.

Scope of Inference

These twins were not randomly selected from general populations of schizophrenic and nonschizophrenic individuals. Tempting as it is to draw inferences to these wider populations, such inferences must be based on an assumption that these individuals are as representative as random samples are. Furthermore, the study is observational, so no causal connection between left hippocampus volume and schizophrenia can be established from the statistics alone. In fact, the researchers had no theories about whether the abnormalities preceded the disease or resulted from it.

2.2 ONE-SAMPLE *t*-TOOLS AND THE PAIRED *t*-TEST

The schizophrenia study used a *paired* t-test, in which measurements taken on paired subjects are reduced to a single set of differences for analysis. This section develops the single population methods for drawing inferences about the population mean from a random sample and at the same time introduces key concepts such as sampling distributions, standard errors, Z-ratios, and t-ratios.

2.2.1 The Sampling Distribution of a Sample Average

A random sample is drawn from a population with the objective of learning about the population's mean. Suppose the average of that sample is written on a piece of paper, which is placed in a box. Then suppose this process is repeated for every one of the equally likely samples that could be drawn. Then the distribution of all the numbers in the box is the *sampling distribution of the average.*

Display 2.3 illustrates a sampling distribution in the conceptual framework of the schizophrenia study. There is an assumed population of twins in which one of the twins has schizophrenia and the other does not. For each set of twins, Y represents the difference between the left hippocampus volumes of the unaffected and the affected twin. The 15 observed differences are assumed to be a random sample from this population. To examine whether there is a structural difference between volumes, one calculates the average of the 15 measurements, $\overline{Y} = 0.20\,\text{cm}^3$, as an estimate of the population mean μ.

Although only the one sample is actually taken, it is important to think about replicating the study—repeatedly collecting 15 sets of twins and repeatedly calculating the average difference. The value of the average varies from sample to sample, and a histogram of its values represents its sampling distribution.

Because, in this case, there is only one average with which to estimate a population mean, it would seem difficult to learn anything about the characteristics of the sampling distribution. Some illuminating facts about the sampling distribution of an average, however, come from statistical theory. If a population has

DISPLAY 2.3 The sampling distribution of the sample average

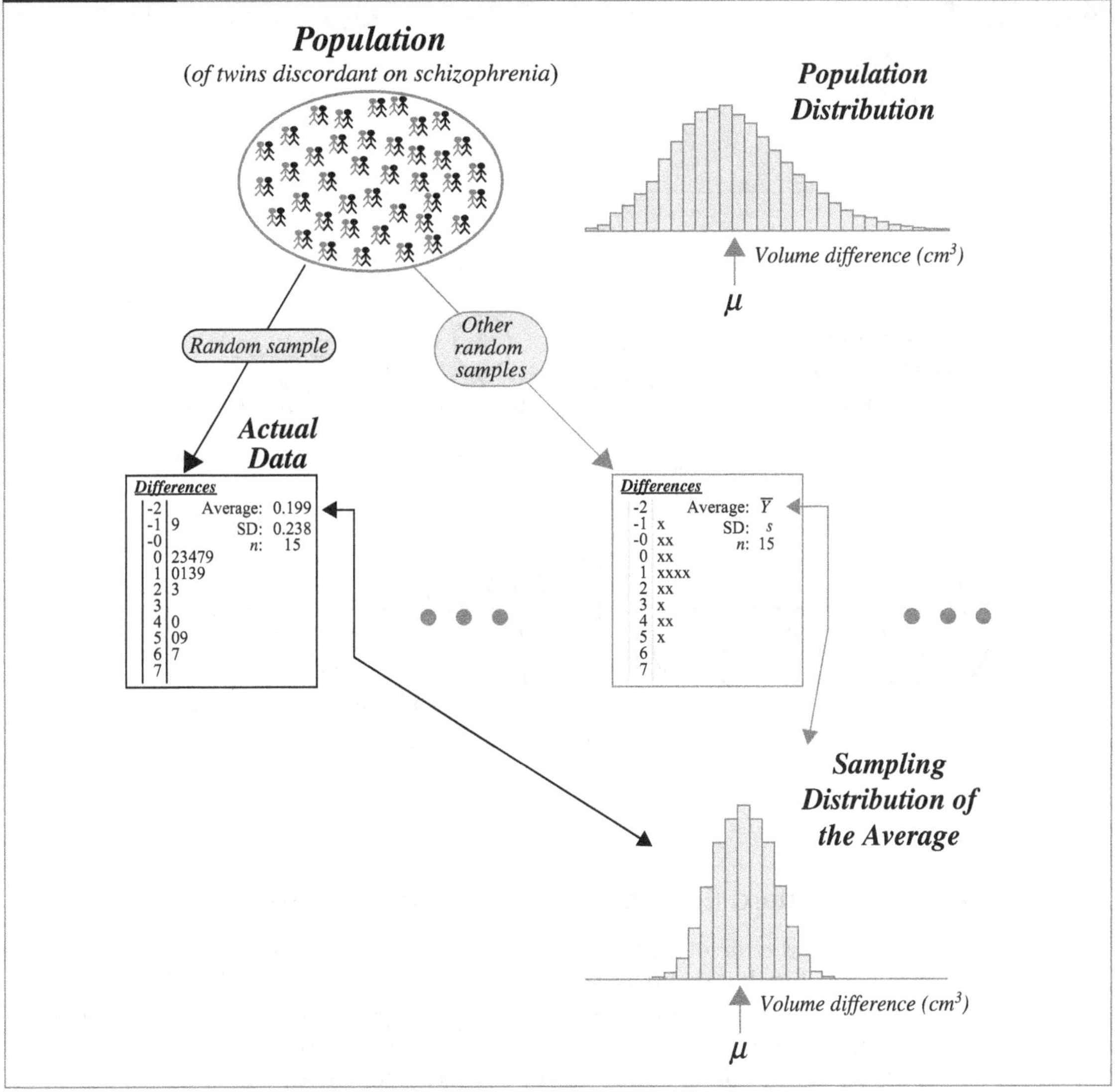

mean μ and standard deviation σ, then—as illustrated in Display 2.4—the mean of the sampling distribution of the average is also μ, the standard deviation of the sampling distribution is $\sigma/\sqrt{n}$, and the shape of the sampling distribution is more nearly normal than is the shape of the population distribution. The last fact comes from the important *Central Limit Theorem.*

The standard deviation in the sampling distribution of an average, denoted by $\text{SD}(\overline{Y})$, is the typical size of $(\overline{Y} - \mu)$, the error in using $\overline{Y}$ as an estimate of μ. This standard deviation gets smaller as the sample size increases.

DISPLAY 2.4 Relationship between the population distribution and the sampling distribution of the average in random sampling

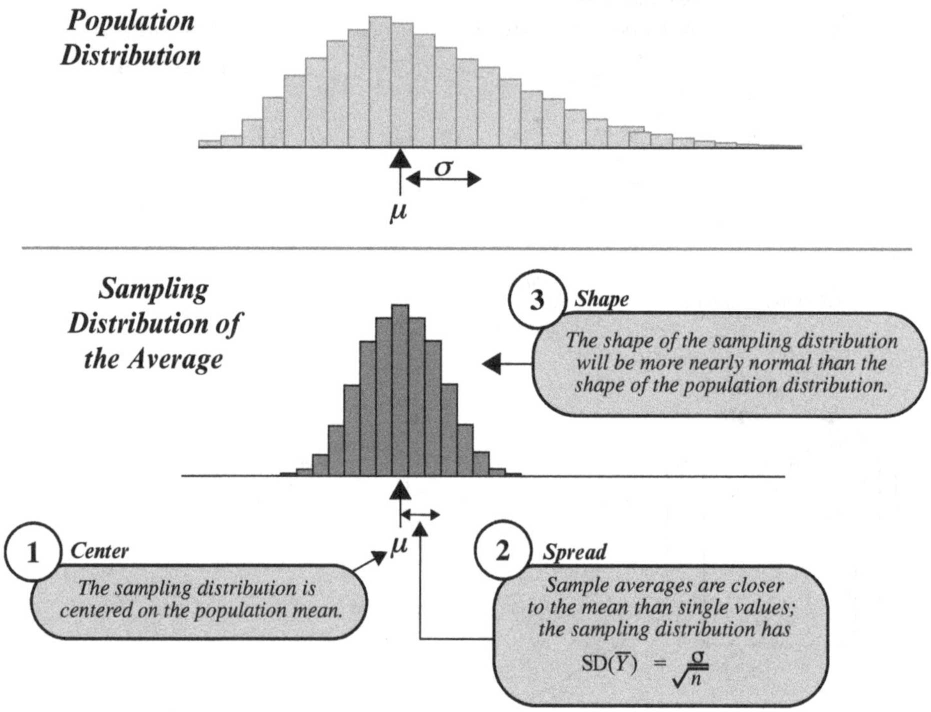
Population Distribution
σ
μ
Sampling Distribution of the Average
3 Shape
The shape of the sampling distribution will be more nearly normal than the shape of the population distribution.
1 Center
The sampling distribution is centered on the population mean.
μ
2 Spread
Sample averages are closer to the mean than single values; the sampling distribution has
SD($\overline{Y}$) = $\frac{\sigma}{\sqrt{n}}$

2.2.2 The Standard Error of an Average in Random Sampling

The *standard error* of any statistic is an estimate of the standard deviation in its sampling distribution. It is therefore the best guess about the likely size of the difference between a statistic used to estimate a parameter and the parameter itself. Standard errors are ordinarily calculated by substituting estimates of variability parameters into the formulas of sampling distribution standard deviations.

Associated with every standard error is a measure of the amount of information used to estimate variability, called its *degrees of freedom*, denoted d.f. Degrees of freedom are measured in units of "equivalent numbers of independent observations." Further information to explain degrees of freedom, more generally, will be provided in future chapters.

Standard Error for a Sample Average

The formula for the standard deviation of the average in a sample of size n is $\sigma/\sqrt{n}$, so if s is the sample standard deviation, the standard error for the average is

$$\mathrm{SE}(\overline{Y}) = \frac{s}{\sqrt{n}}, \qquad \text{d.f.} = (n - 1).$$

The degrees of freedom in a single sample standard deviation are always one less than the sample size.

In the schizophrenia study, the average difference between volumes of unaffected and affected twins is 0.199 cm^3, and the sample standard deviation of the differences is 0.238 cm^3. The standard error of the average is therefore 0.062 cm^3, with 14 degrees of freedom. From this, one makes a preliminary judgment that the population difference is likely to be near the sample estimate, 0.199 cm^3, but that the sample estimate is likely to be somewhere in the neighborhood of 0.062 cm^3 off the mark.

2.2.3 The *t*-Ratio Based on a Sample Average

The ratio of an estimate's error to the anticipated size of its error provides a convenient basis for drawing inferences about the parameter in question.

The Z-Ratio

For any parameter and its sample estimate, its *Z-ratio* is defined as $Z =$ (Estimate–Parameter)/SD(Estimate). If the sampling distribution of the estimate is normal, then the sampling distribution of Z is *standard normal*, where the mean is 0 and the standard deviation is 1. The known percentiles of the standard normal distribution permit an understanding of the likely values of the Z-ratio, even though its value in any single case will not be known. From a computer program with normal percentiles, for example, it can be found that for 95% of all samples the Z-ratio will fall in the interval -1.96 to 1.96. If the standard deviation of the estimate is known, this permits an understanding of the likely size of the estimation error. Consequently, useful statements can be made about the amount of uncertainty with which questions about the parameter can be resolved.

The t-Ratio

When, as is usually the case, the standard deviation of an estimate is unknown, it is natural to replace its value in the Z-ratio with the estimate's standard error. The result is the *t-ratio*,

$$t\text{-ratio} = \frac{(\text{Estimate} - \text{Parameter})}{\text{SE(Estimate)}}$$

Associated with this t-ratio are the same degrees of freedom associated with the standard error of the estimate. The t-ratio does not have a standard normal distribution, because there is extra variability due to estimating the standard deviation. The fewer the degrees of freedom, the greater is this extra variability. Under some conditions, however, the sampling distribution of the t-ratio is known.

If $\overline{Y}$ *is the average in a random sample of size* n *from a normally distributed population, the sampling distribution of its t-ratio is described by a* ***Student's t-distribution*** *on* $n - 1$ *degrees of freedom.* The mathematical formula for the t-distributions was guessed by W. S. Gossett, a scientist who worked at the Guinness Brewery in the early 1900s and who published under the pseudonym "Student." The formula was proved to be correct by R. A. Fisher in 1925.

Histograms for t-distributions are symmetric about zero. For large degrees of freedom (about 30 or more), t-distributions differ very little from the standard normal. For smaller degrees of freedom, they have longer tails than normal. Percentiles of t-distributions are available in statistical computer programs and some pocket calculators.

2.2.4 Unraveling the *t*-Ratio

The average difference between hippocampus volumes for the 15 sets of twins in the schizophrenia study is $0.199\,\text{cm}^3$, and its standard error is $0.0615\,\text{cm}^3$, based on 14 degrees of freedom. The t-ratio is therefore $(0.199 - \mu)/0.0615$, where μ is the mean difference in the population of twins. If it can be assumed that the population distribution is normally distributed, then this t-ratio has a value typical of one drawn at random from a t-distribution with 14 degrees of freedom. A picture of this distribution is shown in Display 2.5.

This distribution indicates the likely values for the t-ratio, which in turn may be used to indicate plausible values of μ.

The Paired t-Test for No Difference

For example, consider the question "Is it plausible, based on the data, that μ could be zero?" If μ were zero, that would imply that the t-ratio was

$$t\text{-ratio (if } \mu \text{ is zero)} = (0.199 - 0)/0.0615 = 3.236.$$

Display 2.5, however, shows that 3.236 is an unusually large value to have come from the t-distribution. More precisely, only 0.003 (i.e., 0.3%) of all random samples from a population in which $\mu = 0$ lead to values of the t-ratio as large as or larger than 3.236. (This value comes from a calculator or from a statistical computer program.) The proportion of random samples that yield t-ratios that are as far or farther from 0 than 3.236 in *either* direction is 0.006 (double the 0.003). So here are your choices: (a) $\mu \neq 0$; or (b) $\mu = 0$ *and the random sampling resulted in a particularly nonrepresentative sample*. You cannot prove that $\mu \neq 0$, but you may infer it from the rarity of the converse. This type of reasoning is the conceptual basis for testing a hypothesis about a parameter, and the measure 0.006 is the (two-sided) p-value based on the t-distribution.

DISPLAY 2.5 Student's t-distribution on 14 degrees of freedom

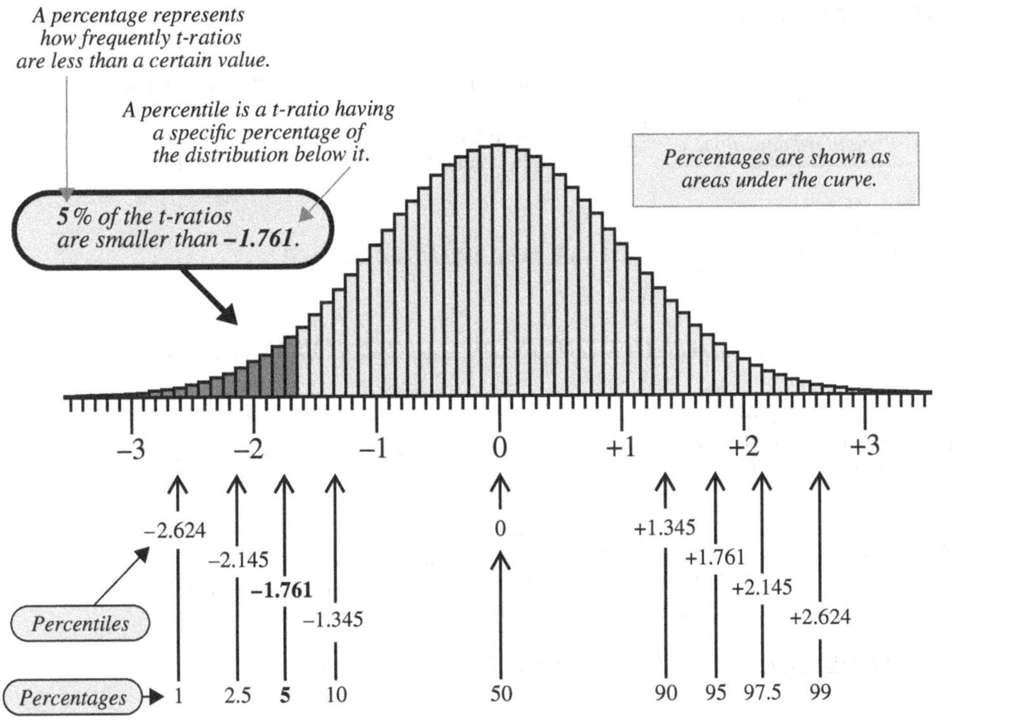
A percentage represents how frequently t-ratios are less than a certain value.
A percentile is a t-ratio having a specific percentage of the distribution below it.
5 % of the t-ratios are smaller than −1.761.
Percentages are shown as areas under the curve.
−3
−2
−1
0
+1
+2
+3
−2.624
−2.145
−1.761
−1.345
0
+1.345
+1.761
+2.145
+2.624
Percentiles
Percentages
1
2.5
5
10
50
90
95
97.5
99

A 95% Confidence Interval for the Mean

Consider also the question "What are plausible values for μ, based on the data?" This can be answered by unraveling the t-ratio in a slightly different way. Display 2.5 shows that the most typical t-ratios are near zero, with 95% of the most likely values being between -2.145 and $+2.145$. If this sample produces one of these 95% most likely t-ratios, then

$$-2.145 < (0.199 - \mu)/0.0615 < +2.145,$$

in which case μ is between 0.067 and 0.331 cm^3. The interval from 0.067 to 0.331 is a *95% confidence interval* for μ.

The Intepretation of a Confidence Interval

A 95% confidence interval will contain the parameter if the t-ratio from the observed data happens to be one of those in the middle 95% of the sampling distribution. Since 95% of all possible pairs of samples lead to such t-ratios, the *procedure* of constructing a 95% confidence interval is successful in capturing the parameter of interest in 95% of its applications. It is impossible to say whether it is successful or not in any particular application.

DISPLAY 2.6 Calculations for the paired t-test and 95% confidence interval for the schizophrenia study

SUMMARY STATISTICS

1. *Compute differences. Obtain their average, $\bar{Y}$, and standard deviation, s.*

Y_i is the difference between the left hippocampus volumes in twin i ($i = 1, \ldots, 15$), unaffected twin minus schizophrenic twin.
Sample average: $\bar{Y} = 0.199\ \text{cm}^3$
Sample standard deviation: $s = 0.238\ \text{cm}^3$; 14 *d.f.*

2. *Compute the standard error of the average: $SE(\bar{Y}) = s/\sqrt{n}$ with its d.f. = n − 1*

$SE(\bar{Y}) = 0.238/\sqrt{15} = 0.0615\ \text{cm}^3$

PAIRED t-TEST FOR THE HYPOTHESIS OF NO DIFFERENCE IN MEAN VOLUMES

3. *Compute the t-statistic: $t = (\bar{Y} - 0)/SE(\bar{Y})$*

t-statistic $= 0.199/0.0615 = 3.236$; 14 *d.f.*

4. *Find the p-value (two-sided here) as the proportion of t-ratios in a t_{n-1} distribution as or more extreme than the t-statistic*

two-sided p-value = 0.006
(from tabulated t-distribution with 14 d.f.)

95% CONFIDENCE INTERVAL FOR THE MEAN VOLUME DIFFERENCE

5. *Find the 97.5th percentile, $t_{n-1}(0.975)$, in the t-distribution with n − 1 d.f.*

$t_{14}(0.975) = 2.145$
(from tabulated t-distribution with 14 d.f.)

6. *The confidence interval: $\bar{Y} \pm t_{n-1}(0.975) \times SE(\bar{Y})$*

$0.199 \pm 2.145 \times 0.0615$
$\Rightarrow$ **0.067 cm³ to 0.331 cm³**

A summary of the calculations for the paired t-test and confidence interval for the schizophrenia study are provided in Display 2.6.

2.3 A t-RATIO FOR TWO-SAMPLE INFERENCE

The preceding discussions provide a conceptual basis for the development of inferential tools based on the t-distributions. This section repeats the discussions more formally, for the structure of independent samples from two normally distributed populations.

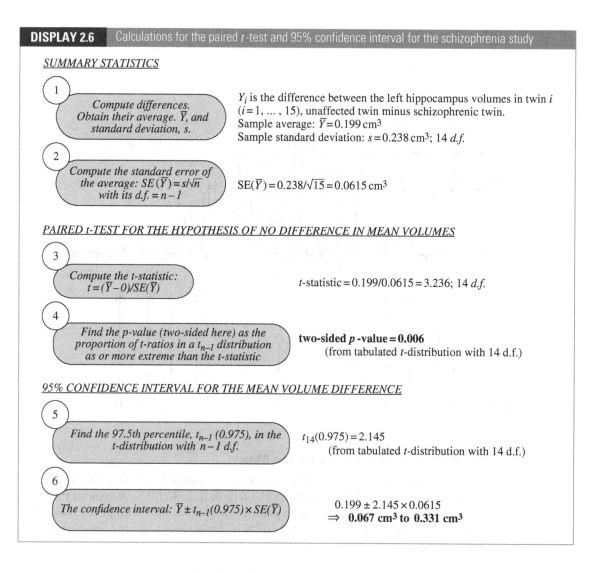

2.3.1 Sampling Distribution of the Difference Between Two Independent Sample Averages

The goal is to draw conclusions about the difference in two population means from the difference in two sample averages. The latter is variable—its value depends on the particular samples that happened to have been selected—and therefore leads to an uncertain conclusion. Fortunately, mathematical theory about the sampling distribution of the difference in averages can be used to simultaneously convey the uncertain conclusion along with an indication of its uncertainty.

Display 2.7 lists the main mathematical results. The top panel shows histograms representing the two unknown population distributions and the bottom panel shows the sampling distribution for the difference in averages, meaning the probability distribution of possible values of the difference in averages that would emerge from (hypothetical) repeated sampling from the two populations. Although only a single sample is selected from each population in practice, mathematical theory based on simple random sampling reveals some useful facts about the sampling distribution, which are indicated in bubbles 1, 2, and 3.

DISPLAY 2.7 Facts about the sampling distribution of the difference of averages from two independent random samples (from statistical theory)

Population Distributions

Post-drought (1978)

σ_2 Beak depth (mm)

μ_2

Pre-drought (1976)

σ_1 Beak depth (mm)

μ_1

Sampling Distribution of the Difference Between Averages

3 *Shape*
The shape of the sampling distribution will be more nearly normal than the shape of the population distributions.

$\mu_2 - \mu_1$

1 *Center*
The sampling distribution is centered on the difference between population means.

2 *Spread*
The sampling distribution has
$$SD(\overline{Y}_2 - \overline{Y}_1) = \sqrt{\frac{\sigma_1^2}{n_1} + \frac{\sigma_2^2}{n_2}}$$

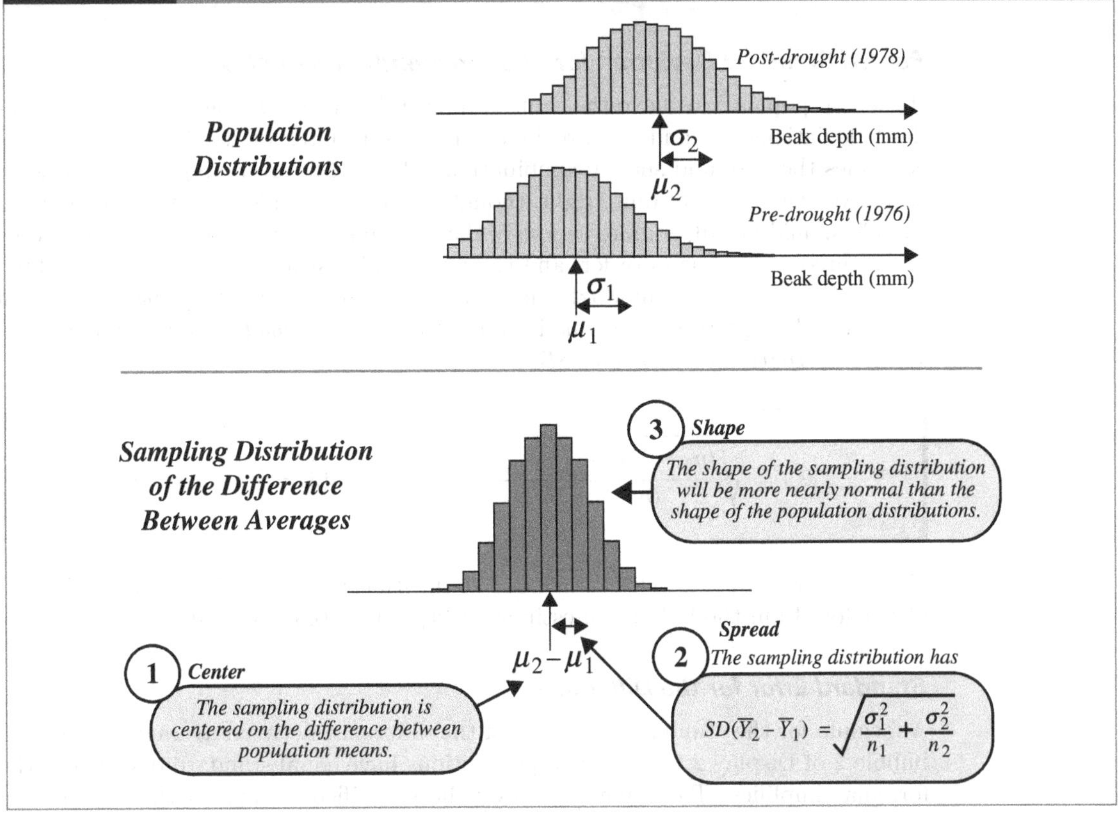

The mathematical theory reveals that the shape of the sampling distribution is approximately normal and that the adequacy of the approximation improves with increasing sample sizes. It also provides a formula for the standard deviation of the sampling distribution, as shown in bubble 3 of Display 2.7. This formula isn't directly useable because it depends on the unknown population standard deviations. As in the one-sample problem, though, an estimate is obtained by replacing the unknown σ's in the formula by estimates. As before, the resulting *estimated* standard deviation of the sampling distribution is called a *standard error.*

2.3.2 Standard Error for the Difference of Two Averages

Statisticians have devised two methods for estimating the standard deviation in the sampling distribution of the difference between two averages. Some prefer an "unequal SD" method in which the two SDs are estimated independently from the two samples. This method will be presented in Chapter 4. Others prefer an "equal SD" method in which the two SDs are assumed equal and a single estimate of the common value is made by pooling information from the two samples. This book focuses on the latter method because it is a fundamental starting point for learning about the more sophisticated tools of regression and analysis of variance, which follow in later chapters. So assume in the following that the two populations have equal standard deviations: $\sigma_1 = \sigma_2 = \sigma$.

Pooled Standard Deviation for Two Independent Samples

If the two populations have the same standard deviation, σ, then the sample standard deviations, s_1 and s_2, are independent estimates of it. A single estimate combines the two, and such a combination is formed by averaging on the variance scale. An average is not quite right, though, since the sample variance from a larger sample should be taken more seriously than a sample variance from a smaller one. A weighted average is in order, and the best single estimate of σ^2 is the weighted average in which the individual sample variances are weighted by their degrees of freedom. The square root of this is called the *pooled estimate of standard deviation*, s_p, or, alternatively, the *pooled SD*:

$$s_p = \sqrt{\frac{(n_1 - 1)s_1^2 + (n_2 - 1)s_2^2}{(n_1 + n_2 - 2)}}, \qquad \text{d.f.} = n_1 + n_2 - 2.$$

The number of degrees of freedom associated with this estimate is the sum of degrees of freedom from the individual estimates: $(n_1 - 1) + (n_2 - 1) = n_1 + n_2 - 2$.

Standard Error for the Difference

A formula for the standard deviation of the difference between averages appears in bubble 2 of Display 2.7. If the two populations have equal standard deviations, the formula simplifies. The following shows the simplified formula with the common standard deviation replaced by the pooled estimate of it. This is the standard error for the difference in sample averages.

$$\mathrm{SE}(\overline{Y}_2 - \overline{Y}_1) = s_p\sqrt{\frac{1}{n_1} + \frac{1}{n_2}}.$$

Display 2.8 shows the standard error calculations from the summary statistics of the two groups of finch beak depths, resulting in a value of 0.1459 mm from 176 degrees of freedom. The standard error may be interpreted as an estimate of the typical size of the deviation of $\overline{Y}_2 - \overline{Y}_1$ from the population quantity of interest, $\mu_2 - \mu_1$. This suggests that the observed difference in averages, 0.6685 mm, might depart by about 0.1459 mm from the quantity of interest. In this way, the standard error can help researchers describe the uncertainty in the uncertain conclusion from their estimate. Even better communication of the uncertainty, though, is accomplished with p-values and confidence intervals.

DISPLAY 2.8 Calculation of the pooled estimate of SD and the standard error for the difference between two sample averages; finch beak data

(1) Summary Statistics

Group	n	Average (mm)	Sample SD (mm)
1: Pre-drought	89	9.4697	1.0353
2: Post-drought	89	10.1382	0.9065

(2) The Pooled SD

$$s_p = \sqrt{\frac{(89-1)(1.0353)^2 + (89-1)(0.9065)^2}{(89+89-2)}}$$

$$= \sqrt{\frac{166.6358}{176}}$$ ← *These are the degrees of freedom associated with the pooled SD.*

$$= \sqrt{0.9468}$$ ← *This is the pooled variance.*

Answer → $s_p = 0.9730$ mm

(3) The Standard Error

$$SE(\overline{Y}_2 - \overline{Y}_1) = 0.9730\sqrt{\frac{1}{89} + \frac{1}{89}}$$

$$= 0.1459 \text{ mm}$$ ← *Answer*

2.3.3 Confidence Interval for the Difference Between Population Means

Inferences about the difference between population means arise from the consideration of a t-ratio, as in Section 2.2.3. The parameter of interest is $\mu_2 - \mu_1$. Its estimate is $\overline{Y}_2 - \overline{Y}_1$. The standard error comes from the previous section, and it has $n_1 + n_2 - 2$ degrees of freedom. In this case the t-ratio is $[(\overline{Y}_2 - \overline{Y}_1) - (\mu_2 - \mu_1)]/\text{SE}(\overline{Y}_2 - \overline{Y}_1)$. If the populations are normally distributed, this t-ratio has a t-distribution with $n_1 + n_2 - 2$ degrees of freedom.

For the finch beak data $\overline{Y}_2 - \overline{Y}_1 = 0.6685$ mm, $\text{SE}(\overline{Y}_2 - \overline{Y}_1) = 0.1459$ mm, and the t-ratio has a sampling distribution described by a t-distribution on 176 degrees of freedom. A statement about the likely values for the t-ratio from this distribution can be translated into a statement about the plausible values for $\mu_2 - \mu_1$.

A computer program that provides percentiles of t-distributions will reveal that 95% of values in a t-distribution with 176 degrees of freedom will fall between -1.9735 and $+1.9735$. A 95% confidence interval can be obtained by supposing that the actual t-ratio is one of these 95% in the center. The extreme t-ratios, -1.9735 and $+1.9735$, are now used to find corresponding endpoints for an interval on the parameter. Setting

$$-1.9735 < \frac{0.6685 - (\mu_2 - \mu_1)}{0.1459} < 1.9735$$

and solving for the parameter value yields the two interval endpoints: $0.3807 \text{ mm} < \mu_2 - \mu_1 < 0.9564 \text{ mm}$.

The preceding interval will contain $\mu_2 - \mu_1$ if the t-ratio for the observed samples is one of those 95% central ones, but not otherwise. The terminology "95% confidence" means that the procedure of constructing a 95% confidence interval is successful in 95% of its applications. It's successful in the 95% of applications for which chance deals a t-ratio from the central part of the t-distribution.

The Mechanics of Confidence Interval Construction

A confidence interval with *confidence level* $100(1 - \alpha)\%$ is the following:

$100(1 - \alpha)\%$ Confidence Limits for the Difference Between Means:

$$(\overline{Y}_2 - \overline{Y}_1) \pm t_{\text{df}}(1 - \alpha/2)\text{SE}(\overline{Y}_2 - \overline{Y}_1).$$

This formula requires that a quantity be subtracted from $\overline{Y}_2 - \overline{Y}_1$ to get the lower endpoint and that the same quantity be added to $\overline{Y}_2 - \overline{Y}_1$ to get the upper endpoint. The symbol $t_{\text{df}}(1-\alpha/2)$ represents the $100(1-\alpha/2)$th percentile of the t-distribution on d.f. degrees of freedom. For example, $t_{176}(0.975)$ represents the 97.5th percentile in the t-distribution with 176 degrees of freedom (which is 1.9735). It may seem strange that the 97.5th percentile is desired in the calculation of a 95% confidence interval, but the 2.5th and the 97.5th percentiles are the ones that divide the middle

DISPLAY 2.9 Construction of a 95% confidence interval for the difference between the mean beak depths in 1978 and 1976

Group	n	Average (mm)	SD (mm)
1: Pre-drought	89	9.4697	1.0353
2: Post-drought	89	10.1382	0.9065

$\overline{Y}_2 - \overline{Y}_1 = 10.1382 - 9.4697 = 0.6685$ mm
$SE(\overline{Y}_2 - \overline{Y}_1) = 0.1459$ mm
Degrees of freedom $= 89 + 89 - 2 = 176$
$t_{176}(0.975) = 1.9735$
Half-width $= (1.9735)(0.1459) = 0.2879$
Lower 95% confidence limit $= 0.6685 - 0.2879 = 0.3807$ mm
Upper 95% confidence limit $= 0.6685 + 0.2879 = 0.9564$ mm

From Display 2.8

From tables of the t-distribution with 176 degrees of freedom

95% of the distribution from the rest; and the 2.5th percentile is always the negative of the 97.5th. Display 2.9 summarizes the solution for the finch beak data.

Factors Affecting the Width of a Confidence Interval

There is a trade-off between the level of confidence and the width of the confidence interval. The level of confidence can be specified to be large by the user (and a high confidence level is good), but only at the expense of having a wider interval (which is bad, since the interval is less specific in answering the question of interest). If the consequences of not capturing the parameter are severe, then it might be wise to use 99% confidence intervals, even though they will be wider and, therefore, less informative. If the consequences of not capturing the parameter are minor, then a 90% interval might be a good choice. Although somewhat arbitrary, a confidence level of 95% is a conventional choice for balancing the trade-off between level of confidence and interval width.

The only way to decrease the interval width without decreasing the level of confidence (or similarly to increase the level of confidence without increasing width) is to get more data or, if possible, to reduce the size of σ. If a guess or initial estimate of σ is available, it is possible to determine the sample size needed to get a confidence interval of a certain width. This is discussed in Chapter 23.

2.3.4 Testing a Hypothesis About the Difference Between Means

To test a hypothesized value for a parameter, a *t-statistic* is formed in the same way as the t-ratio, but supposing the hypothesis is true. The t-distribution permits an evalution of whether the t-statistic is a likely value for a t-ratio and, hence, whether the hypothesis is reasonable. The t-statistic for the difference in means is

$$t\text{-statistic} = \frac{(\overline{Y}_2 - \overline{Y}_1) - [\textit{Hypothesized value for } (\mu_2 - \mu_1)]}{\text{SE}(\overline{Y}_2 - \overline{Y}_1)}.$$

This t-statistic tells how many standard errors the estimate is away from the hypothesized parameter. Its sign tells whether the estimate is above the hypothesized value (+) or below it (−).

The *p-value*, used as a measure of the credibility of the hypothesis, is the proportion of all possible t-ratios that are as far or farther from zero than is the t-statistic.

*The **p-value** for a t-test is the probability of obtaining a t-ratio as extreme or more extreme than the t-statistic in its evidence against the null hypothesis, if the null hypothesis is correct.*

The p-value may be based on a probability model induced by random assignment in a randomized experiment (Section 1.3.2) or on a probability model induced by random sampling from populations, as here.

If the p-value is small, then either the hypothesis is correct—and the sample happened to be one of those rare ones that produce such an unusual t-ratio—or the hypothesis is incorrect. Although it is impossible to know which of these two possibilities is true, the p-value indicates the probability of the first of these results and, therefore, provides a measure of credibility for that interpretation. *The smaller the p-value, the stronger is the evidence that the hypothesis is incorrect.* A large p-value implies that the study is not capable of excluding the null hypothesis as a possible explanation for how the data turned out. A possible wording in this case is "the data are consistent with the hypothesis being true." It is wrong to conclude that the null hypothesis *is true.*

One-Sided and Two-Sided p-Values

In the finch beak example the t-statistic for testing the hypothesis of "no difference" in population means is 4.583. The proportion of t-ratios that are as far or farther from zero than 4.583 is found from the percentiles of the t_{176} distribution. The proportion of t-ratios *greater than or equal* to 4.583 is miniscule, < 0.00001. The t-ratios *as far or farther from zero* than 4.583 are those less than or equal to −4.583 and those greater than or equal to 4.583, and this proportion is twice the proportion greater than 4.583 (because t-distributions are symmetric), but still miniscule.

The proportion of t-ratios farther from zero in one specified direction is referred to as a *one-sided p-value*. The proportion of t-ratios farther from zero than the t-statistic, either positively or negatively, is a *two-sided p-value*.

The choice of one-sided or two-sided depends on how specifically the researcher wishes to declare the alternatives to the hypothesis of equal means. In the finch beak study, the Grants surmised that the response of the species to the sole availability of large, tough seeds would not merely be a change in mean beak sizes but more specifically a tendency toward larger beak sizes. If their intention is to report a conclusion about the evidence that the mean beak size *increased*, regardless of how the data actually turn out, then a one-sided p-value is in order.

Much has been made of whether to report one-sided or two-sided p-values. There are some situations where a one-sided p-value is appropriate, some where a two-sided p-value is appropriate, and many where it is not at all clear which is appropriate. Since the two provide equivalent measures of evidence against the hypothesis of equal means (that is, the two-sided p-value is simply twice the one-sided p-value), the distinction is not terribly important; a reader may convert one to other. There is only one absolute when it comes to reporting: *always report whether the p-value is one- or two-sided.*

2.3.5 The Mechanics of p-Value Computation

The steps required to compute a p-value for the test of the hypothesis that $\mu_2 - \mu_1 = D$ (a specified value, like 0) are as follows.

1. Compute the estimate, $\overline{Y}_2 - \overline{Y}_1$, its standard error, and the degrees of freedom.
2. Compute the t-statistic: $t = [(\overline{Y}_2 - \overline{Y}_1) - D]/\text{SE}(\overline{Y}_2 - \overline{Y}_1)$.
3. Determine the proportion, P, of t-ratios that are less than the t-statistic, using a statistical computer program with t-distribution percentiles with the appropriate degrees of freedom.
4. Determine the p-value based on the proportion, P, and the alternatives of interest. (a) For one-sided alternatives of the form $\mu_2 - \mu_1 > D$, t-ratios larger than t are more extreme, so the one-sided p-value is $1 - P$. (b) For the one-sided alternatives of the form $\mu_2 - \mu_1 < D$, t-ratios smaller than t are more extreme, so the one-sided p-value is P. (c) For two-sided alternatives of the form $\mu_2 - \mu_1 \neq D$, t-ratios that are larger in magnitude than t are more extreme, so the two-sided p-value is $2P$ if $P < 0.5$ or $2(1 - P)$ if $P > 0.5$.
5. Report the hypothesis, the p-value, and whether it is one- or two-sided.

An illustration of this procedure for the finch beak data is shown in Display 2.10.

2.4 INFERENCES IN A TWO-TREATMENT RANDOMIZED EXPERIMENT

Probability models for randomized experiments (Section 1.3.1) are spawned by the chance mechanisms used to assign subjects to treatment groups. Probability models for random sampling (Section 1.4.1) are spawned by the chance mechanisms used to select units from real, finite populations. Chapter 2 has thus far discussed inference procedures whose motivation stems from considerations of random sampling from populations that are conceptual, infinite, and normally distributed. While there seems to be a considerable difference between the situations, it turns out that the t-distribution uncertainty measures discussed in this chapter are useful approximations to both the randomization and the random sampling uncertainty measures for a wide range of problems. The practical consequence is that t-tools

DISPLAY 2.10 The t-test for the hypothesis that the post-drought and pre-drought beak depth means are equal

Group	n	Average (mm)	SD (mm)
1: Pre-drought	89	9.4697	1.0353
2: Post-drought	89	10.1382	0.9065

From Display 2.8

$\overline{Y}_2 - \overline{Y}_1 = 10.1382 - 9.4697 = 0.6685$ mm

$SE(\overline{Y}_2 - \overline{Y}_1) = 0.1459$ mm

Degrees of freedom $= 89 + 89 - 2 = 176$

Hypothesized difference

$$t\text{-statistic} = \frac{0.6685 - \mathbf{0.0}}{0.1459} = 4.583$$

$P = 0.999996$ ← *From tables of the t-distribution with 176 degrees of freedom:*

One-sided p-value = 0.000004

are used for many situations that do not conform to the strict model upon which the t-tools are based, including data from randomized experiments. Conclusions from randomized experiments, however, are phrased in the language of treatment effects and causation, rather than differences in population means and association.

2.4.1 Approximate Uncertainty Measures for Randomized Experiments

Although theoretically motivated by random samples from normal populations, the two-sample t-test can also be applied to data from a two-group randomized experiment. Reconsideration of the creativity study (Section 1.1.1) illustrates the procedure. In the intrinsic group, the average of 24 scores is 19.88, and the SD is 4.44. In the extrinsic group, the average of 23 scores is 15.74, and the SD is 5.25. The pooled estimate of the standard deviation is 4.85, and the standard error for the difference between averages is 1.42 with 45 degrees of freedom. The difference between average scores is 4.14 points.

Hypothesis Test of No Treatment Effect

The t-statistic, $t = (4.14 - 0)/1.42 = 2.92$, can be used to test the hypothesis of no treatment effect in exactly the same way it would be used to test equal population means from two random samples. The probability in a t-distribution on 45 degrees of freedom to the right of 2.92 is 0.0027, so the one-sided p-value for the alternative of a positive difference is 0.0027, and the two-sided p-value is 0.0054. The conclusion is phrased in terms of the additive treatment effect model: "This experiment provides strong evidence that receiving the intrinsic questionnaire caused a higher creativity score (one-sided p-value $= 0.0027$)."

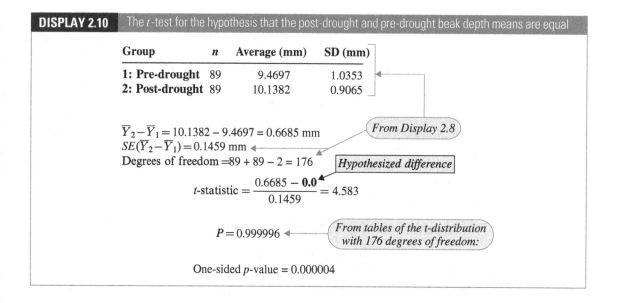

Confidence Interval for a Treatment Effect

Construction of a confidence interval for an additive treatment effect δ is precisely the same as for the difference between population means, $\mu_2 - \mu_1$. The 97.5th percentile in the t-distribution with 45 degrees of freedom is 2.014, so the interval half-width is $(2.014)(1.42) = 2.85$. The interval runs from $4.14 - 2.85 = 1.29$ to $4.14 + 2.85 = 6.99$ points.

Compare this construction with one based on the randomization procedure itself (Section 1.3). The randomization-based procedure relies on a relationship between testing and confidence intervals: *Any hypothesized parameter value should be included or excluded from a* $100(1-\alpha)\%$ *confidence interval according to whether its test yields a two-sided p-value that is greater than or less than* α. Accordingly, to construct a 95% confidence interval for the treatment effect δ, include only those values of δ which, when tested, have two-sided p-values greater than 0.05.

To determine whether $\delta = 5$ should be included in a 95% confidence interval, one must consider the randomization model for $\delta = 5$. If this is the correct value, subtracting 5 from the scores of all persons in the intrinsic group reconstructs the scores they would have had if placed in the extrinsic group. Now all 47 scores are homogeneous, so a randomization test should conclude there is no difference. Perform the randomization test. If the two-sided p-value exceeds (or equals) 0.05, include 5 in the interval. Otherwise, leave it out. That settles the issue for $\delta = 5$, but the limits of the interval must be found by repeating this process to find the smallest and largest δ for which the two-sided p-value is greater than or equal to 0.05.

Approximation of the Randomization Distribution of the t-Statistic

The t-based p-values and confidence intervals are only approximations to the correct p-values and confidence intervals from randomization distributions. To see how good the approximation is in the creativity study, the analysis of Chapter 1 was modified by considering the randomization distribution of the t-statistic rather than the difference in sample averages. A histogram of the t-statistics from 500,000 random regroupings appears in Display 2.11, along with the approximating t-distribution. The observed t-statistic (2.92) was exceeded by only 1,298 of the 500,000 random regroupings, giving an estimated one-sided p-value of $1{,}298/500{,}000 = 0.0026$. The approximation based on the t-distribution (0.0027) is quite good.

2.5 RELATED ISSUES

2.5.1 Interpretation of *p*-Values

A small p-value like 0.005 means either that the null hypothesis is correct—and the randomization or random sampling led by chance to a rare outcome where the t-ratio is quite far from zero—or that the hypothesis is incorrect. The smaller the p-value, the greater is the evidence that the second explanation is the correct one.

DISPLAY 2.11 A histogram of t-statistics from 500,000 random regroupings of the creativity study data with an approximating t-distribution curve, demonstrating that the easier-to-use t-tools can give very good approximations to randomization test inferences, even through they are derived for a quite different model—random samples from two normal populations

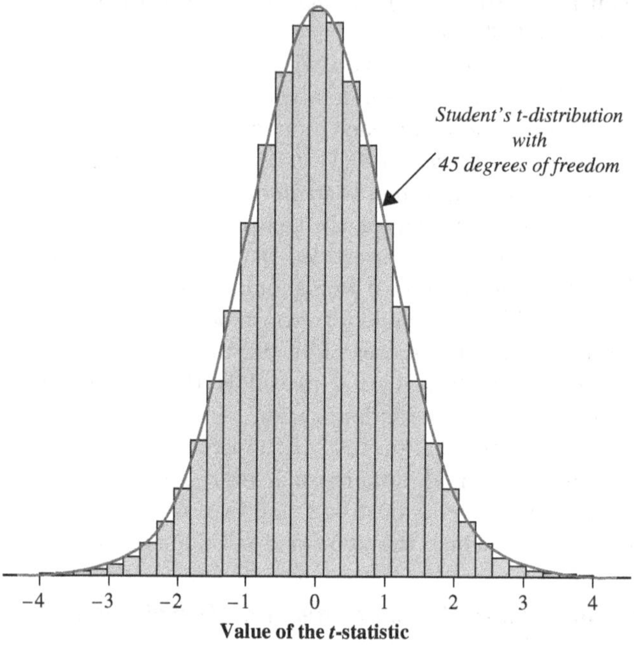

DISPLAY 2.12 Interpreting the size of a p-value

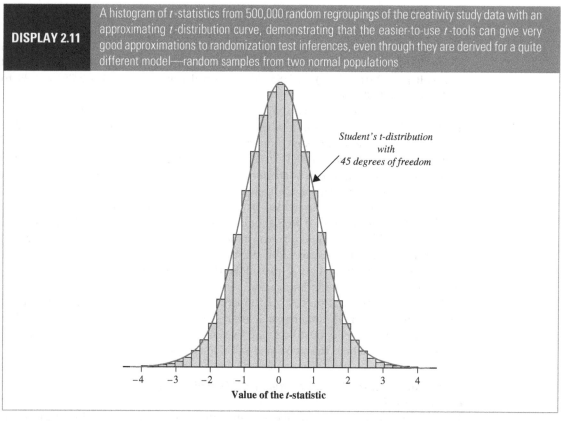

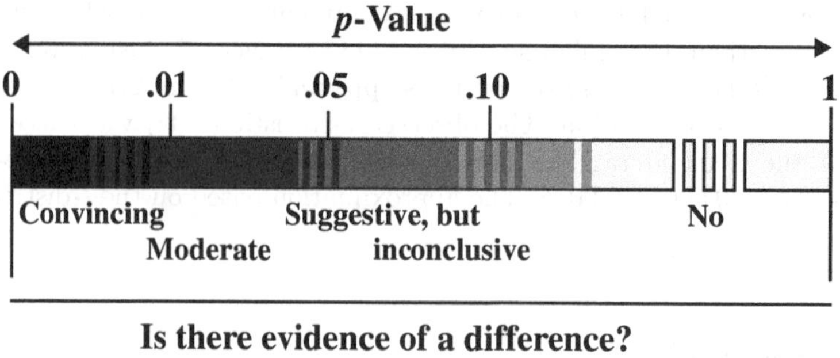

How small is small? It is difficult and unwise to decide on absolute cutoff points for believability to be applied in all situations. Display 2.12 represents a starting point for interpreting p-values.

P-values can be comprehended by comparing them to events whose probabilities are more familiar. For example, at what point does a person flipping a series of heads begin to doubt that the coin is fair? It is not terribly unlikely to get four heads in a row. The probability of doing so with a fair coin is 0.0625. At five heads in a row one might start to get a bit curious. The chance of this, if the coin is fair, is 0.03125. When the sixth toss is also heads, one may start to question the integrity of the coin, even though it is possible that a fair coin could turn up heads six times in a row (with probability 0.015625). Ten heads in a row is convincing evidence that the coin is not of standard issue. The probability of this event, if the chance of heads were in fact one-half, is 0.0009766.

It is tempting to think of a p-value as the probability of the null hypothesis being correct, but this interpretation is technically incorrect and potentially misleading. The hypothesis is or is not correct, and there is no probability associated with that. The probability arises from the uncertainty in the data. So the best technical description is the precise definition (see Section 2.3.4), clumsy as it may sound.

2.5.2 An Example of Confidence Intervals

In 1915 Albert Einstein published four papers in the proceedings of the Prussian Academy of Sciences laying the foundations of general relativity and describing some of its consequences. A paper establishing the field equation for gravity predicted that the arc of deflection of light passing the sun would be twice the angle predicted by Newton's gravity theory. Half the predicted deflection comes directly from Newton's calculations, and the other half comes from the curvature of space near the sun relative to space far away. This is represented by the equation

$$\Delta = (1/2)(1+\gamma)\frac{1.75}{d},$$

where Δ is the deflection of light, d is the distance of the closest approach of the ray to the sun (in solar radii), and γ is the parameter describing space curvature.

The parameter γ, which is predicted by Einstein's general relativity theory to be 1 and by Newtonian physics to be 0, was estimated in 1919 by British astronomers during a total solar eclipse. Since then, measurements have been repeated many times, under various measurement conditions. The efforts are summarized in Display 2.13. (Data from C. M. Will, "General Relativity at 75: How Right Was Einstein?" *Science* 250 (November 9, 1990): 770–75.)

The confidence intervals around the estimates in Display 2.13 reflect uncertainty due to measurement errors. (The actual confidence levels are not given and are not important for this illustration.) After the first relatively crude attempts to measure γ, little improvement was made until the late 1960s and the discovery of quasars. Measurements of light from quasar groups passing near the sun led to dramatic improvement in accuracy, as evident in the narrower intervals with later years.

A Note About the Accumulation of Evidence

This example shows that theories must withstand continual challenges from skeptical scientists. The essence of scientific theory is the ability to predict future

DISPLAY 2.13 Estimates and confidence intervals for γ, the deflection of light around the sun, from 20 experiments

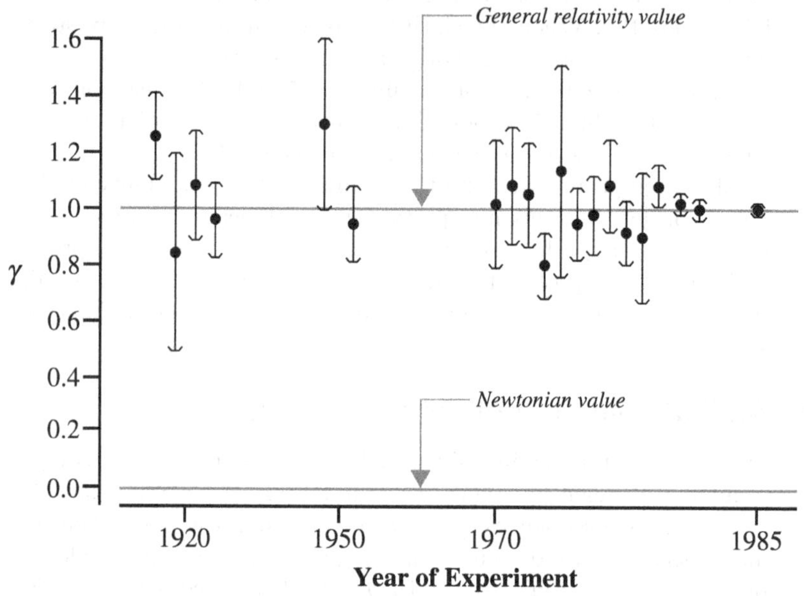

outcomes. Experimental results are typically uncertain. So the fact that some intervals fail to include the value $\gamma = 1$ is not taken to disprove general relativity, but neither would it prove general relativity right if all the intervals did include $\gamma = 1$. When a theory's predictions are consistently denied by a series of experiments—such as the Newtonian prediction of $\gamma = 0$ in this example—scientists agree that the theory is not adequate.

2.5.3 The Rejection Region Approach to Hypothesis Testing

Not long ago, statisticians took a *rejection region* approach to testing hypotheses. A *significance level* of 0.05, say, was selected in advance, and a *p*-value less than 0.05 called for rejecting the hypothesis at the significance level 0.05; otherwise, the hypothesis was accepted, or more correctly, not rejected. Thus *p*-values of 0.048 and 0.0001 both lead to rejection at the 0.05 level, even though they supply vastly different degrees of evidence. On the other hand, *p*-values of 0.049 and 0.051 lead to different conclusions even though they provide virtually identical evidence against the hypothesis. Although important for leading to advances in the theory of statistics, the rejection region approach has largely been discarded for practical applications and *p*-values are reported instead. *P*-values give the reader more information for judging the significance of findings.

2.6 SUMMARY

Many research questions can be formulated as comparisons of two population distributions. Comparison of the distributions' centers effectively summarizes the difference between the parameters of interest when the populations have the same variation and general shape. This chapter concentrated on the difference in means, $\mu_2 - \mu_1$, which is estimated by the difference in sample averages.

The statistical problem is to assess the uncertainty associated with the difference between the estimate (the difference in sample averages) and the parameter (the difference in population means). The sampling distribution of an estimate is the key to understanding the uncertainty. It is represented as a histogram of values of the estimate for every possible sample that could have been selected.

Often with fairly large samples, a sampling distribution has a normal shape. A normal sampling distribution is specified by its mean and its standard deviation. When the populations have common standard deviation σ the difference in sample averages has a sampling distribution with mean $\mu_2 - \mu_1$ and standard deviation

$$\sigma\sqrt{\frac{1}{n_1} + \frac{1}{n_2}}.$$

This could be used to describe uncertainty except that it involves the unknown σ—the common standard deviation in the two populations. In practice, σ is replaced by its best estimate from the data—the pooled standard deviation, s_p, having $n_1 + n_2 - 2$ degrees of freedom. The estimated standard deviation of the sampling distribution is called the standard error.

The standard error alone, however, does not entirely describe the uncertainty in an estimate. More precise statements can be made by using the t-ratio, which has a Student's t-distribution as its sampling distribution (if the ideal normal model applies). This leads directly to confidence intervals and p-values as statistical tools for answering the questions of interest. The confidence interval provides a range of likely values for the parameter, and the confidence level is interpreted as the frequency with which the interval construction procedure gives the right answer. For testing whether a particular hypothesized number could be the unknown parameter, the t-statistic is formed by substituting the hypothesized value for the true value in the t-ratio. The p-value is the chance of getting as extreme or more extreme t-ratios than the t-statistic, and it is interpreted as a measure of the credibility of the hypothesized value.

2.7 EXERCISES

Conceptual Exercises

1. **Finch Beak Data.** Explain why the finch beak study is considered an observational study.

2. For comparing two population means when the population distributions have the same standard deviation, the standard deviation is sometimes referred to as a nuisance parameter. Explain why it might be considered a nuisance.

3. True or false? If a sample size is large, then the shape of a histogram of the sample will be approximately normal, even if the population distribution is not normal.

4. True or false? If a sample size is large, then the shape of the sampling distribution of the average will be approximately normal, even if the population distribution is not normal.

5. Explain the relative merits of 90% and 99% levels of confidence.

6. What is wrong with the hypothesis that $\overline{Y}_2 - \overline{Y}_1$ is 0?

7. In a study of the effects of marijuana use during pregnancy, measurements on babies of mothers who used marijuana during pregnancy were compared to measurements on babies of mothers who did not. (Data from B. Zuckerman et al., "Effects of Maternal Marijuana and Cocaine Use on Fetal Growth," *New England Journal of Medicine* 320(12) (March 1989): 762–68.) A 95% confidence interval for the difference in mean head circumference (nonuse minus use) was 0.61 to 1.19 cm. What can be said from this statement about a p-value for the hypothesis that the mean difference is zero?

8. Suppose the following statement is made in a statistical summary: "A comparison of breathing capacities of individuals in households with low nitrogen dioxide levels and individuals in households with high nitrogen dioxide levels indicated that there is no difference in the means (two-sided p-value $= 0.24$)." What is wrong with this statement?

9. What is the difference between (a) the mean of Y and the mean of $\overline{Y}$? (b) the standard deviation of Y and the standard deviation of $\overline{Y}$? (c) the standard deviation of $\overline{Y}$ and the standard error of $\overline{Y}$? (d) a t-ratio and a t-statistic?

10. Consider blood pressure levels for populations of young women using birth control pills and young women not using birth control pills. A comparison of these two populations through an observational study might be consistent with the theory that the pill elevates blood pressure levels. What tool is appropriate for addressing whether there is a difference between these two populations? What tool is appropriate for addressing the likely size of the difference?

11. The data in Display 2.14 are survival times (in days) of guinea pigs that were randomly assigned either to a control group or to a treatment group that received a dose of tubercle bacilli. (Data from K. Doksum, "Empirical Probability Plots and Statistical Inference for Nonlinear Models in the Two-Sample Case," *Annals of Statistics* 2(1974): 267–77.) (a) Why might the additive treatment effect model (introduced in Section 1.3.1) be inappropriate for these data? (b) Why might the ideal normal model with equal spread be an inadequate approximation?

Computational Exercises

12. Marijuana Use During Pregnancy. For the birth weights of babies in two groups, one born of mothers who used marijuana during pregnancy and the other born of mothers who did not (see Exercise 7), the difference in sample averages (nonuser mothers minus user mothers) was 280 grams, and the standard error of the difference was 46.66 grams with 1,095 degrees of freedom. From this information, provide the following: (a) a 95% confidence interval for $\mu_2 - \mu_1$, (b) a 90% confidence interval for $\mu_2 - \mu_1$, and (c) the two-sided p-value for a test of the hypothesis that $\mu_2 - \mu_1 = 0$.

13. Fish Oil and Blood Pressure. Reconsider the changes in blood pressures for men placed on a fish oil diet and for men placed on a regular oil diet, from Chapter 1, Exercise 12. Do the following steps to compare the treatments.

DISPLAY 2.14 Lifetimes of guinea pigs in two treatment groups

Control (n=64)		*Received bacilli (n=58)*
36,18	0	
91,89,87,86,52,50		76,93,97
49,20,19,18,15,14,14,08,02	1	07,08,13,14,19,36,38,39
89,78,73,67,67,66,65,60		52,54,54,60,64,64,66,68,78,79,81,81,83,85,94,98
16,12,09	2	12,13,16,20,25,25,44
92,79,78,73		53,56,59,65,68,70,83,89,91
41	3	11,15,26,26
82,80,67,55		61,73,73,76,97,98
46,32,21,21	4	06
74,63,55		59,66
46,45,05	5	
90,76,69		92,98
41,38,37,34,21,08,07,03	6	
88,85,63,50		
35,25	7	

Legend: 5 | 98 represents 598 days

(a) Compute the averages and the sample standard deviations for each group separately.
(b) Compute the pooled estimate of standard deviation using the formula in Section 2.3.2.
(c) Compute SE($\overline{Y}_2 - \overline{Y}_1$) using the formula in Section 2.3.2.
(d) What are the degrees of freedom associated with the pooled estimate of standard deviation? What is the 97.5th percentile of the t-distribution with this many degrees of freedom?
(e) Construct a 95% confidence interval for $\mu_2 - \mu_1$ using the formula in Section 2.3.3.
(f) Compute the t-statistic for testing equality as shown in Section 2.3.5.
(g) Find the one-sided p-value (as evidence that the fish oil diet resulted in greater reduction of blood pressure) by comparing the t-statistic in (f) to the percentiles of the appropriate t-distribution (by reading the appropriate percentile from a computer program or calculator).

14. Fish Oil and Blood Pressure. Find the 95% confidence interval and one-sided p-value asked for in Exercise 13(e) and (g) but use a statistical computer package to do so.

15. Auto Exhaust and Lead Concentration in Blood. Researchers took independent random samples from two populations of police officers and measured the level of lead concentration in their blood. The sample of 126 police officers subjected to constant inhalation of automobile exhaust fumes in downtown Cairo had an average blood level concentration of 29.2 μg/dl and an SD of 7.5 μg/dl; a control sample of 50 police officers from the Cairo suburb of Abbasia, with no history of exposure, had an average blood level concentration of 18.2 μg/dl and an SD of 5.8 μg/dl. (Data from A.-A. M. Kamal, S. E. Eldamaty, and R. Faris, "Blood Lead Level of Cairo Traffic Policemen," *Science of the Total Environment* 105(1991): 165–70.) Is there convincing evidence of a difference in the population averages?

16. Motivation and Creativity. Verify the statements made in the summary of statistical findings for the Motivation and Creativity Data (Section 1.1.1) by analyzing the data on the computer.

17. Sex Discrimination. Verify the statements made in the summary of statistical findings for the Sex Discrimination Data (Section 1.1.2) by analyzing the data on the computer.

18. The Grants' Finch Complete Beak Data. The data file ex0218 contains the beak depths (in mm) of all 751 finches captured by Peter and Rosemary Grant in 1976 and all 89 finches captured in 1978 (as described in Section 2.1.1). Use a statistical computer program for parts a–d: (a) Draw side-by-side box plots of the two groups of beak depths. (b) Use the two-sample t-test on these data to find the one-sided p-value for a test of the hypothesis of no difference in means against the alternative that the mean in 1978 is larger. (c) What is the two-sided p-value from the t-test? (d) Provide an estimate and a 95% confidence interval for the amount by which the 1978 mean exceeds the 1976 mean. (e) What is it about the finches in the two populations that might make you question the validity of the independence assumption upon which the two-sample t-test is derived?

19. Fish Oil and Blood Pressure. Reconsider the fish oil and blood pressure data of Chapter 1, Exercise 12. Since the measurements are the reductions in blood pressure for each man, it is of interest to know whether the mean reduction is zero for each group. For the regular oil diet group do the following:

(a) Compute the average and the sample standard deviation. What are the degrees of freedom associated with the sample standard deviation, s_2?

(b) Compute the standard error for the average from this group: $\mathrm{SE}(\overline{Y}_2) = s_2/\sqrt{n_2}$.

(c) Construct a 95% confidence interval for μ_2 as $\overline{Y}_2 + t_d(.975)\mathrm{SE}(\overline{Y}_2)$, where d is the degrees of freedom associated with s_2.

(d) For the hypothesis that μ_2 is zero, construct the t-statistic $\overline{Y}_2/\mathrm{SE}(\overline{Y}_2)$. Find the two-sided p-value as the proportion of values from a t_d-distribution farther from 0 than this value.

20. Fish Oil and Blood Pressure (One-Sample Analysis). Repeat Exercise 19 for the group of men who were given the fish oil diet and then answer these questions: Is there any evidence that the mean reduction for this group is different from zero? What is the typical reduction in blood pressure expected from this type of diet (for individuals like these men)? Provide a 95% confidence interval.

Data Problems

21. Bumpus Natural Selection Data. In 1899, biologist Hermon Bumpus presented as evidence of natural selection a comparison of numerical characteristics of moribund house sparrows that were collected after an uncommonly severe winter storm and which had either perished or survived as a result of their injuries. Display 2.15 shows the length of the humerus (arm bone) in inches for 59 of these sparrows, grouped according to whether they survived or perished. Analyze these data to summarize the evidence that the distribution of humerus lengths differs in the two populations. Write a brief paragraph of statistical conclusion, using the ones in Section 2.1 as a guide, including a

DISPLAY 2.15 Humerus lengths of moribund male house sparrows measured by Hermon Bumpus, grouped according to survival status

Humerus Lengths (inches) of 35 Males That *Survived*

0.687, 0.703, 0.709, 0.715, 0.721, 0.723, 0.723, 0.726, 0.728, 0.728, 0.728, 0.729, 0.730, 0.730, 0.733, 0.733, 0.735, 0.736, 0.739, 0.741, 0.741, 0.741, 0.741, 0.743, 0.749, 0.751, 0.752, 0.752, 0.755, 0.756, 0.766, 0.767, 0.769, 0.770, 0.780

Humerus Lengths (inches) of 24 Males That *Perished*

0.659, 0.689, 0.702, 0.703, 0.709, 0.713, 0.720, 0.720, 0.726, 0.726, 0.729, 0.731, 0.736, 0.737, 0.738, 0.738, 0.739, 0.743, 0.744, 0.745, 0.752, 0.752, 0.754, 0.765

graphical display, a conclusion about the degree of evidence of a difference, and a conclusion about the size of the difference in distributions.

22. Male and Female Intelligence. Males and females tend to exhibit different types of intelligence. Although there is substantial variability between individuals of the same gender, males on average tend to perform better at navigational and spatial tasks, and females tend to perform better at verbal fluency and memory tasks. This is not a controversial conclusion. Some researchers, however, ask whether males and females differ, on average, in their overall intelligence, and that *is* controversial because any single intelligence measure must rely on premises about the types of intelligence that are important. Even if researchers don't make a subjective judgment about a type of intelligence being tested, they are constrained by the available tools for measuring intelligence. Mathematical knowledge is easy to test, for example, but wisdom, creativity, practical knowledge, and social skill are not.

Display 2.16 shows the first five rows of a data set with several intelligence test scores for random samples of 1,306 American men and 1,278 American women between the ages of 16 and 24 in 1981. The column labeled AFQT shows the percentile scores on the Armed Forces Qualifying Test, which is designed for evaluating the suitability of military recruits but which is also used by researchers as a general intelligence test. The AFQT score is a combination of scores from four component tests: word knowledge, paragraph comprehension, arithmetic reasoning, and mathematical knowledge. The data set represented in Display 2.16 includes each individual's score on these components. (The overall AFQT score reported here, officially called AFQT89, is based on a nontrivial combination of the component scores)

DISPLAY 2.16 Armed Forces Qualifying Test (AFQT) score percentile and component test scores in arithmetic reasoning, word knowledge, paragraph comprehension, and mathematical knowledge, for 1,278 women and 1,306 men in 1981; first 5 of 2,584 rows

Gender	Arith	Word	Parag	Math	AFQT
male	19	27	14	14	70.3
female	23	34	11	20	60.4
male	30	35	14	25	98.3
female	30	35	13	21	84.7
female	13	30	11	12	44.5

Analyze the data to summarize the evidence of differences in male and female distributions of AFQT scores. Do they differ? By how much do they differ? Also answer these two questions for each of the four component test scores. Write a statistical report that includes graphical displays and statistical conclusions (like those in the case studies of Section 2.1), and a section of details upon which the conclusions were based (such as a listing of the computer output showing the results of two-sample t-tests and confidence intervals).

Notes about the data: Although these are random samples of American men and women between the ages of 16 and 24 in 1981, they are not simple random samples. The data come from the National Longitudinal Study of Youth (NLSY), which used variable probability sampling (see Section 1.5.4). To estimate the means of the larger populations, more advanced techniques are appropriate. For comparing male and female distributions, the naive approach based on random sampling is not likely to be misleading. These data come from the National Longitudinal Survey of Youth, U.S. Bureau of Labor Statistics, http://www.bls.gov/nls/home.htm (May 8, 2008). Rows with missing values of variables, including variables used in related problems in other chapters, have been omitted.

23. Speed Limits and Traffic Fatalities. The National Highway System Designation Act was signed into law in the United States on November 28, 1995. Among other things, the act abolished the federal mandate of 55-mile-per-hour maximum speed limits on roads in the United States and permitted states to establish their own limits. Of the 50 states (plus the District of Columbia), 32 increased their speed limits either at the beginning of 1996 or sometime during 1996. Shown in Display 2.17 are the percentage changes in interstate highway traffic fatalities from 1995 to 1996. What evidence is there that the percentage change was greater in states that increased their speed limits? How much of a difference is there? Write a brief statistical report detailing the answers to these questions. (Data from "Report to Congress: The Effect of Increased Speed Limits in the Post-NMSL Era," National Highway Traffic Safety Administration, February, 1998; available in the reports library at http://www-fars.nhtsa.dot.gov/.)

DISPLAY 2.17 Number of traffic fatalities in 50 U.S. states and the District of Columbia, and status of speed limit change in the state (retained 55 mph limit or increased speed limit); first 5 of 51 rows

State	Fatalities1995	Fatalities1996	PctChange	SpeedLimit
Alabama	1114	1146	2.87	Inc
Alaska	87	81	−6.9	Ret
Arizona	1035	994	−3.96	Inc
Arkansas	631	615	−2.54	Inc
California	4192	3989	−4.84	Inc

Answers to Conceptual Exercises

1. The birds were *observed* to be in the 1976 or 1978 groups, not *assigned* by the researchers.

2. There is rarely any direct interest in the standard deviation, but it must be estimated in order to clear up the picture regarding means.

3. False.

4. True.

5. There is more confidence that a 99% interval contains the parameter of interest, but the extra confidence comes at the price of the interval being larger and therefore less informative about specific likely values.

6. The hypothesis must be about the population means. A hypothesis must be about the value of an *unknown* parameter. The value of a statistic will be known when the data are collected and analyzed.

7. It is less than 0.05.

8. The statement implies that the null hypothesis is accepted as true. It should be worded as, for example, the data are consistent with the hypothesis that there is no difference. (This issue is partly one of semantics, but it is still important to understand the distinction being made.)

9. (a) The mean of Y is the mean in the population of all individual measurements, and the mean of $\overline{Y}$ is the mean of the sampling distribution of the sample mean. With random sampling, the two have the same value μ.
   (b) The standard deviation of Y is the standard deviation among all observations in the population, and the standard deviation of $\overline{Y}$ is the standard deviation in the sampling distribution

of the sample average. The two are related, but not the same: if the standard deviation of Y is denoted by σ, then the standard deviation of $\overline{Y}$ is $\sigma/\sqrt{n}$.

(c) The standard error is an estimate of the standard deviation in the sampling distribution, obtained by replacing the unknown population standard deviation in the formula by the known sample standard deviation.

(d) The t-ratio is the ratio of the difference between the estimate and the parameter to the standard error of the estimate. It involves the parameter, so you do not generally know what it is. The t-statistic is a trial value of the t-ratio, obtained when a hypothesized value of the parameter is used in place of the actual value.

10. A p-value. A confidence interval.

11. (a) Because the spread of the stem-and-leaf plot is larger for the control group than for the treatment group, it does not appear that the effect of treatment was simply to add a certain number of days onto the lives of every guinea pig. It may have added days for those that would not have lived long anyway, and subtracted days from those that would have lived a long time. (b) The equal variation assumption does not appear to be appropriate.

CHAPTER 3

A Closer Look at Assumptions

Although statistical computer programs faithfully supply confidence intervals and p-values whenever asked, the human data analyst must consider whether the assumptions on which use of the tools is based are met, at least approximately. In this regard, an important distinction exists between the mathematical assumptions on which use of t-tools is exactly justified and the broader conditions under which such tools work quite well.

The mathematical assumptions—such as those of the model for two independent samples from normal populations with the same standard deviation—are never strictly met in practice, nor do they have to be. The two-sample t-tools are often valid even if the population distributions are nonnormal or the standard deviations are unequal. An understanding of the broader conditions, provided by advanced statistical theory and computer simulation, is needed to evaluate the appropriateness of the tools for a particular problem. After checking the actual conditions with graphical displays of the data, the data analyst may decide to use the standard tools, use them but apply the label "approximate" to the inferences, or choose an alternative approach.

An effective alternative is to apply the standard tools after transforming the data. A transformation is useful if the tools are appropriate for the conditions of the transformed data and if the questions of interest are answerable on the new scale. A particularly important transformation is the logarithm, which permits a convenient description of a multiplicative effect.

3.1 CASE STUDIES

3.1.1 Cloud Seeding to Increase Rainfall—A Randomized Experiment

The data in Display 3.1 were collected in southern Florida between 1968 and 1972 to test a hypothesis that massive injection of silver iodide into cumulus clouds can lead to increased rainfall. (Data from J. Simpson, A. Olsen, and J. Eden, "A Bayesian Analysis of a Multiplicative Treatment Effect in Weather Modification," *Technometrics* 17 (1975): 161–66.)

DISPLAY 3.1 Rainfall (acre-feet) for days with and without cloud seeding

Rainfall from Unseeded Days ($n = 26$)

1,202.6	830.1	372.4	345.5	321.2	244.3	163.0	147.8	95.0
87.0	81.2	68.5	47.3	41.1	36.6	29.0	28.6	26.3
26.0	24.4	21.4	17.3	11.5	4.9	4.9	1.0	

Rainfall from Seeded Days ($n = 26$)

2,745.6	1,697.1	1,656.4	978.0	703.4	489.1	430.0	334.1	302.8
274.7	274.7	255.0	242.5	200.7	198.6	129.6	119.0	118.3
115.3	92.4	40.6	32.7	31.4	17.5	7.7	4.1	

On each of 52 days that were deemed suitable for cloud seeding, a random mechanism was used to decide whether to seed the target cloud on that day or to leave it unseeded as a control. An airplane flew through the cloud in both cases, since the experimenters and the pilot were themselves unaware of whether on any particular day the seeding mechanism in the plane was loaded or not (that is, they were *blind* to the treatment). Precipitation was measured as the total rain volume falling from the cloud base following the airplane seeding run, as measured by radar. Did cloud seeding have an effect on rainfall in this experiment? If so, how much?

Box plots of the data in Display 3.2(a) indicate that the rainfall tended to be larger on the seeded days. Both distributions were quite skewed, and more variability occurred in the seeded group than in the control group. The box plots in Display 3.2(b) are drawn from the same data, but on the scale of the natural logarithm of the rainfall measurements. On this scale, the measurements appear to have symmetric distributions, and the variation seems nearly the same.

Statistical Conclusion

It is estimated that the volume of rainfall on days when clouds were seeded was 3.1 times as large as when not seeded. A 95% confidence interval for this multiplicative effect is 1.3 times to 7.7 times. Since randomization was used to determine whether any particular suitable day was seeded or not, it is safe to interpret this as evidence that the seeding caused the larger rainfall amount.

DISPLAY 3.2 Box plots of rainfall amounts on original and transformed scales

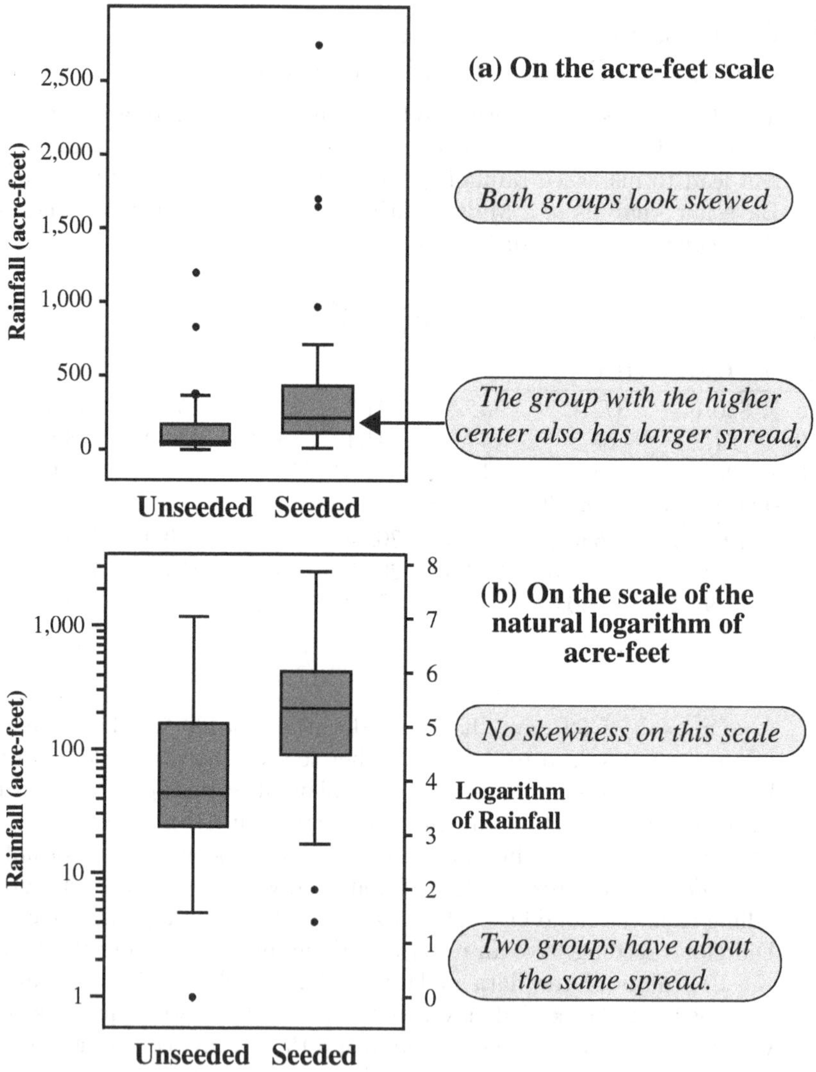

3.1.2 Effects of Agent Orange on Troops in Vietnam—An Observational Study

Many Vietnam veterans are concerned that their health may have been affected by exposure to Agent Orange, a herbicide sprayed in South Vietnam between 1962 and 1970. The particularly worrisome component of Agent Orange is a dioxin called TCDD, which in high doses is known to be associated with certain cancers. Studies have shown that high levels of this dioxin can be detected 20 or more years after heavy exposure to Agent Orange. Consequently, as part of a series of studies, researchers from the Centers for Disease Control compared the current (1987) dioxin levels in living Vietnam veterans to the dioxin levels in veterans who did not serve in Vietnam.

The 646 Vietnam veterans in the study were a sample of U.S. Army combat personnel who served in Vietnam during 1967 and 1968, in the areas that were most heavily treated with Agent Orange. The 97 non–Vietnam veterans entered the Army between 1965 and 1971 and served only in the United States or Germany. Neither sample was randomly selected.

Blood samples from each veteran were analyzed for the presence of dioxin. Box plots of the observed levels are shown in Display 3.3. (Data from a graphical display in Centers for Disease Control Veterans Health Studies, "Serum 2,3,7,8-Tetrachlorodibenzo-*p*-dioxin Levels in U.S. Army Vietnam-era Veterans," *Journal of the American Medical Association* 260 (September 2, 1988): 1249–54.) The question of interest is whether the distribution of dioxin levels tends to be higher for the Vietnam veterans than for the non–Vietnam veterans.

DISPLAY 3.3 Box plots of 1987 dioxin concentrations in 646 Vietnam veterans and 97 veterans who did not serve in Vietnam

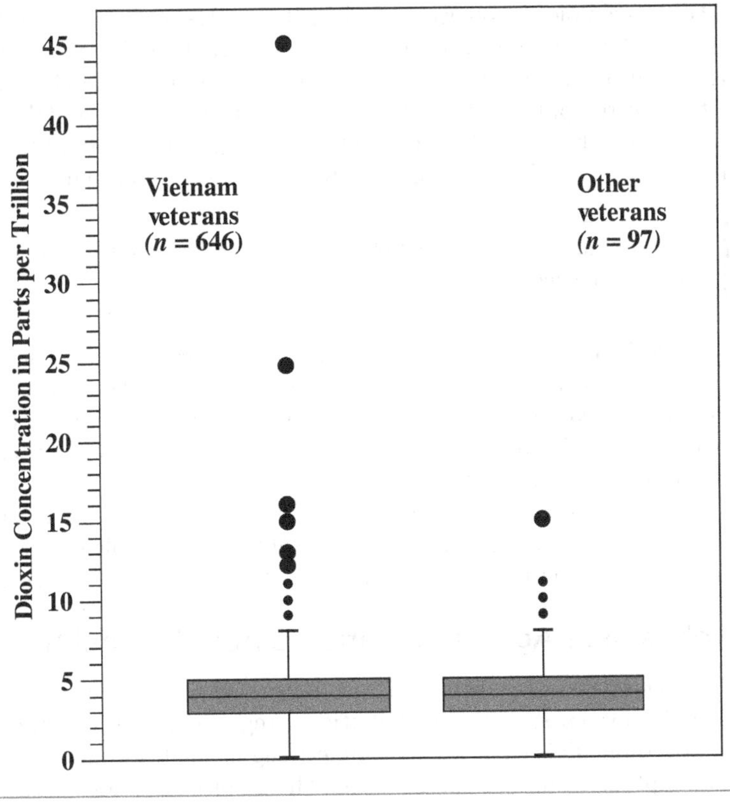

Statistical Conclusion

These data provide no evidence that the mean dioxin level in surviving Vietnam combat troops is greater than that for non–Vietnam veterans (one-sided p-value $= 0.40$, from a two–sample t-test). A 95% confidence interval for the difference in population means is -0.48 to 0.63 parts per trillion.

Scope of Inference

Since the samples were not random, inference to the populations is speculative. Participating veterans may not be representative of their respective groups. For example, nonparticipating Vietnam veterans may have failed to participate because of dioxin-related illnesses. If so, statistical statements about the populations of interest could be seriously biased. It should also be noted that many Vietnam veterans are frustrated and insulted by the prevalence of weak Agent Orange studies, like this one, which appear to address the Agent Orange problem but which actually skirt the main health issues.

3.2 ROBUSTNESS OF THE TWO-SAMPLE *t*-TOOLS

3.2.1 The Meaning of Robustness

The two-sample t-tools were used in the analyses of the Agent Orange study and the cloud seeding study, even though the actual conditions did not seem to match the ideal models upon which the tools are based. In the cloud seeding study, the t-tools were applied after taking the logarithms of the rainfalls. The t-tools could be used for the Agent Orange study, despite a lack of normality in the populations, because of the robustness of the t-tools against nonnormality.

> *A statistical procedure is* ***robust to departures from a particular assumption*** *if it is valid even when the assumption is not met.*

Valid means that the uncertainty measures—the confidence levels and the p-values—are very nearly equal to the stated rates. For example, a procedure for obtaining a 95% confidence interval is valid if it is 95% successful in capturing the parameter. It is robust against nonnormality if it is roughly 95% successful with nonnormal populations.

Robustness of a tool must be evaluated separately for each assumption. The following sections detail the robustness of the two-sample t-tools against departures from the ideal assumptions of the normal, equal standard deviation model.

3.2.2 Robustness Against Departures from Normality

The *Central Limit Theorem* asserts that averages based on large samples have approximately normal sampling distributions, regardless of the shape of the population distribution. This suggests that underlying normality is not a serious issue, as long as sample sizes are reasonably large. The theorem provides only partial reassurance of applicability with respect to t-tools. It states what the sampling distribution of an average should be, but it does not address the effects of estimating a population standard deviation. Many empirical investigations and related theory, however, confirm that the t-tools remain reasonably valid in large samples, with many nonnormal populations.

How large is large enough? That depends on how nonnormal the population distributions are. Because distributions can differ from the normal in infinitely many ways, the question of sample size is difficult to answer. Statistical theory does say something fairly general about the relative effects of skewness and long-tailedness (*kurtosis*):

1. If the two populations have the same standard deviations and approximately the same shapes, and if the sample sizes are about equal, then the validity of the t-tools is affected moderately by long-tailedness and very little by skewness.
2. If the two populations have the same standard deviations and approximately the same shapes, but if the sample sizes are not approximately the same, then the validity of the t-tools is affected moderately by long-tailedness and substantially by skewness. The adverse effects diminish, however, with increasingly large sample sizes.
3. If the skewness in the two populations differs considerably, the tools can be very misleading with small and moderate sample sizes.

Computer simulations can clarify the role of sample sizes. To further investigate the effect of nonnormality on 95% confidence intervals, a computer was instructed to generate samples from the nonnormal distributions shown in Display 3.4. For each pair of generated samples, it computed the 95% confidence interval for the difference in population means and recorded whether the interval actually captured the ***true*** difference in population means. The actual percentage of successful intervals from 1,000 simulations is shown in Display 3.4 for each set of conditions examined. The purpose of the simulation is to identify combinations of sample sizes and nonnormally shaped distributions for which the confidence interval procedure has a success rate of nearly 95%. An actual success rate less than 95% is undesirable because it means the intervals tend to be too short for the given confidence level and they therefore tend to exclude possible parameter values that shouldn't be excluded. The corresponding hypothesis test, in this case, tends to produce more false claims about statistical significance than expected. An actual success rate greater than 95% is also undesirable because it means the intervals tend to be too long and include possible parameter values that shouldn't be included. The corresponding hypothesis test, in this case, tends to miss statistically significant findings that it really should find.

Of the five distributions examined, only the long-tailed distribution appears to have success rates that are poor enough to cause potentially misleading statements—and even those are not too bad. This distribution can be recognized in practice by the presence of outliers. For the skewed distributions, however, the normality assumption does not appear to be a major concern even for small sample sizes, at least as long as the skewness is the same in the two populations and the sample sizes are roughly equal.

DISPLAY 3.4 Percentage of 95% confidence intervals that are successful when the two populations are non-normal (but with same shape and SD, and equal sample sizes) (each percentage is based on 1,000 computer simulations)

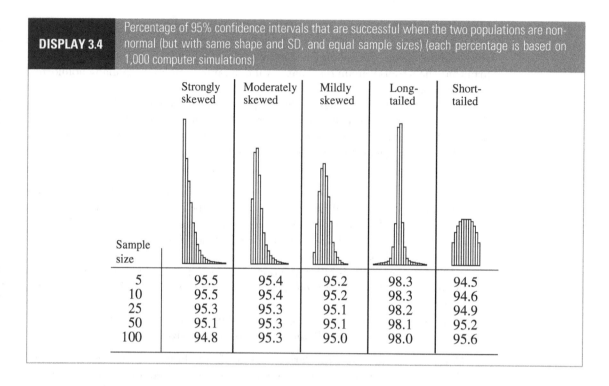

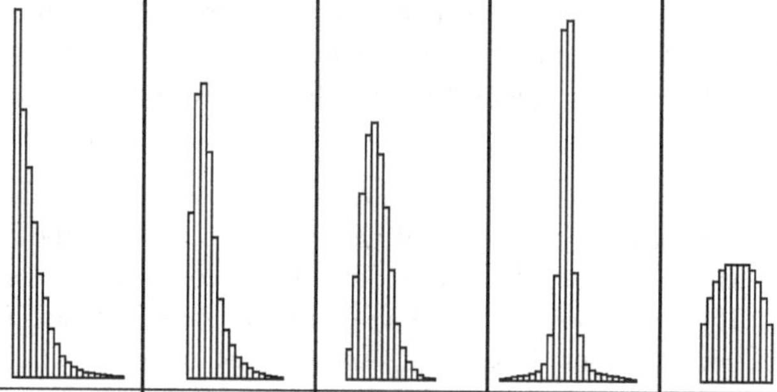

Sample size	Strongly skewed	Moderately skewed	Mildly skewed	Long-tailed	Short-tailed
5	95.5	95.4	95.2	98.3	94.5
10	95.5	95.4	95.2	98.3	94.6
25	95.3	95.3	95.1	98.2	94.9
50	95.1	95.3	95.1	98.1	95.2
100	94.8	95.3	95.0	98.0	95.6

3.2.3 Robustness Against Differing Standard Deviations

More serious problems may arise when the standard deviations of the two populations are substantially unequal. In this case, the pooled estimate of standard deviation does not estimate any population parameter and the standard error formula, which uses the pooled estimate of standard deviation, no longer estimates the standard deviation of the difference between sample averages. As a result, the t-ratio does not have a t-distribution.

Theory shows that the t-tools remain fairly valid when the standard deviations are unequal, as long as the sample sizes are roughly the same. For clarification, a computer was again instructed to generate pairs of samples, this time from two normal populations with different standard deviations, as shown in Display 3.5. It computed 95% confidence intervals for each pair of samples and recorded whether the resulting interval successfully captured the true difference in population means. The actual percentages successful are displayed.

Notice that the success rates for the rows with equal sample sizes ($n_1 = n_2 = 10$ and $n_1 = n_2 = 100$) are very nearly 95%. Thus, as suggested by theory, unequal population standard deviations have little effect on validity if the sample sizes are equal. For substantially different σ's and different n's, however, the confidence intervals are unreliable. The worst situation is when the ratio of standard deviations is much different from 1 and the smaller sized sample is from the population with the larger standard deviation (as, for example, when $n_1 = 100$, $n_2 = 400$, and $\sigma_2/\sigma_1 = 1/4$).

DISPLAY 3.5 Percentage of successful 95% confidence intervals when the two populations have different standard deviations (but are normal) with possibly different sample sizes (each percentage is based on 1,000 computer simulations)

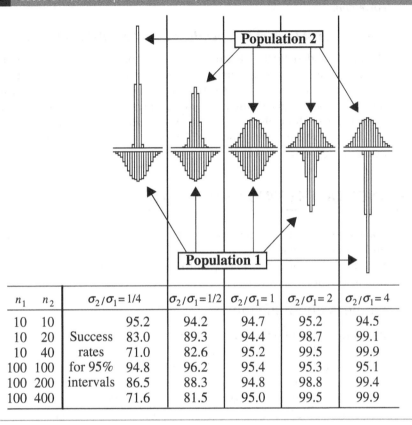

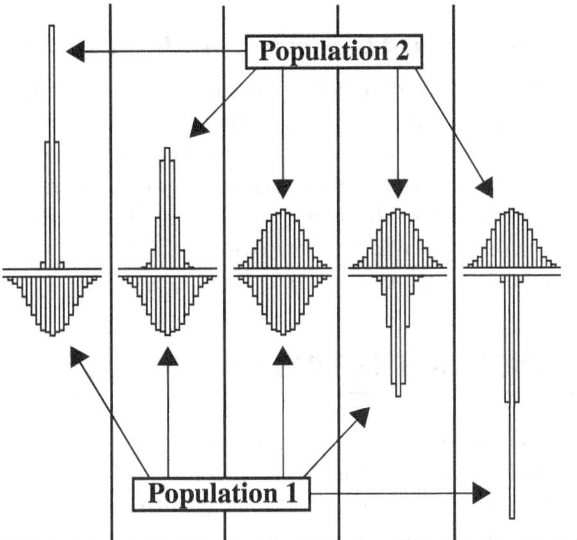

n_1	n_2		$\sigma_2/\sigma_1 = 1/4$	$\sigma_2/\sigma_1 = 1/2$	$\sigma_2/\sigma_1 = 1$	$\sigma_2/\sigma_1 = 2$	$\sigma_2/\sigma_1 = 4$
10	10		95.2	94.2	94.7	95.2	94.5
10	20	Success	83.0	89.3	94.4	98.7	99.1
10	40	rates	71.0	82.6	95.2	99.5	99.9
100	100	for 95%	94.8	96.2	95.4	95.3	95.1
100	200	intervals	86.5	88.3	94.8	98.8	99.4
100	400		71.6	81.5	95.0	99.5	99.9

3.2.4 Robustness Against Departures from Independence

Cluster Effects and Serial Effects

Whenever knowledge that one observation is, say, above average allows an improved guess about whether another observation will be above average, *independence* is lacking. The methods of Chapter 2 may be misleading in this case. Two types of dependence (lack of independence) commonly arise in practical problems.

The first is a *cluster effect*, which sometimes occurs when the data have been collected in subgroups. For example, 50 experimental animals may have been collected from 10 litters and then randomly assigned to one of two treatment groups. Since animals from the same litter may tend to be more similar in their responses than animals from different litters, it is likely that independence is lacking.

The other type of dependence commonly encountered is caused by a *serial effect*, in which measurements are taken over time and observations close together in time tend to be more similar (or perhaps more different) than observations collected at distant time points. This can also occur if measurements are made at different locations, and measurements physically close to each other tend to be more similar than those farther apart. In the latter case the dependence pattern is called *spatial correlation*.

The Effects of Lack of Independence on the Validity of the t-Tools

When the independence assumptions are violated, the standard error of the difference of averages is an inappropriate estimate of the standard deviation of the difference in averages. The t-ratio no longer has a t-distribution, and the t-tools may give misleading results. The seriousness of the consequences depends on the seriousness of the violation. It is generally unwise to use the t-tools directly if cluster or serial effects are suspected. Other methods that adjust for these effects are available (Chapters 9–15).

3.3 RESISTANCE OF THE TWO-SAMPLE *t*-TOOLS

Some practical suggestions will soon be provided for sizing up the actual conditions and choosing a course of action. The effect of outliers on the t-tools is discussed first, however, since decisions about how to deal with the outliers play an important role in the overall strategy.

3.3.1 Outliers and Resistance

An *outlier* is an observation judged to be far from its group average. The effect of outliers on the two-sample t-tools has partially been addressed in the discussion of robustness. In fact, it is evident from the theoretical results and the computer simulations in Display 3.4 that the p-values and confidence intervals may be unreliable if the population distributions are long-tailed. Since long-tailed distributions are characterized by the presence of outliers in the sample, outliers should cause some concern.

Long-tailed population distributions are not the only explanation for outliers, however. The populations of interest may be normal but the sample may be contaminated by one or more observations that do not come from the population of interest. Often it is philosophically difficult and practically irrelevant to distinguish between a natural long-tailed distribution and one that includes outliers that result from contamination, although in some cases the identification of clear contamination may dictate an obvious course of action. For example, if it is discovered that one member of a sample from a population of 25- to 35-year-old women is, in fact, over 50 years old, she should be removed from the sample.

It is useful to know how sensitive a statistical procedure may be to one or two outlying observations. The notion of resistance addresses this issue:

> *A statistical procedure is* ***resistant*** *if it does not change very much when a small part of the data changes, perhaps drastically.*

As an example, consider the hypothetical sample: 10, 20, 30, 50, 70. The sample average is 36, and the sample median is 30. Now change the 70 to 700, and what happens? The sample average becomes 162, but the sample median remains 30. The sample average is not a resistant statistic because it can be severely influenced by the change in a single observation. The median, however, is resistant.

Resistance is a desirable property. A resistant procedure is insensitive to outliers. A nonresistant one, on the other hand, may be greatly influenced by one or two outlying observations.

3.3.2 Resistance of t-Tools

Since t-tools are based on averages, they are not resistant. A small portion of the data can potentially have a major influence on the results. In particular, one or two outliers can affect a confidence interval or change a p-value enough to completely alter a conclusion.

If the outlier is due to contamination from another population, it can lead to false impressions about the population of interest. If the outlier does come from the population of interest, which happens to be long-tailed, the outcome is still undesirable for the following reason. In statistics, the goal is to describe *group* characteristics. An estimate of the center of a distribution should represent the typical value. The estimate is a good one if it represents the typical values possessed by the great majority of subjects; it is a bad one if it represents a feature unique to one or two subjects. Furthermore, a conclusion that hinges on one or two data points must be viewed as quite fragile.

3.4 PRACTICAL STRATEGIES FOR THE TWO-SAMPLE PROBLEM

Armed with information about the broad set of conditions under which the t-tools work well and the effect of outliers, the challenge to the data analyst is to size up the actual conditions using the available data and evaluate the appropriateness of the t-tools. This involves thinking about possible cluster and serial effects; evaluating the suitability of the t-tools by examining graphical displays; and considering alternatives.

In considering alternatives it is important to realize that even though the t-tools may still be valid when the ideal assumptions are not met, an alternative procedure that is more *efficient* (i.e., makes better use of the data) may be available. For example, another procedure may provide a narrower confidence interval.

Consider Serial and Cluster Effects

To detect lack of independence, carefully review the method by which the data were gathered. Were the subjects selected in distinct groups? Were different groups of subjects treated differently in a way that was unrelated to the primary treatment? Were different responses merely repeated measurements on the same subjects? Were observations taken at different but proximate times or locations? Affirmative answers to any of these questions suggest that independence may be lacking.

The principal remedy is to use a more sophisticated statistical tool. Identifiable clusters, which may be planned or unplanned, can be accounted for through analysis of variance (Chapters 13 and 14) or possibly through regression analysis (Chapters 9–12). Serial effects require time series analysis, the topic of Chapter 15.

Evaluate the Suitability of the t-Tools

Side-by-side histograms or box plots of the two groups of data should be examined and departures from the ideal model should be considered in light of the robustness properties of the t-tools. It is important to realize that the conditions of interest, which are those of the populations, must be investigated through graphical displays of the samples.

If the conditions do not appear suitable for use of the t-tools, then some alternative is necessary. A transformation should be considered if the graphical displays of the transformed data appear to be closer to the ideal conditions. (See Section 3.5.) Alternative tools for analyzing two independent samples are the rank-sum procedure, which is resistant and does not depend on normality (Section 4.2); other permutation tests (Section 4.3.1); and the Welch procedure for comparing normal populations that have unequal standard deviations (Section 4.3.2).

A Strategy for Dealing with Outliers

If investigation reveals that an outlying observation was recorded improperly or was the result of contamination from another population, the solution is to correct it if the right value is known or to leave it out. Often, however, there is no way to know how the outliers arose. Two statistical approaches for dealing with this situation exist. One is to employ a resistant statistical tool, in which case there is no compelling reason to ponder whether the offending observations are natural, the result of contamination, or simply blunders. (The rank-sum procedure in Section 4.2 is resistant.) The other approach is to adopt the careful examination strategy shown in Display 3.6. An important aspect of adopting this procedure is that an outlier does not get swept under the rug simply because it is different from the other observations. To warrant its removal, an explanation for why it is different must be established.

Example—Agent Orange

Box plots of dioxin levels in Vietnam and non–Vietnam veterans (Display 3.3) appear again in Display 3.7. The distributions have about the same shape and spread. Although the shape is not normal, the skewness is mild and unlikely to cause any problems with the t-test or the confidence interval. Two Vietnam veterans (#645 and #646) had considerably higher dioxin levels than the others.

From the results listed in Display 3.7 it is evident that the comparison of the two groups is changed very little by the removal of one or both of these outliers. Consequently, there is no need for further action. Even so, it is useful to see what else can be learned about these two, as indicated at the bottom of the display.

Notes

1. It is not useful to give a precise definition for an *outlier*. Subjective examination is the best policy. If there is any doubt about whether a particular observation deserves further examination, give it further examination.

DISPLAY 3.6 Examination strategy

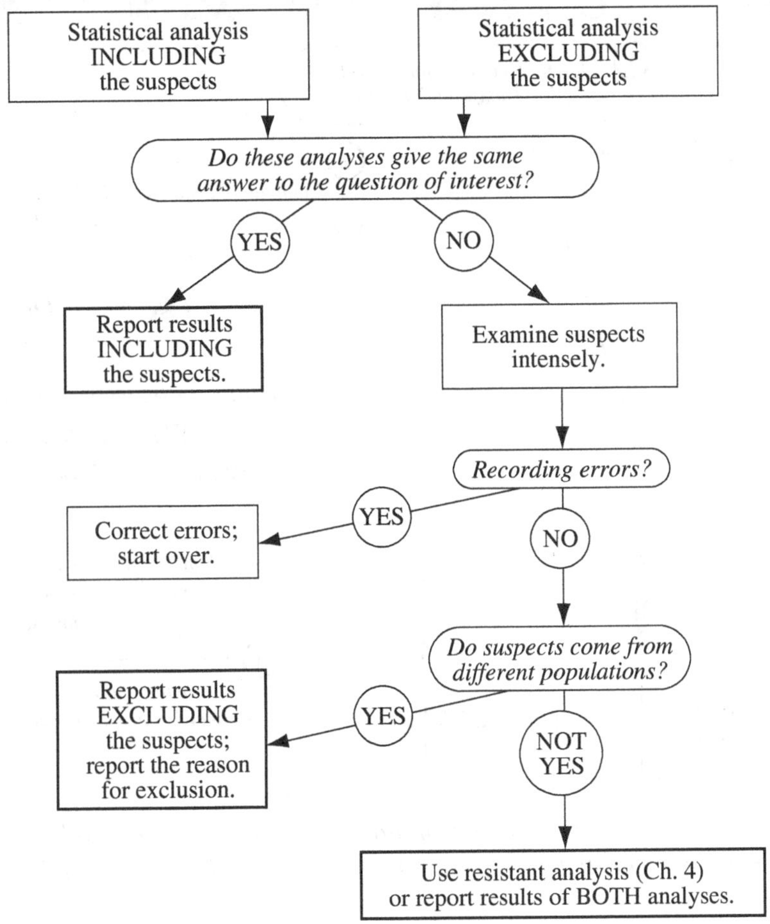

2. It is not surprising that the outliers in the Agent Orange example have little effect, since the sample sizes are so large.
3. The apparent difference in the box plots may be due to the difference in sample sizes. If the population distributions are identical, more observations will appear in the extreme tails from a sample of size 646 than from a sample of size 97.

3.5 TRANSFORMATIONS OF THE DATA

3.5.1 The Logarithmic Transformation

The most useful transformation is the *logarithm* (log) for positive data. The common scale for scientific work is the *natural* logarithm (ln), based on the number

DISPLAY 3.7 Outlier analysis for Agent Orange data: effect of outliers on the p-value, for equal population means

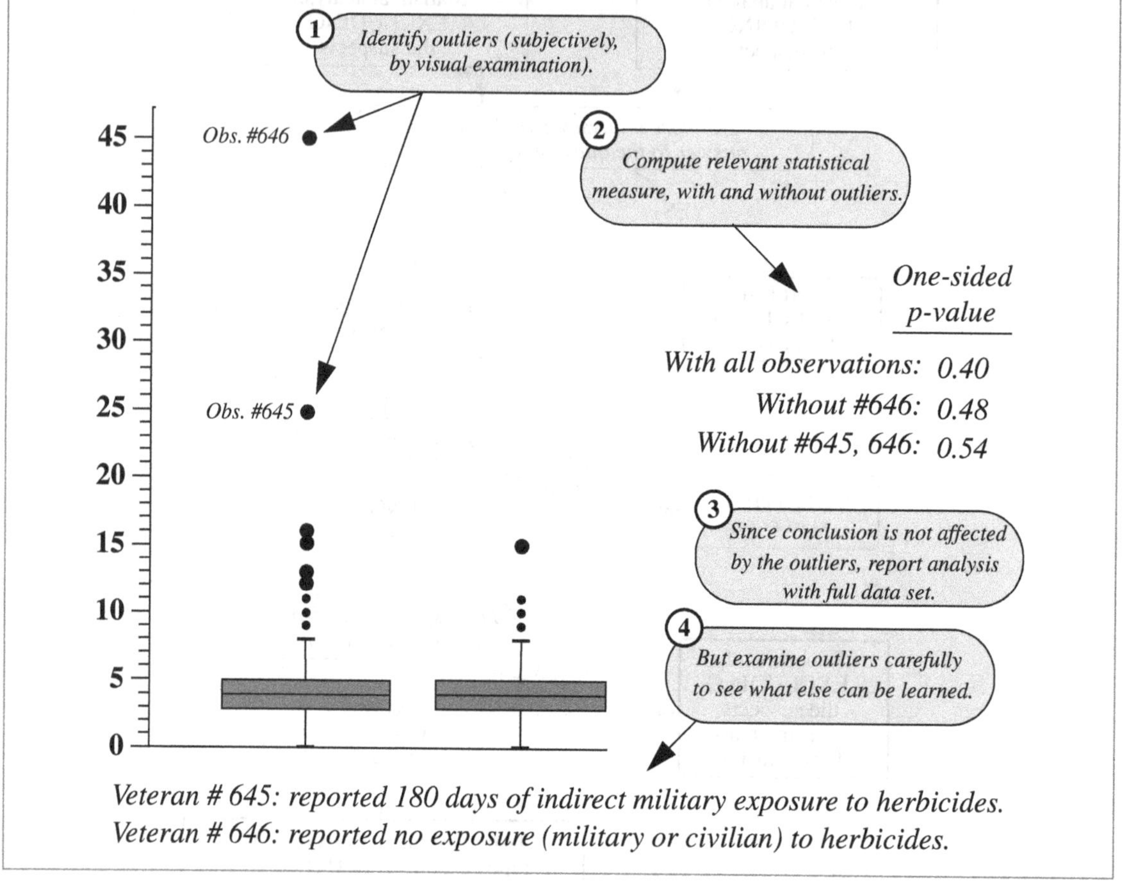

$e = 2.71828\ldots$. The logarithm of e is unity, denoted by $\log(e) = 1$. Also, the log of 1 is 0: $\log(1) = 0$. The general rule for using logarithms is that $\log(e^x) = x$. Another choice is the *common* logarithm based on the number 10, rather than e. Common logs are defined by $\log_{10}(10^x) = x$. Unless otherwise stated, *log* in this book refers to the natural logarithm.

Recognizing the Need for a Log Transformation

The data themselves usually suggest the need for a log transformation. If the ratio of the largest to the smallest measurement in a group is greater than 10, then the data are probably more conveniently expressed on the log scale. Also, if the graphical displays of the two samples show them both to be skewed and if the group with the larger average also has the larger spread (see Display 3.2), the log transformation is likely to be a good choice.

DISPLAY 3.8 The logarithmic transformation used to arrive at favorable conditions for the two-sample t-analysis

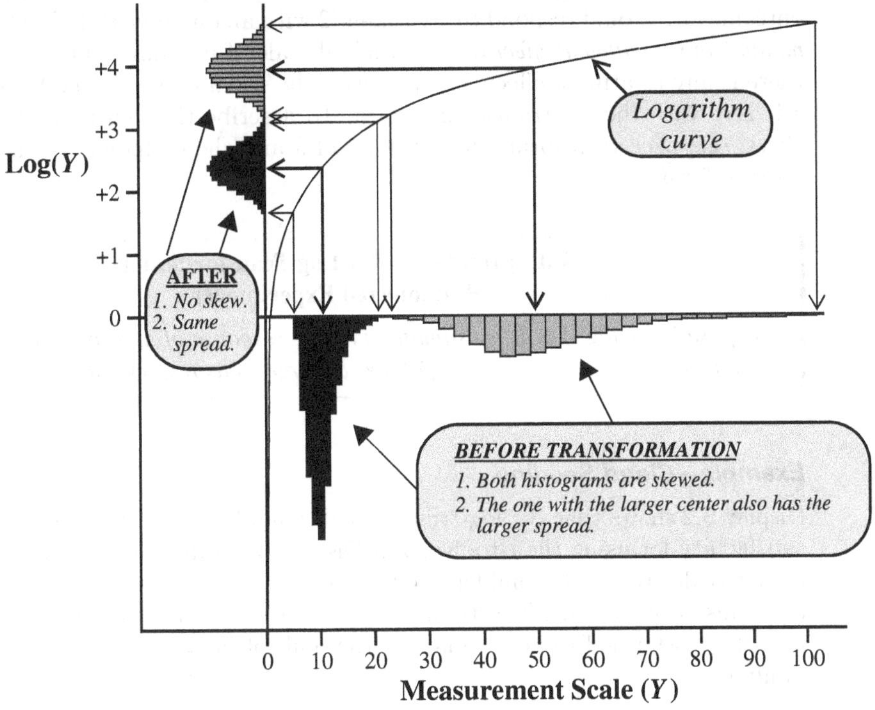

Display 3.8 illustrates the behavior of the log transformation. On the scale of measurement Y the two groups have skewed distributions with longer tails in the positive direction. The group with the larger center also has the larger spread. The measurements on the transformed scale have the same ordering, but small numbers get spread out more, while large numbers are squeezed more closely together. The overall result is that the two distributions on the transformed scale appear to be symmetric and have equal spread—just the right conditions for applying the t-tools.

3.5.2 Interpretation After a Log Transformation

For some measurements, the results of an analysis are appropriately presented on the transformed scale. Most users feel comfortable with the Richter scale for measuring earthquake strength, even though it is a logarithmic scale. Similarly, pH as a measure of acidity is the negative log of ion concentration. In other cases, however, it may be desirable to present the results on the original scale of measurement.

Randomized Experiment Model: Multiplicative Treatment Effect

If the randomized experiment model with additive treatment effect is thought to hold for the log-transformed data, then an experimental unit that would respond to treatment 1 with a logged outcome of $\log(Y)$ would respond to treatment 2 with a logged outcome of $\log(Y) + \delta$. By taking antilogarithms of these two quantities, one finds that an experimental unit that would respond to treatment 1 with an outcome of Y would respond to treatment 2 with an outcome of Ye^{δ}. Thus, e^{δ} is the *multiplicative treatment effect* on the original scale of measurement. To test whether there is any treatment effect, one performs the usual t-test for the hypothesis that δ is zero with the log-transformed data. To describe the multiplicative treatment effect, one back-transforms the estimate of δ and the endpoints of the confidence interval for δ.

Interpretation After Log Transformation (Randomized Experiment)

Suppose $Z = \log(Y)$. *It is estimated that the response of an experimental unit to treatment 2 will be* $\exp(\overline{Z}_2 - \overline{Z}_1)$ *times as large as its response to treatment 1.*

Example—Cloud Seeding

Display 3.2 shows that the log-transformed rainfalls have distributions that appear satisfactory for using the t-tools; so in Display 3.9 a full analysis is carried out on the log scale. Tests and confidence intervals are constructed in the usual way but on the transformed data. The estimate of the additive treatment effect on log rainfall is back-transformed to an estimate of the multiplicative effect of cloud seeding on rainfall.

Population Model: Estimating the Ratio of Population Medians

The t-tools applied to log-transformed data provide inferences about the difference in means of the logged measurements, which may be represented as $\text{Mean}[\log(Y_2)] - \text{Mean}[\log(Y_1)]$, where $\text{Mean}[\log(Y_2)]$ symbolizes the mean of the logged values of population 2. A problem with interpretation on the original scale arises because the mean of the logged values is not the log of the mean. Taking the antilogarithm of the estimate of the mean on the log scale does *not* give an estimate of the mean on the original scale.

If, however, the log-transformed data have symmetric distributions, the following relationships hold:

$$\text{Mean}[\log(Y)] = \text{Median}[\log(Y)]$$

(and since the log preserves ordering)

$$\text{Median}[\log(Y)] = \log[\text{Median}(Y)],$$

where $\text{Median}(Y)$ represents the *population median* (the 50th percentile of the population). In other words, the 50th percentile of the logged values is the log of the 50th percentile of the untransformed values. Putting these two equalities together,

DISPLAY 3.9 Two-sample t-analysis and statement of conclusions after logarithmic transformation—cloud seeding example

(1) **Transform the data.**

Unseeded		Seeded	
Y (acre-ft)	log (Y)	Y (acre-ft)	log (Y)
1202.6	7.092	2745.6	7.918
830.1	6.722	1697.8	7.437
372.4	5.920	1656.0	7.412
345.5	5.845	978.0	6.886
321.2	5.772	703.4	6.556
244.3	5.498	489.1	6.193
163.0	5.094	430.0	6.064
147.8	4.996	334.1	5.811
95.0	4.554	302.8	5.713
87.0	4.466	274.7	5.616
81.2	4.397	274.7	5.616
68.5	4.227	255.0	5.541
47.3	3.857	242.5	5.491
41.1	3.716	200.7	5.302
36.6	3.600	198.6	5.291
29.0	3.367	129.6	4.864
28.6	3.353	119.0	4.779
26.3	3.270	118.3	4.773
26.1	3.262	115.3	4.748
24.4	3.195	92.4	4.526
21.7	3.077	40.6	3.704
17.3	2.851	32.7	3.487
11.5	2.446	31.4	3.447
4.9	1.589	17.5	2.862
4.9	1.589	7.7	2.041
1.0	0.000	4.1	1.411

(2) **Use the two-sample t-tools on the log rainfall.**

Difference in averages = 1.1436 (SE = 0.4495).

Test of the hypothesis of no effect of cloud seeding on log rainfall: one-sided p-value from two-sample t-test = 0.0070 (50 d.f.).

95% confidence interval for additive effect of cloud seeding on log rainfall: 0.2406 to 2.0467.

(3) **Back-transform estimate and confidence interval.**

Estimate = $e^{1.1436}$ = 3.1382

Lower confidence limit = $e^{0.2406}$ = 1.2720.

Upper confidence limit = $e^{2.0467}$ = 7.7425.

(4) **State the conclusions on the original scale.**

Conclusion: *There is convincing evidence that seeding increased rainfall (one-sided p-value = 0.0070). The volume of rainfall produced by a seeded cloud is estimated to be 3.14 times as large as the volume that would have been produced in the absence of seeding (95% confidence: 1.27 to 7.74 times).*

it is evident that the antilogarithm of the mean of the log values is the median on the original scale of measurements.

If $\overline{Z}_1$ and $\overline{Z}_2$ are used to represent the averages of the logged values for samples 1 and 2, then $\overline{Z}_2 - \overline{Z}_1$ estimates $\log[Median(Y_2)] - \log[Median(Y_1)]$, and therefore

$$\overline{Z}_2 - \overline{Z}_1 \text{ estimates } \log\left[\frac{\text{Median}(Y_2)}{\text{Median}(Y_1)}\right]$$

and, therefore,

$$\exp(\overline{Z}_2 - \overline{Z}_1) \text{ estimates } \left[\frac{\text{Median}(Y_2)}{\text{Median}(Y_1)}\right].$$

The point of this is that a very useful multiplicative interpretation emerges in terms of the ratio of population medians. This is doubly important because the median is a better measure of the center of a skewed distribution than the mean. The multiplicative nature of this relationship is captured with the following wording:

Interpretation After Log Transformation (Observational Study)

It is estimated that the median for population 2 is $\exp(\overline{Z}_2 - \overline{Z}_1)$ *times as large as the median for population 1.*

In addition, back-transforming the ends of a confidence interval constructed on the log scale produces a confidence interval for the ratio of medians.

Example (Sex Discrimination)

Although the analysis of the sex discrimination data of Section 1.1.2, was suitable on the original scale of the untransformed salaries, graphical displays of the log-transformed salaries indicate that analysis would also be suitable on the log scale. The average male log salary minus the average female log salary is 0.147. Since $e^{0.147} = 1.16$, it is estimated that the median salary for males is 1.16 times as large as the median salary for females. Equivalently, the median salary for males is estimated to be 16% more than the median salary for females. Since a 95% confidence interval for the difference in means on the log scale is 0.100 to 0.194, a 95% confidence interval for the ratio of population median salaries is 1.11 to 1.21 ($e^{0.100}$ to $e^{0.194}$). With 95% confidence, it is estimated that the median salary for males is between 11% and 21% greater than the median salary for females.

3.5.3 Other Transformations for Positive Measurements

There are other useful transformations for positive measurements with skewed distributions where the means and standard deviations differ between groups. The *square root* transformation $\sqrt{Y}$ applies to data that are counts—counts of bacteria clusters in a dish, counts of traffic accidents on a stretch of highway, counts of red giants in a region of space—and to data that are measurements of area. The *reciprocal* transformation $1/Y$ applies to data that are waiting times—times to failure of lightbulbs, times to recurrence for cancer patients treated with radiation, reaction times to visual stimuli, and so on. The reciprocal of a time measurement can often be interpreted directly as a rate or a speed. The *arcsine square root* transformation, arcsine($\sqrt{Y}$), and the *logit* transformation, $\log[Y/(1-Y)]$, apply when the measurements are proportions between zero and one—proportions of trees infested by a wood-boring insect in experimental plots, proportions of weight lost as a side effect of leukemia therapy, proportions of winning lottery tickets in clusters of a certain size, and so forth.

Only the log transformation, however, gives such ease in converting inferences back to the original scale of measurement. One may estimate the difference in means of $\sqrt{Y_2}$ and $\sqrt{Y_1}$, but the square of this difference does not make much sense on the original scale.

Choosing a Transformation

Formal statistical methods are available for selecting a transformation. Nevertheless, it is recommended here that a trial-and-error approach, with graphical analysis, be used instead. For positive data in need of a transformation, the logarithm should almost always be the first tried. If it is not satisfactory, the reciprocal or the square root transformations might be useful. Keep in mind that the primary goal is to establish a scale where the two groups have roughly the same spread. If several transformations are similar in their ability to accomplish this, think carefully about which one offers the most convenient interpretation.

Caveat About the Log Transformation

Situations arise where presenting results in terms of population medians is not sufficient. For example, the daily emissions of dioxin in the effluent from a paper mill have a very skewed distribution. An agency monitoring the emissions will be interested in estimating the total dioxin load released during, say, a year of operation. The total dioxin load would be the population mean times the population size, and therefore is estimated by the sample average times the population size. It cannot be estimated directly from the median, unless more specific assumptions are made.

3.6 RELATED ISSUES

3.6.1 Prefer Graphical Methods Over Formal Tests for Model Adequacy

Formal tests for judging the adequacy of various assumptions exist. Tests for normality and tests for equal standard deviation are available in most statistical computer programs, as are tests that determine whether an observation is an outlier. Despite their widespread availability and ease of use, these diagnostic tests are not very helpful for model checking. They reveal little about whether the data meet the broader conditions under which the tools work well. The fact that two populations are not exactly normal, for example, is irrelevant. Furthermore, the formal tests themselves are often not very robust against their own model assumptions. Graphical displays are more informative, if less formal. They provide a good indication of whether or not the data are amenable to t-analysis and, if not, they often suggest a remedy.

3.6.2 Robustness and Transformation for Paired t-Tools

The one-sample t-test, of which the paired t-test is a special case, assumes that the observations are independent of one another and come from a normally distributed population. P-values and confidence intervals remain valid for moderate and large sample sizes for nonnormal distributions. For smaller sample sizes skewness can be a problem. When cluster or serial effects are present (see Section 3.2.4), the t-tools may give misleading results. When the observations within each pair are positive, either an apparent multiplicative treatment effect (in an experiment) or a tendency for larger differences in pairs with larger average values suggests the use of a log transformation. The transformation is applied before taking the difference, which is equivalent to forming a ratio within each pair and performing a one-sample analysis on the logarithms of the ratios. If there are n pairs, let $Z_i = \log(Y_{1i}) - \log(Y_{2i})$, which is the same as $\log(Y_{1i}/Y_{2i})$. In an observational study, $\exp(\overline{Z})$ is an estimate of the median of the ratios, Y_1/Y_2. (This is not the same as the ratio of the medians [see Exercise 20].) In a randomized, paired experiment, $\exp(\overline{Z})$ estimates a multiplicative treatment effect on the original scale. In both cases, the statistical work of testing and constructing a confidence interval is done on the log scale. The estimate and associated interval are transformed back to the original scale.

3.6.3 Example—Schizophrenia

In the schizophrenia example of Section 2.1.2, Z_i represents the logarithm of the left hippocampus volume of the unaffected twin divided by the left hippocampus volume of the affected twin in pair i. The average of the 15 log ratios is 0.1285. A one-sample analysis gives a p-value of 0.0065 for the test that the mean is zero and a 95% confidence interval from 0.0423 to 0.2147 for the mean itself. Taking antilogarithms of the estimate and the endpoints of the confidence interval yields the following conclusion: It is estimated that the median of the unaffected-to-affected volume ratios is 1.137. A 95% confidence interval for the median ratio is from 1.043 to 1.239.

3.7 SUMMARY

Cloud Seeding and Rainfall Study

The box plots of the rainfalls for seeded and unseeded days reveal that the two distributions of rainfall are skewed and that the distribution with the larger mean also has the larger variance. This is the situation where log-transformed data behave in accordance with the ideal model. A plot of the data after transformation confirms the adequacy of the transformation. The two-sample t-test can be used as an approximation to the randomization test, and the difference in averages (of log rainfall) can be back-transformed to provide a statement about a multiplicative treatment effect. In the example, it is estimated that the rainfall is 3.1 times as much when a cloud is seeded as when it is left unseeded.

Since randomization is used, the statistical conclusion implies that the seeding causes the increase in rainfall. Since the decision about whether to seed clouds is determined (in this case) by a random mechanism, and since the airplane crew is *blind* to which treatment they are administering, human bias can have had little influence on the result.

Agent Orange Study

Graphical analysis focuses attention on the possibly undue influence of two outliers, but analyses with and without the outliers reveal no such influence, so the t-tools are used on the entire data set. The form of the sampling from the populations of living Vietnam veterans and of other veterans is a major concern in accepting the reliability of the statistical analysis. Protocols for obtaining the samples have not been discussed here, except to note that random sampling is not being used. Conclusions based on the two-sample t-test are supplied, along with the caveat that there may be biases due to the lack of random sampling.

3.8 EXERCISES

Conceptual Exercises

1. Cloud Seeding. What is the experimental unit in the cloud seeding experiment?

2. Cloud Seeding. Randomization in the cloud seeding experiment was crucial in assessing the effect of cloud seeding on rainfall. Why?

3. Cloud Seeding. Why was it important that the airplane crew was unaware of whether seeding was conducted or not?

4. Cloud Seeding. Why would it be helpful to have the date of each observed rainfall?

5. Agent Orange. How would you respond to the comment that the box plots in Display 3.3 indicate that the dioxin levels in the Vietnam veterans tend to be larger since their values appear to be larger?

6. Agent Orange. (a) What course of action would you propose for the statistical analysis if it was learned that Vietnam veteran #646 (the largest observation in Display 3.6) worked for several years, after Vietnam, handling herbicides with dioxin? (b) What would you propose if this was learned instead for Vietnam veteran #645?

7. Agent Orange. If the statistical analysis had shown convincing evidence that the mean dioxin levels differed in Vietnam veterans and other veterans, could one conclude that serving in Vietnam was responsible for the difference?

8. Schizophrenia. In the schizophrenia study in Section 2.1.2, the observations in the two groups (schizophrenic and nonschizophrenic) are not independent since each subject is matched with a twin in the other group. Did the researchers make a mistake?

9. True or false? A statistical computer package will only print out a p-value or confidence interval if the conditions for its validity are met.

10. True or false? A sample histogram will have a normal distribution if the sample size is large enough.

11. A woman who has just moved to a new job in a new town discovers two routes to drive from her home to work. The first Monday, she flips a coin, deciding to take route A if it comes up heads and to take route B if it is tails. The following Monday, she will take the other route. The first Tuesday, she flips the coin again with the same plan. And so on for the first week. At the end of two weeks, she has traveled both routes five times and can compare their average commuting times. Why should she not use the t-tools for two independent samples? What should she use?

12. In which ways are the t-tools more robust for larger sample sizes than for smaller ones (i.e., robust with respect to normality, equal SDs, and/or independence)?

13. **Fish Oil.** Why is a log transformation inappropriate for the fish oil data in Exercise 1.12?

14. Will an outlier from a contaminating population be more consequential in small samples or large samples?

15. What would you suggest as an alternative estimate of the standard deviation of the difference in sample averages when it is clear that the two populations have different SDs? (Check the formula for the standard deviation of the sampling distribution of the difference in averages, in Display 2.6.)

16. A researcher has taken tissue cultures from 25 subjects. Each culture is divided in half, and a treatment is applied to one of the halves chosen at random. The other half is used as a control. After determining the percent change in the sizes of all culture sections, the researcher calculates the standard error for the treatment-minus-control difference using both the paired t-analysis and the two independent sample (Chapter 2) t-analysis. Finding that the paired t-analysis gives a slightly larger standard error (and gives only half the degrees of freedom), the researcher decides to use the results from the unpaired analysis. Is this legitimate?

17. Respiratory breathing capacity of individuals in houses with low levels of nitrogen dioxide was compared to the capacity of individuals in houses with high levels of nitrogen dioxide. From a sample of 200 houses of each type, breathing capacity was measured on 600 individuals from houses with low nitrogen dioxide and on 800 individuals from houses with high nitrogen dioxide. (a) What problem do you foresee in applying t-tools to these data? (b) Would comparing the average *household* breathing capacities avoid the problem?

18. **Trauma and Metabolic Expenditure.** The following data are metabolic expenditures for eight patients admitted to a hospital for reasons other than trauma and for seven patients admitted for multiple fractures (trauma). (Data from C. L. Long, et al., "Contribution of Skeletal Muscle Protein in Elevated Rates of Whole Body Protein Catabolism in Trauma Patients," *American Journal of Clinical Nutrition* 34 (1981): 1087–93.)

Metabolic Expenditures (kcal/kg/day)

Nontrauma patients:	20.1	22.9	18.8	20.9	20.9	22.7	21.4	20.0
Trauma patients:	38.5	25.8	22.0	23.0	37.6	30.0	24.5	

(a) Is the difference in averages resistant? (*Hint*: What happens if 20.0 is replaced by 200?)

(b) Replacing each value with its rank, from the lowest to highest, in the combined sample gives

Metabolic Expenditures (kcal/kg/day)

Nontrauma patients:	3	9	1	4.5	4.5	8	6	2
Trauma patients:	15	12	7	10	14	13	11	

Consider the average of the ranks for the trauma group minus the average of the ranks for the nontrauma group. Is this statistic resistant?

19. In each of the following data problems there is some potential violation of one of the independence assumptions. State whether there is a cluster effect or serial correlation, and whether the questionable assumption is the independence within groups or the independence between groups.

(a) Researchers interested in learning the effects of speed limits on traffic accidents recorded the number of accidents per year for each of 10 consecutive years on roads in a state with speed limits of 90 km/h. They also recorded the number of accidents for the next 7 years on the same roads after the speed limit had been increased to 110 km/hr. The two groups of measurements are the number of accidents per year for those years under study. (Notice that there is also a potential confounding variable here!)

(b) Researchers collected intelligence test scores on twins, one of whom was raised by the natural parents and one of whom was raised by foster parents. The data set consists of test scores for the two groups, boys raised by their natural parents and boys raised by foster parents.

(c) Researchers interested in investigating the effect of indoor pollution on respiratory health randomly select houses in a particular city. Each house is monitored for nitrogen dioxide concentration and categorized as being either high or low on the nitrogen dioxide scale. Each member of the household is measured for respiratory health in terms of breathing capacity. The data set consists of these measures of respiratory health for all individuals from houses with low nitrogen dioxide levels and all individuals from houses with high levels.

Computational Exercises

20. Means, Medians, Logs, Ratios. Consider the following tuitions and their natural logs for five colleges:

College	In-State	Out-of-State	Out/In Ratio	Log(In-State)	Log(Out-of-State)
A	$1,000	$ 3,000	3	6.9078	8.0064
B	$4,000	$ 8,000	2	8.2941	8.9872
C	$5,000	$ 30,000	6	8.5172	10.3090
D	$8,000	$ 32,000	4	8.9872	10.3735
E	$40,000	$ 40,000	1	10.5966	10.5966

(a) Find the average In-State tuition. Find the average log(In-State). Confirm that the log of the average is *not* the same as the average of the logs. (b) Find the median In-State tuition and the median of the logs of In-State tuitions. Verify that the log of the median *is* the same as the median of the logs. (c) Compute the median of the ratios. Compute the differences of logged tuitions—log(Out-of-State) minus log(In-State) and compute the median of these differences. Verify that the median of the differences (of log tuitions) is equal to the natural log of the median of ratios (aside from some minor rounding error).

21. Umpire Life Lengths. When an umpire collapsed and died soon after the beginning of the 1990 U.S. major league baseball season, there was speculation that the stress associated with that job poses a health risk. Researchers subsequently collected historical and current data on umpires to investigate their life expectancies (Cohen et al., " Life Expectancy of Major League Baseball Umpires," *The Physician and Sportsmedicine*, 28(5) (2000): 83–89). From an original list of 441 umpires, data were found for 227 who had died or had retired and were still living. Of these, dates of birth and death were available for 195. Display 3.10 shows several rows of a generated data set based on the study.

DISPLAY 3.10 First 4 rows (of 227) from the umpire data set (Observed is the known lifetime for those umpires who had died by the time of the study [for whom Censored = 0] and the current age of those who had not yet died [for whom Censored = 1]; Expected is the expected life length—from actuarial life tables—for individuals who were alive at the time the person first became an umpire)

Umpire	Observed life length (yr)	Censored (0 if dead)	Expected life length (yr)
1	63	0	70
2	69	0	71
3	58	0	71
4	61	1	70
. . .			

(a) Use a t-test and confidence interval (possibly after transformation) to investigate whether umpires had smaller observed life lengths than expected, using only those with known life lengths (i.e., for whom *Censored* = 0)

(b) What are the potential consequences of ignoring those 214 of the 441 umpires on the original list for whom data was unavailable?

(c) What are the potential consequences of ignoring those 32 umpires in the data set who had not yet died at the time of the study? (See, for example, the survival analysis techniques in S. Anderson et al., *Statistical Methods for Comparative Studies*, New York: Wiley, 1980.)

22. Voltage and Insulating Fluid. Researchers examined the time in minutes before an insulating fluid lost its insulating property. The following data are the breakdown times for eight samples of the fluid, which had been randomly allocated to receive one of two voltages of electricity:

Times (min) at 26 kV:	5.79	1579.52	2323.70		
Times (min) at 28 kV:	68.8	108.29	110.29	426.07	1067.60

(a) Form two new variables by taking the logarithms of the breakdown times: Y_1 = log breakdown time at 26 kV and Y_2 = log breakdown time at 28 kV.

(b) By hand, compute the difference in averages of the log-transformed data: $\overline{Y}_1 - \overline{Y}_2$.

(c) Take the antilogarithm of the estimate in (b): $\exp(\overline{Y}_1 - \overline{Y}_2)$. What does this estimate? (See the interpretation for the randomized experiment model in Section 3.5.2.)

(d) By hand, compute a 95% confidence interval for the difference in mean log breakdown times. Take the antilogarithms of the endpoints and express the result in a sentence.

23. Solar Radiation and Skin Cancer. The data in Display 3.11 are yearly skin cancer rates (cases per 100,000 people) in Connecticut, with a code identifying those years that came two years after higher than average sunspot activity and those years that came two years after lower than average sunspot activity. (Data from D. F. Andrews and A. M. Herzberg, *Data*, New York: Springer-Verlag, 1985.) (a) Is there any reason to suspect that using the two independent sample t-test to compare skin cancer rates in the two groups is inappropriate? (b) Draw scatterplots of skin cancer rates versus year, for each group separately. Are any problems indicated by this plot?

24. Sex Discrimination. With a statistical computer program, reanalyze the sex discrimination data in Display 1.3 but use the log transformation of the salaries. (a) Draw box plots. (b) Find a p-value for comparing the distributions of salaries. (c) Find a 95% confidence interval for the ratio of population medians. Write a sentence describing the finding.

DISPLAY 3.11 Partial listing of Connecticut skin cancer rates (per 100,000 people) from 1938 to 1972, with solar code (1 if there was higher than average sunspot activity and 2 if there was lower than average sunspot activity two years earlier)

Year	Rate	Code
1938	0.8	2
1939	1.3	1
1940	1.4	1
1941	1.2	1
. . .		
1972	4.8	1

DISPLAY 3.12 Proportions of pollen removed and visit durations (in seconds) by 35 bumblebee queens and 12 honeybee workers; partial listing.

Bee	Type	Removed	Duration
1	queen	0.07	2
2	queen	0.10	5
3	queen	0.11	7
4	queen	0.12	11
. . .			
45	worker	0.78	51
46	worker	0.74	64
47	worker	0.77	78

25. Agent Orange. With a statistical computer program, reanalyze the Agent Orange data of Display 3.3 with and without the two largest dioxin levels in the Vietnam veterans group. Verify the one-sided p-values in bubble 2 of Display 3.7.

26. Agent Orange. With a statistical computer package, reanalyze the Agent Orange data of Display 3.3 after taking a log transformation. Since the data set contains zeros—for which the log is undefined—try the transformation log(dioxin + .5). (a) Draw side-by-side box plots of the transformed variable. (b) Find a p-value from the t-test for comparing the two distributions. (c) Compute a 95% confidence interval for the difference in mean log measurements and interpret it on the original scale. (*Note*: Back-transforming does not provide an exact estimate of the ratio of medians since 0.5 was added to the dioxins, but it does provide an approximate one.)

27. Pollen Removal. As part of a study to investigate reproductive strategies in plants, biologists recorded the time spent at sources of pollen and the proportions of pollen removed by bumblebee queens and honeybee workers pollinating a species of lily. (Data from L. D. Harder and J. D. Thompson, "Evolutionary Options for Maximizing Pollen Dispersal of Animal-pollinated Plants," *American Naturalist* 133 (1989): 323–44.) Their data appear in Display 3.12.

(a) (i) Draw side-by-side box plots (or histograms) of the proportion of pollen removed by queens and workers. (ii) When the measurement is the proportion P of some amount, one useful transformation is $\log[P/(1 - P)]$. This is the log of the ratio of the proportion removed to the proportion not removed. Draw side-by-side box plots or histograms on this transformed scale. (iii) Test whether the distribution of proportions removed is the same or different for the two groups, using the t-test on the transformed data.

(b) Draw side-by-side box plots of duration of visit on (i) the natural scale, (ii) the logarithmic scale, and (iii) the reciprocal scale. (iv) Which of the three scales seems most appropriate for use of the t-tools? (v) Compute a 95% confidence interval to describe the difference in means on the chosen scale. (vi) What are relative advantages of the three scales as far as interpretation goes? (vii) Based on your experience with this problem, comment on the difficulty in assessing equality of population standard deviations from small samples.

28. Bumpus's Data. Obtain p-values from the t-test to compare humerus lengths for sparrows that survived and those that perished (Exercise 2.21), with and without the smallest length in the perished group (length = 0.659 inch). Do the conclusions depend on this one observation? What action should be taken if they do?

29. Cloud Seeding—Multiplicative vs. Additive Effects. On the computer, create a variable containing the rainfall amounts for only the unseeded days. (a) Create four new variables by adding 100, 200, 300, and 400 to each of the unseeded day rainfall amounts. Display a set of five box plots to illustrate what one might expect if the effect of seeding were additive. (b) Create four additional variables by multiplying each of the unseeded day rainfall amounts by 2, by 3, by 4, and by 5. Display a set of five box plots to illustrate what could be expected if the effect of seeding were multiplicative. (c) Which set of plots more closely resembles the actual data?

Data Problems

30. Education and Future Income. Display 3.13 shows the first five rows of a data set with annual incomes in 2005 of the subset of National Longitudinal Survey of Youth (NLSY79) subjects (described in Exercise 2.22) who had paying jobs in 2005 and who had completed either 12 or 16 years of education by the time of their interview in 2006. All the subjects in this sample were between 41 and 49 years of age in 2006. Analyze the data to describe the amount (or percent) by which the population distribution of incomes for those with 16 years of education exceeds the distribution for those with 12 years of education. (*Note*: The NLSY79 data set codes all incomes above \$150,000 as \$279,816. To make an exercise version that better matches the actual income distribution, those values have been replaced in the data set by computer-simulated values from a realistic distribution of incomes greater than \$150,000.)

DISPLAY 3.13 Annual incomes in 2005 (in U.S. dollars) of 1,020 Americans who had 12 years of education and 406 who had 16 years of education by the time of their interview in 2006; "Subject" is a subject identification number; first 5 of 1,426 rows

Subject	Educ	Income2005
2	12	5,500
6	16	65,000
7	12	19,000
13	16	8,000
21	16	253,043

31. Education and Future Income II. The data file ex0331 contains a subset of the NLSY79 data set (see Exercise 30) with annual incomes of subjects with either 16 or more than 16 years of education. Analyze the data to describe the amount (or percent) by which the population distribution of incomes for those with more than 16 years of education exceeds the distribution for those with 16 years of education.

32. College Tuition. Display 3.14 shows the first five rows of a data set with 2011–2012 in-state and out-of-state tuitions for random samples of 25 private and 25 public, four-year colleges and universities in the United States. Analyze the data to describe (a) the extent to which out-of-state tuition is more expensive than in-state tuition in the population of public schools, (b) the extent to which private school in-state tuition is more expensive than public school in-state tuition, and (c) the extent to which private school out-of-state tuition is more expensive than public school out-of-state tuition. (Data sampled from College Board: http://www.collegeboard.com/student/ (11 July 2011).)

DISPLAY 3.14 In-state and out-of-state tuitions for 25 public and 25 private colleges and universities in the United States; first 5 of 50 rows

College	Type	InState	OutofState
Albany State University	public	$5,434	$17,048
Appalachian State University	public	$5,175	$16,487
Argosy University: Nashville	private	$19,596	$19,596
Brescia University	private	$18,140	$18,140
Central Connecticut State University	public	$8,055	$18,679

33. Brain Size and Litter Size. Display 3.15 shows relative brain weights (brain weight divided by body weight) for 51 species of mammal whose average litter size is less than 2 and for 45 species of mammal whose average litter size is greater than or equal to 2. (These are part of a larger data set considered in Section 9.1.2.) What evidence is there that brain sizes tend to be different for the two groups? How big of a difference is indicated? Include the appropriate statistical measures of uncertainty in carefully worded sentences to answer these questions.

DISPLAY 3.15 Relative brain sizes, 1,000 × (Brain weight/Body weight), for 96 species of mammals

1,000 × (Brain weight/Body weight) for 51 species with average litter size < 2

0.42	0.86	0.88	1.11	1.34	1.38	1.42	1.47	1.63	1.73	2.17	2.42
2.48	2.74	2.74	2.79	2.90	3.12	3.18	3.27	3.30	3.61	3.63	4.13
4.40	5.00	5.20	5.59	7.04	7.15	7.25	7.75	8.00	8.84	9.30	9.68
10.32	10.41	10.48	11.29	12.30	12.53	12.69	14.14	14.15	14.27	14.56	15.84
18.55	19.73	20.00									

1,000 × (Brain weight/Body weight) for 45 species with average litter size ≥ 2

0.94	1.26	1.44	1.49	1.63	1.80	2.00	2.00	2.56	2.58	3.24	3.39
3.53	3.77	4.36	4.41	4.60	4.67	5.39	6.25	7.02	7.89	7.97	8.00
8.28	8.83	8.91	8.96	9.92	11.36	12.15	14.40	16.00	18.61	18.75	19.05
21.00	21.41	23.27	24.71	25.00	28.75	30.23	35.45	36.35			

Answers to Conceptual Exercises

1. The target clouds on a day that was deemed suitable for seeding.

2. Uncontrollable confounding factors probably explain the variability in rainfall from clouds treated the same way. Randomization is needed to ensure that the confounding factors do not tend to be unevenly distributed in the two groups.

3. Blinding prevents the intentional or unintentional biases of the human investigators from having a chance to make a difference in the results.

4. There may be serial correlation. A plot of rainfall versus date could be used to check.

5. Larger values are to be expected by chance if the populations are the same, since the sample of Vietnam veterans is so much larger than the sample of non–Vietnam veterans.

6. (a) He would not be representative of the target population and should be removed from the data set for analysis. (b) Same thing.

7. No, not from the statistics alone since this is an observational study. It could be said, however, that the data are consistent with that theory.

8. No. The dependence is the result of matching and is desirable. The two-sample t-tools are not appropriate (but the paired t-tools are).

9. False.

10. False. An *average* from a sample will have a sampling distribution that will tend toward normal with large sample sizes, but the sample histogram should mirror the population distribution. As the sample size gets larger, the sample histogram should become a better approximation to the population histogram.

11. There is a cluster effect: the particular day of the week. She should use a paired-t analysis, as will be discussed in Chapter 4.

12. The t-test is robust in validity to departures from normality, especially as the sample size gets large. The robustness with respect to equal standard deviations does not depend much on what the sample sizes are, so long as they are reasonably equal. Sample size does not affect robustness with respect to independence.

13. You cannot take logarithms of negative numbers.

14. It will be more consequential in smaller samples; its effect gets washed out in large ones.

15. Replace the population SDs in the formula (Section 2.2.2) by individual *sample* SDs.

16. No. The paired analysis must be used, even though the inferences may not appear to be as precise. The unpaired analysis is inappropriate.

17. (a) Dependence of measurements on individuals in the same household (cluster effect). (b) Maybe. Getting a single measure for each household may be an easy way out of the dependence problem, but care should be used as these groups also tend to differ in the average number of persons per household.

18. (a) No. (b) Yes.

19. (a) Serial correlation both within and between groups. (Confounding variable is the time at which observations were made.) (b) Cluster effect between groups. (c) Cluster effect (members of the same household should be similar) within groups.

CHAPTER 4

Alternatives to the *t*-Tools

The t-tools have an extremely broad range of application, extending well beyond the strict confines of the ideal model because of robustness. They extend even further when the possibilities of transforming the data and dealing with outliers are taken into account.

Nevertheless, situations arise where the t-tools cannot be applied, because the model assumptions of the t-test are grossly violated. For these situations, a host of other methods, based on different models, may be used. Some are presented in this chapter. Most notable are two distribution-free methods, based on models that do not specify any particular population distributions. The rank-sum test for two independent samples and the signed-rank test for a sample of pairs are useful alternatives, particularly when outliers may be present or when the sample sizes are too small to permit the assessment of distributional assumptions.

4.1 CASE STUDIES

4.1.1 Space Shuttle O-Ring Failures—An Observational Study

On January 27, 1986, the night before the space shuttle *Challenger* exploded, engineers at the company that built the shuttle warned National Aeronautics and Space Administration (NASA) scientists that the shuttle should not be launched because of predicted cold weather. Fuel seal problems, which had been encountered in earlier flights, were suspected of being associated with low temperatures. It was argued, however, that the evidence was inconclusive. The decision was made to launch, even though the temperature at launch time was 29°F.

The data in Display 4.1 are the numbers of O-ring incidents on previous shuttle flights, categorized into those launched at temperatures below 65°F and those launched at temperatures above 65°F. (Data from a graph in Richard P. Feynman, *What Do You Care What Other People Think?* (New York: W. W. Norton, 1988).) Is there a higher risk of O-ring incidents at lower launch temperatures?

DISPLAY 4.1 Numbers of O-ring incidents on 24 space shuttle flights prior to the *Challenger* disaster

Launch temperature	Number of O-ring incidents																			
Below 65°F	1	1	1	3																
Above 65°F	0	0	0	0	0	0	0	0	0	0	0	0	0	0	0	0	0	1	1	2

Statistical Conclusion

These date provide strong evidence that the number of O-ring incidents was associated with launch temperature in these 24 launches. It is highly unlikely that the observed difference of the groups is due to chance (one-sided p-value $= 0.0099$ from a permutation test on the t-statistic).

Scope of Inference

These observational data cannot be used to establish causality, nor is there any broader population of which they are a sample. But the association between temperature and O-ring failure in these particular 24 launches is consistent with the theory that lower temperatures impair the functioning of the O-rings. (At one point in public hearings into the causes of the disaster, Feynman asked for a glass of ice water, placed a small O-ring in it for a time, removed it, and then proceeded to demonstrate that the rubber failed to spring back to its original form.) (*Note*: Other techniques for dealing with count data are given in Chapter 22.)

4.1.2 Cognitive Load Theory in Teaching—A Randomized Experiment

Consider the following problem in coordinate geometry.

> Point A has coordinates $(2, 1)$, point B has $(8, 3)$, and point C has $(4, 6)$. What is the slope of the line that connects C to the midpoint between A and B?

Presenting the solution as a worked problem, a conventional textbook shows a picture of the layout, gives a discussion in the text, and then provides the lines of algebraic manipulation leading to the right answer. (See Display 4.2.) Recent theoretical developments in cognitive science suggest that splitting the presentation into the three distinct units of diagram, text, and algebra imposes a heavy, extraneous cognitive load on the student. The requirement that the student organize and process the separate elements constitutes a cognitive load. The load is extraneous because it is not essential to learning how to solve such problems—indeed, it impedes the learning process by placing heavy demands on cognitive resources that should be used to understand the essentials.

DISPLAY 4.2 Cognitive load experiment: conventional method of instruction (for finding the slope of the line that connects C to the midpoint between A and B)

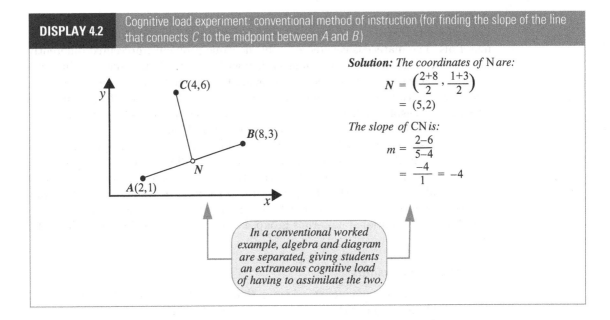

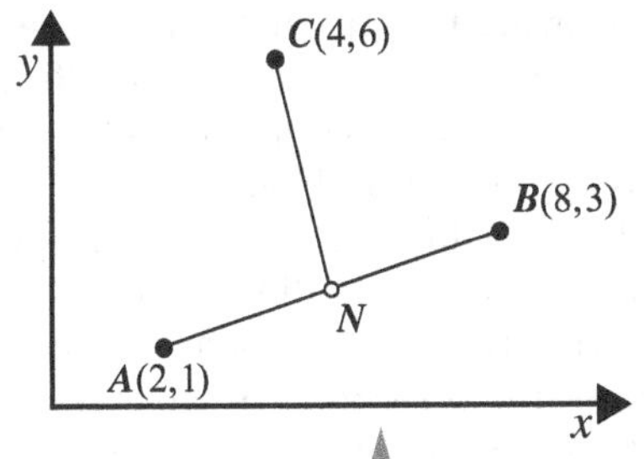

Solution: *The coordinates of* N *are:*

$$N = \left(\frac{2+8}{2}, \frac{1+3}{2}\right) = (5,2)$$

The slope of CN *is:*

$$m = \frac{2-6}{5-4} = \frac{-4}{1} = -4$$

In a conventional worked example, algebra and diagram are separated, giving students an extraneous cognitive load of having to assimilate the two.

In a test of this theory, researchers compared the effectiveness of conventional textbook worked examples to modified worked examples, which present the algebraic manipulations and explanation as part of the graphical display (see Display 4.3). (Data from J. Sweller, P. Chandler, P. Tierney, and M. Cooper, "Cognitive Load as a Factor in the Structuring of Technical Material," *Journal of Experimental Psychology General* 119(2) (1990): 176–92.)

Researchers selected 28 ninth-year students in Sydney, Australia, who had no previous exposure to coordinate geometry but did have adequate mathematics to deal with the problems given. The students were randomly assigned to one of

DISPLAY 4.3 Cognitive load experiment: modified method of instruction (for finding the slope of the line that connects C to the midpoint between A and B)

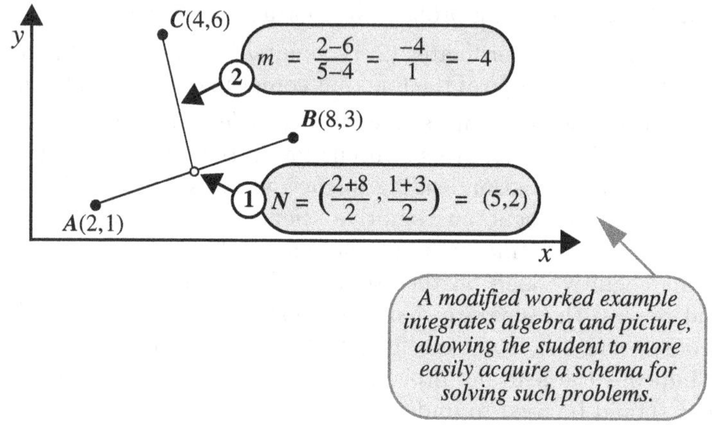

two self-study instructional groups, using conventional and modified instructional materials. The materials covered exactly the same problems, presented differently. Students were allowed as much time as they wished to study the material, but they were not allowed to ask questions. Following the instructional phase, all students were tested with a common examination over three problems of different difficulty. The data in Display 4.4, based on this study, are the number of seconds required to arrive at a solution to the moderately difficult problem.

Both distributions in Display 4.4 are highly skewed. In addition, there were five students in the conventional (control) group who did not come to any solution in the five minutes allotted. Their solution times are considered *censored*—all that is known about them is that they exceed 300 seconds. It appears that the solution times for the "modified instructional materials" group are generally shorter than for the conventional materials group. Is there sufficient evidence to draw this conclusion?

Statistical Conclusions

These data provide convincing evidence that a student could solve the problem more quickly if taught with the modified method than if taught with the conventional method (one-sided p-value $= 0.0013$, from the rank-sum test). The modified instructional materials shortened solution times by an estimated 129 seconds (95% confidence interval for an additive treatment effect: 58 to 159 seconds).

4.2 THE RANK-SUM TEST

The *rank-sum test* is a resistant alternative to the two-sample t-test. It performs nearly as well as the t-test when the two populations are normal and considerably

DISPLAY 4.4 Number of seconds to solution of a problem in coordinate geometry, for students instructed with conventional and modified materials

Modified ($n = 14$)		*Conventional* ($n = 14$)
8	6	
7530	7	
00	8	
	9	
	10	
	11	
	12	
2	13	09
8	14	6
5	15	0
	16	1
	17	7
3	18	
7	19	
6	20	
0	21	
	22	8
	23	
	24	2
	25	
	26	5
	27	
	28	
	29	
	30	+++++

Five students did not complete the problem in the five-minute time allotment.

Legend: | 26 | 5 represents 265 seconds.

better when there are extreme outliers. Its drawbacks are that associated confidence intervals are not computed by most of the statistical computer packages and that it does not easily extend to more complicated situations.

4.2.1 The Rank Transformation

Straightforward transformations of the data were used in Chapter 3 to obtain measurements on a scale where the normality and equal spread assumptions are approximately met. The rank-sum relies on a special kind of transformation that replaces each observation by its rank in the combined sample. It should be noted that this really is a different kind of transformation for two reasons: (1) A single transformed value depends on all the data. So the transformed rank of, say, $Y = 63.925$, may be one value in one problem but a very different value in another problem. (2) There is no reverse transformation, such as the antilog, available for ranks.

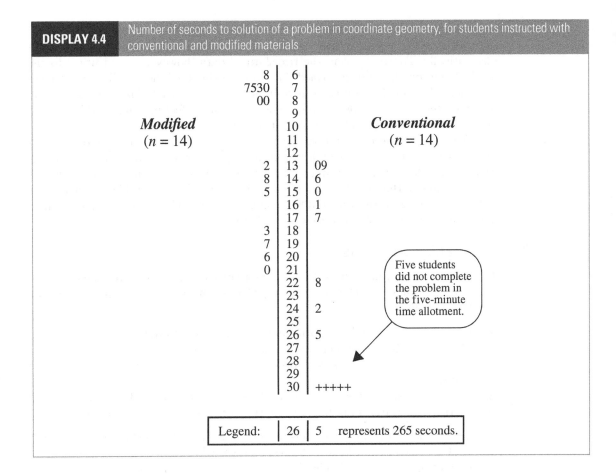

The purpose behind using ranks is not to transform the data to approximate normality but to transform the data to a scale that eliminates the importance of the population distributions altogether. Notice that when the data values are replaced by their square roots or by their logarithms, the sample distributions change shape dramatically. The ranks from the transformed data, however, are identical to the ranks based on the untransformed data, so the shape of the distribution of measurements has no effect on the ranks. In addition, whether the largest observation in a data set is 300 seconds or 3,000 seconds does not affect its rank. In this respect, any statistic based on ranks is resistant to outliers.

A feature of the rank-sum test that makes it an attractive choice for the cognitive load experiment is its ability to deal with *censored observations*, observations that are only known to be greater than (or possibly less than) some number. All that is known about the five students who did not complete the problem is that their solution times are greater than 300 seconds. For the rank-sum test it is enough to know that, in terms of ranks, they were tied for last.

4.2.2 The Rank-Sum Statistic

Calculation of the rank-sum test statistic for the cognitive load experiment is summarized in Display 4.5. The first four steps transform the data to their ranks in the combined sample:

1. List all observations from both samples in increasing order.
2. Identify which sample each observation came from.
3. Create a new column labeled "order," as a straight sequence of numbers from 1 to $(n_1 + n_2)$.
4. Search for ties—that is, duplicated values—in the combined data set. The ranks for tied observations are taken to be the average of the orders for those cases.

Two students tied at 80 seconds, for example. They finished sixth and seventh fastest. One cannot say which of the two deserves which order, so both are assigned the rank of $(6 + 7)/2 = 6.5$. The five students who took the full five minutes were the last five finishers, with orders 24, 25, 26, 27, and 28. So each is assigned rank $(24 + 25 + 26 + 27 + 28)/5 = 26$. Any observation that has a unique value gets its order as its rank.

The test statistic, T, is the sum of all the ranks in one group, called "group 1." Group 1 is conventionally the group with the smaller sample size (because that minimizes computation). The choice, however, is arbitrary.

The impressive feature of the cognitive load experiment is that the early finishers were mostly the students who studied the modified instructional material. This is reflected in the test statistic by a low rank-sum ($T = 137$) for that group.

4.2.3 Finding a *p*-Value by Normal Approximation

The rank-sum procedure is used to test the null hypothesis of no treatment effect from a two-treatment randomized experiment and also to test the null hypothesis of identical population distributions from two independent samples. If the null hypothesis is true in either case, then the sample of n_1 ranks in group 1 is a random sample from the $n_1 + n_2$ available ranks. Both the randomization distribution and the sampling distribution of the rank-sum statistic, T, are represented by a

DISPLAY 4.5 Rank-sum test statistic T for the cognitive load experiment

(1) *List all observations from both samples in increasing order.*

(2) *Identify sample membership of each value.*

(3) *Write down an increasing sequence for list order.*

(4) *Modify the orders, if necessary, by giving the average order to tied values.*

(5) *Add the ranks from all observations in one group (M).*

Y	Group	Order	Rank
68	M	1	1
70	M	2	2
73	M	3	3
75	M	4	4
77	M	5	5
80	M	6	6.5
80	M	7	6.5
130	C	8	8
132	M	9	9
139	C	10	10
146	C	11	11
148	M	12	12
150	C	13	13
155	M	14	14
161	C	15	15
177	C	16	16
183	M	17	17
197	M	18	18
206	M	19	19
210	M	20	20
228	C	21	21
242	C	22	22
265	C	23	23
300	C	24	26
300	C	25	26
300	C	26	26
300	C	27	26
300	C	28	26

$T = 137$

histogram of the rank-sum statistics calculated for each possible regrouping of the $n_1 + n_2$ observations into samples of size n_1 and n_2.

Because conversion to ranks avoids absurd distributional anomalies, the randomization distribution of T can be approximated accurately by a normal distribution in most situations. The exceptions are when at least one sample is small—say under 5—or when large numbers of ties occur. The mean and variance of the permutation distribution, from statistical theory, are shown in Display 4.6.

These facts may be used to evaluate whether the observed rank-sum statistic is unusually small or large. If there is no difference, then the Z-statistic

$$Z\text{-statistic} = [T - \text{Mean}(T)]/\text{SD}(T)$$

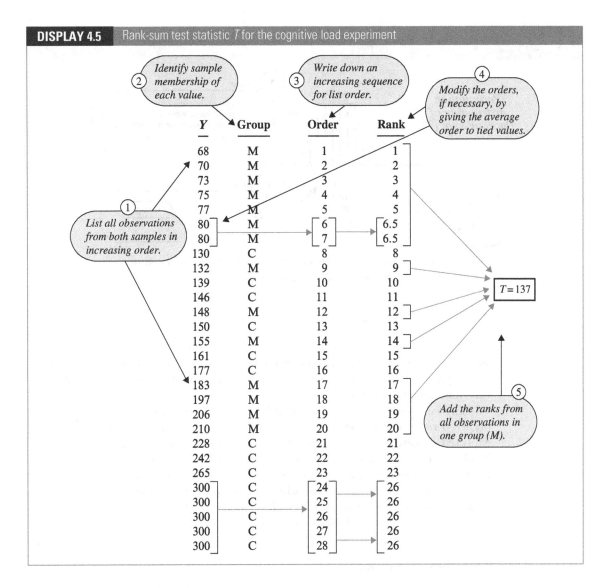

DISPLAY 4.6 Facts about the randomization (or sampling) distribution of the rank-sum statistic—the sum of ranks in group 1—when there is no group difference

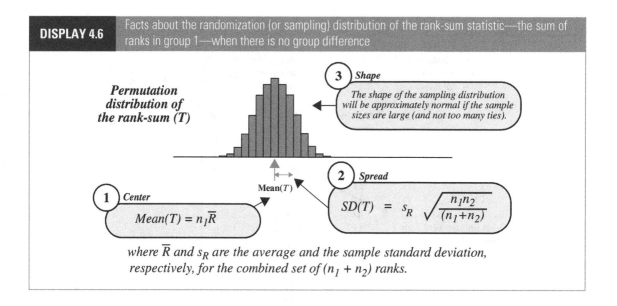

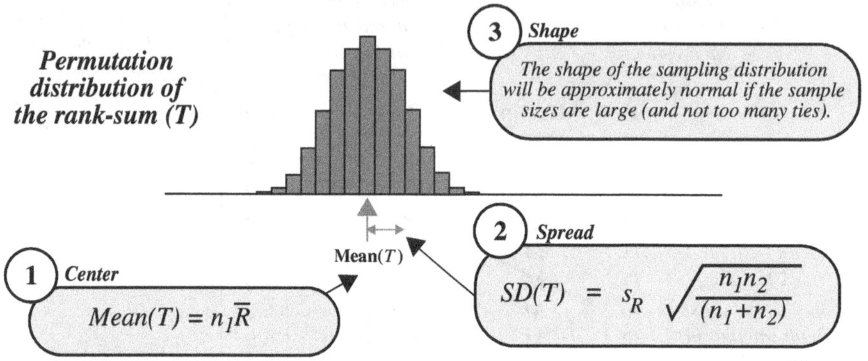
Permutation distribution of the rank-sum (T)
3 Shape
The shape of the sampling distribution will be approximately normal if the sample sizes are large (and not too many ties).
2 Spread
$SD(T) = s_R \sqrt{\frac{n_1 n_2}{(n_1+n_2)}}$
Mean(T)
1 Center
$Mean(T) = n_1\bar{R}$

where $\bar{R}$ and s_R are the average and the sample standard deviation, respectively, for the combined set of $(n_1 + n_2)$ ranks.

should be similar to a typical value from a standard normal distribution. A p-value is the proportion of values from a standard normal distribution that are more extreme than the observed Z-statistic.

The calculations for the cognitive load data are shown in Display 4.7. The observed rank-sum statistic of 137 is smaller than the value of 203 that would be expected if there were no treatment effect. This disparity may be due to the luck involved in the random assignment, with better students placed in the modified instruction group, but the p-value indicates that only about one in a thousand randomizations would produce a disparity as great as or greater than the observed one, if there were no treatment effect. Such a small p-value provides fairly convincing evidence that a treatment effect explains the difference.

Continuity Correction

The statistic T can only take on integer values. Its distribution can be displayed as a histogram with a bar centered on each integer having a height equal to the probability that T equals that integer and a base extending from the integer minus one-half to the integer plus one-half. Thus the bar areas are the probabilities associated with each integer. Probabilities can also be approximated by the normal distribution, which is centered at Mean(T) with a standard deviation of SD(T).

Using the normal distribution to approximate the probability that T is less than or equal to the integer k, one must realize that the desired area under the normal curve is the area less than or equal to $k + 0.5$ in the probability histogram. To get the best approximation, therefore, one must determine the normal distribution area less than or equal to $k + 0.5$.

DISPLAY 4.7 Finding the p-value with the normal approximation to the permutation distribution of the rank-sum statistic, using a continuity correction. Calculations for the cognitive load data are continued from Display 4.5

(1) *Calculate the average and sample standard deviation of the ranks from the combined sample (column 4 of Display 4.5).*

$$\bar{R} = 14.5 \qquad s_R = 8.2023$$

(2) *Compute the theoretical "null hypothesis" mean and standard deviation of T using the formulae in Display 4.6.*

$$\text{Mean}(T) = 14 \times 14.5 = 203 \qquad \text{SD}(T) = 8.2023\sqrt{\frac{14 \times 14}{(14+14)}} = 21.7013$$

(3) *Calculate the Z-statistic using a continuity correction.*

$$Z = \frac{(137.5 - 203)}{21.7013} = -3.0183$$

(4) *Find the p-value from a standard normal table.* **One-sided p-value = 0.0013**

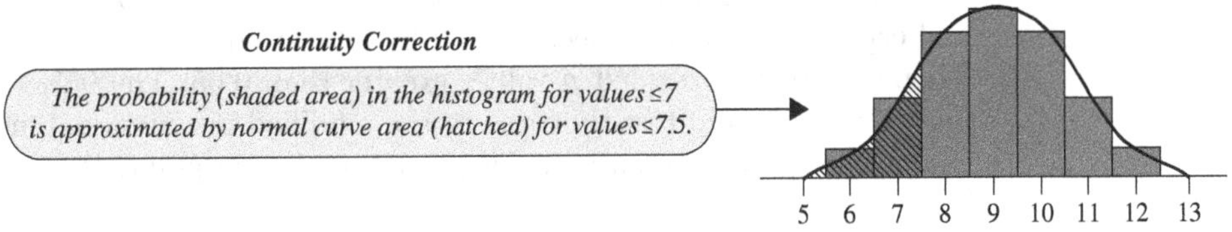
Continuity Correction
The probability (shaded area) in the histogram for values ≤7 is approximated by normal curve area (hatched) for values ≤7.5.
5 6 7 8 9 10 11 12 13

Similarly, the normal approximation to the probability that T is greater than or equal to an integer k is the area under the normal curve to the right of $k - 0.5$ (so that the entire bar above k is included).

This adjustment to the normal approximation is called a *continuity correction* because it corrects the less accurate calculation that simply uses k. The calculations in Display 4.7 include this feature.

Exact p-Values for the Rank-Sum Test

The normal distribution is an easy and adequate approximation to the randomization distribution for most problems. Troublesome situations arise where sample sizes are small or there are large numbers of ties.

It is possible to compute the randomization distribution exactly using the technique described in Section 1.3.2. Published tables with exact percentiles exist (see, for example, Pearson and Hartley 1972, p. 227), but they are only appropriate when there are no ties. Another drawback is that only a few percentiles are published for each combination of sample sizes, so that *p*-values can only be bracketed. Many statistical computer programs can compute the exact *p*-value from the permutation distribution (if the sample sizes aren't too large) or approximate it through repeated sampling from the permutation distribution.

4.2.4 A Confidence Interval Based on the Rank-Sum Test

A 95% confidence interval for any parameter can be constructed by including all hypothesized values that lead to two-sided *p*-values greater than or equal to 0.05. This relation between tests and confidence intervals can be exploited to obtain a confidence interval for an additive treatment effect δ in the cognitive load experiment.

If Y is the time to solution for a student using the conventional study materials, the additive treatment effect model says that $Y - \delta$ is the time to solution for the same student using the modified study materials. The rank-sum test in Display 4.7 indicates that 0 is not very likely as a value for δ. Could δ be, say, 50 seconds? If so, the completion times in the modified group *plus 50 seconds* should be a set of times that are similar to those for the conventional study materials group. This suggests a procedure: (1) add a hypothesized δ to all modified group times, (2) use the rank-sum test to decide whether the two resulting group differences can be explained by chance (in this case whether the two-sided *p*-value exceeds 0.05), and (3) determine—through trial and error—upper and lower limits on δ values satisfying the criterion.

Display 4.8 illustrates the process with a series of proposed values for δ. By trial and error it was determined that all hypothesized values of δ between 58 and 159 seconds lead to two-sided *p*-values greater than 0.05. The 95% confidence interval for the reduction in test time due to the modified instructional method is 58 to 159 seconds. A point estimate for δ is the interval's midpoint, 108 seconds.

DISPLAY 4.8 Using a rank-sum test to construct a confidence interval for an additive treatment effect (cognitive load study)

Hypothesized effect (seconds)	Two-sided *p*-value	Confidence interval inclusion?
50	0.0286	no
60	0.0800	yes
55	0.0403	no
58	0.0502	yes
150	0.1227	yes
160	0.0476	no
155	0.0589	yes
158	0.0530	yes
159	0.0502	yes

Try several hypothesized values for δ to identify those that have two-sided *p*-values ≥ 0.05.

A 95% confidence interval is −159 seconds to −58 seconds.

Notes About the Rank-Sum Procedure

1. Other names for the rank-sum test are the Wilcoxon test and the Mann–Whitney test. The different names refer to originators of different forms of the test statistic. To confuse the issue, there is also a Wilcoxon signed-rank test—which is something entirely different.
2. The rank-sum test is a *nonparametric* or *distribution-free* statistical tool, meaning there are no specific distributional assumptions required.
3. Although the t-test is more efficient when the populations are normal (it makes better use of the available data), the rank-sum test is not that much worse in the normal model, and is substantially better for many other situations, particularly for long-tailed distributions.
4. The theoretical mean of T in Display 4.6 can also be written as $\text{Mean}(T) = n_1(n_1 + n_2 + 1)/2$. If there are no ties, the theoretical standard deviation in the permutation distribution of T is

$$\text{SD}(T) = \sqrt{n_1 n_2 (n_1 + n_2 + 1)/12}$$

The version in bubble 2 of Display 4.6 is correct whether there are ties or not.
5. In its application to the cognitive load problem, the confidence interval method described in Display 4.8 should be modified to reflect the truncation of solution times to 5 minutes. Thus, when the modified group's times are shifted by a certain hypothesized amount, any values exceeding 300 seconds should be replaced by 300 before performing the test. Such a modification makes no change in the lower limit of the confidence interval; but the upper limit changes from 159 to 189 seconds.

4.3 OTHER ALTERNATIVES FOR TWO INDEPENDENT SAMPLES

4.3.1 Permutation Tests

A *permutation test* is any test that finds a p-value as the proportion of regroupings—of the observed $n_1 + n_2$ numbers into two groups of size $n_1 + n_2$—that lead to test statistics as extreme as the observed one. The test statistic may be the difference in group averages, a t-statistic, the sum of ranks in group 1, or any other choice to represent group difference. Permutation tests were previously discussed in the context of interpretation. When used to analyze randomized experiments, for example, permutation tests are called *randomization tests* and provide statistical inferences tied to the chance mechanism in random assignment. For observational studies, they provide no inference to a broader context (except for the special case of the rank-sum test), but may nevertheless be useful for summarizing differences in the data at hand. In the cases so far discussed, the actual calculation of p-values and confidence intervals was based on an approximation to the permutation distribution. Sometimes there is no adequate approximation.

In the O-ring study, for example, the exact rendering of the permutation test is the only method for calculating its p-value. The distribution of the numbers is so nonnormal that a t-distribution approximation is severely inadequate. No transformation helps. The rank-sum test is inadequate because of the large number of ties: 17 of the 24 values are tied at zero. Even though the computational effort for direct calculation of the p-value is considerable, permutation calculations are important because they are always available and require no distributional assumptions or special conditions. The p-value is calculated using the following procedure:

1. Decide on a test statistic, and compute its value from the two samples.
2. List all regroupings of the $n_1 + n_2$ numbers into groups of size n_1 and n_2, and recompute the test statistic for each.
3. Count the number of regroupings that produce test statistics at least as extreme as the observed test statistic from step 1.
4. The p-value is the number found in step 3 divided by the total number of regroupings.

In problems such as the O-ring study, the procedure can be accomplished by counting with *combinatorics*, that is, by using the *combination numbers*

$$C_{n,k} = \frac{n(n-1)\cdots(n-k+1)}{k(k-1)\cdots 1}.$$

The number $C_{n,k}$—read as "n choose k"—is the number of different ways to choose k items from a list of n items. The total number of regroupings for a two-sample problem is $C_{n_1+n_2,n_1}$. Combination numbers can also be used to count the number of regroupings that lead to test statistics as extreme or more extreme than the observed one, as now shown for the O-ring data.

Step 1. The t-statistic was selected as the test statistic (although the difference in averages would be equally good). Its observed value is 3.888.
Step 2. It is only necessary to determine which group 1 outcomes produce t-statistics that are as large as or larger than the 3.888 that came from the observed outcome, $(1,1,1,3)$. After some calculation, it is found that the extreme regroupings are those whose group 1 outcomes are $(1,1,2,3)$, $(1,1,1,3)$, or $(0,1,2,3)$, with t-statistics 5.952, 3.888, and 3.888, respectively.
Step 3. The total number of regroupings is $C_{24,4} = 10{,}626$. For a one-sided p-value it is necessary to count the number of regroupings that produce the outcomes in Step 2. Consider the outcome $(1,1,2,3)$. To get such an outcome, the single 2 and the single 3 must be selected. However, there are five 1's in the full sample, and the indicated outcome would occur with any of the $C_{5,2} = 10$ combinations of 1's. This is illustrated in Display 4.9.

Similarly, the number of regroupings with outcome $(1,1,1,3)$ is the number of ways to select three 1's from the five available in the data set, to go with the obligatory 3. This is $C_{5,3} = 10$. Finally, a regrouping with outcome $(0,1,2,3)$ can be formed by taking any one of the 17 0's in combination with any one of the five 1's, along with 2 and 3. The number of ways to do this is $C_{17,1} \times C_{5,1} = 17 \times 5 = 85$.

DISPLAY 4.9 Counting method for regroupings that give extreme group 1 outcomes

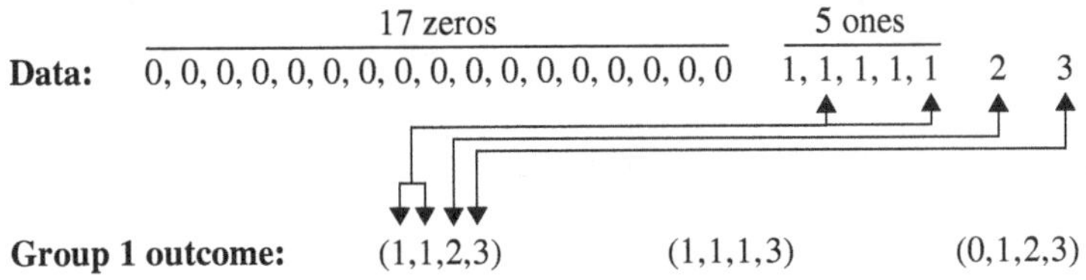

DISPLAY 4.10 A summary of the *t*-statistics calculated from all 10,626 rearrangements of the O-ring data into a Low group of size 4 and a High group of size 20

Number of rearrangements with identical *t*-statistics	*t*-statistic
2,380	−1.188
3,400	−0.463
2,040	0.231
1,530	0.939
855	1.716
316	2.643
95	**3.888**
10	**5.952**

Total number of rearrangements into two groups of size 4 and 20:
10,626

Number of rearrangements with t*-statistics greater than or equal to 3.888:*
105

one-sided p*-value from a permutation test of the* t*-statistic:*
105/10,626 = 0.00988

Summing up, the number of regroupings that lead to *t*-statistics as large as or larger than the observed one is $10 + 10 + 85 = 105$.

Step 4. The one-sided *p*-value is $105/10{,}626 = 0.00988$. This result and the full permutation distribution of the *t*-statistic appear in Display 4.10.

Notes: The combinatorics method is especially useful if there are only a few regroupings that need to be counted, but it might be unmanageable if there are many. A computer can systematically enumerate the regroupings, calculate the test statistic for each, and compute the proportions that have a more extreme test statistic than the observed one. (An alternative is to approximate the *p*-value by the proportion of a *random sample* of all possible regroupings that have a test statistic more extreme than the observed one, as discussed for the discrimination study in Section 1.1.2.)

4.3.2 The Welch *t*-Test for Comparing Two Normal Populations with Unequal Spreads

Welch's t-test employs the individual sample standard deviations as separate estimates of their respective population standard deviations, rather than pooling to obtain a single estimate of a population standard deviation. The result is a different formula for the standard error of the difference in averages:

$$\mathrm{SE}_W(\overline{Y}_2 - \overline{Y}_1) = \sqrt{\frac{s_1^2}{n_1} + \frac{s_2^2}{n_2}}.$$

This becomes the denominator in the t-statistic for comparing the means of populations with different spreads. Even when the populations are normal, however, the exact sampling distribution of the Welch t-ratio is unknown. It can be approximated by a t-distribution with d.f.$_W$ degrees of freedom, known as Satterthwaite's approximation:

$$\text{d.f.}_W = \frac{[\mathrm{SE}_W(\overline{Y}_2 - \overline{Y}_1)]^4}{\dfrac{[\mathrm{SE}(\overline{Y}_2)]^4}{(n_2 - 1)} + \dfrac{[\mathrm{SE}(\overline{Y}_1)]^4}{(n_1 - 1)}}$$

where

$$\mathrm{SE}(\overline{Y}_1) = \frac{s_1}{\sqrt{n_1}} \quad \text{and} \quad \mathrm{SE}(\overline{Y}_2) = \frac{s_2}{\sqrt{n_2}}.$$

The t-test and confidence interval are computed exactly as with the two-sample t-test, except with the modified standard error and the approximate degrees of freedom (rounded down to an integer value).

A Note About the Importance of the Equal-Spread Model

If the two populations have the same shape and the same spread, then the difference in means is an entirely adequate summary of their difference. Any question of interest that requires a comparison of the two distributions can be re-expressed in terms of the single parameter $\mu_2 - \mu_1$.

If, on the other hand, the populations have different means *and different standard deviations*, then $\mu_2 - \mu_1$ may be an inadequate summary. If lifetimes of brand A lightbulbs have a larger mean and a larger standard deviation than lifetimes of brand B lightbulbs, as shown in Display 4.11, then there is a higher proportion of long-life bulbs from brand A, but there may also be a higher proportion of short-life bulbs from brand A. A comparison of brands is not entirely resolved by a comparison of means.

Although some statistical analysts find the unequal variance model more appealing for the two-sample problem, most of the standard methods for more complicated structures make use of a pooled estimate of variance. In that sense, the two-sample t-tools extend to more complicated situations more easily than does Welch's t-test.

DISPLAY 4.11 The conceptual difficulty with comparing population means when population spreads are not the same

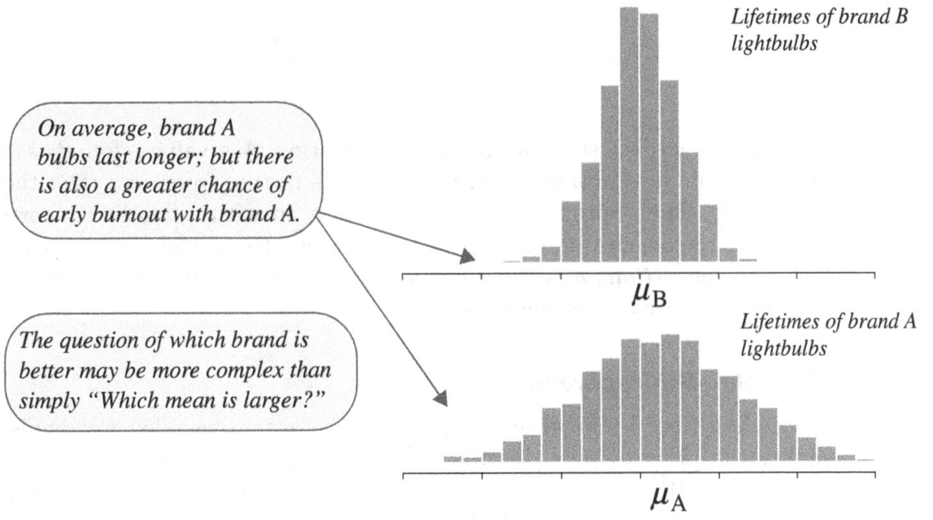
Lifetimes of brand B lightbulbs
On average, brand A bulbs last longer; but there is also a greater chance of early burnout with brand A.
μ_B
Lifetimes of brand A lightbulbs
The question of which brand is better may be more complex than simply "Which mean is larger?"
μ_A

4.4 ALTERNATIVES FOR PAIRED DATA

Two resistant and distribution-free alternatives to the paired t-test are described here. The sign test is a tool for a quick analysis of paired data. If the sign test shows a substantial effect, that may be enough to resolve the answer to the question of interest. But it is not very efficient. If its results are inconclusive, the signed-rank test is a more powerful alternative.

4.4.1 The Sign Test

The *sign test* is a quick, distribution-free test of the hypothesis that the mean difference of a population of pairs is zero (for observational studies) or of the hypothesis that there is no treatment effect in a randomized paired experiment. It counts the number K of pairs where one group's measurement exceeds the other's. In the schizophrenia study (Section 2.1.2), the left hippocampus volume of the unaffected twin was larger than the left hippocampus volume of the affected twin for $K = 14$ out of the $n = 15$ twin pairs. If there were no systematic difference between affected and unaffected individuals, one would expect K to be near $n/2$, as both groups share equal chances for having the larger measurement. The sign test provides evidence against the null hypothesis whenever K is far from $n/2$.

If the null hypothesis is true, the distribution of K is *binomial*, indexed by $n = 15$ trials and probability 0.5 of a positive difference in each trial. The chance of obtaining exactly k positive differences is given by $C_{n,k} \times (1/2)^n$ (with $C_{n,k}$ as defined in Section 4.3.1). The p-value is the sum of these chances for all values of k that are as far or farther from $n/2$ than is the observed value K.

An approximation to the p-value comes from normal approximation to the binomial distribution. If the null hypothesis is true, then the Z-statistic

$$Z\text{-statistic} = \frac{K - (n/2)}{\sqrt{n/4}}$$

has approximately a standard normal distribution. A p-value is found as the proportion of a standard normal distribution that is farther from zero than the observed Z-statistic. The approximation can be improved with a continuity correction, by adding $\frac{1}{2}$ to (or subtracting it from) the numerator of the Z-statistic, to make it closer to zero. (*Note*: n is the number of observations with nonzero differences; we discard ties in using the sign test.)

Example—Schizophrenia

Since 14 of the 15 differences are positive, the exact one-sided p-value is the probability that k is 14 plus the probability that k is 15. This is $C_{15,14} \times (1/2)^{15} + C_{15,15} \times (1/2)^{15} = 15 \times (1/32{,}768) + 1 \times (1/32{,}768) = 0.00049$. To illustrate the normal approximation, the Z-statistic with continuity correction is

$$Z\text{-statistic} = \frac{[14 - (15/2)] - (1/2)}{\sqrt{15/4}} = 3.098,$$

which produces an approximate one-sided p-value of 0.00097.

4.4.2 The Wilcoxon Signed-Rank Test

By retaining only the signs of the difference for each pair, the sign test completely obliterates the effects of outliers, but at the expense of losing potentially useful information about the relative magnitudes of the differences. The *Wilcoxon signed-rank test* uses the ranks of the magnitudes of the differences in addition to their signs. Since ranks are used, the procedure is resistant to outliers.

Computation of the *signed-rank statistic* proceeds as follows.

1. Compute the difference in each of the n pairs.
2. Drop zeros from the list.
3. Order the *absolute differences* from smallest to largest and assign them their ranks $1, \ldots, n$ (or average rank for ties).
4. The signed-rank statistic, S, is the sum of the ranks from the pairs for which the difference is positive.

The computation of S is illustrated in the schizophrenia data in Display 4.12, for the hypothesis that the mean difference is zero.

DISPLAY 4.12 Signed-rank test statistic computations (schizophrenia study)

Pair	Unaffected	Affected	Difference	Ordered magnitude	Order	Rank	+ Ranks	− Ranks
1	1.94	1.27	0.67	0.02 (+)	1	1	1	
2	1.44	1.63	−0.19	0.03 (+)	2	2	2	
3	1.56	1.47	0.09	0.04 (+)	3	3	3	
4	1.58	1.39	0.19	0.07 (+)	4	4	4	
5	2.06	1.93	0.13	0.09 (+)	5	5	5	
6	1.66	1.26	0.40	0.10 (+)	6	6	6	
7	1.75	1.71	0.04	0.11 (+)	7	7	7	
8	1.77	1.67	0.10	0.13 (+)	8	8	8	
9	1.78	1.28	0.50	0.19 (+)	9	9.5	9.5	
10	1.92	1.85	0.07	0.19 (−)	10	9.5		9.5
11	1.25	1.02	0.23	0.23 (+)	11	11	11	
12	1.93	1.34	0.59	0.40 (+)	12	12	12	
13	2.04	2.02	0.02	0.50 (+)	13	13	13	
14	1.62	1.59	0.03	0.59 (+)	14	14	14	
15	2.08	1.97	0.11	0.67 (+)	15	15	15	
							= 110.5	

① *Order the absolute differences and assign ranks to them.*

② *Signed-rank statistic = Sum of ranks for positive differences.*

Exact p-Value

An exact p-value for the signed-rank test is the proportion of all assignments of outcomes within each pair that lead to a test statistic as extreme as or more extreme than the one observed. Assignment refers to switching the group status but keeping the same observations within each pair. Within a single pair there are two possible assignments, so with n pairs there are a total of 2^n possible assignments. The p-value is therefore the number of possible assignments that provide a sum of positive ranks as extreme as or more extreme than the observed one, divided by 2^n.

For the schizophrenia example, there are 15 pairs, so the labels "affected" and "unaffected" can be assigned to observations within pairs in $2^{15} = 32{,}768$ possible ways. The assignments that produce an S greater than or equal to 110.5 are those where the sum of the negative ranks is less than or equal to 9.5.

By systematically examining the possible assignments with no negative ranks, one negative rank, two negative ranks, and so on, the following assignments (indicated by the ranks associated with a negative sign) are found to have a sum of negative ranks less than or equal to 9.5: (none), (1), (2), (3), (4), (5), (6), (7), (8), (9.5), (9.5), (1, 2), (1, 3), (1, 4), (1, 5), (1, 6), (1, 7), (1, 8), (2, 3), (2, 4), (2, 5), (2, 6), (2, 7), (3, 4), (3, 5), (3, 6), (4, 5), (1, 2, 3), (1, 2, 4), (1, 2, 5), (1, 2, 6), (1, 3, 4), (1, 3, 5), and (2, 3, 4). There are 34 assignments with the sum of negative ranks less than or equal to 9.5, and hence 34 assignments with S greater than or equal to 110.5, so the exact one-sided p-value is $34/32{,}768 = 0.00104$.

Normal Approximation

This method of directly counting more extreme cases in order to find the p-value can be replaced by tables or by computer-assisted counting. A normal approximation for convenient computation of an approximate p-value is available. The mean and standard deviation of S in its permutation distribution are

$$\text{Mean}(S) = n(n+1)/4 \qquad \text{and} \qquad \text{SD}(S) = [n(n+1)(2n+1)/24]^{1/2}.$$

Comparing Z-statistic $= [S - \text{Mean}(S)]/\text{SD}(S)$, with a continuity correction, to a standard normal distribution gives a good approximation to the p-value for $n \geq 20$.

For the schizophrenia study, $\text{Mean}(S) = 60$, $\text{SD}(S) = 17.61$, so that Z-stat $= 2.84$. An approximate one-sided p-value is 0.00226.

4.5 RELATED ISSUES

4.5.1 Practical and Statistical Significance

P-values indicate *statistical significance*, the extent to which a null hypothesis is contradicted by the data. This must be distinguished from *practical significance*, which describes the practical importance of the effect in question.

A study may find a statistically significant increase in rainfall due to cloud seeding, but the amount of increase in rainfall, say 10%, may not be sufficient to justify the expense of seeding. A finding that annual male salaries tend to be \$16.32 larger than female salaries, when annual incomes are on the order of \$10,000, does not provide a strong indication of persistent discrimination, even if the difference is statistically significant. A statistically significant reduction in time to complete a test problem may not be practically relevant if the reduction is only from 150 seconds to 147 seconds. These are issues of practical significance. They have little to do with statistics, but they are highly relevant to the interpretation of statistical analyses.

It is important to understand the connection between sample size and statistical significance. If there is a difference in population means (no matter how practically insignificant), large enough samples should indicate the existence of a statistically significant difference. On the other hand, even if there is a practically significant difference in population means (that is, an important difference), a small sample size may fail to indicate the existence of a statistically significant difference between means.

Three practical points deserve attention:

1. p-values are sample-size-dependent.
2. A result with a p-value of 0.08 can have more scientific relevance than one with a p-value of 0.001.

3. Tests of hypotheses by themselves rarely convey the full significance of the results. They should be accompanied by confidence intervals to indicate the range of likely effects and to facilitate the assessment of practical significance.

4.5.2 The Presentation of Statistical Findings

The Statistical Conclusions sections that accompany the case studies in this book are intended as models for communicating results. The following are some general recommendations.

1. State the conclusions as they apply to the question of interest and avoid statistical jargon and symbols, except to include appropriate measures of statistical uncertainty and to state which statistical tools were used.
2. Make clear exactly what is being estimated to answer a question of interest; for example, the difference in population means, a ratio of medians, or an additive treatment effect.
3. Prefer confidence intervals to standard errors, particularly when the estimate does not have a normal sampling (or randomization) distribution.
4. Use graphical displays to help convey important features of the data.
5. Do not include unnecessarily large numbers of digits in estimates or p-values.
6. If transformations have been used, express the results on the original scale of measurement.
7. Comment on the scope of inference. Was random assignment used? Was random sampling used?

4.5.3 Levene's (Median) Test for Equality of Two Variances

Sometimes a question of interest calls for a test of equality of two population variances. The *F-test for equal variances* and its associated confidence interval are available in standard statistical computer packages, but they are not robust against departures from normality. For example, p-values can easily be off by a factor of 10 if the distributions have shorter or longer tails than the normal.

A robust alternative is *Levene's test* (based on deviations from the median). Suppose there are n_1 observations Y_{1i} from population 1, and n_2 observations Y_{2i} from population 2. Let Z_{1i} be the absolute value of the deviation of the ith observation in group 1 from its group median: $|Y_{1i} - \text{median}_1|$, and let Z_{2i} be the absolute value of the deviation of the ith observation in group 2 from its median: $|Y_{2i} - \text{median}_2|$. The typical size of the Z's indicates the degree of variability in each group. The Levene test idea is to perform a two-sample t-test on the Z's to judge equal variability in the two groups. This procedure seems to have good power in detecting nonequal variability yet works well even for nonnormally distributed Y's.

Example—Sex Discrimination

The data of Section 1.1.2 are used here to illustrate Levene's test for the hypothesis that the variance of male salaries is equal to the variance of female salaries. The test compares the 32 absolute deviations from group median of the males to the 61 absolute deviations for the females. The t-statistic is 0.4331 and the two-sided *p*-value, by comparison to a *t*-distribution on 91 degrees of freedom, is 0.67. Thus Levene's test gives no evidence that the variances are unequal.

4.5.4 Survey Sampling

Survey sampling concerns selecting members of a specific (finite) population to be included in a survey and the subsequent analysis. One sampling method already discussed is *simple random sampling*, in which each subset of n items from the population has the same chance of being selected as the sample. This is appealing for its mathematical and conceptual simplicity, but is often practically unrealistic. The prohibitive part of finding a simple random sample of American voters, for example, is first finding a list of all American voters.

Instead of simple random sampling, survey organizations rely on complex sampling designs, which make use of stratification, multistage sampling, and cluster sampling. In *stratified sampling*, the population is divided into strata, based on supplementary information, and separate simple random samples are selected for each stratum. For example, samples of American voters may be taken separately for different states. *Multistage sampling*, as its name suggests, uses sampling at different stages. For example, a random sample of states is taken, then random samples of counties are taken within the selected states, and then random samples of individuals are selected from lists in the chosen counties. In *cluster sampling*, a simple random sample of clusters is selected at some stage of the sampling and all members within the cluster are included in the survey. For example, a random sample of households (the clusters) in a county may be selected and *all* voters in each household surveyed.

Standard Errors from Complex Sampling Designs

The formulas for standard errors so far presented are based on simple random sampling *with replacement*, which means that a member of a population can appear in a sample more than once. Random sampling in sample surveys, however, is conducted *without replacement* and does not permit a member to appear in a sample twice. Consequently, different standard errors must be calculated. The variance of an average from random sampling without replacement differs from that with replacement by a multiplicative factor called the *finite population correction* (FPC). If N is the population size and n the sample size, FPC $= (N - n)/N$. So the difference is small if only a small fraction of the population is sampled.

In addition, the usual standard errors are inappropriate for data from complex sampling designs. The ratio of the sampling variance of an estimate from data based on a particular sampling design to the sampling variance it would have had if it was based on a simple random sample of the same size is called the *design effect*. Design effects for estimates from complex stratified, multistage sampling designs are usually greater than 1. Therefore, application of standard statistical theory to complex sampling designs usually results in an overstatement of the degree of precision. For example, confidence intervals will be narrower than they really should be. Methods for obtaining correct standard errors are discussed in texts on survey sampling.

Nonresponse

Of 979 Vietnam veterans who were deemed to be eligible for the study in Section 3.1.2, only 900 were located, only 871 of those completed the first interview, only 665 of those were available for blood samples, and only 646 of those had valid dioxin readings. Assuming the 979 represented the population of interest, is there any danger in using the available 646 (66%) who responded and provided valid results to represent the population? The answer depends on whether those who were lost along the way differed in their dioxin levels from those who were not. The question that must be asked is whether or not being unavailable or not providing valid results had anything to do with dioxin level. If not, there is little harm in ignoring the nonresponders.

In sample surveys, the issue is usually much more serious, since those individuals who tend to respond to surveys often have much different views from those who do not. As an illustration, consider the popular telephone-in surveys that news programs run after presidential debates. Viewers call in (and have to pay for their calls) to indicate who they think won. Although this is entertaining, the results give an entirely unreliable indication of what all viewers think, since those who respond tend to be a quite different crowd from those who do not.

4.6 SUMMARY

The Space Shuttle O-Ring Study

The main feature of these data is that the two-sample t-tools cannot be relied on to provide valid measures of uncertainty. They are not sufficiently robust with such small, unbalanced samples of discrete data. The permutation test, however, can be applied, because it does not rely on any distributional assumptions. The computations are tedious, and inferences apply only to the 24 launches.

It should be mentioned that some information is lost due to the form in which the data are presented. They have been rather artificially divided into cool temperature launches and warm temperature launches, whereas the actual temperature data for each launch are available. The more advanced method of Poisson log-linear regression is applied to these data in Chapter 22.

The Cognitive Load Study

These data from a randomized experiment could be suitably analyzed with any of a number of techniques if it were not for the censoring of five observations. Without having actual completion times for these five students, it is impossible to take an average. The rank-sum test is convenient in this case since one only needs to know that these five students were tied for last. It can be used to test for an additive treatment effect and, with some effort, to provide a confidence interval for the treatment effect.

It should be mentioned that the rank-sum test is not suitable for censored observations if the rank of the censored observation cannot be determined. Thus, it is limited to handling censoring for data problems like this one, in which a number of censored observations are tied for last.

4.7 EXERCISES

Conceptual Exercises

1. **Cognitive Load.** Suppose that there were two textbooks on coordinate geometry, one written with conventional worked problems and the other with modified worked problems. And suppose it is possible to identify a number of schools in which each of the textbooks is used. If you took random samples of size 14 from schools with each text and obtained exactly the same data as in the example in Section 4.1.2, would the analysis and conclusions be the same?

2. **O-Ring Data.** (a) Is it appropriate to use the two-sample t-test to compare the number of O-ring incidents in cold and warm launches? (b) Is it appropriate to use the rank-sum test? (c) Is it appropriate to use a permutation test based on the t-statistic?

3. **O-Ring Data.** When these data were analyzed prior to the *Challenger* disaster it was noticed that variability was greater in the group with the larger average, so a log transformation was used. Since the log of zero does not exist, all the zeros were deleted from the data set. Does this seem like a reasonable approach?

4. **O-Ring Data.** Explain why the two-sided p-value from the permutation test applied to the O-ring data is equal to the one-sided p-value (see Display 4.10).

5. **O-Ring Data.** If you looked at the source of the O-ring data and found that temperatures for each launch were recorded in degrees F rather than as over/under 65°F, what question would that raise? Would the answer affect your conclusions about the analysis?

6. **Motivation and Creativity.** In what way is the p-value for the motivation and creativity randomized experiment (Section 1.1.1) dependent on an assumed model?

7. Are there occasions when both the two-sample t-test and the rank-sum test are appropriate?

8. Can the rank-sum test be used for comparing populations with unequal variances?

9. Suppose that two drugs are both effective in prolonging length of life after a heart attack. Substantial statistical evidence indicates that the mean life length for those using drug A is greater than the mean life length for those using drug B, but the variation of life lengths for drug A is substantially greater as well. Explain why it is difficult to conclude that drug A is better even though the mean life length is greater.

10. In a certain problem, the randomization test produces an exact two-sided p-value of 0.053, while the t-distribution approximation produces 0.047. One might say that since the p-values are on opposite sides of 0.05, they lead to quite different conclusions and, therefore, the approximation is not adequate. Comment on this statement.

11. What is the difference between a permutation test and a randomization test?

12. Explain what is meant by the comment that there is no single test called a randomization test.

13. What confounding factors are possible in the O-ring failure problem?

Computational Exercises

14. **O-Ring Study.** Find the t-distribution approximation to the p-value associated with the observed t-statistic. Compare this approximation to the (correct) permutation test p-value.

15. Consider these artificial data:

Group 1:	1	5	
Group 2:	4	8	9

The difference in averages $\overline{Y}_1 - \overline{Y}_2$ is -4. What is a one-sided p-value from the permutation distribution of the *difference in averages*? (*Hint*: List the 10 possible groupings; compute the difference in averages for each of these groupings, then calculate the proportion of these less than or equal to -4.)

16. Consider these artificial data:

Group 1:	5	7	12
Group 2:	4	6	

Calculate a p-value for the hypothesis of no difference, using the permutation distribution of the difference in sample averages. (You do not need to calculate the t-statistic for each grouping, only the difference in averages.)

17. O-Ring Study. Suppose the O-ring data had actually turned out as shown in Display 4.13. These are the same 24 numbers as before, but with the 2 and 3 switched. What is the one-sided p-value from the permutation test applied to the t-statistic? (This can be answered by examining Display 4.10.)

DISPLAY 4.13 Hypothetical O-ring data

Launch temperature	Number of O-ring incidents																			
Below 65°F	1	1	1	2																
Above 65°F	0	0	0	0	0	0	0	0	0	0	0	0	0	0	0	0	0	1	1	3

18. Suppose that six persons have an illness. Three are randomly chosen to receive an experimental treatment, and the remaining three serve as a control group. After treatment, a physician examines all subjects and assigns ranks to the severity of their symptoms. The patient with the most severe condition has rank 1, the next most severe has rank 2, and so on up to 6. It turns out that the patients in the treatment group have ranks 3, 5, and 6. The patients in the control group have ranks 1, 2, and 4. Is there any evidence that the treatment has an effect on the severity of symptoms? Use the randomization distribution of the sum of ranks in the treatment group to obtain a p-value. (First find the sum of ranks in the treatment group. Then write down all 20 groupings of the 6 ranks; calculate the sum of ranks in the treatment group for each. What proportion of these give a rank-sum as large as or larger than the observed one?)

19. Bumpus's Study. Use a statistical computer program to perform the rank-sum comparison of humerus lengths in the sparrows that survived and the sparrows that perished (Exercise 2.21). (a) What is the two-sided p-value? (b) Does the statistical computer package report the exact p-value or the one based on the normal approximation? (c) If it reports the one using the normal approximation, does it use a continuity correction to the Z-statistic? (d) How does the p-value from the rank-sum test compare to the one from the two-sample t-test (0.08) and the one from the two-sample t-test when the smallest observation is set aside (see Chapter 3, Exercise 28)? (e) Explain the relative merits of (i) the two-sample t-test using the strategy for dealing with outliers and (ii) the rank-sum test.

20. Trauma and Metabolic Expenditure. For the data in Exercise 18 in Chapter 3: (a) Determine the rank transformations for the data. (b) Calculate the rank-sum statistic by hand (taking the trauma patients to be group 1.) (c) Mimic the procedures used in Display 4.5 and Display 4.7 to compute the Z-statistic. (d) Find the one-sided p-value as the proportion of a standard normal distribution larger than the observed Z-statistic.

21. Trauma and Metabolic Expenditure. Use a statistical computer package to verify the rank-sum and the Z-statistic obtained in Exercise 20. Is the p-value the same? (Does the statistical package use a continuity correction?)

22. Trauma and Metabolic Expenditure. Using the rank-sum procedure, find a 95% confidence interval for the difference in population medians: the median metabolic expenditure for the population of trauma patients minus the median metabolic expenditure for the population of nontrauma patients.

23. Motivation and Creativity. Use a statistical computer package to compute the randomization test's two-sided p-value for testing whether the treatment effect is zero for the data in Section 1.1.1 (file case0101). How does this compare to the results from the two-sample t-test (which is used as an approximation to the randomization test)?

24. Motivation and Creativity. Find a 95% confidence interval for the treatment effect (poem creativity score after intrinsic motivation questionnaire minus poem creativity score after extrinsic motivation questionnaire, from Section 1.1.1 (file case0101)) using the rank-sum procedure. (Use a statistical computer program.) How does this compare to the t-based confidence interval for the treatment effect?

25. Guinea Pig Lifetimes. Use the Welch t-tools to find a two-sided p-value and confidence interval for the effect of treatment on lifetimes of guinea pigs in Chapter 2, Exercise 11. Does the additive treatment effect seem like a sensible model for these data?

26. Schizophrenia Study. (a) Draw a histogram of the differences in hippocampus volumes in Display 4.12. Is there evidence that the population of differences is skewed? (b) Take the logarithms of the volumes for each of the 30 subjects, take the differences of the log volumes, and draw a histogram of these differences. Does it appear that the distribution of differences of log volumes is more nearly symmetric? (c) Carry out the paired t-test on the log-transformed volumes. How does the two-sided p-value compare with the one obtained on the untransformed data? (d) Find an estimate of and 95% confidence interval for the mean difference in log volumes. Back-transform these to get an estimate and confidence interval for the median of the population of ratios of volumes.

27. Schizophrenia Study. Find the two-sided p-value using the signed-rank test, as in Display 4.12, but after taking a log transformation of the hippocampus volumes. How does the p-value compare to the one from the untransformed data? Is it apparent from histograms that the assumptions behind the signed-rank test are more appropriate on one of these scales?

28. Darwin Data. Charles Darwin carried out an experiment to study whether seedlings from cross-fertilized plants tend to be superior to those from self-fertilized plants. He covered a number of plants with fine netting so that insects would be unable to fertilize them. He fertilized a number of flowers on each plant with their own pollen and he fertilized an equal number of flowers on the same plant with pollen from a distant plant. (He did not say how he decided which flowers received which treatments.) The seeds from the flowers were allowed to ripen and were set in wet sand to germinate. He placed two seedlings of the same age in a pot, one from a seed from a self-fertilized flower and one from a seed from a cross-fertilized flower. The data in Display 4.14 are the heights of the plants at certain points in time. (The fertilization experiments were described by Darwin in an 1878 book; these data were found in D. F. Andrews and A. M. Herzberg, *Data* (New York: Springer–Verlag, 1985), pp. 9–12.) (a) Draw a histogram of the differences. (b) Find a two-sided p-value for the hypothesis of no treatment effect, using the paired t-test. (c) Find a 95% confidence interval for the additive treatment effect. (d) Is there any indication from the plot in (a) that the paired

DISPLAY 4.14 Darwin's data: heights (inches) for 15 pairs of plants of the same age, one of which was grown from a seed from a cross-fertilized flower and the other of which was grown from a seed from a self-fertilized flower; first 5 of 15 rows

	Plant height (inches)	
Pair	**Cross-fertilized**	**Self-fertilized**
1	23.5	17.375
2	12	20.375
3	21	20
4	22	20
5	19.125	18.375

t-test may be inappropriate? (e) Find a two-sided p-value for the hypothesis of no treatment effect for the data in Display 4.14, using the signed-rank test.

Data Problems

29. Salvage Logging. When wildfires ravage forests, the timber industry argues that logging the burned trees enhances forest recovery. The 2002 Biscuit Fire in southwest Oregon provided a test case. Researchers selected 16 fire-affected plots in 2004—before any logging was done—and counted tree seedlings along a randomly located transect pattern in each plot. They returned in 2005, after nine of the plots had been logged, and counted the tree seedlings along the same transects. (Data from D.C. Donato et al., 2006. "Post-Wildfire Logging Hinders Regeneration and Increases Fire Risk," *Science*, 311: 352.) The numbers of seedlings in the logged (L) and unlogged (U) plots are shown in Display 4.15.

DISPLAY 4.15 Number of tree seedlings per transect in nine logged (L) and seven unlogged (U) plots affected by the Biscuit Fire, in 2004 and 2005, and the percentage of seedlings lost between 2004 and 2005; first 5 of 16 rows

Plot	Action	Seedlings2004	Seedlings2005	PercentLost
Plot 1	L	298	164	45.0
Plot 2	L	471	221	53.1
Plot 3	L	767	454	40.8
Plot 4	L	576	141	75.5
Plot 5	L	407	217	46.7

Analyze the data to see whether logging has any effect on the distribution of percentage of seedlings lost between 2004 and 2005, possibly using the following suggestions: (a) Use the rank-sum procedure to test for a difference between logged and unlogged plots. Also use the procedure to construct a 95% confidence interval on the difference in medians. (b) Use the t-tools to test for differences in mean percentages lost and to construct a 95% confidence interval. Compare the results with those in (a).

DISPLAY 4.16 Tolerance to sunlight (minutes) for 13 patients prior to treatment and after treatment with a sunscreen; first 5 of 13 rows

	Tolerance to sunlight (minutes)	
Patient	**Pretreatment**	**During treatment**
1	30	120
2	45	240
3	180	480
4	15	150
5	200	480

DISPLAY 4.17 Months of survival after beginning of study for 58 breast cancer patients

Control Patients ($n = 24$)
2, 6, 8, 10, 12, 12, 14, 14, 14, 16, 16, 16, 18, 18, 18, 20, 22, 22, 26, 34, 36, 38, 40, 48

Patients Given Group Therapy for One Year ($n = 34$)
2, 2, 4, 4, 4, 6, 6, 8, 10, 10, 12, 14, 16, 16, 16, 18, 20, 22, 32, 36, 46, 46, 48, 48, 58, 58, 66, 72, 72, 82, 122, 122*, 122*, 122*

*These three patients were still alive at the end of the 122-month study period.

30. Sunlight Protection Factor. A sunscreen sunlight protection factor (SPF) of 5 means that a person who can tolerate Y minutes of sunlight without the sunscreen can tolerate $5Y$ minutes of sunlight with the sunscreen. The data in Display 4.16 are the times in minutes that 13 patients could tolerate the sun (a) before receiving treatment and (b) after receiving a particular sunscreen treatment. (Data from R. M. Fusaro and J. A. Johnson, "Sunlight Protection for Erythropoietic Protoporphyria Patients," *Journal of the American Medical Association* 229(11) (1974): 1420.) Analyze the data to estimate and provide a confidence interval for the sunlight protection factor. Comment on whether there are any obvious potentially confounding variables in this study.

31. Effect of Group Therapy on Survival of Breast Cancer Patients. Researchers randomly assigned metastatic breast cancer patients to either a control group or a group that received weekly 90-minute sessions of group therapy and self-hypnosis, to see whether the latter treatment improved the patients' quality of life. The group therapy involved discussion and support for coping with the disease, but the patients were not led to believe that the therapy would affect the progression of their disease. Surprisingly, it was noticed in a follow-up 10 years later that the group therapy patients appeared to have lived longer. The data on number of months of survival after beginning of the study are shown in Display 4.17. (Data from a graph in D. Spiegel, J. R. Bloom, H. C. Kraemer, and E. Gottheil, "Effect of Psychosocial Treatment on Survival of Patients with Metastatic Breast Cancer," *Lancet* (October 14, 1989): 888–91.) Notice that three of the women in the treatment group were still alive at the time of the follow-up, so their survival times are only known to be larger than 122 months. Is there indeed evidence of an effect of the group therapy treatment on survival time and, if so, how much more time can a breast cancer patient expect to live if she receives this therapy? Analyze the data as best as possible and write a brief report of the findings.

32. Therapeutic Marijuana. Nausea and vomiting are frequent side effects of cancer chemotherapy, which can contribute to the decreased ability of patients to undergo long-term chemotherapy

DISPLAY 4.18 Number of vomiting and retching episodes for 15 chemotherapy-receiving cancer patients, under placebo and marijuana treatments; first 5 of 15 rows

	Total number of vomiting and retching episodes	
Subject number	**Marijuana**	**Placebo**
1	15	23
2	25	50
3	0	0
4	0	99
5	4	31

schedules. To investigate the capacity of marijuana to reduce these side effects, researchers performed a double-blind, randomized, crossover trial. Fifteen cancer patients on chemotherapy schedules were randomly assigned to receive either a marijuana treatment or a placebo treatment after their first three chemotherapy sessions, and then "crossed over" to the opposite treatment after their next three sessions. The treatments, which involved both cigarettes and pills, were made to appear the same whether in active or placebo form. Shown in Display 4.18 are the number of vomiting and retching episodes for the 15 subjects. Does marijuana treatment reduce the frequency of episodes? By how much? Analyze the data and write a statistical summary of conclusions. (Data from A. E. Chang et al., "Delta-9-Tetrahydrocannabinol as an Antiemetic in Cancer Patients Receiving High-Dose Methotrexate," *Annals of Internal Medicine*, Dec. 1979. The order of the treatments is unavailable.)

Answers to Conceptual Exercises

1. The analysis would be the same. The conclusions would be very different. You could infer a real difference in the solution times of the two groups, but you could not attribute it to the different text types because of the possibility of a host of confounding factors.

2. (a) No, the extent of the nonnormality in these small and unequally sized samples is more than can be tolerated by the two-sample t-test. (b) Probably not. The spreads apparently are not equal. (c) Yes. The permutation test for significance requires no model or assumptions.

3. No! Observations cannot be deleted simply because the transformation does not work on them. In this case, a major portion of the data was deleted, leaving a very misleading picture.

4. There are 105 groupings that lead to t-statistics greater than or equal to 3.888 and no groupings that lead to t-statistics less than or equal to -3.888.

5. Was the 65°F cutoff chosen to maximize the apparent difference in the two groups? If so, the p-value would not be correct. Why? Because the p-value represents the chance of getting evidence as damaging to the hypothesis when there is no difference. The "chance" is the frequency of occurrence in replications of the study, *using the same statistical analysis*. The p-value calculation assumes that the 65°F cutoff will always define the groups. If the choice of cutoff was part of the statistical analysis used on this data set, the calculation was not correct. To get a correct p-value would require that you allow for a different cutoff to be chosen in each replication. Further discussion of data snooping is given in Chapter 6.

6. The p-value is based on the two-sample t-test but it is now understood that this p-value serves as an approximation to the p-value from the exact randomization test. For this approximation to be valid, the histograms of creativity scores should be reasonably normal (which they are).

7. Yes. Since the population model for the t-tools requires that the populations be normal with equal spread they will necessarily have the same shape and spread. Therefore, the assumptions for the rank-sum test are also satisfied.

8. Yes, but the meaning may be unclear if the variances are substantially unequal.

9. Generally, it is hard to make use of the difference in the centers of two distributions when the spreads are quite different. Specifically, the mean life length for drug A may be longer, but more people who use it may die sooner than for drug B.

10. The statement takes the rejection region approach too literally. There is very little difference in the degree of evidence against the null hypothesis in p-values of 0.053 and 0.047, so the approximation is pretty good.

11. A randomization test is a permutation test applied to data from a randomized experiment.

12. There is a different permutation distribution for each statistic that can be calculated from the data.

13. Perhaps workers tended to make more mistakes in cold weather or wind stress was greater on days with cold weather.

CHAPTER 5

Comparisons Among Several Samples

The issues and tools associated with the analysis of three or more independent samples (or treatment groups in a randomized experiment) are very similar to those for comparing two samples. Notable differences, however, stem from the particular kinds of questions and the greater number of them that may be asked.

An initial question, often asked in a preliminary stage of the analysis, is whether all of the means are equal. An easy-to-use F-test is available for investigating this. A typical analysis, however, goes beyond the F-test and explores linear combinations of the means to address particular questions of interest. A simple example of a linear combination of means is $\mu_3 - \mu_1$, the difference between means in populations 3 and 1. Inferences about this parameter can be made with t-tools just as for the two-independent-sample problem, with the important difference that the pooled estimate of standard deviation is from all groups, not just from those being compared.

This chapter discusses the use of the pooled estimate of variance for specific linear combinations of means and the one-way analysis of variance F-test for testing the equality of several means. The next chapter looks at linear combinations more generally and the problem of compounded uncertainty from the multiple, simultaneous comparisons of means.

5.1 CASE STUDIES

5.1.1 Diet Restriction and Longevity—A Randomized Experiment

A series of studies involving several species of animals found that restricting caloric intake can dramatically increase life expectancy. In one such study, female mice were randomly assigned to one of the following six treatment groups:

1. **NP:** Mice in this group ate as much as they pleased of a nonpurified, standard diet for laboratory mice.
2. **N/N85:** This group was fed normally both before and after weaning. (The slash distinguishes the two periods.) After weaning, the ration was controlled at 85 kcal/wk. This, rather than NP, serves as the control group because caloric intake is held reasonably constant.
3. **N/R50:** This group was fed a normal diet before weaning and a reduced-calorie diet of 50 kcal/wk after weaning.
4. **R/R50:** This group was fed a reduced-calorie diet of 50 kcal/wk both before and after weaning.
5. **N/R50 lopro:** This group was fed a normal diet before weaning, a restricted diet of 50 kcal/wk after weaning, and had dietary protein content decreased with advancing age.
6. **N/R40:** This group was fed normally before weaning and was given a severely reduced diet of 40 kcal/wk after weaning.

Display 5.1 shows side-by-side box plots for the lifetimes, measured in months, of the mice in the six groups. Summary statistics and sample sizes are reported in

DISPLAY 5.1 Lifetimes of female mice fed on six different diet regimens

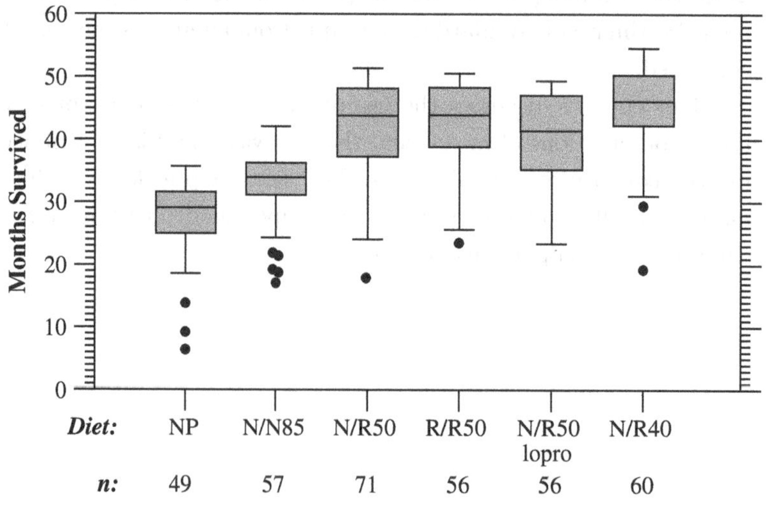

DISPLAY 5.2 Summary statistics for lifetimes of mice on six different diet regimens

Group	n	Range (months)	Average	SD	95% CI for mean
NP	49	6.4–35.5	27.4	6.1	25.6–29.2
N/N 85	57	17.9–42.3	32.7	5.1	31.3–34.1
N/R50	71	18.6–51.9	42.3	7.8	40.5–44.1
R/R50	56	24.2–50.7	42.9	6.7	41.1–44.7
N/R50 lopro	56	23.4–49.7	39.7	7.0	37.8–41.6
N/R40	60	19.6–54.6	45.1	6.7	43.4–46.8

Display 5.2. (Data from R. Weindruch, R. L. Walford, S. Fligiel, and D. Guthrie, "The Retardation of Aging in Mice by Dietary Restriction: Longevity, Cancer, Immunity and Lifetime Energy Intake," *Journal of Nutrition* 116(4) (1986): 641–54.)

The questions of interest involve specific comparisons of treatments as diagrammed in Display 5.3. Specifically, (a) Does lifetime on the 50 kcal/wk diet exceed

DISPLAY 5.3 Structure of planned comparisons among groups in the diet restriction study

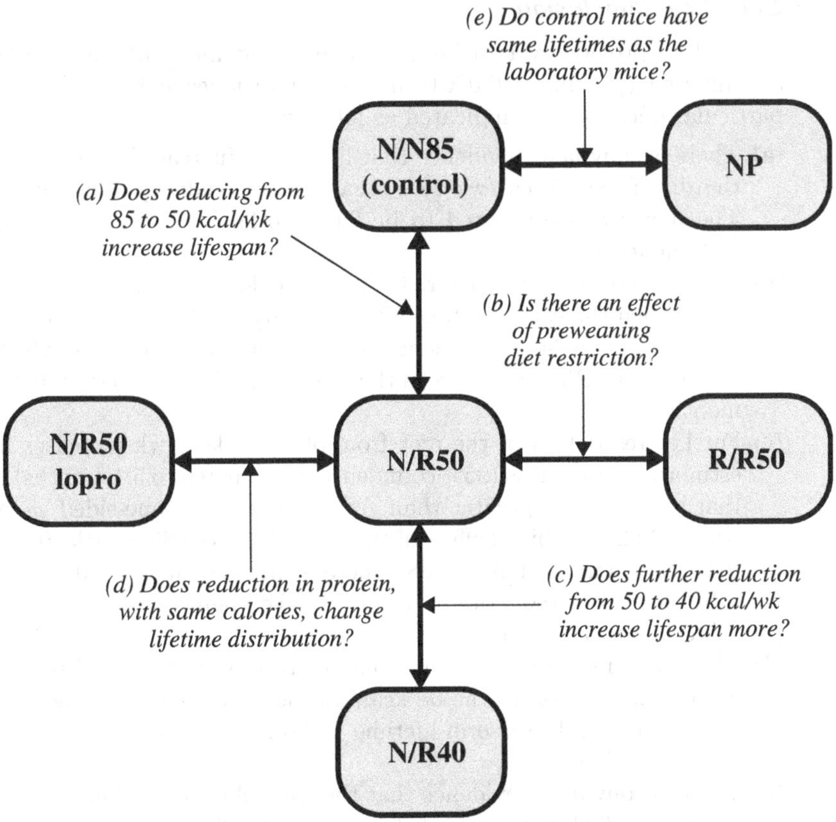

the lifetime on the 85 kcal/wk diet? If so, by how much? (This calls for a comparison of the N/R50 group to the N/N85 group.) (b) Is lifetime affected by providing a reduced calorie diet before weaning, given that a 50 kcal/wk diet is provided after weaning? (This calls for a comparison of the R/R50 group to the N/R50 group.) (c) Does lifetime on the 40 kcal/wk diet exceed the lifetime on the 50 kcal/wk diet? (This calls for a comparison of the N/R40 group to the N/R50 group.) (d) Given a reduced calorie diet of 50 kcal/week, is there any additional effect on lifetime due to decreasing the protein intake? (This calls for a comparison of the N/R50 lopro diet to the N/R50 diet.) (e) Is there an effect on lifetime due to restriction at 85 kcal/week? This would indicate the extent to which the 85 kcal/wk diet served as a proper control and possibly whether there was any effect of restricting the diet, even with a standard caloric intake. (This calls for a comparison of the N/N85 group to the NP group.) Comparisons other than those indicated by the arrows are not directly meaningful because the group treatments differ in more than one way. The N/R50 lopro and the N/R40 groups, for example, differ in both the protein composition and the total calories in the diet, so a difference would be difficult to attribute to a single cause.

Statistical Conclusion

These data provide overwhelming evidence that mean lifetimes in the six groups are different (p-value < 0.0001; analysis of variance F-test). Answers to the five particular questions are indicated as follows:

(a) There is convincing evidence that lifetime is increased as a result of restricting the diet from 85 kcal/wk to 50 kcal/wk (one-sided p-value < 0.0001; t-test). The increase is estimated to be 9.6 months (95% confidence interval: 7.3 to 11.9 months).

(b) There is no evidence that reducing the calories before weaning increased lifetime, when the calorie intake after weaning is 50 kcal/wk (one-sided p-value $=$ 0.32; t-test). A 95% confidence interval for the amount by which the lifetime under the R/R50 diet exceeds the lifetime under the N/R50 diet is -1.7 to 2.9 months.

(c) Further restriction of the diet from 50 to 40 kcal/wk increases lifetime by an estimated 2.8 months (95% confidence interval: 0.5 to 5.1 months). The evidence that this effect is greater than zero is moderate (one-sided p-value $=$ 0.017; t-test). (The combined effect of the reduction from 85 to 40 kcal/wk is estimated to be 12.4 months. This is a 38% increase in mean lifetime. If extended to human subjects, a 50% reduction in caloric intake might increase typical lifetimes—of 80 years—to 110 years.)

(d) There was moderate evidence that lifetime was *decreased* by the lowering of protein in addition to the 50 kcal/wk diet (two-sided p-value $=$ 0.024; t-test). The estimated decrease in lifetime is 2.6 months (95% confidence interval: 0.3 to 4.9 months).

(e) There is convincing evidence that the control mice live longer than the mice on the nonpurified diet (one-sided p-value < 0.0001).

5.1.2 The Spock Conspiracy Trial—An Observational Study

In 1968 Dr. Benjamin Spock was tried in United States District Court of Massachusetts in Boston on charges of conspiring to violate the Selective Service Act by encouraging young men to resist being drafted into military service for Vietnam. The defense in that case challenged the method by which jurors were selected, claiming that women—many of whom had raised children according to popular methods developed by Dr. Spock—were underrepresented. In fact, the Spock jury had no women.

Boston area juries are selected in three stages. From the City Directory, the Clerk of the Court selects at random 300 names for potential jury duty. Before a trial, a *venire* of 30 or more jurors is selected from the 300 names, again—according to the law—at random. An actual jury is selected from the venire in a nonrandom process allowing each side to exclude certain jurors for a variety of reasons.

The Spock defense pointed to the venire for their trial, which contained only one woman. That woman was released by the prosecution, making an all-male jury. Defense argued that the judge in the trial had a history of venires in which women were systematically underrepresented, contrary to the law. They compared this district judge's recent venires with the venires of six other Boston area district judges. The percents of women in those venires are presented in Display 5.4 as stem-and-leaf diagrams. (Data from H. Zeisel and H. Kalven, Jr., "Parking Tickets, and Missing Women: Statistics and the Law," in J. M. Tanur, F. Mosteller, W. H. Kruskal, R. F. Link, R. S. Pieters, and G. R. Rising, eds., *Statistics: A Guide to the Unknown*, San Francisco: Holden-Day, 1972.)

DISPLAY 5.4 Percents of women in 30-juror venires for Boston area U.S. District Court trials, grouped according to the judge presiding

	Spock Trial Judge	Other Boston Area U.S. District Court Judges					
		A	B	C	D	E	F
0	64,87						
1	33,36,50,52,77,86	68				77,97	65
2	31		70,89	10,34,75,75	43,97	15,79	07,35,64,67,95,98
3		08,36	20,27,55	05,19,25,38,38		48	19,62
4		05,89	56			02	

Legend: 4 | 89 represents a venire with 48.9% women.

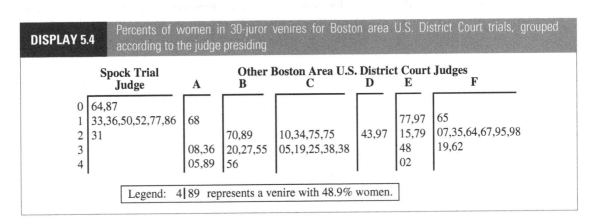

There are two key questions: (1) Is there evidence that women are underrepresented on the Spock judge's venires compared to the venires of the other judges? and (2) Is there any evidence that there are differences in women's representation in the venires of the other six judges? The first question addresses the key issue, but the second question has considerable bearing on the interpretation of the first. If the other judges all had about the same percentage of women on their venires while the Spock judge had significantly fewer women, this would make a strong statement about that particular judge. But if the percentages of women in the venires of the other judges are all different, this would put a very different perspective on any difference that the Spock judge's venires would have from the average of the other judges' venires.

Statistical Conclusion

As evident from the box plots in Display 5.5, the percentages of women on Spock's judge's venires (with an average of 15%) were substantially lower than those of the other judges (with an average of 30%). The one-sided p-value from a two-sample t-test comparing the mean percentage of Spock's judge to the mean percentage of all others combined is less than 0.000001. This choice of comparison, which combines the venires from the six other judges into a single group, is supported by a lack of significant differences among the other judges (p-value $= 0.32$ from a one-way analysis of variance F-test of whether the mean percentage is the same for judges A–F). It is estimated that the mean percentage of women on Spock judge's venires is 15% less than the mean of other venires (with a 95% confidence interval of 10% to 20%). (*Note:* A separate approach that does not combine venires from judges A–F into a single group is detailed in Section 6.2.3, and yields essentially the same conclusion.)

Scope of Inference

There is no indication that the venires in this observational study were randomly selected, so inference to some population is speculative. Thinking of the p-values as approximate p-values for permutation tests, however, leads one to conclude that the Spock judge did have a lower proportion of females on his venires than did the other judges—more so than can be explained by chance.

DISPLAY 5.5 Percentages of women on venires of the seven Boston area judges

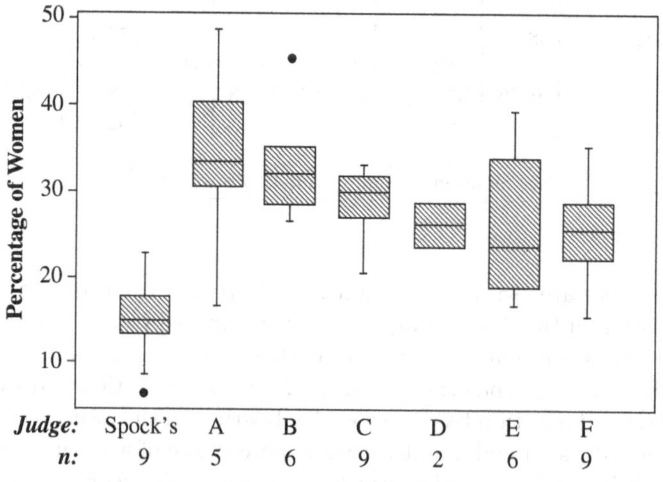

5.2 COMPARING ANY TWO OF THE SEVERAL MEANS

When subjects in a study are divided into distinct experimental or observational categories, the study itself is said to be a *several-group* problem, or a *one-way classification* problem. Mice were divided into six experimental groups; samples of venires were obtained for seven judges. A typical analysis of several-group data involves graphical exploration (like side-by-side box plots), consideration of transformations, initial screening to evaluate differences among all groups, and inferential techniques chosen to specifically address the questions of interest.

In the two-sample problem the questions of interest require inference about $\mu_2 - \mu_1$. In the several-sample problem the questions of interest may involve a few pairwise differences of means, like $\mu_2 - \mu_1$ and $\mu_3 - \mu_1$; all possible pairwise differences of means; or specific linear combinations of means, like

$$[-1 \times \mu_1] + [(1/2) \times \mu_2] + [(1/2) \times \mu_3].$$

This aspect of the data structure requires careful attention to how the questions of interest can be addressed through model parameters. If there are multiple questions, as, for example, "Does group 1 differ from group 2, does group 1 differ from group 3, and does group 2 differ from group 3?" then attention to interpretations of multiple, simultaneous statistical inferences is important. This *multiple comparison* problem is discussed further in the next chapter. For now, the discussion focuses on the single comparison of any two means.

5.2.1 An Ideal Model for Several-Sample Comparisons

The ideal population model, upon which the standard tools are derived, is a straightforward extension of the normal model for two-sample comparisons: (1) The populations have normal distributions. (2) The population standard deviations are all the same. (3) Observations within each sample are independent of each other. (4) Observations in any one sample are independent of observations in other samples.

As in the two-sample problem, the equal standard deviation model is entertained not because all data sets with several groups of measurements necessarily fit the description but because (1) it is conceptually difficult to compare populations with unequal variability, (2) it is statistically difficult as well, (3) for many problems a treatment is associated with an effect on the mean but not on the standard deviation, and (4) for many problems with unequal standard deviations it is possible to transform the data in such a way that the equal-spread model is valid on the transformed scale.

Notation

The symbols for population parameters are the same as before. A population mean is denoted by the Greek letter μ with a subscript indicating its group. The Greek letter σ is used to represent the standard deviation common to all the sampled populations. The number of samples will be represented by I, and the number of observations within the ith sample will be represented by n_i. The total number of observations from all groups combined will be $n = n_1 + n_2 + \cdots + n_I$. There are $I + 1$ parameters in the ideal model—the I means $\mu_1, \mu_2, \ldots, \mu_I$, and the single standard deviation σ.

Treatment Effects for Randomized Experiments

The discussion that follows in this chapter will continue to use the terminology of samples rather than of treatment groups, even though the methods also apply to data from randomized experiments with I treatment groups. An additive treatment effect model asserts that an experimental unit that would produce a response of Y_1 on treatment 1 would produce a response of $Y_1 + \delta_1$ on treatment 2, $Y_1 + \delta_2$ on treatment 3, and so on. As before, exact randomization tests are available for inferences about the δ's, but approximations based on the tools developed for random samples from populations are usually adequate. The practical upshot is that data from randomized experiments will be analyzed in exactly the same way as samples from populations, but concluding statements will be worded in terms of treatment effects rather than differences in population means.

5.2.2 The Pooled Estimate of the Standard Deviation

The mean from the ith population, μ_i, is estimated by the average from the ith sample, $\overline{Y}_i$. The variance σ^2 is estimated separately by ${s_i}^2$ from each of the I samples. If all the sample sizes are equal, then the best single estimate of σ^2 is their average. When the samples have different numbers of observations, a *weighted average* is more appropriate, where the estimates from larger samples are given more weight than those from smaller samples. The *pooled estimate of variance*, s_p^2, is a weighted average of sample variances in which each sample variance receives the weight of its degrees of freedom:

$$s_p^2 = \frac{(n_1 - 1)s_1^2 + (n_2 - 1)s_2^2 + \cdots + (n_I - 1)s_I^2}{(n_1 - 1) + (n_2 - 1) + \cdots + (n_I - 1)}.$$

Its associated degrees of freedom are the sum of degrees of freedom from all samples, $n - I$. An illustration of the calculations for the diet restriction data is shown in Display 5.6.

5.2.3 *t*-Tests and Confidence Intervals for Differences of Means

If $\overline{Y}_i$ is the average based on a sample from the population with mean μ_i and variance σ_i^2, and if $\overline{Y}_j$ is the average based on an independent sample from the population with mean μ_j and variance σ_j^2, then the sampling distribution of $\overline{Y}_i - \overline{Y}_j$ has variance $\sigma_i^2/n_i + \sigma_j^2/n_j$. If the two population variances are equal, this reduces to $\sigma^2[(1/n_i) + (1/n_j)]$. This may be estimated by replacing the unknown σ^2 by

DISPLAY 5.6 Pooled estimate of standard deviation; diet restriction data

Group	n	Sample SD
NP	49	6.1
N/N 85	57	5.1
N/R50	71	7.8
R/R50	56	6.7
N/R50 lopro	56	7.0
N/R40	60	6.7

① *Calculate the pooled estimate of variance, s_p^2.*

$$s_p^2 = \frac{(49-1)(6.1)^2 + (57-1)(5.1)^2 + (71-1)(7.8)^2 + (56-1)(6.7)^2 + (56-1)(7.0)^2 + (60-1)(6.7)^2}{(49-1)+(57-1)+(71-1)+(56-1)+(56-1)+(60-1)}$$

$$= \frac{15,313.90}{343} = 44.647; \quad s_p = \sqrt{44.647} = 6.68$$

② *s_p is the square root.*

$d.f. = 343$

③ *d.f. is the denominator.*

its best estimate. The important aspect of this in the one-way classification is that if the variances from all I populations can be assumed to be equal, then the best estimate is the pooled estimate of variance from all groups. So, for example, the standard error of $\overline{Y}_3 - \overline{Y}_2$ is

$$\mathrm{SE}(\overline{Y}_3 - \overline{Y}_2) = s_p\sqrt{\frac{1}{n_3} + \frac{1}{n_2}},$$

where s_p is the pooled estimate from *all* groups, with $(n - I)$ degrees of freedom.

The theory leading to confidence intervals and tests is the same as for the two-independent-sample problem. In this case the t-ratio has a t distribution on $n - I$ degrees of freedom. A 95% confidence interval for $\mu_3 - \mu_2$ is $(\overline{Y}_3 - \overline{Y}_2) \pm t_{n-I}(0.975) \times \mathrm{SE}(\overline{Y}_3 - \overline{Y}_2)$ and a t-statistic for testing the hypothesis that $\mu_3 - \mu_2$ equals zero is $(\overline{Y}_3 - \overline{Y}_2)/\mathrm{SE}(\overline{Y}_3 - \overline{Y}_2)$. These are illustrated in Display 5.7.

5.3 THE ONE-WAY ANALYSIS OF VARIANCE *F*-TEST

One question often asked, possibly in a first stage of the analysis, is "are there differences between *any* of the means?" The *analysis of variance* (ANOVA) F-test provides evidence in answer to this question. The term *variance* should not mislead; this is most definitely a test about *means*. It assesses mean differences by comparing the amounts of variability explained by different sources.

DISPLAY 5.7 A confidence interval for $\mu_3 - \mu_2$ and a test that $\mu_3 - \mu_2 = 0$ (diet restriction data)

(1) ***Get averages, sample sizes, and pooled estimate of standard deviation.***

Group	3: N/R50	2: N/N85
Sample size	71	57
Average (months)	42.3	32.7

Pooled estimate of σ: $s_p = 6.68$ months; d.f. = 343 (from Display 5.6)

(2) ***Compute the estimate of $\mu_3 - \mu_2$ and its standard error.***

$$\text{Estimate: } \overline{Y}_3 - \overline{Y}_2 = 42.3 - 32.7 = 9.6 \text{ months}$$

$$\text{SE}(\overline{Y}_3 - \overline{Y}_2) = 6.68\sqrt{\frac{1}{71} + \frac{1}{57}} = 1.2 \text{ months}$$

(3) ***95% confidence interval for $\mu_3 - \mu_2$.***

$$t_{343}(0.975) = 1.96$$

95% CI: $9.6 \pm (1.96)(1.2) =$ 7.3 months, 11.9 months

(4) ***Test the hypothesis that $\mu_3 - \mu_2 = 0$.***

$$t\text{-stat} = \frac{9.6}{1.2} = 8.08 \longrightarrow \text{two-sided } p\text{-value} < 0.0001$$

5.3.1 The Extra-Sum-of-Squares Principle

One model for the Spock data is that the percentage of women on venires for judge i comes from a normal distribution with mean μ_i and variance σ^2, for i from 1 to 7. A hypothesis for initial exploration is that all seven means are equal. That is, the null hypothesis is H: $\mu_1 = \mu_2 = \cdots = \mu_7$, and the alternative is that at least one of the means differs from the others.

The term *extra-sum-of-squares* refers to a general idea for hypothesis testing and is fundamental to a large class of tests. This section introduces the general notion of extra-sum-of-squares and the associated F-test, but with particular attention to testing the above hypothesis in the one-way classification problem.

Full and Reduced Models

Testing a hypothesis is formulated as a problem of comparing two models for the response means. A *full model* is a general model that is found to adequately describe the data. The *reduced model* is a special case of the full model obtained by imposing the restrictions of the null hypothesis.

For comparing equality of all means in the several-sample problem, the full model is the one that has a separate mean for each group. The reduced model, obtained from the full model by supposing the hypothesis of equal means is true, specifies a single mean for all populations. For the Spock data, the means from the two models are the following:

Group:	1	2	3	4	5	6	7
Full (separate-means) model:	μ_1	μ_2	μ_3	μ_4	μ_5	μ_6	μ_7
Reduced (equal-means) model:	μ	μ	μ	μ	μ	μ	μ

The terminology of *full* and *reduced* models provides a framework for the general procedure called the *extra-sum-of-squares* F-test. For any particular application, the models will have different names that refer more specifically to the problem at hand. As a test of equality of means in one-way classification, it makes more sense to call the full model the *separate-means* model and the reduced model the *equal-means* model.

Fitting the Models

The idea behind analysis of variance is to estimate the parameters in both the full and reduced models and to see whether the variability of responses about the estimated means is comparable in the two models. The *estimated* means for each group are different for the two models:

Group:	1	2	3	4	5	6	7
Full (separate-means) model:	$\overline{Y}_1$	$\overline{Y}_2$	$\overline{Y}_3$	$\overline{Y}_4$	$\overline{Y}_5$	$\overline{Y}_6$	$\overline{Y}_7$
Reduced (equal-means) model:	$\overline{Y}$	$\overline{Y}$	$\overline{Y}$	$\overline{Y}$	$\overline{Y}$	$\overline{Y}$	$\overline{Y}$

where $\overline{Y}$ is the average of all observations, called the *grand average*.

Residuals

Associated with each observation in the data set is an estimated group mean based on the full model and a different estimated group mean based on the reduced model. Also associated with each observation is a residual for each model. A residual is the observation value minus its estimated mean. So, for observation Y_{ij} (the percentage of women on the jth venire for judge i), the residual from the full model is $Y_{ij} - \overline{Y}_i$ and the residual from the reduced model is $Y_{ij} - \overline{Y}$. Display 5.8 shows the sets of estimated means and residuals for the Spock trial data in both the full (separate-means) and the reduced (equal-means) models. Notice that the residuals tend to be larger for the equal-means model.

If the null hypothesis is correct, then the two models should be about equal in their ability to explain the data, and the magnitudes of the residuals should be about the same. If the null hypothesis is incorrect, the magnitudes of the residuals from the equal-means model will tend to be larger.

DISPLAY 5.8 Estimated means and residuals from two models for mean percentage of women (%*W*) in venires: Spock trial data

Large residuals indicate that the model fits poorly.

		Equal means		Separate means	
Judge	**%W**	**Est.**	**Res.**	**Est.**	**Res.**
Spock	6.4	26.6	−20.2	14.6	−8.2
Spock	8.7	26.6	−17.9	14.6	−5.9
Spock	13.3	26.6	−13.3	14.6	−1.3
Spock	13.6	26.6	−13.0	14.6	−1.0
Spock	15.0	26.6	−11.6	14.6	0.4
Spock	15.2	26.6	−11.4	14.6	0.6
Spock	17.7	26.6	−8.9	14.6	3.1
Spock	18.6	26.6	−8.0	14.6	4.0
Spock	23.1	26.6	−3.5	14.6	8.5
A	16.8	26.6	−9.8	34.1	−17.3
A	30.8	26.6	4.2	34.1	−3.3
A	33.6	26.6	7.0	34.1	−0.5
A	40.5	26.6	13.9	34.1	6.4
A	48.9	26.6	22.3	34.1	14.8
B	27.0	26.6	0.4	33.6	−6.6
B	28.9	26.6	2.3	33.6	−4.7
B	32.0	26.6	5.4	33.6	−1.6
B	32.7	26.6	6.1	33.6	−0.9
B	35.5	26.6	8.9	33.6	1.9
B	45.6	26.6	19.0	33.6	12.0

In the equal-means model, estimated means are equal to the grand average.

In the separate-means model, estimated means are the group averages.

		Equal means		Separate means	
Judge	**%W**	**Est.**	**Res.**	**Est.**	**Res.**
C	21.0	26.6	−5.6	29.1	−8.1
C	23.4	26.6	−3.2	29.1	−5.7
C	27.5	26.6	0.9	29.1	−1.6
C	27.5	26.6	0.9	29.1	−1.6
C	30.5	26.6	3.9	29.1	1.4
C	31.9	26.6	5.3	29.1	2.8
C	32.5	26.6	5.9	29.1	3.4
C	33.8	26.6	7.2	29.1	4.7
C	33.8	26.6	7.2	29.1	4.7
D	24.3	26.6	−2.3	27.0	−2.7
D	29.7	26.6	3.1	27.0	2.7
E	17.7	26.6	−8.9	27.0	−9.3
E	19.7	26.6	−6.9	27.0	−7.3
E	21.5	26.6	−5.1	27.0	−5.5
E	27.9	26.6	1.3	27.0	0.9
E	34.8	26.6	8.2	27.0	7.8
E	40.2	26.6	13.6	27.0	13.2
F	16.5	26.6	−10.1	26.8	−10.3
F	20.7	26.6	−5.9	26.8	−6.1
F	23.5	26.6	−3.1	26.8	−3.3
F	26.4	26.6	−0.2	26.8	−0.4
F	26.7	26.6	0.1	26.8	−0.1
F	29.5	26.6	2.9	26.8	2.8
F	29.8	26.6	3.2	26.8	3.0
F	31.9	26.6	5.3	26.8	5.1
F	36.2	26.6	9.6	26.8	9.4

Residual Sums of Squares

A single summary of the magnitude of the residuals for a particular model is the *residual sum of squares* for that model (i.e., the sum of the squared residuals). By adding the squares of the two sets of residuals in Display 5.8 separately, one finds that the residual sum of squares for the equal-means model is 3,791.53 and the residual sum of squares from the separate-means model is 1,864.45. Is this large difference (3,791.53 − 1,864.45 = 1,927.08) due to the relatively poor fit of the equal-means model, or can it be explained by sampling variability? The F-test answers this question precisely.

General Form of the Extra-Sum-of-Squares F-Statistic

The *extra sum of squares* is the single number that summarizes the difference in sizes of residuals from the full and reduced models. As just calculated above,

Extra sum of squares =
Residual sum of squares (reduced) – Residual sum of squares (full).

A residual sum of squares measures the variability in the observations that remains unexplained by a model, so the extra sum of squares measures the amount of unexplained variability in the reduced model that *is* explained by the full model.

The extra sum of squares is a test statistic for judging the plausibility of the null hypothesis. For practical use, however, it must be converted into a form that can be compared to a known probability distribution. The particular form is the F-statistic:

$$F\text{-statistic} = \frac{(\text{Extra sum of squares})/(\text{Extra degrees of freedom})}{\hat{\sigma}^2_{\text{full}}},$$

where the *extra degrees of freedom* are the number of parameters in the mean for the full model minus the number of parameters in the mean for the reduced model, and $\hat{\sigma}^2_{\text{full}}$ is the estimate of σ^2 based on the full model. Thus, the F-statistic is the extra sum of squares per extra degree of freedom, scaled by the best estimate of variance.

The F-Test

Large F-statistics are associated with large differences in the size of residuals from the two models. They supply evidence against the hypothesis of equal means and in favor of a model with different means. The test is summarized by its p-value, the chance of finding an F-statistic as large as or larger than the observed one when all means are, in fact, equal.

F-Distributions

If all means are equal, the sampling distribution of the F-statistic is that of an *F-distribution*, which depends on two known parameters: the *numerator degrees of freedom* and the *denominator degrees of freedom.* For each degrees of freedom pair, there is a separate F-distribution. The letter F was given to this class of distributions by George Snedecor, honoring the British statistician Sir Ronald Fisher.

Theoretical histograms for four of the F-distributions are shown in Display 5.9. Notice that F-values in the general range of 0.5 to 3.0 are fairly typical. An F-statistic in this range would not be considered as very strong evidence of unequal means for most degree of freedom combinations. An F-statistic in the range from

DISPLAY 5.9 Four *F*-distributions, having different degrees of freedom

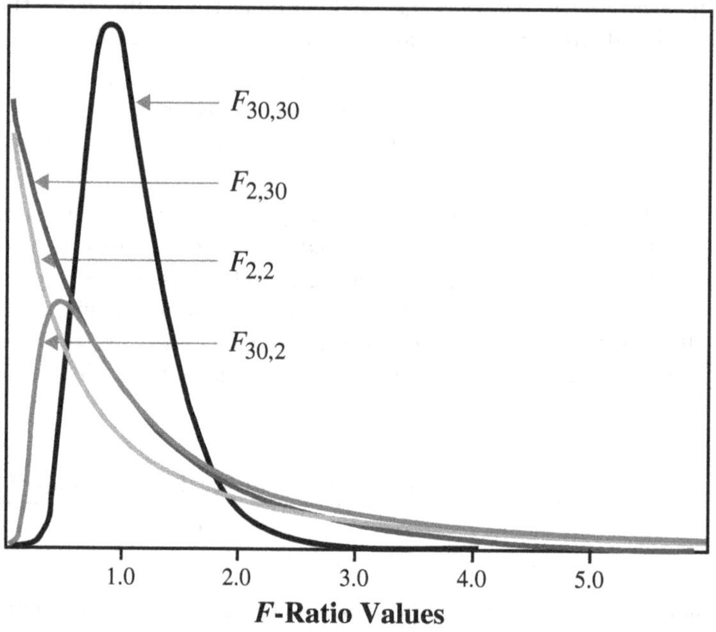

3.0 to 4.0 would be highly unlikely with large degrees of freedom in both numerator and denominator but would be only moderately suggestive of differences with smaller degrees of freedom. An F-statistic larger than 4.0 is strong evidence against equal means except for the smallest degrees of freedom, particularly in the denominator. (*Note:* All the curves have the same area below them. This means that the $F_{2,2}$ and $F_{30,2}$ curves must have considerable area, spread thinly, off the graph to the right.)

Statistical computer programs and some calculators can provide tail area probabilities of F-distributions for specified F-values and, conversely, can provide the F-values corresponding to specified tail area probabilities. In both cases, the user must also specify the particular F-distribution of interest by supplying the numerator and denominator degrees of freedom. The p-value indicated by bubble 7 in Display 5.10, showing the F-test results for the Spock trial data, was obtained with a statistical computer package.

5.3.2 The Analysis of Variance Table for One-Way Classification

The analysis of variance table organizes and displays the calculations used in the F-test. Analysis of variance tables extend to more complicated structures, where several F-tests simultaneously evaluate how well different models fit the data. The ANOVA table for the Spock data appears in Display 5.10.

DISPLAY 5.10 Analysis of variance table: a test for equal mean percentages of women in venires of seven judges (Spock data)

Source of Variation	Sum of Squares	d.f.	Mean Square	*F*-Statistic	*p*-Value
Between Groups	1,927.08	6	321.18	6.72	0.000061
Within Groups	1,864.45	39	47.81		
Total	3,791.53	45			

(1) *Sum of squared residuals from fitting the full (separate-means) model.*

(2) *Sum of squared residuals from fitting the reduced (equal-means) model.*

(3) *degrees of freedom.* ($n-I$; $n-1$)

(4) *Extra sum of squares is Total minus within (d.f. also obtained by subtraction).*

(5) *A mean square is the ratio of a sum of squares to its degrees of freedom.*

(6) *The F-statistic is the ratio of the Between MS to the Within MS.*

(7) *The p-value comes from an F-distribution with 6 and 39 d.f.*

***Note:* *This is* s_p^2.**

The table is organized so that all calculations follow a fixed pattern after entry of the residual sums of squares from the full (separate-means) and reduced (equal-means) models and their degrees of freedom. The between-groups sum of squares and its degrees of freedom are found by subtraction. Each mean square equals its sum of squares divided by its degrees of freedom. The F-statistic equals the ratio of the between-groups mean square to the within-groups mean square; and the evidence is summarized by the p-value, equal to the upper tail area from the F-distribution, whose numerator degrees of freedom are those associated with between groups and whose denominator degrees of freedom are those associated with within groups. The p-value of 0.000061 provides convincing evidence that at least one of the judges' means differs from one of the others' means.

5.4 MORE APPLICATIONS OF THE EXTRA-SUM-OF-SQUARES *F*-TEST

5.4.1 Example: Testing Equality in a Subset of Groups

The one-way analysis of variance F-test shown in Display 5.10 provided convincing evidence that the means were not equal for all seven judges. The next step in the analysis is to see whether the six other judges (not Spock's) have the same mean percentage of women on their venires. Although this step does not answer the main question of interest, it establishes an important context for addressing the main question. If there are no systematic differences among judges A through F in their inclusion of women, then the comparison of them to Spock's judge has greater relevance (in the court proceedings) than if they do differ among each other.

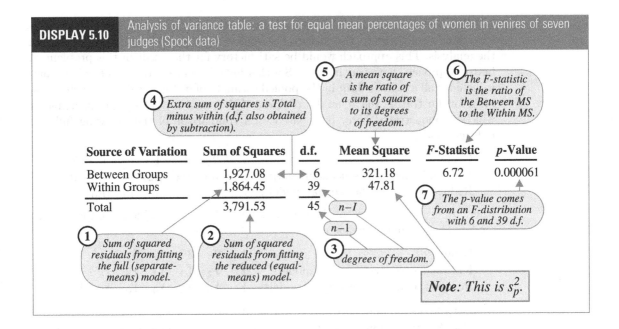

One way to test the hypothesis $H: \mu_2 = \mu_3 = \cdots = \mu_7$ is to perform a one-way analysis of variance F-test with the venires from Spock's judge excluded from the analysis. This approach would be satisfactory for the needs of this problem. A better approach, however, includes Spock's judge's venires. It is "better" because it includes all available data in the pooled estimate of σ. Since this approach is also ideal for illustrating the general usefulness of the extra-sum-of-squares principle, it will be demonstrated here. It is based on a comparison of the following full and reduced models:

Group:	1	2	3	4	5	6	7
Full model (separate-means model):	μ_1	μ_2	μ_3	μ_4	μ_5	μ_6	μ_7
Reduced model (others-equal model):	μ_1	μ_0	μ_0	μ_0	μ_0	μ_0	μ_0

where μ_0 is used to represent the common mean of the last six judges in the reduced model. The estimated means are:

Group:	1	2	3	4	5	6	7
Full model (separate-means model):	$\overline{Y}_1$	$\overline{Y}_2$	$\overline{Y}_3$	$\overline{Y}_4$	$\overline{Y}_5$	$\overline{Y}_6$	$\overline{Y}_7$
Reduced model (others-equal model):	$\overline{Y}_1$	$\overline{Y}_0$	$\overline{Y}_0$	$\overline{Y}_0$	$\overline{Y}_0$	$\overline{Y}_0$	$\overline{Y}_0$

where $\overline{Y}_0$ is the average percentage among all venires for the other six judges.

The F-test for comparing these reduced and full models is not automatically computed in the analysis of variance program. Since the full model in this hypothesis is the separate-means model, its sum of squared residuals is available as the within-groups sum of squares in the analysis of variance table in Display 5.10 (1,864.45 on 39 degrees of freedom). The sum of squared residuals from the reduced model can be found by performing a second one-way classification analysis with just the two groups: *Spock* and *others*. The resulting residual sum of squares is 2,190.90, with $46 - 2 = 44$ degrees of freedom.

The extra-sum-of-squares F-statistic is $[(2{,}190.90 - 1{,}864.45)/(44 - 39)]/(1{,}864.45/39) = 1.37$, and the p-value is the proportion of an F-distribution on 5 and 39 degrees of freedom that exceeds 1.366, which is 0.26. There is consequently no evidence from these data of differences in means among the six other judges.

The next step in the analysis, assuming the six other judges do have equal means, is a test of the hypothesis that the Spock judge's mean is equal to the common mean of the other six. An F-test compares a full model with two means (one for the Spock judge and one for the other six) to a reduced model with a single mean:

Group:	1	2	3	4	5	6	7
Full model (others-equal model):	μ_1	μ_0	μ_0	μ_0	μ_0	μ_0	μ_0
Reduced model (equal-means model):	μ	μ	μ	μ	μ	μ	μ

Notice that the previous step in the analysis has indicated the appropriateness of the two-parameter model, with a common mean for the other judges. This then becomes the full model for the inferential question: Is the Spock judge's mean different from the mean common to the six others, $H: \mu_1 = \mu_0$? The reduced model, consequently, is the equal-means model. The sum of squared residuals from this reduced model is the total sum of squares from the one-way analysis of variance table: 3,791.53 on 45 degrees of freedom. The sum of squared residuals from this full model is 2,190.90 on 44 degrees of freedom. The F-statistic is [(3,791.53 − 2,190.90)/(45 − 44)]/(2,190.90/44), which is equal to 32.14. By comparison to an F-distribution on 1 and 44 degrees of freedom, the p-value is found to be 0.000001. There is convincing evidence that the Spock judge's mean differs from the others.

5.4.2 Summary of ANOVA Tests Involving More Than Two Models

The *others-equal* model (using two parameters to describe the means) lies intermediate between the *equal-means* model (with one parameter) and the *separate-means* model (with seven parameters). It is a simplification of the *separate-means* model, while the *equal-means* model is a simplification of it.

Display 5.11 illustrates the additive nature of the extra sums of squares for comparing these three nested models. The residual sum of squares in each model is viewed as a measure of the unexplained variability from that model, or as a *distance* from the model to the data. The residual sum of squares for the equal-means model is 3,791.53. By including one extra parameter to allow the Spock judge

DISPLAY 5.11 Residual sums of squares as distances from the data to proposed models for the means (Spock trial example)

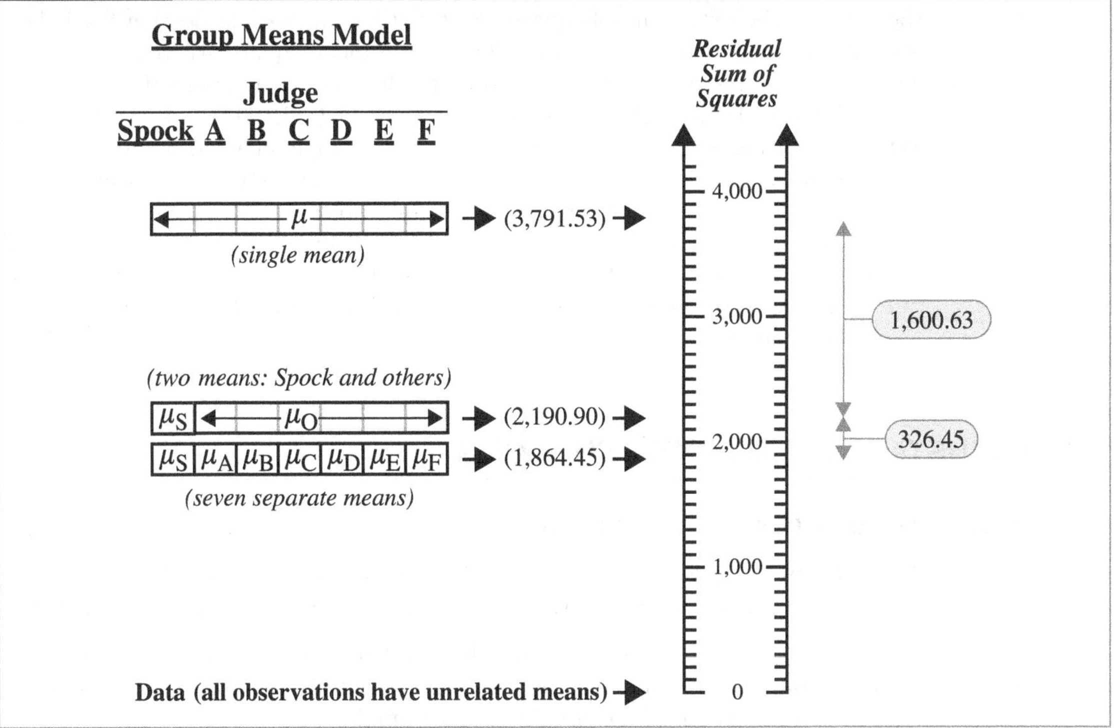

DISPLAY 5.12 Complete analysis of variance table for three tests involving the mean percents of women in venires of seven judges

Source of variation	Sum of squares	d.f.	Mean square	F-statistic	p-value
Between groups	1,927.08	6	321.18	6.72	0.000061
Spock vs. others	1,600.63	1	1,600.63	33.48	0.000001
Among others	326.45	5	65.29	1.37	0.26
Within groups	1,864.45	39	47.81		
Total	3,791.53	45			

to have a different mean (the others-equal model), the unexplained variability is reduced by 1,600.63. By including an additional five extra parameters (the separate-means model), the unexplained variability is further reduced by 326.45. Notice that the extra sum of squares in the comparison of the separate-means model to the equal-means model (i.e., the between-group sum of squares) is the sum of the two-component sums of squares (1,927.08 = 1,600.63 + 326.45).

The two tests in the Spock trial data could be combined into a single analysis of variance table, as in Display 5.12. In this table, the ***Between groups*** sum of squares has been decomposed into two pieces corresponding to extra sums of squares for the two individual tests.

The F-statistic for "Spock vs. others" presented in Display 5.12 is *not* exactly the same as the extra-sum-of-squares F-statistic computed in Section 5.4.1, however. It has as its denominator the *Within groups* mean square, 47.81 (with 39 d.f). The actual extra-sum-of-squares F-statistic has as its denominator the mean squared residuals from the *two-group* model, 49.79 (=2,190.90/44, with 44 d.f.). When presenting several tests simultaneously in a single analysis of variance table it is customary to use, as the denominator of the F-statistics, the estimate of σ^2 from the fullest model fit. There are some philosophical differences between these two approaches. For now, we encourage the student to apply the extra-sum-of-squares idea directly to address specific questions of interest (as in Section 5.4.1), but to be aware of differences in computer output due to the type of convention used in Display 5.12.

5.5 ROBUSTNESS AND MODEL CHECKING

5.5.1 Robustness to Assumptions

The robustness of t-tests, F-tests, and confidence intervals to departures from model assumptions can be described essentially in the same way as in Chapter 3:

1. Normality is not critical. Extremely long-tailed distributions or skewed distributions coupled with different sample sizes present the only serious distributional problems, particularly if sample sizes are small.

2. The assumptions of independence within and across groups are critical. If lacking, different analyses should be attempted.
3. The assumption of equal standard deviations in the populations is crucial.
4. The tools are not resistant to severely outlying observations.

The robustness with respect to equal standard deviations in the populations requires further discussion. It is important to pool together the estimates of variability from all the groups in order to make the most powerful comparisons possible. If one of the populations has a very different standard deviation, however, serious problems may result, even if the comparisons do not involve the mean from that population.

A computer was used to simulate situations in which three samples from normal populations were selected, with the aim of obtaining a 95% confidence interval for the difference in means from the first two. Six different configurations of population standard deviations were used with each of four sample size combinations. Based on 2,000 simulated data sets, the results appear in Display 5.13. The procedure is robust against unequal standard deviations for those situations where success rates are approximately equal to 95%.

DISPLAY 5.13 Success rates for 95% confidence intervals for $\mu_1 - \mu_2$ from samples simulated from normal populations with possibly different SDs

			$\sigma_2=\sigma_1$			$\sigma_2=2\sigma_1$		
n_1	n_2	n_3	$\sigma_3=\sigma_1$	$\sigma_3=2\sigma_1$	$\sigma_3=4\sigma_1$	$\sigma_3=\sigma_1$	$\sigma_3=2\sigma_1$	$\sigma_3=4\sigma_1$
10	10	10	95.4	98.9	99.9	91.9	96.8	99.6
20	10	10	95.5	98.7	99.8	84.8	91.7	98.9
10	20	10	94.1	98.7	99.9	97.0	98.8	99.8
10	10	20	95.6	99.6	99.9	90.4	97.5	99.9

The simulations suggest that if σ_3 is quite different from σ_1 and σ_2, then the actual success rates of 95% confidence intervals for $\mu_1 - \mu_2$ can be quite different from 95%. Unlike the result for the two-sample tools, the effect of unequal standard deviations can be serious even if the three sample sizes are equal.

5.5.2 Diagnostics Using Residuals

Initial graphical examination of the data by stem-and-leaf diagrams, box plots, or histograms is important. If there is a large number of groups, then side-by-side box plots are particularly useful, since they eliminate much of the clutter contained in other displays. As in the two-sample problem, initial assessment helps identify (1) the centers, (2) the relative spreads, (3) the general shapes of the distributions, and (4) the presence of outliers. If the spreads are quite different, transforming the data to a different scale should be considered.

DISPLAY 5.14 Residual plot: lifetimes of mice fed six different diets

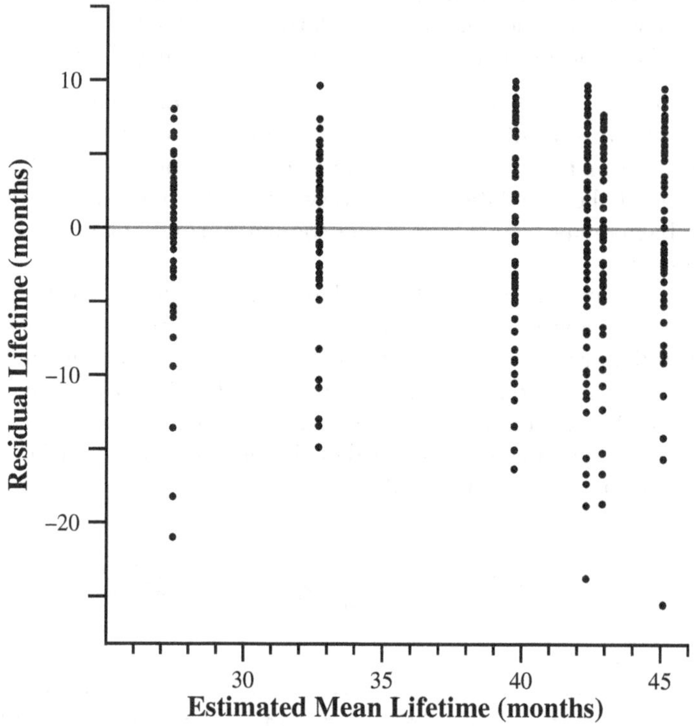

An important tool, which extends to almost all statistical methods in this book, is a *residual plot*. Because the residuals $\overline{Y}_{ij} - \overline{Y}_i$ are the original observations with their group averages subtracted out, they exhibit the variation of the observations without the visual interference caused by differences between group means. A scatterplot of these residuals versus the group averages can reveal a relationship between the spread and the group means, which may be exploited to improve the analysis.

Display 5.14 shows the residual plot from the diet restriction and longevity data. The features to look for in such a plot are (1) an increase in the spread from left to right, in a *funnel-shaped* pattern (which would suggest the need for a log or some other transformation), or (2) seriously outlying observations. The residual plot in Display 5.14 has neither feature, suggesting that analysis on the natural scale is adequate. It is evident that the distributions of lifetimes are skewed. For these large samples, however, there should be no problem in relying on the inferential tools derived from the normal model.

Finally, if the data were collected over time, a plot of residuals versus the time or order of data collection will help to reveal any serial effects that might be present. A pattern of generally increasing or generally decreasing residuals indicates a time trend, which may be accounted for using regression techniques. A pattern in which

DISPLAY 5.15 Some important patterns in residual plots

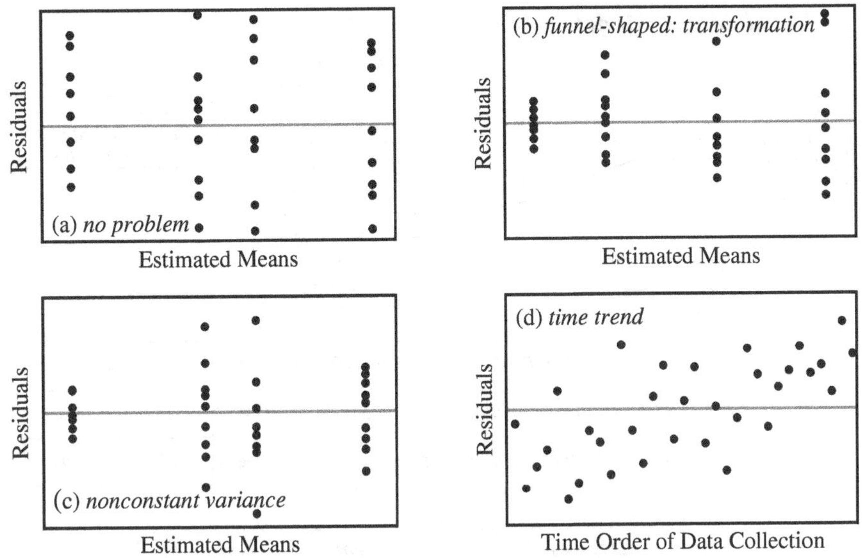

residuals close to each other tend to be more alike (or perhaps more different) than any two arbitrarily chosen residuals may indicate serial correlation. Formal investigation into serial correlation, and methods that account for it, are provided in Chapter 15.

Display 5.15 shows patterns in residual plots that would indicate (a) no obvious problems; (b) nonconstant variance, particularly a variance that is increasing with increasing mean (a log transformation might be in order); (c) nonconstant variance, particularly a variance that is smaller for small and large means (this might be the case if the response is restricted to be between 0 and 1, and a logit transformation might help: $\log[Y/(1-Y)]$); and (d) a relationship between the response and time order of data collection.

5.6 RELATED ISSUES

5.6.1 Further Illustration of the Different Sources of Variability

The analysis of variance is a general method of partitioning total variation into several components. This section attempts to shed further light on that partitioning.

A computer was used to generate independent random samples of size four each from nine normal populations having the same mean and the same standard deviation. The samples are illustrated in Display 5.16. The smooth curve above is the common population histogram, centered at the mean, μ. The horizontal axes

DISPLAY 5.16 Three sources of variation for data simulated from the equal-means model

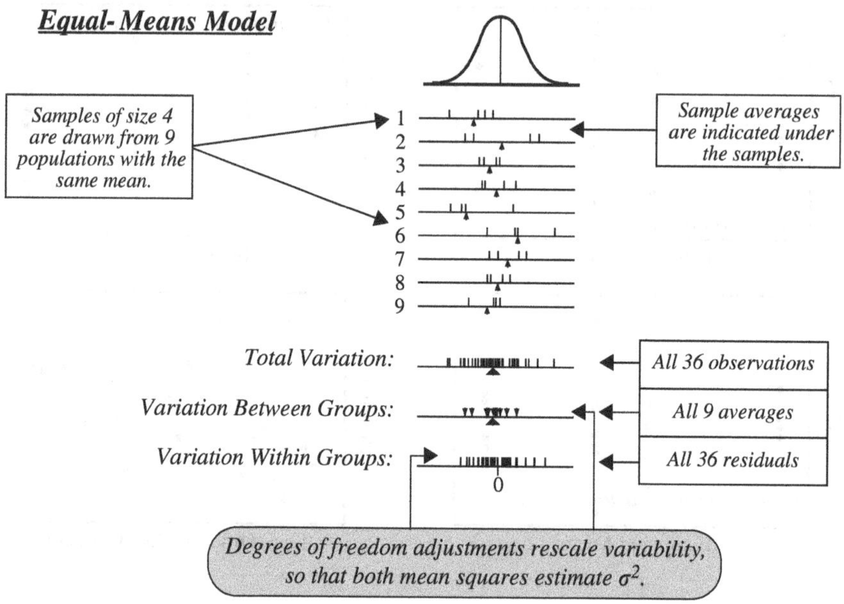
Equal-Means Model
Samples of size 4 are drawn from 9 populations with the same mean.
1
2
3
4
5
6
7
8
9
Sample averages are indicated under the samples.
Total Variation:
Variation Between Groups:
Variation Within Groups:
0
All 36 observations
All 9 averages
All 36 residuals
Degrees of freedom adjustments rescale variability, so that both mean squares estimate σ².

below the histogram locate the separate samples. The ticks above each axis locate the four sample values, and the single arrow below each axis locates the sample's average.

Three additional axes appear below the nine samples, labeled according to the three sources of variation in the analysis of variance table. The *Total* axis shows all 36 sample values on one common axis. Sample values are ticked on top of the axis and the grand average of all 36 is marked below the axis. On the ***Between Groups*** axis, bullets locate the nine sample averages on top, and their average—also the grand average—is marked by the arrowhead on the bottom. The ***Within Groups*** axis displays all 36 ***residuals*** from the separate-means model. These are the observations minus their group averages.

Variation When All Means Are Equal

Variation is different on the three different summary axes for two basic reasons: (1) ***Observations are always closer to their sample average than to their population mean.*** This is easily visible in sample #1, where all four observations fall below the population mean and where the sample average follows them. And (2) ***sample averages are less variable than individual sample values.*** Reason (2) explains why the between-group variation is visibly less than the total variation. Reason (1) explains why the within-group variation is also visibly less than the total variation.

The means of the sampling distributions of the average squared distance of a tick from the axis center are available from theory. The average of the squared residuals is $(1/36)\sum\sum(Y_{ij} - \overline{Y}_i)^2$. The mean of its sampling distribution is $(27/36)\sigma^2$.

DISPLAY 5.17 Variations in the several-group problem for data simulated from the separate-means model

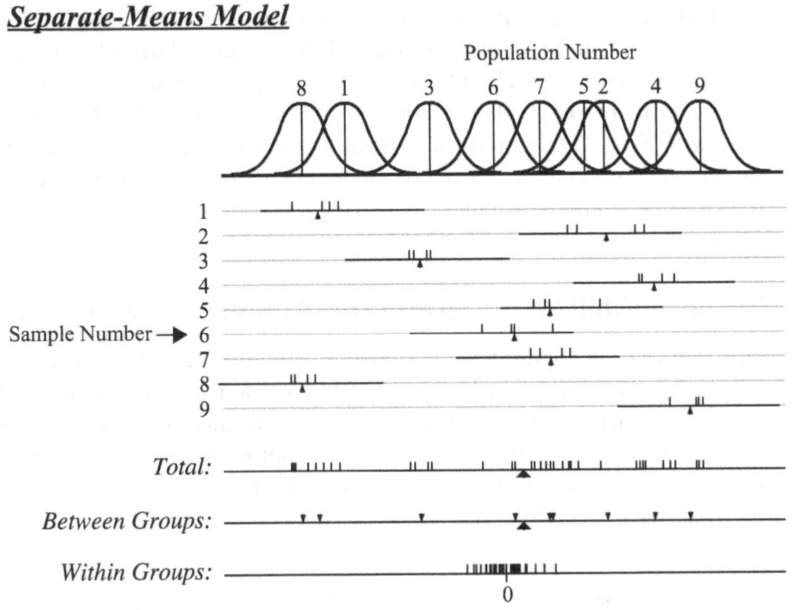

This is less than σ^2 because of reason (1) mentioned above, and illustrates the need for a degrees of freedom adjustment: The mean square of the residuals, $[1/(36-9)]\sum\sum(Y_{ij}-\overline{Y}_i)^2$ is an unbiased estimate of σ^2. (Its sampling distribution has mean σ^2.)

Since the populations have the same mean, the nine $\overline{Y}_i$'s are like a sample of size 9 from a normal population with mean μ and variance $\sigma^2/4$. The quantity $\sum(\overline{Y}_i-\overline{Y})^2/9$ is an unbiased estimate of $(8/9)(\sigma^2/4)$. It is not surprising then that the sample variance of the sample of averages, $\sum(\overline{Y}_i-\overline{Y})^2/8$, is an unbiased estimate of $\sigma^2/4$. It follows that the between-group mean square, which is $\sum n_i(\overline{Y}_i-\overline{Y})^2/8$, with $n_i = 4$ for all i in this case, is an unbiased estimate of σ^2.

If the hypothesis of equal population means is correct, the numerator and the denominator of the F-statistic (the between-groups and within-groups mean squares) are both unbiased estimates of σ^2, so the F-statistic should be close to 1. As shown in the next display, if the population means are, in fact, unequal, then the numerator of the F-statistic is estimating something larger than σ^2 and the F-statistic will tend to be substantially larger than 1.

Variation When the Means Are Different

Display 5.17 depicts simulated samples drawn from populations with *different* means. The samples are the same as in Display 5.16, but shifted to their population means. Notice that the within-group variation is unchanged, whereas the difference in population means results in a larger between-group variation. Hence the F-statistic will be larger than in the equal-means case.

Data that turn out like the bottom three axes in Display 5.17 show strong evidence that the between-group variability is much larger than that expected from the equal means model. The one-way analysis of variance F-test formalizes the comparison. In particular, while the denominator of the F-statistic is always an estimate of σ^2, the numerator is estimating something larger than σ^2. The amount by which it exceeds σ^2 depends on how different the means are. An F-statistic much larger than 1 provides strong evidence of differences among the population means.

5.6.2 Kruskal–Wallis Nonparametric Analysis of Variance

One method for coping with seriously outlying observations is to replace all observation values by their ranks in a single combined sample and then apply a one-way analysis of variance F-test on the rank-transformed data. The Kruskal–Wallis test, which is available in many statistical computer packages, is similar in its approach but takes advantage of the known variance of the ranks.

The Kruskal–Wallis test statistic is

$$KW = 1/[\sigma_R^2] \times \text{Between Group Sum of Squares (of ranks)},$$

where σ_R^2 is the variance of all n ranks (using an $n-1$ divisor) and where n is the total number of observations in all groups. A p-value is found as the proportion of a chi-squared distribution on $(I-1)$ degrees of freedom that is larger than this test statistic.

Display 5.18 shows the rank-transformed data from the Spock trial example. Notice that Spock's judge has venires that rank quite low. The between-group sum of squares (which could be obtained from an analysis of variance table based on the data in Display 5.18) is 3,956.325. The variance of the ranks is 180.122. The value of the Kruskal–Wallis test statistic is therefore 21.96, and p-value is 0.0012 from a chi-square distribution with 6 degrees of freedom. Testing equality of the

DISPLAY 5.18 Spock trial data, rank-transformed

Judge	Rank of venire from smallest (1) to largest (46) percentage of women							
Spock's	1	2	3	4	5	6	9.5	11	16
A	8	31	37	44	46				
B	22	26	34	36	41	45			
C	14	17	23.5	23.5	30	32.5	35	38.5	38.5
D	19	28							
E	9.5	12	15	25	40	43			
F	7	13	18	20	21	27	29	32.5	42

judges other than Spock's can be accomplished by performing a Kruskal–Wallis test ignoring the venires of the Spock judge. The Spock judge can be compared to the other judges combined using the rank-sum test.

5.6.3 Random Effects

Rationale for the Random Effects Model

It has so far been assumed that there is direct interest in the particular groups chosen. Sometimes, however, the groups are selected as representative of some broader population, and an inference is to be drawn to that population. A distinction is made between the *fixed effects model*, in which the group means are fixed and the *random effects model*, in which the group means are a random sample from a population of means. For the case studies in this chapter there is direct interest in the particular groups, so fixed effects models were used.

To illustrate when each model is appropriate, suppose that measurements are taken on the yield of a machine operated by each of several operators. An analysis of variance may be used to compare the mean yields under the different operators. A fixed effects model would be appropriate if there was interest in only those particular operators. (They may constitute all the operators at the plant.) A random effects model would be appropriate if those operators were just a sample and if the question of interest pertained to the population of operators from which they were sampled. There may be interest, for example, in estimating the proportion of the yield variance that could be explained by between-operator variability in a plant with a large number of operators.

Is the random effects model the right one to use? There are two pertinent questions: (1) Is inference desired to a larger set from which these groups are a sample? (2) Are the groups (operators) truly a random sample from the larger set? A yes answer to the first question would ordinarily prompt a user to use the random effects model. Statistical inference to the larger population would only be justified, however, if there was also a yes answer to the second question. If there was no random sampling to obtain the particular operators, then the usual warnings about potential biases due to using nonrandom samples apply.

The Random Effects Model

In the fixed effects model, observed sample i is thought to be a random sample from a normal population with mean μ_i and variance σ^2. There are $I + 1$ parameters in the fixed effects model: the I means and the single variance σ^2.

In the random effects model the μ_i's themselves are thought to be a random sample from a normal population with mean μ and variance σ_μ^2. The random effects model has three parameters: the overall mean μ, the within-group variance σ^2, and the between-group variance σ_μ^2.

Analysis of the one-way classification random effects model involves a test of whether σ_μ^2 is zero and an estimate of the ratio $\sigma_\mu^2/(\sigma_\mu^2 + \sigma^2)$. Notice that this ratio is between 0 and 1. It is 0 when there is no between-group variance, and it is 1 when there is no within-group variance. Since the denominator describes the total variance of the measurements, the ratio may be interpreted as the proportion of the total variance of the measurements that is explained by between-group variability. It is also called the *intraclass correlation.*

Estimation and Testing

The overall mean μ is estimated by the grand average $\overline{Y}$. The estimates of the two variances σ^2 and σ_μ^2 are often found by equating the mean squares in the analysis of variance table to the means of their sampling distributions, under the random effects model. In particular, letting MS(W) and MS(B) represent the within-group and between-group means squares, respectively, one obtains

$$\text{Mean}\{\text{MS(W)}\} = \sigma^2$$

$$\text{Mean}\{\text{MS(B)}\} = \sigma^2 + \frac{1}{n(I-1)}\left(n^2 - \sum_{i=1}^{I} n_i^2\right)\sigma_\mu^2,$$

so the estimates of the variances are

$$\hat{\sigma}^2 = \text{MS(W)}$$

and

$$\hat{\sigma}_\mu^2 = \frac{n(I-1)[\text{MS(B)} - \text{MS(W)}]}{n^2 - \sum_{i=1}^{I} n_i^2},$$

with the modification that the latter is set to zero if the numerator turns out to be negative.

It is sometimes desired to test the hypothesis $H{:}\, \sigma_\mu^2 = 0$, against the alternative that it is greater than zero. It should be evident that this is analogous to the hypothesis that the means are all equal in the fixed effects model. The usual F-test for the fixed effects model (Section 5.3.2) is appropriate for this hypothesis as well.

The bottom line is that testing for between-group differences may be carried out in the usual way with an analysis of variance procedure. An additional, useful summary for the random effects model, however, is $\hat{\sigma}_\mu^2/(\hat{\sigma}_\mu^2 + \hat{\sigma}^2)$.

Example—Spock Trial Data

Although the questions of interest in the Spock example focus on the specific judges, the data set can be used as an example to demonstrate random effects, by ignoring Spock's judge and thinking of the six other judges as representative of some large population. The parameters in the random effects model are μ, the overall mean percentage of women on venires, the variance σ^2 of percentages about the judge mean, and the variance σ_μ^2 of the population of judge means. A test of whether the between-judge variance is zero ($H{:}\, \sigma_\mu^2 = 0$) is the F-test from the standard analysis of variance. The p-value is 0.32, so the data are consistent with there being no between-judge variability. The estimates of σ_μ^2 and σ^2 are 1.96 and 53.6, so it can be said that the proportion of variability that is due to differences between judges is $1.96/(1.96 + 53.6) = 0.035$. The square root of this number is also interpreted as the intraclass correlation—the correlation that percentages from two venires have if the venires come from the same judge. Of course, these inferential statements are speculative since the six judges were not, in fact, a random sample from a population of judges.

5.6.4 Separate Confidence Intervals and Significant Differences

Published research articles often present graphs of estimated means with separate confidence intervals for each. If a reader wishes to know whether two means are different, there is a fairly close—but not exact—relationship between the overlap in the confidence intervals and the result of a test of equal means. *The proper course of action for judging whether two means are equal is to carry out a* t*-test directly.* The comments here apply to a situation where either the reader is looking for a quick approximate answer or where the article fails to provide enough information to conduct the test.

Four categories of results are possible (Display 5.19). *Case 1*: If the intervals do not overlap, it is safe to infer that the means are different. Some readers incorrectly assume this is the only case that provides strong evidence of a difference. *Case 2*, however, also shows strong evidence. Even though the intervals overlap, the best estimate for each mean lies outside the confidence interval for the other mean.

DISPLAY 5.19 Separate confidence intervals for two group means: Are the means different?

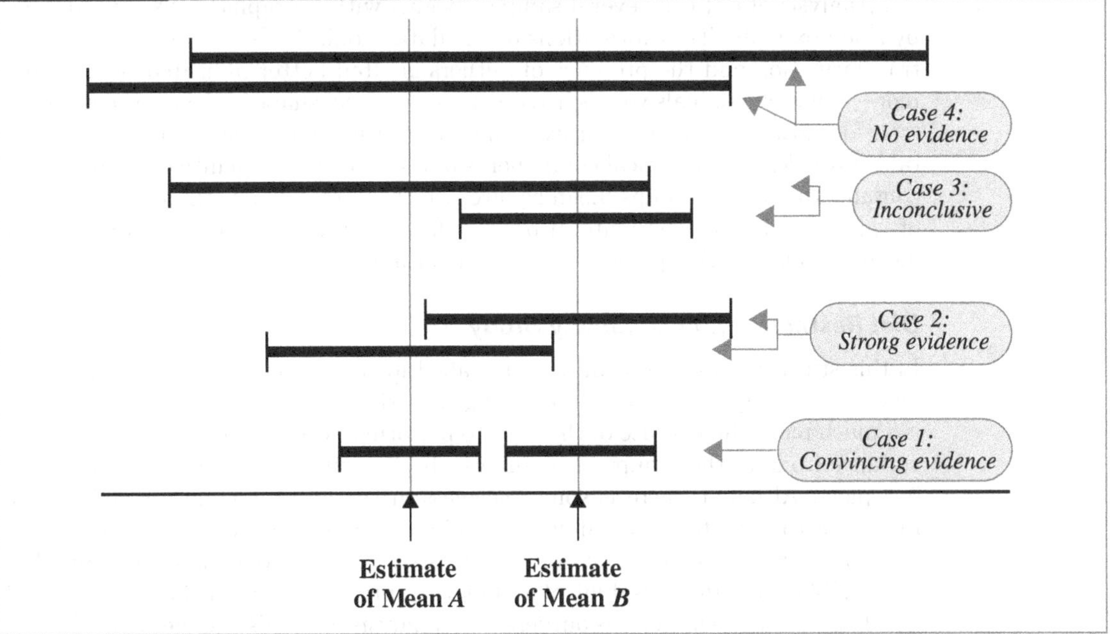

Case 3, where one estimate lies within the confidence interval for the other mean, but the second estimate lies outside the first interval, is difficult to judge. But in *Case 4*, where the best estimate of each mean lies inside the confidence interval for the other, there is no evidence of any difference.

Finally, it must be mentioned that the discussion of this section applies to the comparison of two confidence intervals only. If there are more than two confidence intervals, then it may be quite misleading to compare the two most disparate ones, unless some adjustment for multiple comparisons is made. This topic is discussed in the next chapter.

5.7 SUMMARY

The term *analysis of variance* is often initially confusing as it seems to imply a comparison of variances. It is most definitely a method for comparing means, however, and the name derives from the approach for doing so—assessing variability from several sources. The analysis of variance F-test is used for assessing equality of several means.

Another point of confusion arises from the mistaken belief—due to the prevalence of the F-test in textbooks and computer programs—that the F-test necessarily plays a central role in the analysis of several samples. Usually it does not. It offers a convenient approach for detection of group differences, but it does not ordinarily provide answers to particular questions of interest. Tests and confidence intervals for pairs of means or linear combinations of means (discussed in the next chapter) provide much more specific information.

Analysis of data in several samples begins with a graphical display, like side-by-side box plots. Transformations of the data should be considered. The need for transformation and the presence of outliers is often better indicated by a residual plot—a plot of residuals versus fitted values. A funnel shape indicates the need for a transformation like the log, for example, for positive data. The analysis of variance table provides the numerical components of the F-test for equality of means. It also contains the within-groups mean square, which exactly equals the pooled estimate of variance, the best estimate of σ^2. Confidence intervals and t-tests for pairs of means should use this pooled estimate of variance from all groups.

Diet Restriction and Longevity Study

In this study, the questions of interest called for five specific pairwise comparisons among the groups. It might be tempting to perform five two-sample t-tests, but it is a much more efficient use of the data to perform t-tests using a pooled estimate of variance from all the groups. The analysis begins with examination of side-by-side box plots. Although there is some skewness in the data, it is not enough to warrant concern—the tests are sufficiently robust against this type of departure from normality—and no transformation is suggested. A closer look at possible problems is available through a residual plot, but the spreads appear to be approximately equal and there are no serious outliers. The analysis proceeds, therefore, with t-tests and confidence intervals in the usual way, but using the pooled estimate of variance from all groups.

Spock Trial Study

The stem-and-leaf plots in Display 5.4 and box plots in Display 5.5 are useful for suggesting some answers to the question of interest and for indicating the appropriateness of the tools based on the standard one-way classification model. An analysis of variance F-test confirms the strong evidence of some differences between means. An application of the extra-sums-of-squares F-test for comparing equality of the six other judges shows no evidence of a difference in mean percentages of women on their venires. Assuming that the six other means are equal, a further F-test shows overwhelming evidence that the Spock judge mean is different from the mean of the other six. Since this is a test for equality of two means, a t-test could be used. In fact, the F-test is equivalent to a two-sided t-test when $I = 2$. The actual p-value reported in the summary of statistical findings comes from a different test, not based on the assumption of equal means among the other six judges, and is discussed in the next chapter.

5.8 EXERCISES

Conceptual Exercises

1. Spock Trial. Why is it important to obtain a pooled estimate of variance in the Spock trial study? Is it ever a mistake to obtain a pooled estimate of variance in a comparison involving several groups?

2. Four methods of growing wheat are to be compared on five farms. Four plots are used on each farm and each method is applied to one of the plots. Five measurements are therefore obtained on yield per acre for each of the four growing methods. Are the methods of this chapter appropriate for analyzing the results?

3. Diet Restriction. Is there any explanation for why the distribution of lifetimes of mice in Display 5.1 are all negatively skewed?

4. Diet Restriction. For comparing group 3 to group 2, explain why it is better to use the t-tools presented in Section 5.2.3 (using s_p from all six groups) than to use the Chapter 2 t-tools (using s_p from only the two groups involved).

5. Spock Trial. Should Spock's accusers question the defense on how the venires were selected for their study?

6. Spock Trial. Why is it useful to test whether the six judges other than Spock's have equal mean percentages of women on their venires?

7. Why is s_p^2 not simply taken as the average of the I sample variances?

8. Diet Restriction. If the longevity study was a planned experiment, why are the sample sizes different?

9. If s_p is zero, what must be true about the residuals?

10. Explain the role of degrees of freedom of the F-distribution associated with the F-statistic. How are degrees of freedom related to how far the F-statistic is likely to be from 1?

11. What does it mean if the F-statistic is so *small* that the chance of getting an F-statistic that small or smaller is only, say, 0.0001?

12. Flycatcher Species Identification. One of the most challenging field identification problems for North American ornithologists is to distinguish the 10 species of flycatchers in the genus *Empidonax*. Many articles have appeared in popular and scientific journals suggesting different morphological clues to proper identification. F. Rowland ("Identifying *Empidonax* Flycatchers: The Ratio Approach," 2009, *Birding* 41 (2): 30–38) asserted that the relative size of wing length to tail length is the appropriate physical characteristic for distinguishing the species in the field. This conclusion was based on the average values of the wing length minus the tail length for 24 birds in each species, as shown in the following table.

Species:	Yellow-bellied	Acadian	Alder	Willow	Least	Hammond's	Gray	Dusky	Pacific-slope	Cordilleran
Average wing–tail (mm; n=24)	13.6	15.4	14.7	12.4	9.2	13.7	10.3	7.0	9.5	9.5

Explain why a conclusion that this measurement tends to differ in the 10 species cannot be made from the averages alone. What additional piece of information is needed to test for group differences and to evaluate the extent to which individuals from different species can be distinguished?

Computational Exercises

13. Spock Trial. By examining Display 5.8, answer the following:

(a) What is the average percentage of women from all 46 venires?

(b) For how many of the 9 Spock judge's venires is the percentage of women less than the grand average from all 46 venires?

(c) For how many of the 9 Spock judge venires is the percentage of women less than the Spock judge's average?

14. Spock Trial. Use the following summary statistics to (a) compute the pooled estimate of the standard deviation and (b) carry out a t-test for the hypothesis that the Spock judge's mean is equal to the mean for judge A.

Judge:	Spock	A	B	C	D	E	F
Average % women:	14.62	34.12	33.61	29.10	27.00	26.97	26.80
SD of % women:	5.039	11.942	6.582	4.593	3.818	9.010	5.969
Sample size:	9	5	6	9	2	6	9

15. Spock Trial. (a) Use a calculator or statistical package to get the sample variance for the percentage of women on all 46 venires treated as one sample. (b) Multiply this by 45 to get the residual sum of squares for the equal-means model. (c) Multiply s_p^2 found in Exercise 14(a) above by (46−7) to get the residual sum of squares for the separate-means model. (d) Use these to construct an analysis of variance table, including the F-statistic for the hypothesis of equal means. Compare the result with Display 5.10.

16. Spock Trial. Use a statistical computer package to obtain the analysis of variance table in Display 5.10.

17. Display 5.20 shows the start of an analysis of variance table. Fill in the whole table from what is given here. How many groups were there? Is there evidence that the group means are different?

DISPLAY 5.20 Incomplete ANOVA table for Exercise 17

Source	d.f.	Sum of squares	Mean square	F-statistic	p-value
Between groups	?	?	?	?	?
Within groups	24	35,088	?		
Total	31	70,907			

18. Fatty Acid. The data in Display 5.21 were obtained from a randomized experiment to estimate the effect of a certain fatty acid (CPFA) on the level of a certain protein in rat livers. Only one level of the CPFA could be investigated in a day's work, so a control group (no CPFA) was investigated each day as well. (Data from Donald A. Pierce.)

DISPLAY 5.21 Levels of protein (×10) found in rat livers

	Treatment					
Day	CPFA 50	CPFA 150	CPFA 300	CPFA 450	CPFA 600	Control
1	154, 177, 174					157, 165, 150
2		164, 192, 159				186, 206, 195
3			157, 159, 124			192, 202, 216
4				160, 152, 141		190, 187, 160
5					147, 152, 158	191, 188, 199

(a) Obtain estimated means for the model with six independent samples, one for each treatment. Determine the residuals and plot them versus the estimated means. Plot the residuals versus the day on which the investigation was conducted. Is there any indication that the methods of this chapter are not appropriate?

(b) Obtain estimated means for the model with 10 independent samples, one from each treatment-day combination. Calculate the ANOVA F-test to see whether these 10 groups have equal means.

(c) Use (a) and (b) and the methods of Section 5.4.1, to test whether the means for the control groups on different days are different. That is, compare the model with 10 different means to the model in which there are 6 different means.

19. Cavity Size and Use. Biologists freely discuss the concept of competition between species, but it is difficult to measure. In one study of competition for nesting cavities in Southeast Colorado, Donald Youkey (Oregon State University Dept. of Fisheries & Wildlife) located nearly 300 cavities occupied by a variety of bird and rodent species. Display 5.22 shows box plots of the entrance area measurements from cavities chosen by nine common nesting species. The general characteristics—positive skewness, larger spreads in the groups with larger means—suggest the need for a transformation. On the logarithmic scale, the spreads are relatively uniform, and the summary statistics appear in Display 5.23. Are the species competing for the same size cavities? Or, are there differences in the cavity sizes selected by animals of different species? It appears that there are two very different sets of species here. The first six selected relatively small cavities while the last three selected larger ones. Is that the only significant difference?

DISPLAY 5.22 Box plots for areas of entrances to cavities used by different species

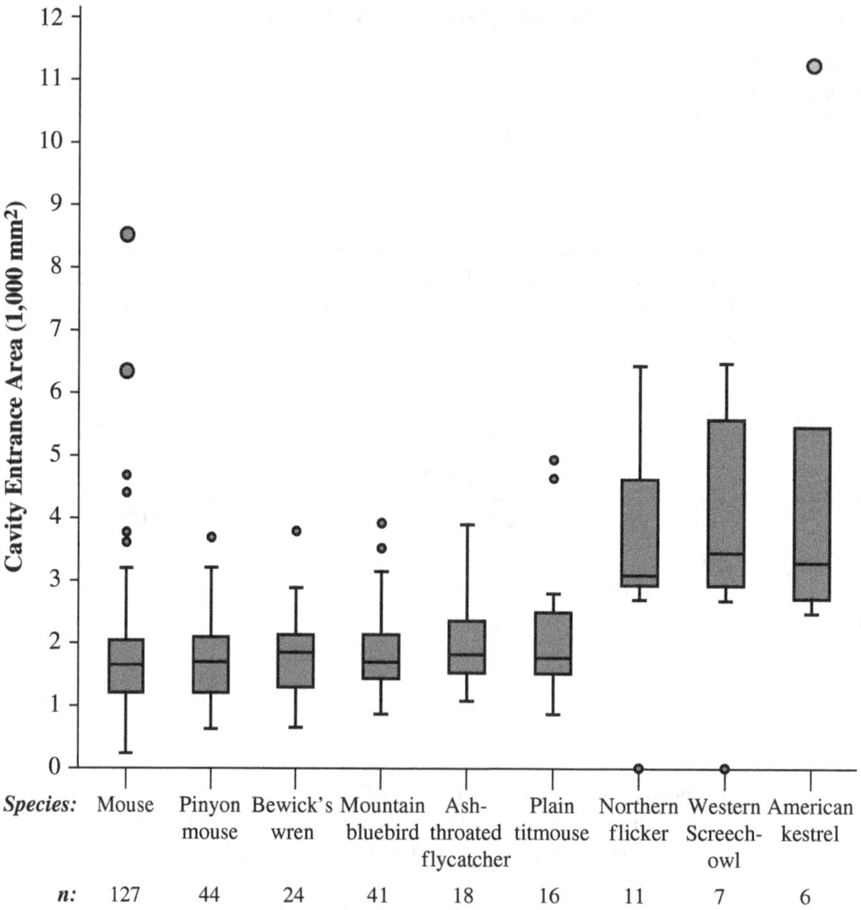

(a) Compute the pooled estimate of variance.

(b) Construct an analysis of variance table to test for species differences. (The sample standard deviation of all 294 observations as one group is SD = 0.4962.) Perform the F-test.

(c) Verify that the analysis of variance method for calculating the between-group sum of squares yields the same answer as the formula

$$\text{Between-group SS} = \sum_{i=1}^{I} n_i \overline{Y}_i^{\,2} - n\overline{Y}^2.$$

(d) Fit an intermediate model in which the first six species have one common mean and the last three species have another common mean. Construct an analysis of variance table with F-tests to compare this model with (i) the equal-means model and (ii) the separate-means model. Perform the F-test.

DISPLAY 5.23 Summary statistics for areas of cavity entrances (logarithmic scale)

Species	n	Mean	Sample SD
Mouse	127	7.347	0.4979
Pinyon mouse	44	7.368	0.4235
Bewick's wren	24	7.418	0.3955
Mountain bluebird	41	7.487	0.3183
Ash-throated flycatcher	18	7.563	0.3111
Plain titmouse	16	7.568	0.4649
Northern flicker	11	8.214	0.2963
Western Screech-owl	7	8.272	0.3242
American kestrel	6	8.297	0.5842

20. Flycatcher Species Identification. Consider the table of averages (of wing lengths minus tail lengths) from 24 birds in each of 10 species of flycatcher in Exercise 12. If it is assumed that the 11 populations all have the same mean and same population standard deviation, what is an estimate of the population standard deviation?

21. A robust test for equality of several population variances is *Levene's test*, which was previously discussed in Section 4.5.3 for the case of two variances. This procedure carries out the usual one-way analysis of variance F-test on the absolute values of the differenes of observations from their group medians. For practice, carry out Levene's test on the Spock data.

22. Equity in Group Learning. Several studies have demonstrated that engaging students in small learning groups increases student performances on subsequent tests. However, N. M. Webb and her colleagues argue that this raises a question of equity: Does the quality of the learning depend on the composition of the group? They chose students from five 7th and 8th grade classes in the Los Angeles school system. Based upon a science and language pretest, they classified each student's ability level as Low, Low-Medium, Medium-High, or High. They formed study groups consisting of three students each. The students were given a problem involving the setting up of two electrical circuits that would produce different brightness in a standard lightbulb. Each group was given the equipment to work with and time to discuss the problem and come to a solution. Afterward, each student was tested on the basics of the problem and its solution.

The table in Display 5.24 shows the results of the scores on this final test of the students whose ability level was Low in pretest. The students are grouped in the table according to the highest level of ability of a member in their study group. (Data from N. M. Webb, K. M. Nemer, A. W. Chizhik, and G. Sugrue "Equity Issues in Collaborative Group Assessment: Group Composition and Performance," *American Educational Research Journal* 35(4): (1998) 607–51.)

DISPLAY 5.24 Achievement test scores of Low ability students who worked in different study groups

	Highest ability level in the study group			
	Low	Low-medium	Medium-high	High
Average:	0.26	0.37	0.36	0.47
St. Dev.:	0.14	0.21	0.17	0.21
n:	17	24	25	14

DISPLAY 5.25 Achievement test scores of High ability students who worked in different study groups

	Lowest ability level in the study group			
	Low	Low-medium	Medium-high	High
Average:	0.75	0.77	0.72	0.85
St. Dev.:	0.16	0.11	0.12	0.10
n:	13	22	42	28

(a) How strong is the evidence that at least one group mean differs from the others?
(b) Display 5.25 shows a companion table. How strong is the evidence from this table that at least one mean differs from the others?
(c) The study groups apparently were not formed using random assignment. How might this affect any conclusions you might draw from the analysis?

Data Problems

23. Was Tyrannosaurus Rex Warm-Blooded? Display 5.26 shows several measurements of the oxygen isotopic composition of bone phosphate in each of 12 bone specimens from a single *Tyrannosaurus rex* skeleton. It is known that the oxygen isotopic composition of vertebrate bone phosphate is related to the body temperature at which the bone forms. Differences in means at different bone sites would indicate nonconstant temperatures throughout the body. Minor temperature differences would be expected in warm-blooded animals. Is there evidence that the means are different for the different bones? (Data from R. E. Barrick, and W. J. Showers, "Thermophysiology of *Tyrannosaurus rex*; Evidence from Oxygen Isotopes," *Science* 265 (1994): 222–24.)

24. IQ and Future Income. Display 5.27 shows the first five rows of a data set with annual incomes in 2005 for 2,584 Americans who were selected in the National Longitudinal Study of Youth 1979, who were available for re-interview in 2006, and who had paying jobs in 2005, along with the quartile of their AFQT (IQ) test score taken in 1981 (see Exercise 2.22). How strong is the evidence that the

DISPLAY 5.26 Measurements of oxygen isotopic composition of vertebrate bone phosphate (per mil deviations from SMOW) in 12 bones of a single *Tyrannosaurus rex* specimen

Bone	Oxygen isotopic composition					
Rib 16	11.10	11.22	11.29	11.49		
Gastralia	11.32	11.40	11.71			
Gastralia	11.60	11.78	12.05			
Dorsal vertebra	10.61	10.88	11.12	11.24	11.43	
Dorsal vertebra	10.92	11.20	11.30	11.62	11.70	
Femur	11.70	11.79	11.91	12.15		
Tibia	11.33	11.41	11.62	12.15	12.30	
Metatarsal	11.32	11.65	11.96	12.15		
Phalange	11.54	11.89	12.04			
Proximal caudal	10.93	11.01	11.08	11.12	11.28	11.37
Mid-caudal	11.35	11.43	11.50	11.57	11.92	
Distal caudal	11.95	12.01	12.25	12.30	12.39	

DISPLAY 5.27 Annual income in 2005 and test score quartile for an IQ test taken in 1981 for 2,584 Americans in the NLSY79 survey; first 5 of 2,584 rows

Subject	IQquartile	Income2005
2	1stQuartile	5,500
6	4thQuartile	65,000
7	2ndQuartile	19,000
8	2ndQuartile	36,000
9	3rdQuartile	65,000

distributions of 2005 annual incomes differ in the four populations? By how many dollars or by what percent does the distribution of 2005 incomes for those within the highest (fourth) quartile of IQ test scores exceed the distribution for the lowest (first) quartile?

25. Education and Future Income. The data file ex0525 contains annual incomes in 2005 of a random sample of 2,584 Americans who were selected for the National Longitudinal Survey of Youth in 1979 and who had paying jobs in 2005 (see Exercise 22 in Chapter 2). The data set also includes a code for the number of years of education that each individual had completed by 2006: <12, 12, 13–15, 16, and >16. How strong is the evidence that at least one of the five population distributions (corresponding to the different years of education) is different from the others? By how many dollars or by what percent does the mean or median for each of the last four categories exceed that of the next lowest category?

Answers to Conceptual Exercises

1. To make comparisons, one must estimate variation. There are not many venires for any particular judge, so pooling the information gives better precision to the variance estimate. But if the groups have very different spreads, pooling is a bad idea.

2. Not appropriate. You should not expect the measurements from plots on the same farm to be independent of each other.

3. Perhaps there is something like an upper bound, a maximum possible lifetime for each group, and healthy mice all tend to get close to it. Unhealthy mice, however, die off sooner and at very different ages.

4. If the variances in all populations are equal, s_p from all groups uses much more data to estimate σ, resulting in a more precise estimator.

5. Yes. Perhaps these are just as good as random samples of all venires for each judge. If there was any bias in the selection, however—for example, if the nine venires for Spock's judge were chosen because they did not have many women—the results would be misleading.

6. Spock's lawyers will have a stronger case if they can show that Spock's judge is particularly different from *all others* in having low representation of women.

7. It is, if the sample sizes are all equal. Otherwise, it gives more weight to estimates from larger samples.

8. It is unusual for experimenters to purposefully plan on unequal sample sizes. In this study it is likely that the larger number of mice in the N/R50 group was planned, because that was the major experimental group. Inequalities in the other group sample sizes are likely the result of losing mice to factors unrelated to the experiment.

9. All the residuals would have to be identically zero for this to happen.

10. The larger the degrees of freedom in either the numerator or denominator, the less variability there is in their sampling distributions. With smaller degrees of freedom in either, sampling variability can result in an F-ratio which is considerably different from 1, even when the null hypothesis is true.

11. That would suggest that the sample averages are closer to each other than one would expect in the course of natural sampling from identical populations. You may want to check out the independence assumption.

12. There are two important quantities: (1) the within-mean square is s_p^2, and (2) the p-value allows for judging group differences.

CHAPTER 6

Linear Combinations and Multiple Comparisons of Means

The F-test for equality of several means gives a reliable result with any number of groups. Its weakness is that it neither tells which means are different from which others nor accounts for any structure possessed by the groups. Consequently, its role is mainly to act as an initial screening device.

If the groups have a structure, or if the research requires a specific question of interest involving several groups, a particular *linear combination* of the means may address the question of interest. This chapter shows how to make inferences about linear combinations of means and how to choose linear combinations for some important kinds of problems.

When no planned comparison is called for by the questions of interest or the group structure, one may compare all means with each other. The large number of comparisons, however, compounds the statistical uncertainty in the statements of evidence. Some methods of adjustment to account for this *multiple comparisons* problem are provided and discussed here.

6.1 CASE STUDIES

6.1.1 Discrimination Against the Handicapped—A Randomized Experiment

The U.S. Vocational Rehabilitation Act of 1973 prohibited discrimination against people with physical disabilities. The act defined a handicapped person as any individual who had a physical or mental impairment that limits the person's major life activities. Approximately 44 million U.S. citizens fit that definition. In 1984, 9 million were in the labor force, and these individuals had an unemployment rate of 7%, compared to 4.5% in the nonimpaired labor force.

One study explored how physical handicaps affect people's perception of employment qualifications. (Data from S. J. Cesare, R. J. Tannenbaum, and A. Dalessio, "Interviewers' Decisions Related to Applicant Handicap Type and Rater Empathy," *Human Performance* 3(3) (1990): 157–71.) The researchers prepared five videotaped job interviews, using the same two male actors for each. A set script was designed to reflect an interview with an applicant of average qualifications. The tapes differed only in that the applicant appeared with a different handicap. In one, he appeared in a wheelchair; in a second, he appeared on crutches; in another, his hearing was impaired; in a fourth, he appeared to have one leg amputated; and in the final tape, he appeared to have no handicap.

Seventy undergraduate students from a U.S. university were randomly assigned to view the tapes, fourteen to each tape. After viewing the tape, each subject rated the qualifications of the applicant on a 0- to 10-point applicant qualification scale. Display 6.1 shows the results. The question is, do subjects systematically evaluate qualifications differently according to the candidate's handicap? If so, which handicaps produce the different evaluations?

DISPLAY 6.1 Stem-and-leaf diagrams of applicant qualification scores given to applicants simulating five different handicap conditions

	None	Amputee	Crutches	Hearing	Wheelchair
0					
1	9	9		4	7
2	5	56		149	8
3	06	268	7	479	5
4	129	06	033	237	78
5	149	3589	18	589	03
6	17	1	0234	5	1124
7	48	2	445		246
8			5		
9					

Legend: 7 | 4 represents a score of 7.4 on the Applicant Qualification Scale.

Statistical Conclusion

The evidence that subjects rate qualifications differently according to handicap status is moderately strong, but not convincing (F-test p-value $= 0.030$). The difference between the average qualification scores given to the *crutches* candidate and to the *hearing-impaired* candidate is difficult to attribute to chance. The difference is estimated to be 1.87 points higher for the *crutches* tape, with a 95% confidence interval from 0.14 to 3.60 points based on the Tukey–Kramer procedure. The strongest evidence supports a difference between the average scores given to the *wheelchair* and *crutches* handicaps and the average scores given to the *amputee* and *hearing* handicaps (t-statistic $= 3.19$ for a linear contrast). None of the average qualification scores from the various feigned handicaps differ significantly from the no-handicap control. (The protected least significant differences all have two-sided p-values > 0.05.)

Scope of Inference

Although the evidence suggests that differences exist among some of the handicap categories, the overall picture is made difficult by the location of the control in the middle of the groups. Any inference statements must also be qualified by the fact that the subjects used in this study may not accurately represent the population of employers making hiring decisions.

6.1.2 Pre-Existing Preferences of Fish—A Randomized Experiment

Charles Darwin proposed that sexual selection by females could explain the evolution of elaborate characteristics in males that appear to decrease their capacity to survive. In contrast to the usual model stressing the co-evolution of the female preference with the preferred male trait, A. L. Basolo proposed and tested a selection model in which females have a pre-existing bias for a male trait, even before males of the same species possess it. She studied a Central American genus of small fish. The males in some species of the genus develop brightly colored swordtails at sexual maturity. For her study, Basolo selected one species in the genus—the Southern Platyfish—whose males do not naturally develop the swordtails.

Six pairs of males were surgically given artificial, plastic swordtails. One male of each pair received a bright yellow sword, while the other received a transparent sword. The males in a pair were placed in closed compartments at opposite ends of a fish tank. One at a time, females were placed in a central compartment, where they could choose to engage in courtship activity with either of the males by entering a side compartment adjacent to it (see Display 6.2). Of the total time spent by each female engaged in courtship during a 20-minute observation period, the percentages of time spent with the yellow-sword male were recorded. These appear in Display 6.3. (Data from A. L. Basolo, "Female Preference Predates the Evolution of the Sword in Swordtail Fish," *Science* 250 (1990): 808–10.) Did these females show a preference for the males that were given yellow swordtails?

DISPLAY 6.2 Experimental tank allowing female fish to choose between males

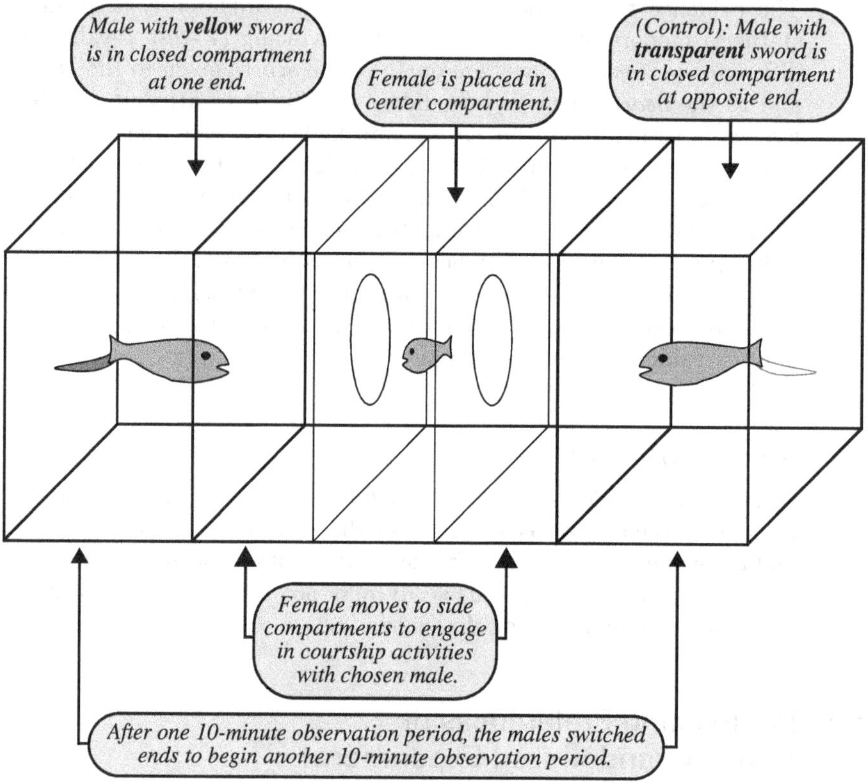

Statistical Conclusion

These data provide convincing evidence that the females tended to spend a higher percentage of time with the yellow-sword male than with the transparent-sword male (one-sided p-value < 0.0001 from a one-sample test that the mean percentage of time spent with the yellow-sword male is 50%). The estimated mean percentage of time with the yellow-sword male was 62% (95% confidence interval: 59% to 65%). The data provide no evidence that the mean percentage differed among the six pairs (p-value $= 0.56$, from a one-way analysis of variance F-test). There was also no evidence of a linear association between mean percentage of time spent with the yellow-sword male and the males' body size (p-value $= 0.32$ from a linear contrast).

6.2 INFERENCES ABOUT LINEAR COMBINATIONS OF GROUP MEANS

6.2.1 Linear Combinations of Group Means

Questions of interest sometimes involve comparing only two group means. Each question in the diet and lifetime study (Section 5.1.1) had this feature. Examining

DISPLAY 6.3 Percentage of courtship time spent by 84 females with the yellow-sword male; body sizes of the males are shown in parentheses

	Pair 1 (35 mm)	Pair 2 (31 mm)	Pair 3 (33 mm)	Pair 4 (34 mm)	Pair 5 (28 mm)	Pair 6 (34 mm)
	43.7	52.5	91.0	72.2	78.3	33.4
	54.0	65.6	62.0	58.5	66.0	42.2
	49.8	68.5	10.0	51.0	47.7	35.6
	65.5	45.9	83.8	56.8	77.5	79.9
	53.1	80.2	91.3	92.4	58.3	59.0
	53.0	67.0	56.3	55.3	61.1	58.1
	62.3	73.0	83.6	59.3	65.1	64.2
	49.4	71.7	53.3	42.0	62.9	82.8
	45.7	55.0	36.5	68.5	61.0	75.7
	56.6	70.0	65.4	78.4		66.3
	59.0	63.2	48.1	69.6		56.3
	67.8	39.6	50.6	89.2		84.5
	73.3	41.0	40.4	67.3		61.1
	43.8	59.2	90.6	77.5		87.6
	67.4		74.9			
	58.1		56.0			
			67.5			
Average:	56.41	60.89	62.43	67.00	64.21	63.34
SD:	9.02	12.48	22.29	14.33	9.41	17.68
***n*:**	16	14	17	14	9	14

differences between the corresponding sample averages answers such questions. But this situation is uncommon in complex studies.

More typically, questions of interest involve several group means. For example, the study of qualification scores given to handicapped applicants might focus on a comparison of two handicaps—*crutches* and *wheelchair*—with the two handicaps—*hearing* and *amputee*. If $\mu_1, \mu_2, \mu_3, \mu_4$, and μ_5 are the mean scores in the *none*, *amputee*, *crutches*, *hearing*, and *wheelchair* groups, respectively, that question can be explored by studying the difference between the average of mean responses, $\gamma = (\mu_3 + \mu_5)/2 - (\mu_2 + \mu_4)/2$.

The parameter γ introduced here is called a *linear combination* of the group means. Linear combinations have the form

$$\gamma = C_1\mu_1 + C_2\mu_2 + \cdots + C_I\mu_I.$$

in which the coefficients $C_1, C_2, \ldots, C_I$ are chosen by the researcher to measure specific features of interest. In the handicap example, $C_1 = 0, C_2 = C_4 = -1/2$, and $C_3 = C_5 = +1/2$. These particular coefficients add to zero, which gives this linear combination the special designation of being a *contrast*.

6.2.2 Inferences About Linear Combinations of Group Means

The Estimate of a Linear Combination and Its Sampling Distribution

The same linear combination of sample averages, called g, is the natural estimate of the parameter γ. It is

$$g = C_1\overline{Y}_1 + C_2\overline{Y}_2 + \cdots + C_I\overline{Y}_I.$$

The sampling distribution of this estimate has mean γ. The standard deviation in the sampling distribution is given by the formula

$$\mathrm{SD}(g) = \sigma\sqrt{\frac{C_1{}^2}{n_1} + \frac{C_2{}^2}{n_2} + \cdots + \frac{C_I{}^2}{n_I}},$$

which depends on the nuisance parameter σ. This assumes that the equal-spread model applies. The shape of the sampling distribution is normal if the individual populations are normal, and it is approximately normal more generally.

Standard Errors for Estimated Linear Combinations

The standard error for g is obtained by substituting the pooled estimate for σ in the formula for the standard deviation of g:

$$\mathrm{SE}(g) = s_p\sqrt{\frac{C_1^2}{n_1} + \frac{C_2^2}{n_2} + \cdots + \frac{C_I^2}{n_I}}.$$

Two-sample comparisons are a special case. To compare the mean score of the crutches ratings with the mean score of the no handicaps ratings, for example, the investigator chooses $C_3 = +1$ and $C_1 = -1$, with $C_2 = C_4 = C_5 = 0$. The expression under the square root in the standard error reduces to the familiar sum of the reciprocals of the two group sample sizes. The SD is estimated from information in all groups, even when only two groups are being compared.

Inferences Based on the t-Distributions

The t-ratio, $t = (g - \gamma)/\mathrm{SE}(g)$, has an approximate Student's t-distribution with degrees of freedom equal to that of the pooled SD: d.f. $= (n_1 + n_2 + \cdots + n_I - I)$. The t-ratio may now be used as before either to construct a confidence interval or to test a hypothesized value for γ.

Example—Handicap Study

A computer can produce the averages and the pooled estimate of variability, but hand calculations are usually required from there. Display 6.4 illustrates all the steps

DISPLAY 6.4 Confidence interval construction for the linear combination $\gamma = (\mu_3 + \mu_5)/2 - (\mu_2 + \mu_4)/2$ in the handicap study

(1) *Summary statistics* → $s_p = 1.6329$; 65 *d.f.*

	Feigned Handicap				
	None	**Amputee**	**Crutches**	**Hearing**	**Wheelchair**
n:	14	14	14	14	14
Average:	4.900	4.4286	5.9214	4.0500	5.3429
C:	0	−1/2	+1/2	−1/2	+1/2

(2) *Specify the coefficients for the linear combination.*

(3) *Estimate the linear combination.*

$$g = \frac{(\overline{Y}_3 + \overline{Y}_5)}{2} - \frac{(\overline{Y}_2 + \overline{Y}_4)}{2} = \frac{(5.9214 + 5.3429)}{2} - \frac{(4.4286 + 4.0500)}{2}$$
$$= 1.3929$$

(4) *Find the standard error of the estimate.*

$$\text{SE}(g) = 1.6329\sqrt{\frac{(0)^2}{14} + \frac{(-1/2)^2}{14} + \frac{(+1/2)^2}{14} + \frac{(-1/2)^2}{14} + \frac{(+1/2)^2}{14}}$$
$$= 0.4364$$

(5) *Construct the 95% confidence interval.*

$t_{65}(0.975) = 1.9971$ ← *from the t-distribution with 65 d.f.*

$1.3929 \pm (1.9971) \times (0.4364)$ → **from 0.521 to 2.264**

involved in finding a confidence interval for the contrast between the *wheelchair* and *crutches* means and the average of the *amputee* and *hearing* means. A 95% confidence interval for the parameter $\gamma = (\mu_3 + \mu_5)/2 - (\mu_2 + \mu_4)/2$ extends from 0.522 to 2.264.

6.2.3 Specific Linear Combinations

Here are some examples of linear combinations that arise frequently in practical situations.

Comparing Averages of Group Means

One common problem has two *sets of groups* distinguished by a specific factor; a comparison of the two sets is of interest. The preceding comparison is a typical example. Another example, from Section 5.1.2, involves comparing the Spock judge's mean percentage women with the average of the means from the other six judges.

If J groups in one set are to be compared to K groups in the second set, the relevant parameter is

$$\gamma = \frac{(\mu_{1,1} + \cdots + \mu_{1,J})}{J} - \frac{(\mu_{2,1} + \cdots + \mu_{2,K})}{K}.$$

The coefficients will be $+1/J, -1/K$, or zero, depending on whether a group is in the first set, the second set, or neither. *Important note:* One group cannot belong to both sets.

Comparing Rates

In problems like the diet restriction study in Section 5.1.1, where groups are structured according to levels of a quantitative explanatory variable, it may be desirable to report results as *rates* of increase in the mean response associated with changes in the explanatory variable. A comparison of the increase in mean lifetime associated with the reduction from 50 to 40 kcal/wk to the increase in mean lifetime associated with the reduction from 85 to 50 kcal/wk, for example, is best made on the basis of increases associated with one-unit changes in the caloric intake. Thus one would inquire whether the rate of increase in lifetime is the same in the reduction from 50 to 40 kcal/wk as in the reduction from 85 to 50 kcal/wk.

The rate of increase in mean lifetime associated with the reduction from 50 to 40 kcal/wk is rate2 $= (\mu_6 - \mu_3)/(50 - 40)$, which is estimated to be

$$\text{est. rate2} = \frac{(\text{Average lifetime on } N/\text{R40}) - (\text{Average lifetime on } N/\text{R50})}{(50 - 40)}$$

$$= \frac{(45.1 - 42.3)}{10} = 0.2800 \text{ months/[kcal/wk]}.$$

Similarly, the rate of increase associated with the reduction from 85 to 50 kcal/wk is rate1 $= (\mu_3 - \mu_2)/(85 - 50)$, which is estimated to be:

$$\text{est. rate1} = \frac{(\text{Average lifetime on } N/\text{R50}) - (\text{Average lifetime on } N/\text{N85})}{(85 - 50)}$$

$$= \frac{(42.3 - 32.7)}{35} = 0.2743 \text{ months/[kcal/wk]}.$$

Reducing caloric intake increased longevity in both instances, and the rates of increase appear to be about the same—about 0.28 month of extra lifetime for each kcal/wk reduction in caloric intake.

A formal comparison of the two rates will resolve whether the difference is real. The difference between the rates is estimated to be:

$$(\text{est. rate1} - \text{est. rate2}) = 0.2743 - 0.2800 = -0.0057 \text{ mo/[kcal/wk]}.$$

This is a linear combination of only three means, because μ_3 occurs in both rates. To get the correct coefficients for calculating the standard error, reduce the comparison as follows:

$$\gamma = (\text{rate1} - \text{rate2}) = \frac{(\mu_3 - \mu_2)}{35} - \frac{(\mu_6 - \mu_3)}{10}$$

$$= -\frac{1}{35}\mu_2 + \frac{9}{70}\mu_3 - \frac{1}{10}\mu_6.$$

This is a linear combination of three averages, so the standard error may be computed with the general formula of Section 6.2.2:

$$\text{SE}(g) = (6.68)\sqrt{\frac{\left[-\frac{1}{35}\right]^2}{57} + \frac{\left[+\frac{9}{70}\right]^2}{71} + \frac{\left[-\frac{1}{10}\right]^2}{60}}$$

$$= 0.1359 \text{ mo/[kcal/wk]}.$$

The estimate of σ, 6.68, is the pooled estimate from all six groups and has 343 degrees of freedom. The resulting t-statistic, $0.0057/0.1359 = 0.04$, provides no evidence that the two rates differ (two-sided p-value $= 0.97$). It might therefore be appropriate to estimate a common rate for increased lifetime over the entire 85 to 40 kcal/wk range.

Linear Trends

Sometimes the group means are associated with quantitative levels of an additional, *explanatory* variable. In the platyfish preference study, for example, the six groups correspond to different body sizes of the male pairs. A particular linear combination of group means may be used to assess the evidence for a linear trend in means as a function of the explanatory variable (for example, a linear trend in time with the yellow-sword male as a function of male body size).

Let X_i be the value of the explanatory variable associated with group i. Then the particular linear combination for linear trend happens to be the one that has $C_i = (X_i - \overline{X})$ as the coefficient of μ_i. The inference based on the linear combination will be unchanged if all the C_i's are multiplied by the same constant. It is tidier, therefore, to use as C_i some multiple of $(X_i - \overline{X})$, where the multiplier is chosen to make all coefficients integers or to express the linear combination in convenient units of measurement. Display 6.5 demonstrates the linear combination for comparing the model of linear trend (in percentage time with the yellow-sword male), against the more general model in which the means are unrestricted. In this case a convenient multiplier to make all coefficients integers is 2, so that C_i is $2(X_i - \overline{X})$.

DISPLAY 6.5 Analysis of the pre-existing preference example: F-test for differences in mean percentage of time spent with yellow-sword male, and t-test for linear effect of male body size

ANOVA F-Test

Source of variation	Sum of squares	d.f.	Mean square	F-statistic	p-value
Between male groups	938.75	5	187.75	0.786	0.56
Within groups	18,636.68	78	238.93		
Total	19,575.43	83			

Conclusion: *There is no evidence that the group means are different for different pairs of males (p-value = 0.56, from ANOVA F-statistic).*

t-Test for linear effect of body size

Group	n	Average (%)	Standard deviation	Male body size (mm)	Coefficient
Pair 1	16	56.41	9.02	35	5
Pair 2	14	60.89	12.48	31	−3
Pair 3	17	62.43	22.29	33	1
Pair 4	14	67.00	14.33	34	3 ← C's
Pair 5	9	64.21	9.41	28	−9
Pair 6	14	63.34	17.68	34	3
Pooled	84	62.13	15.46	Average = 32.5	

(1) *Calculate the coefficients for the linear combination.*

$$C_i = 2*(X_i - 32.5)$$

(2) *Calculate the effect's estimate*

$$g = (5)(56.41) + (-3)(60.89) + (1)(62.43) + (3)(67.00) + (-9)(64.21) + (3)(63.34) = -25.06$$

and its standard error.

$$\text{SE}(g) = (15.46)\sqrt{\frac{(5)^2}{16} + \frac{(-3)^2}{14} + \frac{(1)^2}{17} + \frac{(3)^2}{14} + \frac{(-9)^2}{9} + \frac{(3)^2}{14}} = 54.77$$

(3) *Calculate the t-statistic and determine the p-value.*

$$t\text{-statistic} = \frac{-25.06}{54.77} = -0.458$$

one-sided p-value = 0.32 (from t-distribution with 78 d.f.)

Conclusion: *There is no evidence of a linear association between group means and male body size (one-sided p-value = 0.32).*

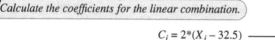

There are two tests shown in Display 6.5. The analysis of variance F-test indicates the evidence against the reduced, single-mean model as a special case of the full, six-mean model. The t-test for linear effect indicates the evidence against the reduced model in which the slope of a straight line function of body size is zero, as a special case of the model in which the slope is unrestricted.

Important note: It is useful to understand the linear combination for testing linear effect. For practical purposes, however, ***regression methods*** introduced in Chapter 7 provide a more complete data analytic process for investigating this type of structure.

Averages

Given the absence of any apparent differences between the mean percentages of times spent with the various yellow-sword males, the question of whether female platyfish prefer yellow-sword males becomes sensible globally. The average of the group means is another linear combination (but not a contrast), and the appropriate null hypothesis is that $(\mu_1 + \mu_2 + \cdots + \mu_6)/6 = 0.5$. The average of the group sample averages is 62.38%. Its standard error, 1.72%, is calculated in the same way from the general formula. This leads to the conclusive statement that the females were not dividing their time evenly between the males.

Caveat concerning average: The overall average just described differs slightly from the grand average of all female percentages (treated as a single group), because it gives equal weight to each male pair instead of equal weight to each female percentage. The conclusions will, in most cases, agree.

6.3 SIMULTANEOUS INFERENCES

A 95% confidence interval procedure is successful in capturing its parameter in 95% of its applications. When several 95% confidence intervals are considered simultaneously, they constitute a *family* of confidence intervals. The relative frequency with which all of the intervals in a family simultaneously capture their parameters is smaller than 95%. Because this rate is often of interest, the following distinction is drawn.

> ***Individual confidence level*** *is the success rate of a procedure for constructing a single confidence interval.*

> ***Familywise confidence level*** *is the success rate of a procedure for constructing a family of confidence intervals, where a "successful" usage is one in which all intervals in the family capture their parameters.*

If the family consists of k confidence intervals, each with individual confidence level 95%, the familywise confidence level can be no larger than 95% and no smaller than $100(1 - 0.05k)\%$. The actual familywise confidence level depends on the degree of dependence between the intervals.

The lower limit decreases rapidly with k. With a family of 10 confidence intervals, the familywise level could be as low as 50%, indicating the strong possibility that at least one of the intervals fails to capture its parameter. In other words, one should not suppose that all the confidence intervals capture their parameters, especially when a large number of intervals are being considered simultaneously.

The issue here is *compound uncertainty*—the increased chance of making at least one mistake when drawing more than one direct inference. Compound uncertainty also arises when many *tests* are considered simultaneously. The greater the number of tests performed, the higher the chance that a low *p*-value will be found for at least one of them, even in the absence of group differences. Consequently, the researcher is likely to find group differences that are not really there.

Multiple Comparisons

Multiple comparison procedures have been developed as ways of constructing individual confidence intervals so that the familywise confidence level is controlled (at 95%, for example). The important issue for the researcher to consider is whether to control the individual confidence levels or the overall confidence level.

Planned Comparisons, Unplanned Comparisons, and Data Snooping

Consider a one-way classification with 100 groups. One researcher may be particularly interested in comparing groups 23 and 78 because the comparison answers a research question directly. The researcher knows which groups are involved before seeing the data, and the comparison will be reported regardless of its statistical and practical significance. This constitutes a *planned comparison*. The individual confidence level should be controlled for planned comparisons.

Another researcher may examine differences between all possible pairs of groups—4,950 confidence intervals in all. As a result of these efforts, the researcher finds that groups 36 and 44 and groups 27 and 90 suggest actual group differences. Only these pairs are reported as significant. They exemplify *unplanned comparisons*. The familywise confidence level should be controlled for unplanned comparisons, since the uncertainty measure must incorporate the process of searching for important comparisons.

A third researcher notices that group 36 has the largest average and group 44 has the smallest, and presents only the single confidence interval—the one comparing group 36 to group 44. This is an instance of *data snooping*, in which the particular hypothesis or comparison chosen originates from looking at the data. The familywise confidence level should be controlled, for the same reason as in the second case.

The Spock trial, the diet restriction, and the platyfish preference studies all involved planned comparisons. The handicap study had no prespecified comparisons, so any comparisons reported in it should be treated as unplanned.

6.4 SOME MULTIPLE COMPARISON PROCEDURES

Confidence intervals for differences between pairs of means are centered at the difference between sample averages. Interval half-widths are computed as follows:

Interval half-width = (Multiplier) × (Standard error).

As usual the standard error of the difference is the pooled standard deviation times the square root of the sum of reciprocals of sample sizes.

There are many multiple comparison procedures, and these differ in their multipliers. The two highlighted in the ensuing subsections offer strict control over the familywise confidence levels for two important families.

6.4.1 Tukey–Kramer Procedure and the Studentized Range Distributions

The Tukey–Kramer procedure utilizes the unique structure of the multiple comparisons problem by selecting a multiplier from the *studentized range distributions* rather than from the t-distributions. The idea is to incorporate the search for the two most divergent sample averages directly into the statistical procedure.

Consider the case where all group means are equal and where all sample sizes are equal. The standard errors for all comparisons are the same and are equal to SE, say. A confidence interval is successful when it includes zero, so success occurs for a particular comparison when the magnitude of the difference between sample averages, $|\overline{Y}_i - \overline{Y}_j|$, is small. All such differences are less than $M \times \text{SE}$ if $(\overline{Y}_{\text{max}} - \overline{Y}_{\text{min}})$, the range of sample averages, is less than $M \times \text{SE}$. By selecting M in such a way that the chance of getting

$$(\overline{Y}_{\text{max}} - \overline{Y}_{\text{min}}) \leq M \times \text{SE}$$

is 95%, one guarantees that all intervals include zero 95% of the time. That is, the overall familywise confidence level is set at 95%.

A studentized range distribution describes values for the ratio of the range in I sample averages to the standard error of a single sample average, given that all samples are drawn from the same normal population. Tables of the studentized range distributions provide the $100 \times (1-\alpha)$th percentile, $q_{I,\text{d.f.}}(1-\alpha)$, depending on the number of groups (I) and the degrees of freedom (d.f.) for estimating σ. (In the one-way classification problem, d.f. $= n - I$.) The procedure originally proposed by Tukey—called Tukey's HSD, for "honest significant difference"—assumed an ideal normal model with equal spreads and also assumed equal sample sizes in all groups. The modification for unequal sample sizes provides confidence intervals with approximately the correct confidence levels, and goes by the name of the Tukey–Kramer procedure. The multiplier used in the interval half-width calculation is $[q_{I,n-I}(1-\alpha)]/\sqrt{2}$.

In the handicap study, $I = 5$ groups and $(n - I) = 65$ degrees of freedom. The 95th percentile in the corresponding studentized range distribution (interpolated between 60 and 120 d.f. in Table A.5) is 3.975, so the multiplier for constructing 95% confidence intervals is 2.8107.

6.4.2 Dunnett's Procedure

Dunnett proposed a multiple comparison procedure for comparing every other group to a reference group. This is often appropriate, such as for randomized experiments that compare several other treatments to a control. Dunnett realized that the t-statistics for comparing $I - 1$ other groups to a reference group were correlated due to the common appearance of the reference group average. To achieve a familywise error rate, the Dunnett procedure replaces the usual t-distribution with a multivariate t-distribution, which accounts for that correlation. Since there are fewer comparisons in this family than in the family of all possible pairwise comparisons, the Dunnett-adjusted confidence intervals of a given confidence level will be narrower than those from the Tukey–Kramer procedure.

Tables for the multivariate t-distribution aren't readily available, but are incorporated in computer routines for multiple comparisons with the Dunnett procedure. Such a routine was used to find the multiplier for 95% confidence intervals for comparing every other group to "None" (i.e., control) in Case Study 6.1.1. to be 2.5032. Notice that this falls between the unadjusted t-multiplier, 1.9971, and the Tukey–Kramer multiplier, 2.8107.

6.4.3 Scheffé's Procedure

Scheffé proposed the multiplier

$$\sqrt{(I-1)F_{(I-1),\text{d.f.}}(1-\alpha)},$$

where $F_{(I-1),\text{d.f.}}(1-\alpha)$ is the $(1-\alpha)$th percentile of the F-distribution with $I - 1$ and d.f. degrees of freedom. Here, $(I - 1)$ represents the between-group degrees of freedom, and d.f. is the within-group degrees of freedom. Scheffé's multiplier controls the overall confidence level for the family of parameters consisting of all possible linear contrasts among group means. When applied to the smaller family of differences between pairs of group means, the overall confidence level is *at least* $100(1-\alpha)\%$, and generally is higher. The Scheffé method finds a more appropriate application in providing intervals for regression curves, as will be seen in later chapters.

In the handicap study, $I - 1 = 4$, d.f. $= n - I = 65$, and the 95th percentile in the F-distribution is 2.513. The resulting multiplier is 3.1705.

6.4.4 Other Multiple Comparisons Procedures

The multiple comparisons procedures in this section present a range of options for balancing individual and familywise confidence levels.

The LSD

The familiar choice for a multiplier is the $100(1 - \alpha/2)\%$ critical value in the Student's t-distribution with degrees of freedom equal to those associated with the pooled SD. The resulting interval half-width is called the *least significant difference*, or LSD. The terminology arises naturally because any difference that exceeds the LSD in size is significant in a $100\alpha\%$-level hypothesis test. The multiplier for the handicap study is $t_{65}(0.975) = 1.9971$.

F-Protected Inferences

The method known as "protected LSD" is a simple and widely used alternative for testing unplanned comparisons. It is a two-step procedure, as follows:

1. Perform the ANOVA F-test.

2. **(a)** If the p-value from the F-test is large (>0.05, say), do not declare any individual difference significant, even though some differences appear large enough to be declared real.

(b) If the p-value from the F-test is small, proceed with individual comparisons, as in Section 5.2, using t-tests or confidence intervals with the t-multiplier.

In the handicap study, the p-value from the ANOVA test was 0.03, so the F-protected comparison plan would proceed to step 2(b).

Although one should compute confidence intervals whether or not the F-test's p-value is small, there is no convenient method for F-protecting confidence intervals. It is tempting to use a t-multiplier in 2(b) and to substitute either the Tukey–Kramer or Scheffé multipler in 2(a), depending on the family. That practice, however, would control neither the individual nor the familywise success rate at 95%.

Bonferroni

If the confidence level for each of k individual comparisons is adjusted upward to $100(1 - \alpha/k)\%$, the chance that all intervals succeed simultaneously is at least $100(1-\alpha)\%$. This result is an application of the Bonferroni inequality in probability theory. Using the Student's t-multiplier $t_{\text{d.f.}}(1-\alpha/2k)$ allows the user to be "at least $100(1-\alpha)\%$ confident" that all intervals succeed. In a multiple comparisons problem involving I groups, there are $k = I(I-1)/2$ pairs of means to be compared. The exact confidence level for the Bonferroni intervals is not generally as predictable as for the Tukey–Kramer intervals in the multiple comparisons problem. Bonferroni intervals may be used in a wider range of problems, however, including some situations where the Tukey–Kramer approach is not appropriate. With the five groups in the handicap study, $k = 10$, so the t-multiplier is the 99.75th percentile in the t-distribution with 65 d.f., or 2.9060.

Others

The *Newman–Keuls* procedure also employs studentized range distributions, but with different multipliers for different ranges. *Duncan's multiple range* procedure

DISPLAY 6.6 Summary of 95% confidence interval procedures for differences between treatment means in the handicap study

Group	Average	Difference with . . . Hearing	Amputee	Control	Wheelchair
Crutches	5.921	1.871	1.492	1.021	0.578
Wheelchair	5.343	1.293	0.914	0.443	
Control	4.900	0.850	0.471		
Amputee	4.429	0.379			
Hearing	4.050				

A confidence interval is centered at a difference with half-width given by one of the procedures.

Procedure	95% interval half-width
LSD	1.233
Dunnett	1.545 (for comparisons with control only)
Tukey–Kramer	1.735
Bonferroni	1.794
Scheffé	1.957

extends the Newman–Keuls procedure, with a Bonferroni protective correction to the nominal level.

6.4.5 Multiple Comparisons in the Handicap Study

The pooled estimate of the standard deviation of the data in Display 6.1 is 1.633. All groups have the same sample size (14) so the standard error for any and all differences between sample averages is

$$\mathrm{SE}(\overline{Y}_i - \overline{Y}_j) = 1.633\sqrt{\frac{1}{14} + \frac{1}{14}} = 0.6172.$$

The other relevant information is as follows: There are $I = 5$ groups, so there are $k = 10$ different comparisons, and the degrees of freedom for the standard error are 65.

Display 6.6 summarizes the ten 95% confidence intervals computed according to the multiple comparisons methods described earlier. Since sample sizes are all the same, confidence intervals have the same width for all comparisons under each method. The upper part of Display 6.6 shows the centers of the confidence intervals. The lower part shows the interval half-widths.

If the researchers' intent is to compare each of the handicap groups to the control group, then the Dunnett procedure is appropriate. In this case, none of the 95% confidence intervals for the handicap minus control difference exclude zero, so the data provide no evidence of a handicap effect. If, however, their intent is to compare every group to every other group, then the Tukey–Kramer procedure is appropriate. It suggests one difference—between the *hearing* and *crutches* groups.

Recall, however, that comparing the combined *crutches* and *wheelchair* group with the combined *amputee* and *hearing* group (Display 6.4) revealed a very clear difference. Making that comparison was suggested by an examination of the data; by analogy, the interval here should also be widened by using the Scheffé multiplier in place of the t-multiplier. When this is done, the confidence interval for the contrast is from 0.011 to 2.775, which still excludes zero. Because the Scheffé method incorporates the search among linear contrasts for the most significant, one should conclude that strong evidence exists that this difference is real.

6.4.6 Choosing a Multiple Comparisons Procedure

The LSD is the most liberal procedure (narrowest confidence intervals), and the Scheffé procedure is the most conservative (widest confidence intervals). Bonferroni and Tukey–Kramer procedures occupy intermediate positions. Aside from conducting these general comparisons, the best approach is to think carefully about whether it is desirable to control the familywise confidence level and, if so, what the appropriate family of comparisons includes. If the answer to the first question is no, then standard t-tools apply. If the family of comparisons includes all differences of other groups with a reference group, the Dunnett method gives the appropriate control. If the family of comparisons includes pairwise differences of all group means, the Tukey–Kramer method gives precise control. If the family includes a large number of comparisons and no other method seems feasible, the Bonferroni method—although conservative—can always be applied.

6.5 RELATED ISSUES

6.5.1 Reasoning Fallacies Associated with Statistical Hypothesis Testing and p-Values

p-values indicate the ***strength of evidence*** from data in support of an alternative to a null hypothesis. Although p-value reasoning is both natural and logical, widespread misinterpretations and misuses cause many scientists and statisticians to advocate their elimination from scientific discourse. The abuses are easily understood, though. For the sake of preserving a tool that is very useful when used correctly, it is important for students of statistics to recognize and avoid the misinterpretations. To this end, Display 6.7 lists the important ones. Scientists that understand this display are unlikely to error in their interpretations of p-values.

6.5.2 Example of a Hypothesis Based on How the Data Turned Out

Although not a one-way classification problem, the following example demonstrates the need to incorporate data snooping into the assessment of uncertainty. The letters in Display 6.8 represent 2,436 mononucleotides in a DNA molecule. Mononucleotides come in four varieties—A, C, G, and T—and their sequence along the DNA strand forms a molecule's genetic code. DNA molecules break, drift for a time, and

DISPLAY 6.7 Common fallacies of reasoning from statistical hypothesis tests

Fallacy name	The fallacy	Avoiding the fallacy
False Causality Fallacy	Incorrectly interpreting statistical significance (i.e., a small *p*-value) from an observational study as evidence of causation	Use the word *association* to indicate a relationship that is not necessarily a causal one.
Fallacy of Accepting the Null	Incorrectly interpreting a lack of statistical evidence that a null hypothesis is false (i.e., a large *p*-value) as statistical evidence that the null hypothesis is true	Avoid this incorrect wording: "the study provides evidence that there is no difference." Say instead: "there is no evidence from this study of a difference." Also, report a confidence interval to emphasize the many possible hypothesized values (in addition to 0) that are consistent with the observed data.
Confusing Statistical for Practical Significance	Interpreting a "statistically significant" effect (which has to do with the strength of evidence that there's an effect) as a practically important one, which it may or may not be	If you must use the term *statistically significant*, don't abbreviate it. Also, report a confidence interval so that the *size* of an effect can be evaluated for its practical importance.
Data Dredging (Fishing for Significance, Data Snooping)	Incorrectly drawing conclusions from an unadjusted *p*-value that emerged from a process of sifting through many possible *p*-values Note: "Publication Bias" is the de facto data dredging that results if journals only accept research papers with statistically significant findings.	For multiple comparisons of means, use the adjustments in this chapter. For identifying a few from many possible predictor variables, use the variable selection methods in Chapter 12. For tests based on many different response variables (data mining), use the False Discovery Rate methods of Chapter 16.
Good Statistics from Bad Data	Incorrectly accepting conclusions based on sound statistical technique when there are problems with data collection, such as biased sampling or data contamination	Critically evaluate the potential biases from non–randomly selected samples.

DISPLAY 6.8 2,436 mononucleotides along a DNA molecule. All 40 occurrences of the trinucleotide TGG appear in boldface. Eleven breaks occurred in the string, at the positions indicated by dashes.

TAAAGAAACATAAATGCCGATATTTGTTAATACTGTGTACTGTAAGAATATATTAGCATTGT
CTATGACTAAGAATTCAAAACAATTATTGATGCTATAGGG**TGG**CAATATAATAGTCAATTC
TACGATATTGAAAAAGTTATCTCCTTACTTTCGCACACATTTACGTCAAAAATACACGAAA
ATAAAGATCCAGTTACT**TGG**GTTTGTCTAGACCTTGACATTCACAGTTTAACTTCTATAGTT
ATTTACTCGTATAC**TGG**AAAGGTATATATAATAGTCATAACGTCGTCAATTTATTA-CGTGC
TTCTATATTAACCTCTGTAGAATTTATCATCTACACTTGTATAAACTTTATCTTACGAGATTT
TAGAAAGGAATATTGTGTCGAGTGTTACATGA**TGG**G–TATATAATACGGACTATCCAATCTC
TTATGTCATACTAAAAACTTTATTGCCAAACACTTTT**TGG**AAC**TGG**AAGATGACATCATAG
ACAATTTTGATTATCTATCTATGAAACTTATTCTAGAAAGCGATGAACTAAATGTTCCAGAT
GAGGATTATGTAGTTGATTTTGTCATTAAG**TGG**TATATAAAGCGAAGAAATAAATTAGGAA
ATCTGCTACTCCTTATCAAAAATGTAATCAGGTCAAATTATCTTTCTCCCAGAGGTATAAAT
AATGTAAAA**TGG**ATACTAGACTGTACCA–AAATATTTCATTGTGATAAACAACCACGCAAA
TCATACAAGTATCCATTCATAGAGTATCCTATGAACA**TGG**ATCAAATTATAGATATATCCA
TATGTGTACAAGTACTCATGT**TGG**AGAAGTAGTATATCTCATCGG**T–GG**A**TGG**ATGAACAA
TGAAATACATAACAATGCTATAGCGGTAAATTATATATCAAACAAT**TGG**AT–TCCAATTCCT
CCGATGAATAGCCCCAGACTGTATGCTAGCGGGATACCCGCTAACA–ATAAATTATACGTAG
TAGGAGGTCTACCAAATCCCACATCTGTTGAGCGT**TGG**T–TCCACGGGGATGCTGCT**TGG**G
TTAATATGCCGAGTCTTCTGAAACCTAGATGTAATCCAGCAG**TGG**C–ATCCATAAACAATGT
TATATACGTAA**TGG**GAGGACATTCTGAAACTGATACAACTACAGAATATTTGCTACCCAAT
CATGATCAG**TGG**CAGTT**TGG**ACCATTCCACTTATTATCCTCATTATAAATCATGCGCGTTAG
TGTTCGGTAGAAGGTTATTCT**TGG**T**TGG**TAGAAATGCGGAATTTTATTGTGAATCCAGCAA
TACA**TGG**CTCTGATAGATGATCCTATTTATCCGAGGGATAATCCAGAATTGATCATAG**TGG**
ATAATAAACTGCTATTGATAGGAGGATTTAATCGTGCATCGTATATAGATACTATAGAAGT
GTACATCATCACACTTATTCA**TGG**AATATA**TGG**GA**TGG**TAAATAATTTTGAAATAAAATAT
TAGTTTTATGTTCAACATGAATATTAAC–TCACCAGTTAGATTTGTTAAGGAAACTAACAGA
GCTAAATCTCCTACTAGGCAATCACCTTACGCCGCCGGATATGATTTATATAGCGCTTACGA
TTATACTATCCCTCCAGGAGAACGACAGTTAATTAAGACAGATATTAGTATGTCCATGCCT
AAGTTCTGCTA**TGG**TAGAATAGCTCCTAGGTC**TGG**TCTGTCCCTAAAAGGCATTGATATAG
GAGGCGGTGTAATAGACGAAGATTATAGGGGAAACATAGGAGTCATTCTTATTAATAA**TG**
GAAAATGTACGTTTAATGTAAATAC**TGG**AGATACAATAGCTCAGCTAATCTATCAACGTAT
ATATTATCCAGAAC**TGG**AAGAAGTACAATCTCTAGATAGTACAAATAGAGGAGATCAAGG
GTT**T–GG**ATCAACAGGACTTAGATAATAAACAATAGTATGTTGTCGATGTTTATAGTGTAAT
AATATCGTAGATTATGTAGATGATATAGATAA**TGG**TATAGTACAGGATATAGAAGATGAG
GCTAGCAATAATGTTGATCACGACTATGTATATCCACTTCCAGAAAATA**TGG**TATATAGAT
TTGACAAGTCCACTAACATACTCGATTATCTATCAACGGAACGGGACCATGTAATGA**TGG**C
TGTTCGATACTATATGAGTAAACAACGTTTAGACGACTTGTATAGACAGTTGCCCACAAAG
ACTAGATCATATATAGATATTATCAACATATATTGTGATAAAGTTAGTAATGATTATAATAG
GGACATGAATATCATGTATGA–TA**TGG**CATCTACAAAATCATTTACAGTTTATGACATAAAT
AACGAAGTTAATACTATACTAA**TGG**ATAACAAGGGGT**TGG**GTGTAAGAT**TGG**CGACAATT
TCATTTATAACCGAAT**TGG**GTAGACGATGTATGA

then recombine with other strands to form new molecules. The molecule shown in Display 6.8 has broken in 11 places, indicated by the dashes between two mononucleotides. Three consecutive mononucleotides form a trinucleotide, which functions as a genetic word. In this molecule, the 40 occurrences of the trinucleotide TGG are shown in boldface. In line 7, a break point has appeared shortly after an occurrence of TGG, which is of interest because TGG may indicate an increased likelihood of a break to follow.

If a break occurs between any of the mononucleotides of TGG plus the four following, the break is said to be *downstream* from TGG. And in this particular molecule, 6 of the 11 breaks were downstream from a TGG trinucleotide. The question is, does this evidence support TGG as a precursor of breaks in the DNA?

Summary of Statistical Analysis

Two different analyses are possible, depending on whether the hypothesis that TGG was specifically indicated arose prior to viewing this molecule (planned comparison) or whether TGG was actually suggested by examination of this molecule (data snooping).

1. Of the 2,435 possible break positions in the entire string, 235 (9.65%) are downstream from TGG trinucleotides. If breaks occurred at *random* positions, about 9.65% of the 11 breaks (that is, one) would be downstream from TGG. But in fact, 6 of the 11 (54.55%) occurred downstream from TGG. That would seem to be too many to have happened by chance. If breaks occurred at random positions, the possibility that six or more would occur at positions downstream from TGG trinucleotides is precisely 0.000243. This is very small, showing strong evidence of an association between the occurrences of breaks and TGG.

The p-value of 0.000243 assumes, however, that, in every trial determining 11 breaks, the number downstream from TGG is counted. What if the focus on TGG were the result of a search on this molecule for the trinucleotide occurring most frequently upstream of the existing breaks? *If the search for the trinucleotide was an integral part of the statistical analysis of this data set, the series of trials used to evaluate the evidence must also include the search procedure.*

2. The computer was programmed to conduct a trial incorporating the search. It randomly selected 11 break points from the 2,435 possible, and searched upstream from all the breaks to find the trinucleotide appearing most frequently. Then it recorded the frequency of the occurrence of this trinucleotide. The computer ran 1,000 of these trials. Of course, the most frequently occurring trinucleotide differed considerably from trial to trial. The key piece of evidence is the distribution of the frequency of the most frequent upstream trinucleotide, which appears in Display 6.9. In 320 of the 1,000 trials, some trinucleotide was found upstream from the 11 breaks 6 times or more. In one trial, the same trinucleotide occurred upstream 11 times. Thus, having some trinucleotide upstream from many of the breaks in a molecule appears to be rather common! Consequently, if the search was conducted on this molecule, the p-value for an observed highest frequency of 6 is $p = 0.32$.

DISPLAY 6.9 Simulated estimate of the distribution of the highest frequency of occurrence of any trinucleotide upstream of 11 randomly selected breaks

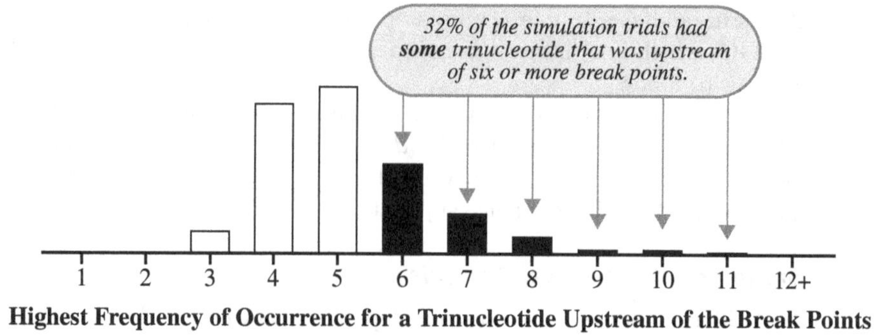

Scope of Inference

This example illustrates the tremendous difference between evaluating the strength of evidence about a preplanned hypothesis and evaluating the strength of evidence about a hypothesis suggested by the data themselves. Correct evaluation of *post hoc* hypotheses is possible, but it must incorporate the actual procedure by which the hypothesis was formulated. (Data from M. B. Slabaugh, N. A. Roseman, and C. K. Mathews, "Amplification of the Ribonucleotide Reductase Small Subunit Gene: Analysis of Novel Joints and the Mechanism of Gene Duplication in Vaccinia Virus," Biochemistry and Biophysics Department Report, Oregon State University, Corvallis, Oregon (1989).)

How valuable is the correct p-value? These researchers did "discover" TGG from this molecule. Had they accepted the strength of evidence 0.000243, they might have wasted considerable effort devising explanations for what turned out to be a nonrepeatable phenomenon.

6.5.3 Is Choosing a Transformation a Form of Data Snooping?

The Statistical Sleuth emphasizes examining graphical displays for clues to possible data transformation. When a stem-and-leaf diagram suggests a logarithmic transformation, should the resulting inferences take the data-selected transformation into account? The answer is yes. Uncertainty about which transformation (if any) is best is part of the problem, so a measure of uncertainty regarding the question of interest should incorporate uncertainty about the form of the statistical model.

In practice, however, most researchers do not incorporate model uncertainty into uncertainty measures—largely because doing so is extremely difficult. In later chapters (in particular, Chapter 12), *The Statistical Sleuth* will investigate some methods that have been devised for solving this problem. For now, the scope of inference should be limited to replicates where the transformation is selected. To caution a reader about additional uncertainty, the statistical summary should explain clearly all the steps taken to select the transformation.

6.6 SUMMARY

Many researchers routinely examine an analysis of variance F-statistic for group differences; if it is significant, they proceed directly to multiple comparison methods to search for differences. In most cases, this is a mistake. Usually the groups have an inherent structure related to levels of specific factors set by the researcher. Linear combinations of group means can be devised to answer questions related to this structure.

A host of studies fall within the general class of one-way classifications. Yet the statistical analysis is not the same for all. A major distinction must be made between studies in which the group structure calls for specific planned comparisons and those in which the only question of interest is which means differ from which others. Selecting an appropriate statistical procedure requires the researcher to evaluate honestly whether hypotheses were clearly stated prior to data collection or whether the data themselves guided the formulation of hypotheses. In the latter case, proper statistical evaluation must acknowledge the data-snooping process.

This chapter demonstrates some statistical tools for assessing uncertainty when a family of inferences is desired. Selecting an appropriate tool requires some introspection about the nature of the family examined. This chapter also introduced a powerful tool—computer simulation—that can help evaluate evidence about more complex hypotheses suggested by the data.

6.7 EXERCISES

Conceptual Exercises

1. Handicap Study. (a) Is it possible that the applicant's handicap in the videotape is confounded with the actor's performance? (b) Is there a way to design the study to avoid this confounding?

2. Mate Preference of Platyfish. If $\mu_1, \mu_2, \ldots, \mu_6$ represent the mean percentage of time spent by females with the yellow-sword male, for the six pairs of males, (a) state the null and alternative hypotheses that are tested by (i) the analysis of variance F-test and (ii) the t-test for the hypothesis that the linear contrast (for the linear effect of male body size) is zero. (b) Say why it is possible that the second test might find evidence that the means are different even if the first does not.

3. Mate Preference of Platyfish. For the test that the mean percent of time females spent with yellow-sword males is 50%, a one-tailed p-value was reported. Why?

4. An experimenter takes 20 samples of bark from each of 10 tree species in order to estimate the differences between fuel potentials. The data give 10 species averages, the lowest being 1.6 Btu/lb and the highest 3.8 Btu/lb for a range of 2.2 Btu/lb. A colleague suggests that another species be included, so the experimenter plans to gather 20 samples from that species and calculate its average potential. Which of the following is true about the range that the 11 species averages will have when the new species is included? (a) The range will equal the old range, 2.2 Btu/lb. (b) The range will be larger than 2.2 Btu/lb. (c) The range will be smaller than 2.2 Btu/lb. (d) The range cannot be smaller than 2.2 Btu/lb. (e) The range cannot be larger than 2.2 Btu/lb. (f) It is not possible to say that any of the above options is true until the average is known.

5. O-Ring Data. The case study in Section 4.1.1 involved the numbers of O-ring events on U.S. space shuttle flights launched at temperatures above and below 65°F. In the context of this chapter, is anything suspicious about that data? (*Hint*: Is there a possibility of data snooping?)

6. Does a confidence interval for the difference between two groups use information about variability from other groups? Why? or Why not?

7. What is the distinction between planned and unplanned comparisons?

8. Does a planned comparison always consist of estimating the difference between the means in two groups?

9. In comparing 10 groups a researcher notices that $\overline{Y}_7$ is the largest and $\overline{Y}_3$ is the smallest, and then tests the hypothesis that $\mu_7 - \mu_3 = 0$. Why should a multiple comparison procedure be used even though there is only one comparison being made?

10. If the analysis of variance screening test shows no significant evidence of any group differences, does that end the issue of there being any differences to report?

DISPLAY 6.10 Test scores for the experimental CAD instruction course

Group	Logo	Teaching method	n	Average	SD
1	L+D	Lecture and discussion	9	30.20	3.82
2	R	Programmed text	9	28.80	5.26
3	R+L	Programmed text with lectures	9	26.20	4.66
4	C	Computer instruction	9	31.10	4.91
5	C+L	Computer instruction with lectures	9	30.20	3.53

11. When choosing coefficients for a contrast, does the choice of $\{C_1, C_2, \ldots, C_I\}$ give a different t-ratio than the choice of $\{3C_1, 3C_2, \ldots, 3C_I\}$?

Computational Exercises

12. Handicap Study. Consider the groups *amputee*, *crutches*, and *wheelchair* to be handicaps of mobility and *hearing* to be a handicap affecting communication. Use the appropriate linear combination to test whether the average of the means for the mobility handicaps is equal to the mean of the communication handicap.

13. Handicap Study. Use the Bonferroni method to construct simultaneous confidence intervals for $\mu_2 - \mu_3, \mu_2 - \mu_5$, and $\mu_3 - \mu_5$ (to see whether there are differences in attitude toward the mobility type of handicaps).

14. Handicap Study. Examine these data with your available statistical computer package. See what multiple comparison procedures are available within the one-way analysis of variance procedure. Verify the 95% confidence interval half-widths in Display 6.6.

15. Comparison of Five Teaching Methods. An article reported the results of a planned experiment contrasting five different teaching methods. Forty-five students were randomly allocated, nine to each method. After completing the experimental course, a one-hour examination was administered. Display 6.10 summarizes the scores on a 10-minute retention test that was given 6 weeks later. (Data from S. W. Tsai and N. F. Pohl, "Computer-Assisted Instruction Augmented with Planned Teacher/Student Contacts," *Journal of Experimental Education*, 49(2) (Winter 1980–81): 120–26.)

(a) Compute the pooled estimate of the standard deviation from these summary statistics.
(b) Determine a set of coefficients that will contrast the methods using programmed text as part of the method (groups 2 and 3) with those that do not use programmed text (1, 4, and 5).
(c) Estimate the contrast in (b) and compute a 95% confidence interval.

16. A study involving 36 subjects randomly assigned six each to six treatment groups gives an ANOVA F-test with p-value $= 0.0850$. What multipliers are used to construct 95% confidence intervals for treatment differences with the following methods: (i) LSD, (ii) F-protected LSD, (iii) Tukey–Kramer, (iv) Bonferroni, and (v) Scheffé?

17. Adder Head Size. Red Riding Hood: "My, what big teeth you have!" Big Bad Wolf: "The better to eat you with, my dear." Are predators morphologically adapted to the size of their prey? A. Forsman studied adders on the Swedish mainland and on groups of islands in the Baltic Sea to determine if there was any relationship between their relative head lengths (RHL) and the body size of their main prey, field voles. (Data from A. Forsman, "Adaptive Variation in Head Size in *Vipera berus* L. Populations," *Biological Journal of the Linnean Society* 43 (1991): 281–96.) Relative head length is head length adjusted for overall body length, determined separately for males and females. Field vole body size is a combined measure of several features, expressed on a standardized scale.

DISPLAY 6.11 Average relative head lengths of adders from seven μ Swedish localities with their distances to the mainland and the body sizes of prey

Locality	Sample size	Average relative head length	Distance (km) to mainland	Field vole body size
Uppsala	21	−6.98	0	−1.75
In-Fredeln	34	−4.24	25.1	
Inre Hamnskär	20	−2.79	13.4	−0.16
Norrpada	25	2.22	14.7	1.31
Kärringboskär	7	1.27	10.0	
Ängskär	82	1.88	22.7	1.67
Svenska Hägarna	48	4.98	39.6	2.17

The data appear in Display 6.11. The pooled estimate of standard deviation of the RHL measurements was 11.72, based on 230 degrees of freedom.

(a) Determine the half-widths of 95% confidence intervals for all 21 pairwise differences among means for the seven localities, using (i) the LSD method and (ii) the Tukey–Kramer method.

(b) Using a linear contrast on the groups for which vole body size is available, test whether the locality means (of relative head length) are equal, against the alternative that they fall on a straight line function of vole body size, with nonzero slope.

(c) Repeat (b) for the pattern of distances to the mainland rather than vole body size.

18. Nest Cavities. Using the nest cavity data in Exercise 5.19, estimate the difference between the average of the mean entry areas for flickers, screech-owls, and kestrels and the average of the mean entry areas for the other six animals (on the transformed scale). Use a contrast of means.

19. Diet Restriction. For the data in Display 5.1 (and the summary statistics in Display 5.2), obtain a 95% confidence interval for the difference $\mu_3 - \mu_2$ using the Tukey–Kramer procedure. How does this interval differ from the LSD interval? Why is the Tukey–Kramer procedure the wrong thing to use for this problem?

20. Equity in Group Learning. [Continuation of Exercise 5.22.] (a) To see if the performance of low-ability students increases steadily with the ability of the best student in the group, form a linear contrast with increasing weights: -3 = Low, -1 = Low–Medium, $+1$ = Medium–High, and $+3$ = High. Estimate the contrast and construct a 95% confidence interval. (b) For the High-ability students, use multiple comparisons to determine which group composition differences are associated with different levels of test performance.

21. Education and Future Income. Reconsider the data problem of Exercise 5.25 concerning the distributions of annual incomes in 2005 for Americans in each of five education categories. (a) Use the Tukey–Kramer procedure to compare every group to every other group. Which pairs of means differ and by how many dollars (or by what percent)? (Use p-values and confidence intervals in your answer.) (b) Use the Dunnett procedure to compare every other group to the group with 12 years of education. Which group means apparently differ from the mean for those with 12 years of education and by how many dollars (or by what percent)? (Use p-values and confidence intervals in your answer.)

22. Reconsider the measurements of oxygen composition in 12 dinosaur bones from Exercise 5.23. Using a multiple comparisons procedure in a statistical computer package, find 95% confidence intervals for the difference in means for all pairs of bones (a) without adjusting for multiple comparisons and (b) using the Tukey–Kramer adjusted intervals. (c) How many of the unadjusted intervals exclude zero? (d) How many of the Tukey–Kramer adjusted intervals exclude zero? (e) By how much does the width of the adjusted interval exceed the width of the unadjusted interval, as a percentage, for comparing bone 1 to bone 2?

Data Problems

23. **Diet Wars.** To reduce weight, what diet should one combine with exercise? Most studies of human dieting practices have faced problems with high dropout rates and questionable adherence to diet protocol. Some studies comparing low-fat and low-carbohydrate diets have found that low-carb diets produced weight loss early, but that the loss faded after a short time. In an attempt to exert more control on subject adherence, a team of researchers at Ben-Gurion University in Negev, Israel, conducted a trial within a single workplace where lunch—the main meal of the day—was provided by the employer under the guidance of the research team. The team recruited 322 overweight employees and randomly assigned them to three treatment groups: a low-fat diet, a low-carb diet (similar to the Atkins diet), and a Mediterranean diet. Trans-fats were discouraged in all three diets. Otherwise, the restrictions and recommendations were as follows:

	Low-fat ($n = 104$)	**Mediterranean** ($n = 109$)	**Low-carbohydrate** ($n = 109$)
Calorie/day restriction	Women: 1,500 kcal Men: 1,800 kcal	Women: 1,500 kcal Men: 1,800 kcal	(Not Specified)
Percentage of calories from fat	30%	35%	(not specified)
Carbohydrates/day	(not specified)	(not specified)	20 g at start; increasing to 120 g
Percentage of calories from saturated fat	10%	(not specified)	(not specified)
Cholesterol/day	300 mg	(not specified)	(not specified)
Recommended:	low-fat grains vegetables fruits legumes	30–45 g olive oil 5–7 nuts < 20 g vegetables fish and poultry	get fat and protein from vegetables
Discouraged:	added fats	beef and lamb sweets high-fat snacks	

The study ran for two years, with 272 employees completing the entire protocol. Display 6.12 shows some of a data set simulated to match the weight losses (kg) of the participants at the study's conclusion. Is there evidence of differences in average weight loss after two years among these diets? If so, which diets appear to be better than which others? (Notice the consequences of controlling the family-wise confidence level on the widths of 95% confidence intervals.)

24. **A Biological Basis for Homosexuality.** Is there a physiological basis for sexual preference? Following up on research suggesting that certain cell clusters in the brain govern sexual behavior, Simon LeVay (data from S. LeVay, "A Difference in Hypothalamic Structure Between Heterosexual and Homosexual Men," *Science*, 253 (August 30, 1991): 1034–37) measured the volumes of four cell groups in the interstitial nuclei of the anterior hypothalamus in postmortem tissue from 41 subjects at autopsy from seven metropolitan hospitals in New York and California. The volumes of one cell

DISPLAY 6.12 Partial listing of data from a diet and weight loss experiment, showing subject number (there were 272 subjects), diet treatment group (there were 3), and weight loss (in kg) after 24 months

Subject	Group	WtLoss24
1	Low-Fat	2.2
2	Low-Fat	−4.8
3	Low-Fat	2.9
. . .		
105	Mediterranean	10.8
106	Mediterranean	6.4
107	Mediterranean	−0.3
. . .		
214	Low-Carbohydrate	3.4
215	Low-Carbohydrate	4.8
216	Low-Carbohydrate	10.9

DISPLAY 6.13 Volumes of INAH3 (1,000 × mm^3) cell clusters from 41 human subjects at autopsy, by sex, sexual orientation, and cause of death

Males				Females	
Heterosexual		Homosexual		Heterosexual	
AIDS death	Non-AIDS death	AIDS death		AIDS death	Non-AIDS death
12	20	1	34	12	10
105	37	7	39		19
105	103	12	41		29
118	129	15	46		105
119	135	18	66		155
161	140	18	86		
	161	23	128		
	175	26	142		
	179	29	193		
	209	32			

cluster, INAH3, are re-created in Display 6.13. The numbers are 1,000 times volumes in mm^3. Subjects are classified into five groups according to three factors: gender, sexual orientation, and cause of death. One male classified as a homosexual who died of AIDS (volume 128) was actually bisexual. LeVay used the term ***presumed*** heterosexual to indicate the possibility of misclassifying some subjects. Do heterosexual males tend to differ from homosexual males in the volume of INAH3? Do heterosexual males tend to differ from heterosexual females? Do heterosexual females tend to differ from homosexual males? Analyze the data and write a brief statistical report including a summary of statistical findings, a graphical display, and a details section describing the details of the particular methods used. Also describe the limitations of inferences that can be made. (*Hint:* What linear combination of the five means can be used to test whether cause of death can be ignored? If cause of death can be ignored, what linear combinations of the resulting three means are appropriate for addressing the questions above?)

Answers to Conceptual Exercises

1. (a) Yes, it is possible that the acting performance of the actor portrayed a more competent worker in the *crutches* role, even though the script was held constant. (b) With two pairs of actors and twice as many groups, the handicap effect could be isolated from the actor effect.

2. (a) (i) Hypothesis: all means are equal; alternative: at least one is different from the others. (ii) Hypothesis: all means are equal: alternative: the means are not equal but fall on a straight line function of male body size (with nonzero slope). (b) The alternative in (ii) is more specific. For a given set of data there is more power in detecting differences if the (correct) specific alternative can be investigated. (This has previously been noticed by the fact that a one-tailed p-value is smaller than a two-tailed.)

3. The researcher had reason to believe, because of other species of fish in the same genus, that the colored tail would be more attractive to the females.

4. If the new species average is somewhere between 1.6 and 3.8 Btu/lb, the range of the set of means is unchanged. If the new species average is either less than 1.6 or greater than 3.8 Btu/lb, the range is larger. Those are the only possibilities. So the answer is (d): the range cannot be smaller than the old range (but it could be larger). The range increases as the number of groups increase.

5. Where did the 65°F cutoff come from? If the analyst chose that cutoff because it produced the most dramatic difference between the two groups, the search procedure should be included in the assessment of evidence.

6. Yes, it does. It is important to pool information about variability because the population SD is difficult to estimate from small samples.

7. A planned comparison is one of a few specific comparisons that is designed to answer a question of interest. An unplanned comparison is one (of a large number) of comparisons that is suggested by the data themselves.

8. No. More complex comparisons can be made by examining linear combinations of group means.

9. The more groups there are, the larger the difference one expects between the smallest and largest averages. To incorporate the selection of this hypothesis on the basis of how the data turned out, the appropriate statistical measure of uncertainty is the same as the one that is appropriate for comparing every mean to every other mean.

10. Not necessarily. If there are no planned comparisons, it may be best to report no evidence of differences (protected LSD procedure). But the evidence about planned comparisons to answer questions of interest should be assessed on its own. It is possible that a planned comparison shows something when the F-test does not.

11. No. The parameter changes from γ to 3γ, the estimate changes from g to $3g$, and the standard error also changes from $\mathrm{SE}(g)$ to $\mathrm{SE}(3g) = 3\mathrm{SE}(g)$. So the t-ratio is not changed at all. This is why one can take the convenient step of multiplying a set of coefficients by a common factor to make all coefficients into integers, if desirable.

CHAPTER 7

Simple Linear Regression: A Model for the Mean

Chapters 5 and 6 advocated the investigation of specific questions of interest for the several-sample problem by paying attention to the structure of the grouped data. When the different groups correspond to different levels of a quantitative explanatory variable, the idea can be extended with the *simple linear regression* model, in which the means fall on a straight line function of the explanatory variable.

Such a model has several advantages when it is appropriate. It offers a concise summary of the mean of the response variable as a function of the explanatory variable through two parameters: the slope and the intercept of the line. For a surprisingly large number of data problems this model is appropriate (possibly after transformation) and the questions of interest can be conveniently reworded in terms of the two parameters. The parameters are estimable even without replicate values at each distinct level of the explanatory variable.

This chapter presents the model and some associated inferential tools. The next chapter takes a closer look at the simple linear regression model assumptions and how to measure departures from them.

7.1 CASE STUDIES

7.1.1 The Big Bang—An Observational Study

Edwin Hubble used the power of the Mount Wilson Observatory telescopes to measure features of nebulae outside the Milky Way. He was surprised to find a relationship between a nebula's distance from earth and the velocity with which it was going away from the earth. Hubble's initial data on 24 nebulae are shown as a *scatterplot* in Display 7.1. (Data from E. Hubble, "A Relation Between Distance and Radial Velocity Among Extra-galactic Nebulae," *Proceedings of the National Academy of Science* 15 (1929): 168–73.) The horizontal axis measures the recession velocity, in kilometers per second, which was determined with considerable accuracy by the red shift in the spectrum of light from a nebula. The vertical scale measures distance from the earth, in megaparsecs: 1 megaparsec is 1 million parsecs, and 1 parsec is about 30.9 trillion kilometers. Distances were measured by comparing mean luminosities of the nebulae to those of certain star types, a method that is not particularly accurate. The data are shown later in this chapter as Display 7.8.

DISPLAY 7.1 Scatterplot of measured distance versus velocity for 24 extra-galactic nebulae

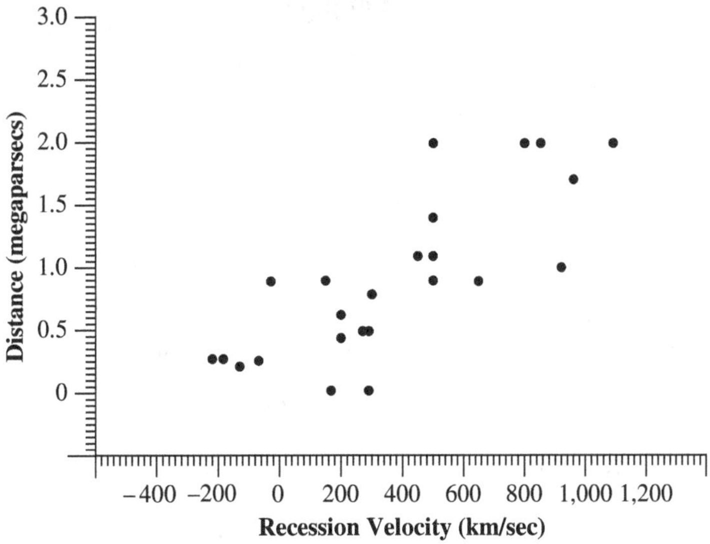

The apparent statistical relationship between distance and velocity led scientists to consider how such a relationship could arise. It was proposed that the universe came into being with a Big Bang, a long time ago. The material in the universe traveled out from the point of the Big Bang, and scattered around the surface of an expanding sphere. If the material were traveling at a constant velocity (V) from the point of the bang, then the earth and any nebulae would appear as in Display 7.2.

DISPLAY 7.2 Big Bang theory model for distance–velocity relationship of nebulae

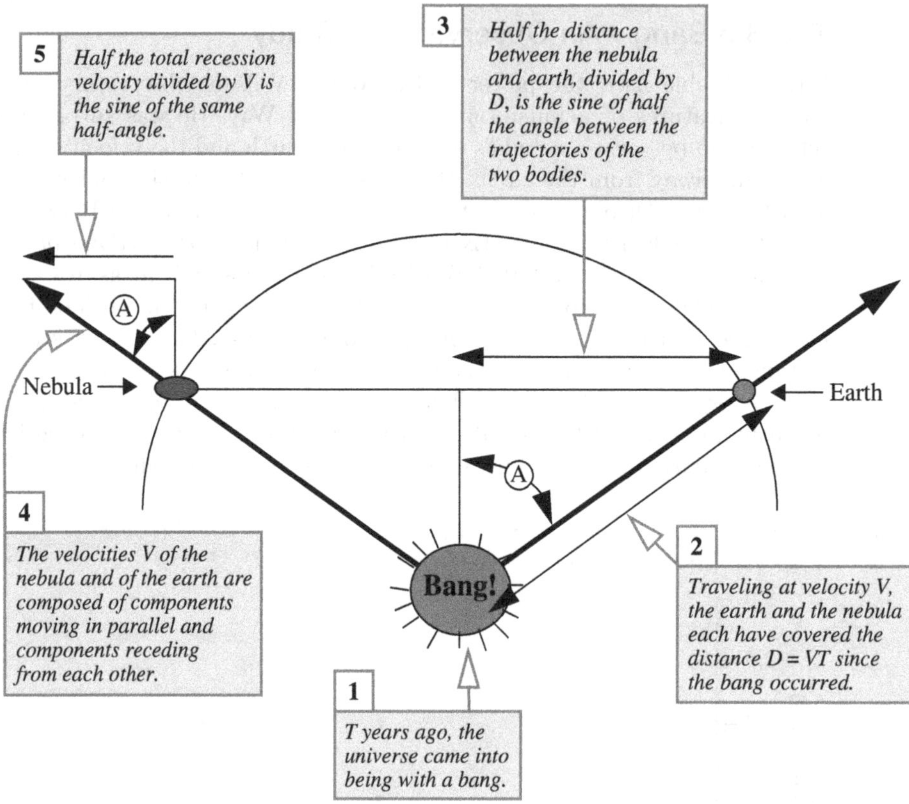

The distance (Y) between them and the velocity (X) at which they appear to be going away from each other satisfy the relationship

$$(Y/2)/VT = (X/2)/V = \sin(A),$$

where A is half the angle between them. In that case,

$$Y = TX$$

is a straight line relationship between distance and velocity. The points in Display 7.1 do not fall exactly on a straight line. It might be, however, that the *mean* of the distance measurements is TX. The slope parameter T in the equation $\text{Mean}\{Y\} = TX$ is the time elapsed since the Big Bang (that is, the age of the universe).

Several questions arise. Is the relationship between distance and velocity indeed a straight line? Is the y-intercept in the straight line equation zero, as the Big Bang theory predicts? How old is the universe?

Statistical Conclusion

If the theory is taken as correct, then the estimated age of the universe is 0.001922 megaparsecs-seconds per kilometer, or about 1.88 billion years (estimate of slope in simple linear regression through the origin). A 95% confidence interval for the age is 1.50 to 2.27 billion years. However, the data are not consistent with the Big Bang theory as proposed. Although the relationship between mean measured distance and velocity approximates a straight line, the value of the line at velocity zero is apparently not zero, as predicted by the theory (two-sided p-value $= 0.0028$ for a test that the intercept is zero).

Scope of Inference

These data are not a random sample, so they do not necessarily represent what would result from taking measurements from other nebulae. This analysis assumes that there is an exact linear relationship between distance and velocity, but that the measured distances do not fall exactly on a straight line because of measurement errors. The confidence interval above summarizes the uncertainty that comes from errors in distance measurements. Uncertainty due to errors in measuring velocities is not included in the p-values and confidence coefficients. Such errors are a potential source of pronounced bias. (*Note:* The Big Bang theory is still intact. Astronomers now estimate the universe to be 13.7 billion years old.)

7.1.2 Meat Processing and pH—A Randomized Experiment

A certain kind of meat processing may begin once the pH in postmortem muscle of a steer carcass decreases to 6.0, from a pH at time of slaughter of around 7.0 to 7.2. It is not practical to monitor the pH decline for each animal, so an estimate is needed of the time after slaughter at which the pH reaches 6.0. To estimate this time, 10 steer carcasses were assigned to be measured for pH at one of five times after slaughter. The data appear in Display 7.3. (Data from J. R. Schwenke and

DISPLAY 7.3 pH of 10 steer carcasses measured at five different times after slaughter

Steer	Time after slaughter (hours)	pH
1	1	7.02
2	1	6.93
3	2	6.42
4	2	6.51
5	4	6.07
6	4	5.99
7	6	5.59
8	6	5.80
9	8	5.51
10	8	5.36

DISPLAY 7.4 Meat processing data with estimated regression line (from the simple linear regression of pH on log time after slaughter) and a 95% prediction band

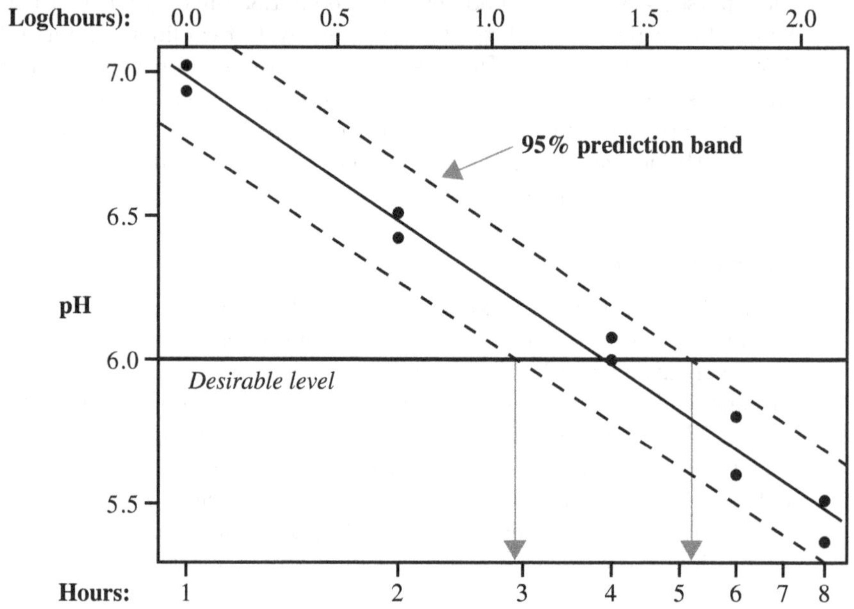

G. A. Milliken, "On the Calibration Problem Extended to Nonlinear Models," *Biometrics* 47 (1991): 563–74.)

Statistical Conclusion

The estimated relationship between postmortem muscle pH and time after slaughter is summarized in Display 7.4. The solid line shows the estimated mean pH as a function of the logarithm of time after slaughter. The dotted lines are the upper and lower endpoints of 95% prediction intervals for the pH of steer carcasses at times after slaughter in the range of 1 to 8 hours. It is estimated that the mean pH at 3.9 hours is 6. It is predicted that at least 95% of steer carcasses will reach a pH of 6.0 sometime between 2.94 and 5.10 hours after slaughter (95% calibration interval).

7.2 THE SIMPLE LINEAR REGRESSION MODEL

7.2.1 Regression Terminology

Regression analysis is used to describe the distribution of values of one variable, the *response*, as a function of other—*explanatory*—variables. This chapter deals with *simple* regression, where there is a single explanatory variable.

Think of a series of subpopulations of responses, one for each value of the explanatory variable, such as the subpopulation of pH values of steer carcasses, 4 hours after slaugher. The *regression of the response variable on the explanatory variable* is a mathematical relationship between the *means* of these subpopulations and the explanatory variable. The *simple linear regression* model specifies that this relationship is a straight line function of the explanatory variable. As evident in Display 7.4, a useful model for the meat processing data has the means of the subpopulations of pH falling on a straight line function of the logarithm of time after slaughter.

Let Y and X denote, respectively, the response variable and the explanatory variable. The notation $\mu\{Y|X\}$ will represent the regression of Y on X, and it should be read as "the mean of Y as a function of X." When a specific value, $X = x$, is given to the explanatory variable, the same expression should be read as "the mean of Y when $X = x$." The simple linear regression model is

$$\mu\{Y|X\} = \beta_0 + \beta_1 X.$$

The equation involves two statistical parameters. β_0 is the *intercept* of the line, measured in the same units as the response variable. β_1 is the *slope* of the line, equal to the rate of change in the mean response per one-unit increase in the explanatory variable. The units of β_1 are the ratio of the units of the response variable to the units of the explanatory variable.

A simple linear regression model for the statistical relationship between Y = Distance and X = Velocity in the Big Bang example is:

$$\mu\{\text{Distance}|\text{Velocity}\} = \beta_0 + \beta_1 \text{Velocity},$$

in which β_0 and β_1 are the intercept and slope of the regression line. The intercept is in units of megaparsecs and the slope is in units of megaparsecs per (km/sec).

A similar notation is $\sigma\{Y|X\}$ for the standard deviation of Y as a function of X. The equal-SD assumption specifies that $\sigma\{Y|X\} = \sigma$, for all X.

The complete picture of a probability model emerges if, as depicted in Display 7.5, each of the subpopulations has a normal distribution. This model has only three unknown parameters—β_0, β_1, and σ. Notice that the model describes the distributions of responses, but it says nothing about the distribution of the explanatory variable. In various applications, X-values may be chosen by the researcher, or they may themselves be random.

The independence assumption means that each response is drawn independently of all other responses from the same subpopulation and independently of all responses drawn from other subpopulations. The common types of violations are cluster effects and serial effects, as before.

7.2.2 Interpolation and Extrapolation

The simple linear regression model makes it possible to draw inference about any mean response within the range of the explanatory variable. Statements about the

DISPLAY 7.5 The ideal normal, simple linear regression model

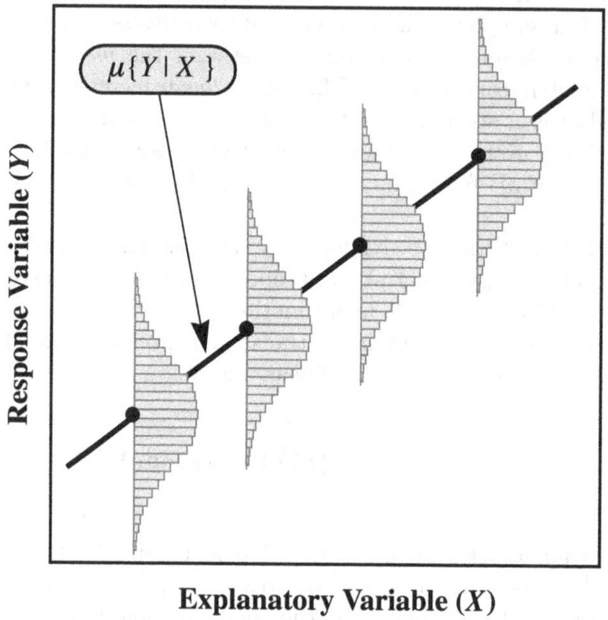

Model Assumptions

1. **There is a normally distributed subpopulation of responses for each value of the explanatory variable.**
2. **The means of the subpopulations fall on a straight line function of the explanatory variable.**
3. **The subpopulation standard deviations are all equal (to σ).**
4. **The selection of an observation from any of the subpopulations is independent of the selection of any other observation.**

mean at values of X not in the data set, but within the range of the observed explanatory variable values, are called *interpolations*. The ability to interpolate is a strong advantage to regression analysis. Making statements for values outside of this range—*extrapolation*—is potentially dangerous, however, because the straight line model is not necessarily valid over a wider range of explanatory variable values.

The regression of a response variable on an explanatory variable is not necessarily a straight line. The simple regression model is very useful, though, because the regression is often well approximated by a straight line in the region of interest or is approximated by a straight line after transformation.

The remainder of this chapter assumes the simple regression model is appropriate for the case studies. Model checking, transformations, and alternatives will be discussed in the next chapter.

7.3 LEAST SQUARES REGRESSION ESTIMATION

The method of least squares is one of many procedures for choosing estimates of parameters in a statistical model. The method was described by Legendre in 1805, but the German mathematician Karl Gauss had been using it since 1795 and is generally regarded as its originator. The method provides the foundation for nearly all regression and analysis of variance procedures in modern statistical computer software.

7.3.1 Fitted Values and Residuals

The hat notation (^) denotes an estimate of the parameter under the hat, so $\hat{\beta}_0$ and $\hat{\beta}_1$ are symbols for estimates of β_0 and β_1, respectively. Once these estimates are determined, they combine to form an estimated mean function,

$$\hat{\mu}\{Y|X\} = \hat{\beta}_0 + \hat{\beta}_1 X.$$

This general expression may be calculated for each X_i in the data set to estimate the mean of the distribution from which the corresponding response Y_i arose. The estimated mean is called the *fitted value* (or sometimes *predicted value*) and is represented by fit_i. The difference between the observed response and its estimated mean is the *residual*, denoted by res_i:

$$fit_i = \hat{\mu}\{Y_i|X_i\} = \hat{\beta}_0 + \hat{\beta}_1 X_i.$$

$$res_i = Y_i - fit_i$$

These definitions are illustrated in Display 7.6 for a simple example.

DISPLAY 7.6 Illustration of the residual and fitted value for observation (X_4, Y_4) in a hypothetical data set of size 4

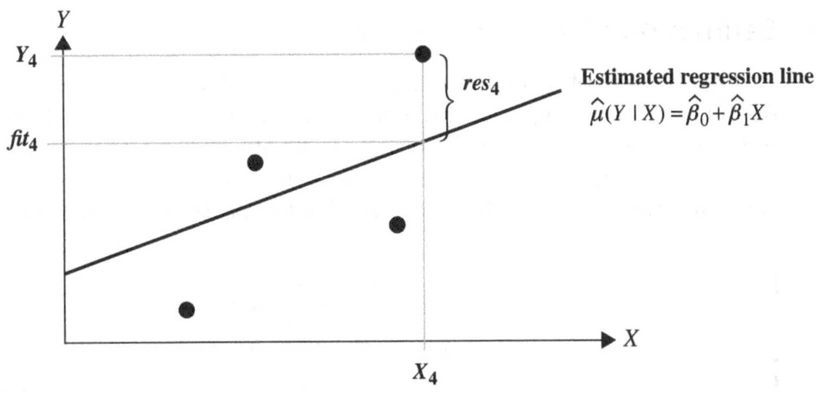

The magnitude of a single residual measures the distance between the corresponding response and its fitted value according to the estimated mean function. A good estimate should result in a small distance. A measure of the distance between *all* responses and their fitted values is provided by the *residual sum of squares.*

7.3.2 Least Squares Estimators

Least squares estimates of β_0 and β_1 are those values for the intercept and slope that minimize the sum of squared residuals (hence "least squares"). The solution to this fitting criterion is found by calculus to be:

$$\hat{\beta}_1 = \frac{\sum_{i=1}^{n}(X_i - \overline{X})(Y_i - \overline{Y})}{\sum_{i=1}^{n}(X_i - \overline{X})^2}, \qquad \hat{\beta}_0 = \overline{Y} - \hat{\beta}_1\overline{X}$$

(the interested reader is referred to Exercise 26). Statistical computer packages carry out the arithmetic, as shown for the Big Bang data in Display 7.9.

7.3.3 Sampling Distributions of the Least Squares Estimators

The sampling distributions of $\hat{\beta}_0$ and $\hat{\beta}_1$ are the distributions of least squares estimates from hypothetical repeated sampling of response measurements at the same (fixed) values of the explanatory variable. In the meat processing example, a single repetition consists of obtaining pH readings from two new steers at each of the times 1, 2, 4, 6, and 8 hours after slaughter. If the normal, simple linear regression model applies, then Display 7.7 summarizes what statistical theory concludes about the sampling distribution.

7.3.4 Estimation of σ from Residuals

The parameters of primary interest are β_0 and β_1, but the sampling distributions of their estimators depend on the parameter σ. Therefore, even if there is no direct interest in the variation of the observations about the regression line, an estimate of σ must be obtained in order to assess the uncertainty about estimates of the slope and the intercept. Residuals provide the basis for an estimate:

$$\hat{\sigma} = \sqrt{\frac{\textit{Sum of all squared residuals}}{\textit{Degrees of freedom}}},$$

DISPLAY 7.7 Facts about the sampling distributions of the least squares estimators of slope and intercept in the ideal normal model (from statistical theory)

Sampling of Distribution of $\hat{\beta}_1$

3 | SHAPE
The shape of the sampling distribution is normal.

1 | Center
The mean of the sampling distribution is β_1.

β_1

2 | Spread
$$SD(\hat{\beta}_1) = \sigma\sqrt{\frac{1}{(n-1)s_X^2}}$$
where s_X^2 is the sample variance of the X's.

Sampling Distribution of $\hat{\beta}_0$

3 | Shape
The shape of the sampling distribution is normal.

1 | Center
The mean of the sampling distribution is β_0.

β_0

2 | Spread
$$SD(\hat{\beta}_0) = \sigma\sqrt{\frac{1}{n}+\frac{\overline{X}^2}{(n-1)s_X^2}}$$

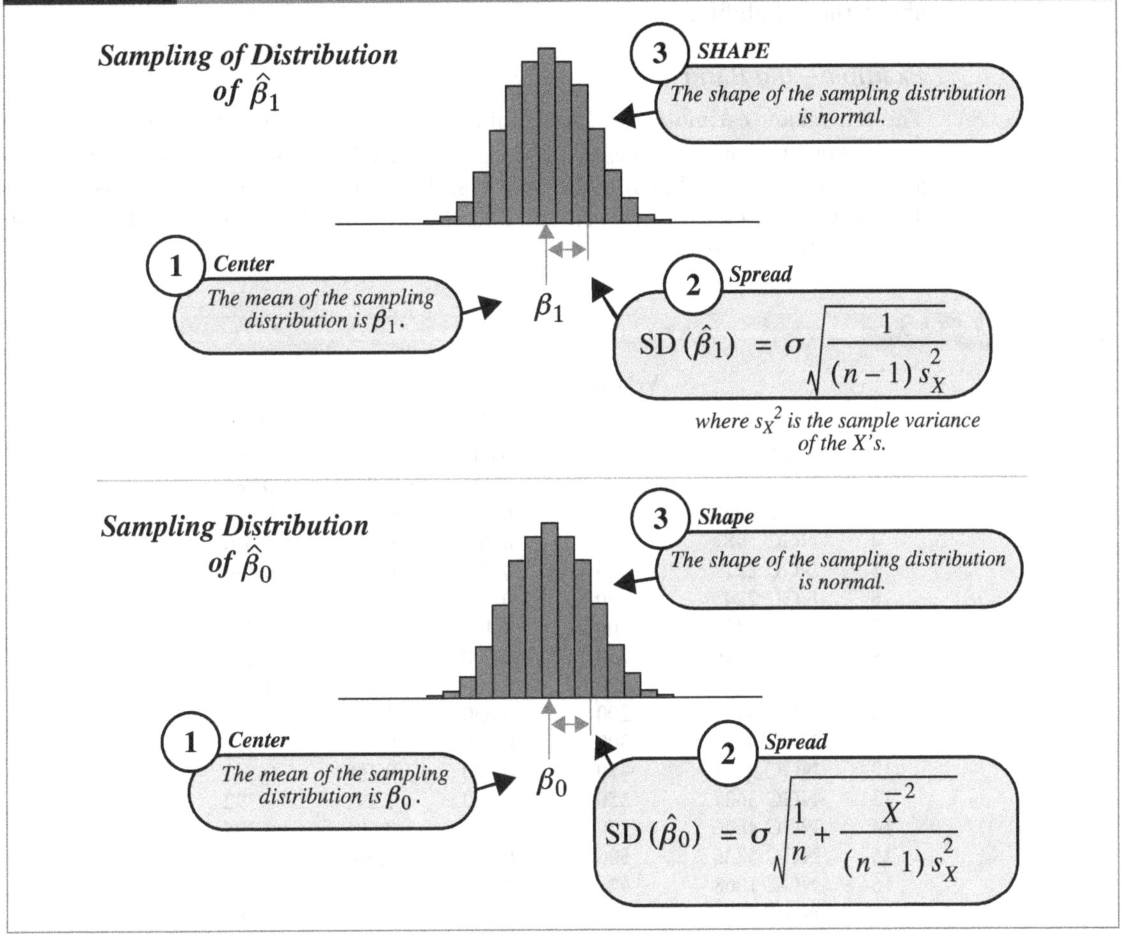

where the degrees of freedom are the sample size minus *two*, $n-2$. This is another instance of applying the general rule for finding the degrees of freedom associated with residuals from estimating a model for the means.

> **Rule for Determing Degrees of Freedom**
> *Degrees of freedom associated with a set of residuals =*
> *(Number of observations) −*
> *(Number of parameters in the model for the means).*

To understand the rule's application to simple linear regression, it is helpful to think about the case where $n = 2$. The least squares estimated line connects the two points, and both residuals are zero. There are no degrees of freedom for estimating variance. Only by adding a third or more observations is there any information about the variability.

Example—Big Bang

The parameter estimates for the Hubble data are: $\hat{\beta}_0 = 0.3991$ megaparsecs and $\hat{\beta}_1 = 0.001373$ megaparsecs/(km/sec). The fitted values and residuals from the least squares fit are shown in Display 7.8. The sum of the 24 squared residuals is 3.6085, with 22 degrees of freedom. The estimate of σ^2 is 0.164, and the estimate of σ is 0.405 megaparsecs.

DISPLAY 7.8 Fitted values and residuals for the Big Bang study

i	Nebula	Velocity X_i	Distance Y_i	fit_i	res_i
1	S. Mag.	170	0.032	0.632	−0.600
2	L. Mag.	290	0.034	0.797	−0.763
3	NGC 6822	−130	0.214	0.221	−0.007
4	NGC 598	−70	0.263	0.303	−0.040
5	NGC 221	−185	0.275	0.145	0.130
6	NGC 224	−220	0.275	0.097	0.178
7	NGC 5457	200	0.450	0.674	−0.224
8	NGC 4736	290	0.500	0.797	−0.297
9	NGC 5194	270	0.500	0.770	−0.270
10	NGC 4449	200	0.630	0.674	−0.044
11	NGC 4214	300	0.800	0.811	−0.011
12	NGC 3031	−30	0.900	0.358	0.542
13	NGC 3627	650	0.900	1.292	−0.392
14	NGC 4626	150	0.900	0.605	0.295
15	NGC 5236	500	0.900	1.086	−0.186
16	NGC 1068	920	1.000	1.662	−0.662
17	NGC 5055	450	1.100	1.017	0.083
18	NGC 7331	500	1.100	1.086	0.014
19	NGC 4258	500	1.400	1.086	0.314
20	NGC 4151	960	1.700	1.717	−0.017
21	NGC 4382	500	2.000	1.086	0.914
22	NGC 4472	850	2.000	1.566	0.434
23	NGC 4486	800	2.000	1.497	0.503
24	NGC 4649	1090	2.000	1.896	0.104

7.3.5 Standard Errors

Standard errors for the intercept and slope estimates are obtained by replacing σ with $\hat{\sigma}$ in the formulas for their standard deviations (see Display 7.7). The degrees of freedom assigned to these standard errors are $n - 2$, the same as those attached to the residual estimate of the standard deviation:

$$\text{SE}(\hat{\beta}_1) = \hat{\sigma}\sqrt{\frac{1}{(n-1){s_X}^2}}, \quad \text{d.f.} = n-2$$

and

$$\text{SE}(\hat{\beta}_0) = \hat{\sigma}\sqrt{\frac{1}{n} + \frac{\overline{X}^2}{(n-1){s_X}^2}}, \quad \text{d.f.} = n-2,$$

where ${s_X}^2$ is the sample variance of the X's.

Example—Big Bang

Display 7.9 shows a table that is typical of computer output for regression. Separate rows show the estimates, standard errors, and t-tests for the hypotheses that parameter values are zero. The first row corresponds to β_0, the coefficient of the oxymoronic *constant variable*. The second row corresponds to β_1, the coefficient of the variable *velocity*.

DISPLAY 7.9 Regression parameter estimates for the Big Bang study

Variable	Coefficient	Standard error	t-statistic	p-value
Constant	0.3991	0.1185	3.369	0.0028
Velocity	0.001373	0.000227	6.036	0.0000045

Estimate of $\sigma = 0.4050$ (22 d.f.)

Ratios of coefficient estimates to their standard errors

For hypotheses that the coefficients = 0

Another recommended way to report the results of regression analysis is to present the estimated equation with standard errors in parentheses below the corresponding parameter estimates:

$$\begin{aligned} \text{Estimated Mean Distance} = {} & \underset{(0.1185)}{0.3991} + \underset{(0.000227)}{0.001373}\ (\text{Velocity}) \\ \text{Estimated SD of Distances} = {} & 0.405 \text{ megaparsec (22 d.f.)} \end{aligned}$$

7.4 INFERENTIAL TOOLS

7.4.1 Tests and Confidence Intervals for Slope and Intercept

Inferences are based on the t-ratios, $(\hat{\beta}_0 - \beta_0)/\mathrm{SE}(\hat{\beta}_0)$ and $(\hat{\beta}_1 - \beta_1)/\mathrm{SE}(\hat{\beta}_1)$, which have t-distributions with $n-2$ degrees of freedom under the conditions of the normal simple linear regression model. As usual, *tests* are appropriate when the question is whether the data are consistent with some particular hypothesized value for the parameter, as in "Is 7 a possible value for β_0?" *Confidence intervals* are appropriate for presenting a range of values for a parameter that are consistent with the data, as in "What are the possible values of β_0?"

The p-values for the tests that β_0 and β_1 are individually equal to zero are included as standard output from statistical computer packages, as shown in Display 7.9. Although the test that β_1 is zero is often very important—so as to judge whether the mean of the response is unrelated to the explanatory variable—neither of the reported tests necessarily answers a relevant question. To obtain p-values for testing other hypothesized values or to construct confidence intervals, the user must perform other calculations based on the estimates and their standard errors. Some examples follow.

Does Hubble's Data Support the Big Bang Theory? (Test That β_0 Is 0)

According to the theory discussed in Section 7.1.1, the true distance of the nebulae from earth is a constant times the recession velocity, so, if Y is measured distance, $\mu\{Y|X\} = \beta_1 X$. Thus, if the simple theory is right, the linear regression of measured distance on velocity will have intercept zero. Since the relationship does indeed appear to be a straight line, a check on the theory is accomplished by a test that β_0 is 0. From Display 7.9 the two-sided p-value for this test is reported as 0.0028, providing convincing evidence that these data do not support the theory as stated.

What Is the Age of the Universe According to the Big Bang Theory? (Confidence Interval for β_1)

According to the Big Bang theory, $\mu\{Y|X\} = \beta_1 X$, and β_1 is the age of the universe. To estimate β_1 with the value (0.001373) reported in Display 7.9 would be wrong, because that estimate refers to a different model—one with an intercept parameter—in which the slope parameter has a somewhat different meaning. A least squares fit to the Big Bang model—without the intercept parameter—gives an estimate for β_1 of 0.001922 megaparsec-second per km, with a standard error of 0.000191 based on 23 degrees of freedom. Since $t_{23}(0.975) = 2.069$, the confidence interval for β_1 is $0.001922 \pm (2.069)(0.000191)$, or from 0.001527 to 0.002317, or 1.50 to 2.27 billion years. The best estimate is 1.88 billion years. (One megaparsec-second per kilometer is about 979.8 billion years.)

7.4.2 Describing the Distribution of the Response Variable at Some Value of the Explanatory Variable

At some specified value, $X = X_0$, for the explanatory variable, the response distribution has mean $\beta_0 + \beta_1 X_0$ and standard deviation σ. For example, the distribution of pH levels of steers 4 hours after slaughter will have mean $\beta_0 + \beta_1 \log(4)$. Not all steers will have this pH; the variability of individual pH's around this mean is represented by the standard deviation σ. With least squares estimates, $\hat{\beta}_0 = 6.98$, $\hat{\beta}_1 = -0.7257$, and $\hat{\sigma} = 0.0823$, it can be estimated that the distribution of pH's at 4 hours has a mean of $6.98 - 0.7257 \log(4) = 5.98$ and an SD of 0.0823.

The Standard Error for the Estimated Mean

The standard error of $\hat{\beta}_0 + \hat{\beta}_1 X_0$, for some specific number X_0, is

$$\mathrm{SE}[\hat{\mu}\{Y|X_0\}] = \hat{\sigma}\sqrt{\frac{1}{n} + \frac{(X_0 - \overline{X})^2}{(n-1)s_X{}^2}}, \quad \text{d.f.} = n = 2.$$

Once this standard error is computed, a confidence interval or a test for the mean of Y at X_0 follows according to standard t-tools. In addition to $\hat{\sigma}$, the formula requires the average and the sample variance of the explanatory variable values in the data set. The calculations, as demonstrated on the meat processing example in Display 7.10, are straightforward, but inconvenient to do by hand.

DISPLAY 7.10 95% confidence interval for the mean pH of steers 4 hours after slaughter (from the estimated regression of pH on log(time) after slaughter for the meat processing data)

$$\hat{\mu}\{Y \mid 1.386\} = 6.9836 - 0.7257 \times 1.386 = 5.98$$

$\hat{\sigma}$; $X_0 = \log(4)$; $\overline{X}$

$$\mathrm{SE}[\hat{\mu}\{Y|1.386\}] = 0.08226\sqrt{\frac{1}{10} + \frac{(1.386 - 1.190)^2}{9\,(0.6344)}}$$

$$= 0.0269$$

n ; $(n-1)$; $s_X{}^2$; $t_8(0.975)$

Upper limit: $5.98 + 2.306 \times 0.0269 = 6.04$

Lower limit: $5.98 - 2.306 \times 0.0269 = 5.92$

Computer Centering Trick

A computer trick can be used to bypass the hand calculations. First, create an artificial explanatory variable, $X^* = X - X_0$, by subtracting X_0 from all explanatory variable values. This centers X at X_0, in the sense that the value $X = X_0$ becomes the value $X^* = 0$. Next, fit the simple linear regression of Y on X^*. The *intercept* in the linear regression model $\mu\{Y|X^*\}$ is the mean of Y at $X^* = 0$, which is exactly the mean of Y at $X = X_0$. Therefore, you may now read the standard error for the estimated mean directly off the computer output as the standard error for the intercept parameter. This trick extends to many more complicated situations.

Additional Notes About the Estimated Mean at Some Value of X

1. *SE for estimated intercept.* Notice that for $X_0 = 0$, the formula for the standard error of the estimated mean, above, reduces to the formula for the standard error of the estimated intercept.
2. *Precision in estimating $\mu\{Y|X\}$ is not constant for all values of X.* The formula for the standard error indicates that the precision of the estimated mean response decreases as X_0 gets farther from the sample average of X's.
3. *Compound uncertainty in estimating several means simultaneously.* This section has addressed the situation where there is one single value of X_0 at which the mean response is desired. The construction of a 95% confidence interval is such that 95% of repetitions of the sampling process result in intervals that include the correct mean response *at that specified* X_0. But if one constructs different 95% confidence intervals for the mean responses at two values of X, the proportion of repetitions that result in both intervals correctly covering the respective means is something less than 95%. This is another case of compound uncertainty. To account for compound uncertainty in estimating mean responses at k different values of X, the Bonferroni procedure (Section 6.4.3) can be used. If k is very large, or if it is desired to obtain a procedure that protects against an unlimited number of comparisons, the Scheffé method can be used to construct a confidence band, as discussed next.
4. *Where is the regression line? (The Workman–Hotelling procedure)* Suppose the question is, "What is the mean response at all X-values within the observational range?" That is, where is the entire line, as opposed to where is one specific value on the line? By the device of replacing the t-multiplier in confidence intervals with a Scheffé multiplier based on an F-percentile, the confidence interval for the mean response can be converted to a *confidence band*, having the property that at least 95% of the repetitions mentioned above produce bands that include the correct mean response *everywhere*. If the 95th percentile in the F-distribution based on 2 and $(n-2)$ degrees of freedom is $F_{2,n-2}(0.95)$, then the 95% confidence band has upper and lower bounds at each X given by

$$\hat{\mu}\{Y|X\} \pm \sqrt{2 \times F_{2,n-2}(0.95)} \times \text{SE}[\hat{\mu}\{Y|X\}].$$

A 95% confidence band for the regression line in the Hubble example uses $F_{2,22}(0.95) = 3.443$, so the multiplier for the standard error is 2.624 instead of the t-multiplier of 2.074. Band limits are calculated for different values of X and the confidence band is obtained by connecting all the upper endpoints and all lower endpoints, as shown in Display 7.11. Notice that the confidence band is not substantially wider than the intervals for individual means. The *prediction band* in the display is discussed in Section 7.4.3.

DISPLAY 7.11 The 95% confidence band on the population regression line, the 95% confidence interval band for single mean estimates, and a 95% prediction interval band for the Big Bang example

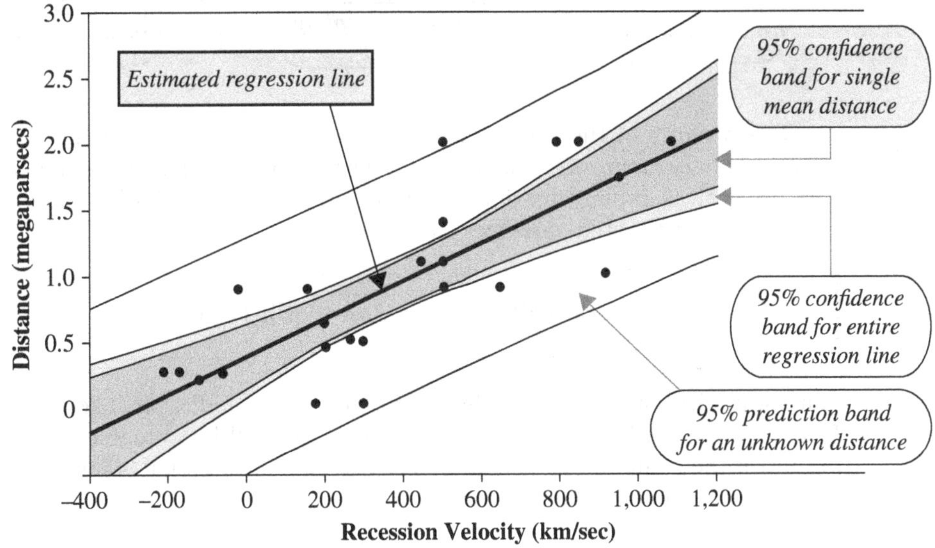

7.4.3 Prediction of a Future Response

Confidence intervals indicate likely values for parameters. Another inferential tool is a *prediction interval*, which indicates likely values for a future value of a response variable at some specified value of the explanatory variable.

Notice the difference between these two questions: (1) What is the *mean pH* 4 hours after slaughter? (2) What will be the *pH of a particular steer carcass* 4 hours after slaughter? The first question is about a parameter and the second is about an individual within the population. Even if the mean and standard deviation of pH at 4 hours after slaughter are known exactly, it is impossible to predict exactly the pH of an individual steer.

The best single prediction of a future response at X_0, denoted by $\text{Pred}\{Y|X_0\}$, is the estimated mean response at X_0:

$$\text{Pred}\{Y|X_0\} = \hat{\mu}\{Y|X_0\} = \hat{\beta}_0 + \hat{\beta}_1 X_0.$$

A *prediction interval* is an interval of likely values, along with a measure of the likelihood that the interval will include the future response value. This measure must account for two independent sources of uncertainty: uncertainty about the location of the subpopulation mean, and uncertainty about where the future value will be in relation to its mean. The error in using the predictor above is

$$Y - \text{Pred}\{Y|X_0\} = Y - \hat{\mu}\{Y|X_0\}$$

$$= [Y - \mu\{Y|X_0\}] - [\hat{\mu}\{Y|X_0\} - \mu\{Y|X_0\}].$$

$$\boxed{\text{Prediction error}} = \boxed{\text{Random sampling error}} + \boxed{\text{Estimation error}}$$

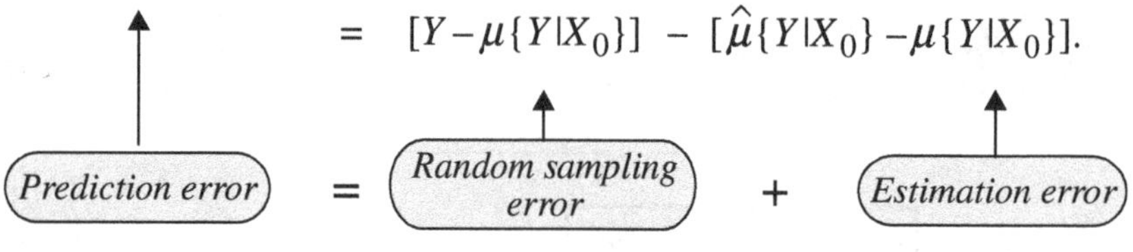

The variance of the random sampling error, as evident from the definition of the simple linear regression model, is σ^2. The variance of the estimation error is the variance of the sampling distribution of $\hat{\beta}_0 + \hat{\beta}_1 X_0$. The standard error of prediction combines the estimated variances as

$$\text{SE}[\text{Pred}\{Y|X_0\}] = \sqrt{\hat{\sigma}^2 + \text{SE}[\hat{\mu}\{Y|X_0\}]^2}.$$

Some statistical programs provide standard errors of prediction at requested values. If not, both quantities under the square root sign will be available from computer output, if the computer centering trick of the previous section is invoked. As a last resort, the formula for hand calculation is

$$\text{SE}[\text{Pred}\{Y|X_0\}] = \hat{\sigma} \sqrt{1 + \frac{1}{n} + \frac{(X_0 - \overline{X})^2}{(n-1)s_X^2}},$$

where, again, s_X^2 is the sample variance of the X's.

Prediction intervals may be obtained from this standard error and a multiplier from the t-distribution with $n - 2$ degrees of freedom. Display 7.12 shows the construction of a 95% prediction interval for the pH's of steers at 4 hours after slaughter. The 95% prediction interval succeeds in capturing the future Y in 95% of its applications. The prediction bands in Displays 7.4 and 7.11 are the curves obtained by connecting the upper endpoints and connecting the lower endpoints of prediction intervals at many different X-values.

7.4.4 Calibration: Estimating the X That Results in $Y = Y_0$

Sometimes it is desired to estimate the X that produces a specific response Y. This happens to be the case for the meat processing example: "At what time after slaughter will the pH in any particular steer be 6?" This is a *calibration* problem, also known as *inverse prediction*.

DISPLAY 7.12 95% prediction interval for the pH of a steer carcass 4 hours after slaughter (from the estimated regression of pH on log time after slaughter)

$X_0 = \log(4)$

$$\text{Pred}\{\,Y \mid 1.386\,\} = 6.9836 - 0.7257 \times 1.386 = 5.98$$

$\hat{\sigma}$ n $\overline{X}$

$$\text{SE}[\text{Pred}\{Y \mid 1.386\}] = 0.08226\sqrt{1 + \frac{1}{10} + \frac{(1.386 - 1.190)^2}{9\,(0.6344)}}$$

$$= 0.0865$$

$t_8(0.975)$ $(n-1)$ $s_X{}^2$

Upper limit: $5.98 + 2.306 \times 0.0865 = 6.18$
Lower limit: $5.98 - 2.306 \times 0.0865 = 5.78$

In a typical calibration problem, an experiment is used to estimate the regression of an imprecise measurement (Y) of some quantity on its actual value (X), as determined by a more precise, but more expensive measuring instrument. The regression provides a calibration equation, so that the accurate measurement can be predicted from the cheaper, inaccurate one. It may seem appropriate to call the precise value Y, to use the imprecise value as X, and to employ prediction as in the previous section. This is not possible, however, when the precise values are controlled by the researcher, because the distributional model makes no sense.

The simplest method for calibration is to invert the prediction relationship. If Pred$\{Y|X_0\}$ is taken to be $\hat{\beta}_0 + \hat{\beta}_1 X_0$, then Pred$\{X|Y_0\}$ is $(Y_0 - \hat{\beta}_0)/\hat{\beta}_1$. This is the value of X at which the predicted Y is Y_0. A graphical method for obtaining a calibration interval (inverse prediction interval) for the X at some Y_0 is to plot prediction bands (the ones for predicting Y from X), draw a horizontal line at Y_0, then draw vertical lines down to the X-axis from the points where the horizontal line intersects the upper and lower prediction limits. These two points are the limits of the calibration interval. This procedure is shown for the meat processing data in Display 7.4.

The uncertainty in a calibration interval has to do with the prediction of new Y's. For the meat processing data: at 2.94 hours after slaughter (i.e., when log hours = 1.08), it is predicted that 95% of steer carcasses will have a pH between 6.0 and 6.3. At 5.10 hours (when log hours = 1.63) it is predicted that 95% of steer carcasses will have a pH between 5.7 and 6.0. According to these statements, only 2.5% of carcasses will have a pH less than 6.0 at time 2.94 hours, and only 2.5% of carcasses will have a pH above 6.0 at time 5.10 hours.

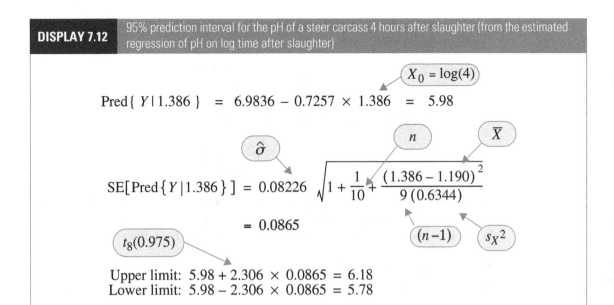

A similar method directly estimates (or predicts) $\hat{X} = (Y_0 - \hat{\beta}_0)/\hat{\beta}_1$ with an approximate standard error of either

$$\mathrm{SE}(\hat{X}) = \frac{\mathrm{SE}(\hat{\mu}\{Y|\hat{X}\})}{|\hat{\beta}_1|}$$

(to estimate the X at which the mean of Y is Y_0) or

$$\mathrm{SE}(\hat{X}) = \frac{\mathrm{SE}(\mathrm{Pred}\{Y|\hat{X}\})}{|\hat{\beta}_1|}$$

(to find the X at which the value of Y is Y_0). An interval is centered at $\hat{X}$, with a half-width equal to a t-multiplier (d.f. $= n - 2$) times the standard error. (*Note*: When the regression line is too flat, these methods are not satisfactory.)

Example—Meat Processing

The log time at which the mean pH has decreased to 6.0 is estimated to be $\hat{X} = 1.3554$. Its standard error is 0.0367, so a 95% confidence interval is from 1.2708 to 1.4400. Hence the estimate of time required is 3.88 hours, with an approximate 95% confidence interval from 3.56 to 4.22 hours. To predict the time when the pH of a single steer might decline to 6.0, the standard error must account for the pH not being at its average. The second standard error is 0.1191, so the interval for $\log(\hat{X})$ is from 1.0806 to 1.6301. The estimate of the time is the same, 3.88 hours, but the approximate 95% confidence interval is from 2.95 to 5.10 hours. The interval is similar to the one determined graphically in Display 7.4.

7.5 RELATED ISSUES

7.5.1 Historical Notes About Regression

Regression analysis was first used by Sir Francis Galton, a 19th-century English scientist studying genetic similarities in parents and offspring. In describing the mean of child height as a function of parent height, Galton found that children of tall parents tended to be tall, but on average not as tall as their parents. Similarly, children of short parents tended to be short, but on average not as short as their parents. He described this as "regression towards mediocrity." Others using his method to analyze statistical relationships referred to the method as Galton's "regression line," and the term *regression* stuck. (See Exercise 7.26.)

The Regression Effect

Galton's phenomenon of regression toward the mean appears in any test–retest situation. Subjects who score high on the first test will, as a group, score closer to the average (lower) on the second test, while subjects who score low on the first test will, as a group, also score closer to the average (higher) on the second test. The

DISPLAY 7.13 Test–retest scores, illustrating the regression effect

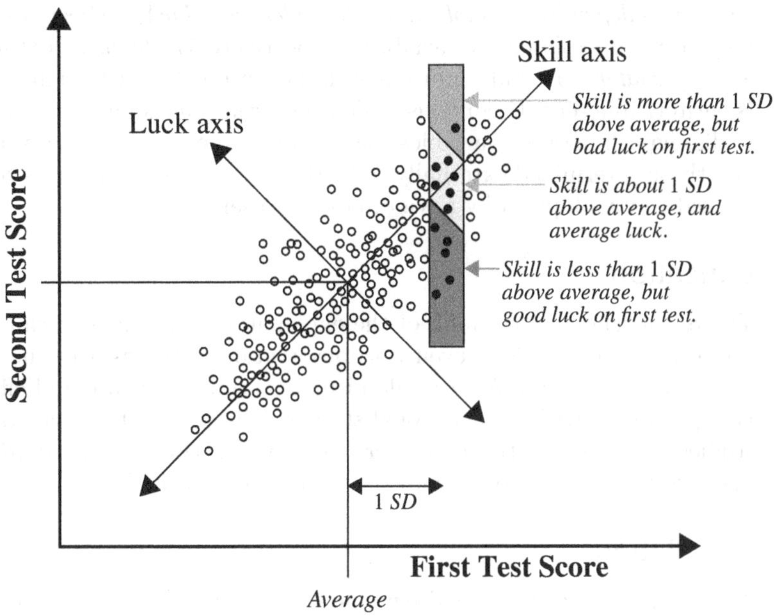

reason is illustrated in Display 7.13. Each point shows the scores of a single subject on the two tests. The horizontal axis contains the first test score, and a vertical box isolates a set of subjects who scored about one standard deviation above average on the first test. Subjects in the set are there (1) because their overall skill is about 1 SD above average, and they performed at their skill level; (2) because their overall skill level was lower than 1 SD above average, and they performed somewhat above their overall skill level on this test; or (3) because their overall skill level was higher than 1 SD above average, and they performed somewhat below their overall skill level on this test. Performances on single tests that differ from one's overall skill level might be called luck or chance.

As is apparent in Display 7.13, more subjects fall into category (2) than into category (3), and they bring the average of the second test scores to a level lower than 1 SD above average. This is the *regression effect*, and it is real. The reason for the regression effect is that there are many more individuals in the whole set with skill levels near average than there are with skill levels farther than one standard deviation from it. So, by chance, more appear in the strip whose skill level is closer to the average (and their skill level tends to catch up with them on the second test).

The regression effect is therefore a natural phenomenon in any test–retest situation. It is often mistakenly interpreted as something more, suggesting some theory about the attitude toward the second test of the test takers based on their first test scores. The term *regression fallacy* is used to describe the mistake of attaching some special meaning to the regression effect.

7.5.2 Differing Terminology

The terms *dependent variable* and *independent variable* have been used in the past for response and explanatory variables, respectively. *The Statistical Sleuth* uses *response* and *explanatory* to avoid confusion with the notion of statistical independence, which is entirely different. (Economists often use *endogenous* and *exogenous*, respectively, to distinguish between variables that are determined within the system under study and those determined externally to it. However, the merits of this terminology have not led to its adoption by a wider circle of users.)

7.5.3 Causation

There is a tendency to think of the explanatory variable as a causative agent and the response as an effect, even with observational studies. But the same conditions apply here as before. With randomized experiments, cause-and-effect terminology is appropriate; with observational studies, it is not. With data from observational studies, the most satisfactory terminology describes an *association* between the mean response and the value of the explanatory variable.

7.5.4 Correlation

The *sample correlation coefficient* for describing the degree of linear association between any two variables X and Y is

$$r_{XY} = \frac{\sum_{i=1}^{n}(X_i - \overline{X})(Y_i - \overline{Y})/(n-1)}{s_X s_Y},$$

where s_X and s_Y are the sample standard deviations. Unlike the estimated slope in regression, the sample correlation coefficient is symmetric in X and Y, so it does not depend on labeling one of them the response and one of them the explanatory variable.

A correlation is dimension-free. It falls between -1 and $+1$, with the extremes corresponding to the situations where all the points in a scatterplot fall exactly on a straight line with negative and positive slopes, respectively. Correlation of zero corresponds to the situation where there is no linear association.

Although the sample correlation coefficient is always a useful summary measure, inferences based on it only make sense when the pairs (X, Y) are selected randomly from a population. This is more restrictive than the conditions for inference from the regression model, where X need not be random. It is a common mistake to base conclusions on correlations when the X's are not random.

If the data are randomly selected pairs, then the sample correlation coefficient estimates a population correlation, $\rho_{XY} = \text{Mean}\{(X - \mu_X)(Y - \mu_Y)\}/(\sigma_X \sigma_Y)$. Conveniently, a test of the hypothesis that $\rho_{XY} = 0$ turns out to be identical to the test of the hypothesis that the slope in the linear regression of Y on X is zero.

Tests for nonzero hypothesized values are also available, but they are quite different and are not pursued here.

Correlation only measures the degree of *linear* association. It is possible for there to be an exact relationship between X and Y and yet the sample correlation coefficient is zero. This would be the case, for example, if there was a parabolic relationship with no linear component.

7.5.5 Planning an Experiment: Replication

Replication means applying exactly the same treatment to more than one experimental unit. With the 10 steers available for the meat processing experiment, the researchers decided to administer five treatment levels—the five different times after slaughter—with two replicate observations at each level. Although they could have investigated 10 different times after slaughter, the benefits of replication can outweigh the extra information that would come from more explanatory variable values. The main benefit is that replication allows for an estimate of σ^2 that does not depend on any model being correct (the pooled estimate of the sample variances). This is sometimes referred to as the *pure error* estimate of σ^2. This, in turn, permits an assessment of model adequacy that would not otherwise be available. (The details of a formal test for model adequacy are shown in Section 8.5.3.) A secondary benefit is a clearer picture of the relationship between the variance and the mean, which may permit an easier assessment of the need for and the type of transformation for the response variable.

Even though the benefits of replication are many, the practical decision about allocating a fixed number of experimental units to treatment levels is difficult. The researcher must weigh the benefits of replication against the benefits of further treatment levels. It would not be good in the meat processing study, for example, to place all 10 steers in only two treatment levels unless the researchers were certain that the regression was a straight line.

7.6 SUMMARY

In the simple linear regression model the mean of a response variable is a straight line function of an explanatory variable. The slope and intercept of the regression line can be estimated by the method of least squares. The variance about the regression is estimated by the sum of squared residuals divided by the degrees of freedom $n-2$. Inferential tools discussed in this chapter are t-tests and confidence intervals for slope and intercept, t-tests and confidence intervals for the mean of the response at any particular value of the explanatory variable, prediction intervals for a future response at any particular value of the explanatory variable, and calibration intervals for predicting the X associated with a specified Y.

Big Bang Analysis

The analysis begins with a scatterplot of distance versus velocity. It is apparent from this (Display 7.1) that the regression of distance on velocity is at least approximately a straight line, as predicted by the Big Bang theory. The least squares method provides a way to estimate the slope and intercept in this regression. Two questions are asked: is the intercept zero, as predicted by the theory, and what is the slope; which, according to the theory, is the age of the universe? The evidence in answer to the first question is expressed through a p-value for the test that β_0 is zero. The evidence in answer to the second is expressed through a confidence interval for β_1.

Meat Processing Analysis

Even though the question of interest calls for finding the value of time after slaughter at which pH is 6, it is necessary to specify pH to be the response and time (log hours after slaughter) to be the explanatory variable, since the latter was controlled by the researchers. The regression of pH on log time is well approximated by a straight line, at least for times up to 8 hours. The calibration-like question of interest requires the data analyst to find the value of X (log time) at which Y (pH) is predicted to be 6.

7.7 EXERCISES

Conceptual Exercises

1. Big Bang Data. Can the estimated regression equation be used to make inferences about (a) the mean distance of nebulae whose recession velocities are zero? (b) the mean distance of nebulae whose recession velocities are 1,000 km/sec? (c) the mean distance of nebulae whose velocities are 2,000 km/sec? (See Display 7.1.)

2. Big Bang Data. Explain why improved measurement of distance would lead to more precise estimates of the regression coefficients.

3. Meat Processing Data. By inspecting Display 7.4, describe the distribution of pH's for steer carcasses 1.65 hours after slaughter (where $X = \log(1.65) = 0.5$).

4. What is wrong with this formulation of the regression model: $Y = \beta_0 + \beta_1 X$?

5. Consider a regression of weight (kg) on height (cm) for a sample of adult males. What are the units of measurement of (a) the intercept? (b) the slope? (c) the SD? (d) the correlation?

6. Explain the differences between the following terms: regression, regression model, and simple linear regression model.

7. What assumptions are made about the distribution of the explanatory variable in the normal simple linear regression model?

8. A group of children were all given a dexterity test in their fifth grade physical education class. The teacher noted a small group who performed exceptionally well, and she informed the grade six teacher as to which children were in the group. When the same children were given a similar dexterity test the next year, they performed reasonably well, but not as well as the sixth grade teacher had expected. What might have caused this?

9. Consider the regression of weight on height for a sample of adult males. Suppose the intercept is 5 kg. (a) Does this imply that males of height 0 weigh 5 kg, on average? (b) Would this imply that the simple linear regression model is meaningless?

10. (a) At what value of X will there be the most precise estimate of the mean of Y? (b) At what value of X will there be the most precise prediction of a future Y?

11. What is the standard error of prediction as the sample size approaches infinity?

Computational Exercises

12. Suppose that the fit to the simple linear regression of Y on X from 6 observations produces the following residuals: −3.3, 2.1, −4.0, −1.5, 5.1, 1.6. (a) What is the estimate of σ^2? (b) What is the estimate of σ? (c) What are the degrees of freedom?

13. Big Bang Data. Using the results shown in Display 7.9, find a 95% confidence interval for the intercept in the regression of measured distance on recession velocity.

14. Big Bang Data. Using the results in Display 7.9, find a 95% confidence interval for the slope in the regression of measured distance on recession velocity.

15. Pollen Removal. Reconsider the data in Exercise 3.27. (a) Draw a scatterplot of proportion of pollen removed versus duration of visit, for the bumblebee queens. (b) Fit the simple linear regression of proportion of pollen removed on duration of visit. Draw the estimated regression line on the scatterplot. (Do this with the computer, if possible; otherwise draw the line with pencil and ruler.) Is there any indication of a problem with this fit? (This problem will be continued in the next chapter.)

16. Most computational procedures utilize the following identities:

$$\sum_{i=1}^{n}(X_i - \overline{X})^2 = \sum_{i=1}^{n} X_i{}^2 - \frac{1}{n}\left(\sum_{i=1}^{n} X_i\right)^2$$

and

$$\sum_{i=1}^{n}(X_i - \overline{X})(Y_i - \overline{Y}) = \sum_{i=1}^{n} X_i Y_i - \frac{1}{n}\left(\sum_{i=1}^{n} X_i\right)\left(\sum_{i=1}^{n} Y_i\right).$$

The left-hand expressions are needed for calculating $\hat{\beta}_1$, but the right-hand equivalents require less effort. Using the meat processing data, calculate the left-hand versions. Then calculate the right-hand versions. Did you get the same answers? Which version requires fewer steps? Which version requires less storage in a computer?

17. Meat Processing. (a) Enter the data from Display 7.3 into a computer and find the least squares estimates for the simple linear regression of pH on log hours. (b) Noting that the average and sample standard deviation of X are provided in Display 7.12, calculate the estimated mean pH at 5 hours after slaughter, and the standard error of the estimated mean (using the formula in Section 7.4.2). (c) Find the standard error of the estimated mean pH at 5 hours after slaughter using the computer trick described in Section 7.4.2.

18. Meat Processing. (a) Find the standard error of prediction for the prediction of pH at 5 hours after slaughter. (b) Construct a 95% prediction interval at 5 hours after slaughter.

19. Meat Processing. Compute a 95% calibration interval (using the graphical approach) for the time at which pH of steer carcasses should be 7.

20. Meat Processing (Sample Size Determination). The standard error of the estimated slope based on the 10 data points is 0.0344. Using the formula for SE in Section 7.3.5, and supposing that the spread of the X's and the estimate of σ will be about the same in a future study, calculate how large the sample size would have to be in order for the SE of the estimated slope to be 0.01.

21. Planetary Distances and Order from the Sun. The first three columns in Display 7.14 show the distances from the sun (scaled so that earth = 1) and the order from the sun of the 8 planets in

DISPLAY 7.14 Orders and distances from the sun (in astronomical units, so that the distance from earth to the sun is 1) of planets in our solar system; without and with the asteroid belt

Name	Order	Distance	Name 2	Order 2	Distance 2
Mercury	1	0.387	Mercury	1	0.387
Venus	2	0.723	Venus	2	0.723
Earth	3	1	Earth	3	1
Mars	4	1.524	Mars	4	1.524
Jupiter	5	5.203	(asteroids)	5	2.9
Saturn	6	9.546	Jupiter	6	5.203
Uranus	7	19.2	Saturn	7	9.546
Neptune	8	30.09	Uranus	8	19.2
Pluto	9	39.5	Neptune	9	30.09
			Pluto	10	39.5

our solar system and the dwarf planet, Pluto. The last three columns are the same but also include the asteroid belt. Using the first three columns, (a) draw a scatterplot of log of distance versus order, (b) include the least squares estimated simple linear regression line on the plot, (c) find the estimate of σ from the least squares fit, and (d) draw a scatterplot of the residuals versus the fitted values from this fit. Using the last three columns, (e) draw a scatterplot of log of distance versus order, (f) include the least squares estimated simple linear regression line on the plot, (g) find the estimate of σ from the least squares fit, and (h) draw a scatter plot of the residuals versus the fitted values from this fit. (i) Does it appear that the simple linear (straight line) regression model fits better to the first set of 9 planets or the second set of 10 "planets"? Explain.

22. Crab Claw Size and Force. As part of a study of the effects of predatory intertidal crab species on snail populations, researchers measured the mean closing forces and the propodus heights of the claws on several crabs of three species. Their data (read from their Figure 3) appear in Display 7.15. (Data from S. B. Yamada and E. G. Boulding, "Claw Morphology, Prey Size Selection and Foraging Efficiency in Generalist and Specialist Shell-Breaking Crabs," *Journal of Experimental Marine Biology and Ecology*, 220 (1998): 191–211.)

(a) Estimate the slope in the simple linear regression of log force on log height, separately for each crab species. Obtain the standard errors of the estimated slopes.

(b) Use a t-test to compare the slopes for *C. productus* and *L. bellus*. Then compare the slopes for *C. productus* and *H. nudus*. The standard error for the difference in two slope estimates from independent samples is the following:

$$\mathrm{SE}[\hat{\beta}_{1(1)} - \hat{\beta}_{1(2)}] = \sqrt{[\mathrm{SE}(\hat{\beta}_{1(1)})]^2 + [\mathrm{SE}(\hat{\beta}_{1(2)})]^2},$$

where $\hat{\beta}_{1(j)}$ represents the estimate of slope from sample j. Use t-tests with the sum of the degrees of freedom associated with the two standard errors. What do you conclude? (*Note*: A better way to perform this test, using multiple regression, is described in Chapter 9.)

23. For Those Literate in Calculus and Linear Algebra. The least squares problem is that of finding estimates of β_0 and β_1 that minimize the sum of squares,

$$\mathrm{SS}(\beta_0, \beta_1) = \sum_{i=1}^{n} (Y_i - \beta_0 - \beta_1 X_i)^2.$$

DISPLAY 7.15 Closing force (in newtons) and propodus heights (in mm) in three predatory crab species

Hemigrapsus nudus ($n = 14$)		Lophopanopeus bellus ($n = 12$)		Cancer productus ($n = 12$)	
Force	Height	Force	Height	Force	Height
3.2	5.0	2.1	5.1	5.0	6.7
6.4	6.0	8.7	5.9	7.8	7.1
2.0	6.4	2.9	6.6	14.6	11.2
2.0	6.5	6.9	7.2	16.8	11.4
4.9	6.6	8.7	8.6	17.7	9.4
3.0	7.0	15.1	7.9	19.8	10.7
2.9	7.9	14.6	8.1	19.6	13.1
9.5	7.9	17.6	9.6	22.5	9.4
4.0	8.0	20.6	10.2	23.6	11.6
3.4	8.2	19.6	10.5	24.4	10.2
7.4	8.3	27.4	8.2	26.0	12.5
2.4	8.8	29.4	11.0	29.4	11.8
4.0	12.1				
5.2	12.2				

(a) Setting the partial derivatives of $SS(\beta_0, \beta_1)$ with respect to each parameter equal to zero, show that β_0 and β_1 must satisfy the *normal equations*:

$$\beta_0 n + \beta_1 \sum_{i=1}^{n} X_i = \sum_{i=1}^{n} Y_i$$

$$\beta_0 \sum_{i=1}^{n} X_i + \beta_1 \sum_{i=1}^{n} X_i^2 = \sum_{i=1}^{n} X_i Y_i.$$

(b) Show that $\hat{\beta}_0$ and $\hat{\beta}_1$ given in Section 7.3.1 satisfy the normal equations. (c) Verify that the solutions provide a minimum to the sum of squares.

24. Decline in Male Births. Display 7.16 shows the proportion of male births in Denmark, The Netherlands, Canada, and the United States for a number of years. (Data read from graphs in Davis et al., "Reduced Ratio of Male to Female Births in Several Industrial Countries," *Journal of the American Medical Association*, 279 (1998): 1018–23.) Also shown are the results of least squares fitting to the simple linear regression of proportion of males on year, separately for each country, with standard errors of estimated coefficients in parentheses.

- **(a)** With a statistical computer package and the data in the file ex0725, obtain the least squares fits to the four simple regressions, individually, to confirm the estimates and standard errors presented in Display 7.17.
- **(b)** Obtain the t-statistic for the test that the slopes of the regressions are zero, for each of the four countries. Is there evidence that the proportion of male births is truly declining?
- **(c)** Explain why the United States can have the largest of the four t-statistics (in absolute value) even though its slope is only the third largest (in absolute value).
- **(d)** Explain why the standard error of the estimated slope is smaller for the United States than for Canada, even though the sample size is the same.

DISPLAY 7.16 Proportions of male births in four countries; and simple regression statistics

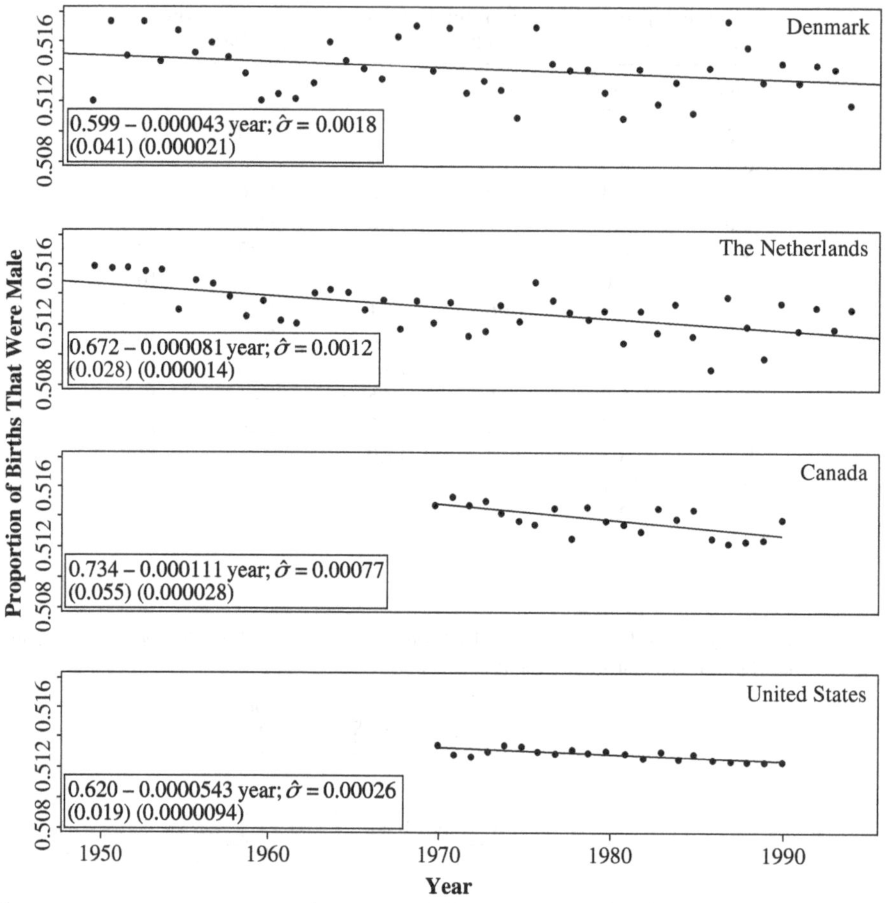

(e) Can you think of any reason why the standard deviations about the regression line might be different for the four countries? (*Hint*: The proportion of males is a kind of average, i.e., the average number of births that are male.)

Data Problems

25. **Big Bang II.** The data in Display 7.17 are measured distances and recession velocities for 10 clusters of nebulae, much farther from earth than the nebulae reported in Section 7.1.1. (Data from E. Hubble and M. Humason, "The Velocity–Distance Relation Among Extra-Galactic Nebulae," *Astrophysics Journal* 74 (1931): 43–50.) If Hubble's theory is correct, then the mean of the measured distance, as a function of velocity, should be β_1 Velocity, and β_1 is the age of the universe. Are the data consistent with the theory (that the intercept is zero)? What is the estimated age of the universe? (*Note*: The slope here is in units of megaparsecs-seconds per kilometer. Multiply by 979.8 to get an answer in billions of years. You should find out how to fit simple linear regression through the

DISPLAY 7.17 Measured distance (million parsecs) and recession velocity (km/sec) for 10 clusters of nebulae

Cluster	Distance	Velocity
Virgo	1.8	890
Pegasus	7.25	3,810
Pisces	7.00	4,638
Cancer	9.00	4,820
Perseus	11.00	5,230
Coma	13.80	7,500
Ursa Major	22.00	11,800
Leo	32.00	19,600
Isolated nebulae I	4.20	2350
Isolated nebulae II	2.15	630

DISPLAY 7.18 Galton's data on heights (in inches) of adult children and their parents; first 5 of 933 rows

Gender	Family	Height	Father	Mother
Male	1	73.2	78.5	67
Female	1	69.2	78.5	67
Female	1	69	78.5	67
Female	1	69	78.5	67
Male	2	73.5	78.5	66.5

origin—that is, how to drop the intercept term—with your statistical computer package.) To what extent is the relationship shown by these far-away nebulae clusters similar to and different from the relationship indicated in Case Study 7.1.1? (Analyze the data and write a brief statistical report including a summary of statistical findings, a graphical display, and a details section describing the details of the particular methods used.)

26. Origin of the Term *Regression*. Motivated by the work of his cousin, Charles Darwin, the English scientist Francis Galton studied the degree to which human traits were passed from one generation to the next. In an 1885 study, he measured the heights of 933 adult children and their parents. Display 7.18 shows the first five rows of his data set. Galton multiplied all female heights by 1.08 to convert them to a male-equivalent height. He estimated the simple linear regression line of child's height on average parent's height and, upon finding that the slope was positive but less than 1, concluded that children of taller-than-average parents tended to also be taller than average but not as tall as their parents; and, similarly, children of shorter-than-average parents tended to be shorter than average, but not as short as their parents. He labeled this "regression towards mediocrity" because of the apparent regression (i.e., reversion) of children's height toward the average. Other scientists began to refer to the line as "Galton's regression line". Although the term was not intended to describe the model for the mean, the name stuck. Reproduce Galton's analysis by converting females' heights to their male-equivalents (multiply them by 1.08). Do the same for mothers' heights. Compute the parent height by taking the average of the father's height and the converted mother's height. Then fit the simple linear regression of child height on parent height. (For now, do as Galton did and ignore the fact that the heights of individuals from the same family are probably not independent.) Find the predicted height and a 95% prediction interval for the adult height of a boy whose average parent height is 65 inches. Repeat for a boy whose average parent height is 76 inches. (These are the raw data used in the paper by F. Galton, "Regression towards Mediocrity in Hereditary Stature" in the *Journal of the Anthropological Institute* in 1886, as described in "'Transmuting' Women into Men: Galton's Family Data On Human Stature" by James Henley in *The American Statistician*, 58 (2004): 237–43.)

27. Male Displays. Black wheatears, *Oenanthe leucura*, are small birds of Spain and Morocco. Males of the species demonstrate an exaggerated sexual display by carrying many heavy stones to nesting cavities. This 35-gram bird transports, on average, 3.1 kg of stones per nesting season! Different males carry somewhat different sized stones, prompting a study of whether larger stones may be a signal of higher health status. M. Soler et al. ("Weight Lifting and Health Status in the Black Wheatear," *Behavioral Ecology* 10(3) (1999): 281–86) calculated the average stone mass (g) carried by each of 21 male black wheatears, along with T-cell response measurements reflecting their immune systems' strengths. The data in Display 7.19 were taken from their Figure 1. Analyze the data and write a statistical report summarizing the evidence supporting whether health—as measured by T-cell response—is associated with stone mass, and quantifying the association.

DISPLAY 7.19 Mass of stones carried and immune system strength for 21 wheatear birds; first 5 of 15 rows

Bird	**Mean stone mass** (g)	**T-cell response** (mm)
1	3.33	0.252
2	4.62	0.263
3	5.43	0.251
4	5.73	0.251
5	6.12	0.183

28. Brain Activity in Violin and String Players. Studies over the past two decades have shown that activity can effect the reorganization of the human central nervous system. For example, it is known that the part of the brain associated with activity of a finger or limb is taken over for other purposes in individuals whose limb or finger has been lost. In one study, psychologists used magnetic source imaging (MSI) to measure neuronal activity in the brains of nine string players (six violinists, two cellists, and one guitarist) and six controls who had never played a musical instrument, when the thumb and fifth finger of the left hand were exposed to mild stimulation. The researchers felt that stringed instrument players, who use the fingers of their left hand extensively, might show different behavior in the brain—as a result of this extensive physical activity—than individuals who did not play stringed instruments. Display 7.20 shows a neuron activity index from the MSI and the years that the individual had been playing a stringed instrument (zero for the controls). (Data based on a graph in Elbert et al., "Increased Cortical Representation of the Fingers of the Left Hand in String Players," *Science* 270 (13 October, 1995) 305–7.) Is the neuron activity different in the stringed musicians and the controls? Is the amount of activity associated with the number of years the individual has been playing the instrument?

29. Sampling Bias in Exit Polls. Exit pollsters predict election results before final counts are tallied by sampling voters leaving voting locations. The pollsters have no way of selecting a random sample, so they instruct their interviewers to select every third exiting voter, or fourth, or tenth, or some other specified number. Some voters refuse to participate or avoid the interviewer. If the refusers and avoiders have the same voting patterns as the rest of the population, then it shouldn't matter; the sample, although not random, wouldn't be biased. If, however, one candidate's voters are more likely to refuse or avoid interview, the sample would be biased and could lead to a misleading conclusion.

DISPLAY 7.20 Years that the individual has been playing a stringed instrument and neuronal activity index ("D5 dipole strength, in nA·m") for nine stringed musicians and six controls

Years playing	Neuron activity index
0	5
0	6
0	7.5
0	9
0	9.5
0	11
5	16
6	16.5
8	11.5
10	16
12	25
13	25.5
17	25.5
18	23
19	26.5

On November 4, 2004, exit pollsters incorrectly predicted that John Kerry would win the U.S. presidential election over George W. Bush. The exit polls overstated the Kerry advantage in 42 of 50 states. No one expects exit polls to be exact, but chance alone cannot reasonably explain this discrepancy. Although fraud is a possibility, the data are also consistent, with Bush supporters being more likely than Kerry supporters to refuse or avoid participation in the exit poll.

In a postelection evaluation, the exit polling agency investigated voter avoidance of interviewers. Display 7.21 shows the average Kerry exit poll overestimate (determined after the actual counts were available) for a large number of voting precincts, grouped according to the distance of the interviewer from the door. If Bush voters were more likely to avoid interviewers in general, one might also expect a greater avoidance with increasing distance to the interviewer (since there is more opportunity for escape). A positive relationship between distance of the interviewer from the door and amount of Kerry overestimate, therefore, would lend credibility to the theory that Bush voters were more likely to avoid exit poll interviews. How strong is the evidence that the mean Kerry overestimate increases with increasing distance of interviewer from

DISPLAY 7.21 Exit poll error in favor of Kerry and distance of exit poll interviewer from the voting precinct door, in the 2004 U.S. presidential election between George W. Bush and John Kerry

Overestimate	Distance
5.3	0
6.4	5
5.6	17
7.6	37
9.6	75
12.3	100

DISPLAY 7.22 Average exit poll interview refusal rates for precincts grouped according to the approximate age of the interviewer in the 2004 U.S. presidential election between George W. Bush and John Kerry

Age	Refusal
22	0.39
30	0.38
40	0.35
50	0.32
60	0.31
65	0.29

the door? (Data from Evaluation of Edison/Mitofsky Election System 2004 prepared by Edison Media Research and Mitofsky International for the National Election Pool (NEP), January 15, 2005. http://abcnews.go.com/images/Politics/EvaluationofEdisonMitofskyElectionSystem.pdf (accessed May 9, 2008).)

30. Sampling Bias in Exit Polls 2. This exercise is about differential interview *refusal* rates in the exit polls conducted in the 2004 U.S. presidential election. Display 7.22 shows the average proportion of voters who refused to be interviewed at precincts grouped according to the approximate age of the interviewer. What evidence do these data provide that the mean refusal rate decreased with increasing age of interviewer? An affirmative answer to this question doesn't provide any direct evidence of a difference between Kerry and Bush voters, but is consistent with an undercount of Bush votes in the exit polls if, as one might speculate based on the relative conservativeness of Bush supporters, Bush voters were more likely to avoid younger interviewers. (See Exercise 29.)

Answers to Conceptual Exercises

1. (a) Yes. (b) Yes. (c) Such an extrapolation would be risky.

2. The standard deviation σ about the regression reflects measurement error variation. Making this smaller will cause the standard deviations of the sampling distributions of the least squares estimates to be smaller (see Display 7.7).

3. The model says that the distribution is normal. The estimated mean pH is about 6.6. The prediction limits will be about 2 SDs up and down, so the SD is about 0.1 pH units.

4. This implies an exact relationship between Y and X. The model should be for the *mean* of Y as a function of X.

5. (a) kg; (b) kg/cm; (c) kg; (d) none.

6. Regression refers to the mean of a response variable as a function of an explanatory variable. A regression model is a function used to describe the regression. The simple linear regression model is a particular regression model in which the regression is a straight-line function of a single explanatory variable.

7. None.

8. This is the regression effect. It is exactly what you can expect to happen.

9. (a) No. Height = 0 is outside the range of observed values, so the model may not extend to that situation. (b) No. It may be useful for answering questions pertaining to the regression of weight on height for heights in a certain range.

10. (a) At the sample average of the X's used in the estimation. (b) Same as (a).

11. σ.

CHAPTER 8

A Closer Look at Assumptions for Simple Linear Regression

The inferential tools of the previous chapter are based on the normal simple linear regression model with constant variance. Since real data do not necessarily conform to this model, the data analyst must size up the situation and choose a course of action based on an understanding of the robustness of the tools to model violations.

This chapter presents some informal graphical tools and a formal test for assessing the lack of fit. As before, the graphical procedures are used to find suitable transformations to scales where the simple linear regression model seems appropriate. The lack-of-fit test, on the other hand, looks specifically at the issue of whether the straight line assumption is plausible.

8.1 CASE STUDIES

8.1.1 Island Area and Number of Species—An Observational Study

Biologists have noticed a consistent relation between the area of islands and the number of animal and plant species living on them. If S is the number of species and A is the area, then $S = CA^{\gamma}$ (roughly), where C is a constant and γ is a biologically meaningful parameter that depends on the group of organisms (birds, reptiles, or grasses, for example). Estimates of this relationship are useful in conservation biology for predicting species extinction rates due to diminishing habitat.

The data in Display 8.1 are the numbers of reptile and amphibian species and the island areas for seven islands in the West Indies. (Data on species from E. O. Wilson, *The Diversity of Life*, New York: W. W. Norton, 1992; areas from *The 1994 World Almanac*, Mahwah, N.J.: Funk & Wagnalls, 1993.) These are typical of data used to estimate the area effect. Researchers wish to estimate γ in the species–area equation for this group of animals and to summarize the effect of area on the typical number of species.

DISPLAY 8.1 Island area and number of reptile and amphibian species for seven islands in the West Indies

Island	Area (square miles)	Number of species
Cuba	44,218	100
Hispaniola	29,371	108
Jamaica	4,244	45
Puerto Rico	3,435	53
Montserrat	32	16
Saba	5	11
Redonda	1	7

Statistical Conclusion

A second graph in Display 8.2 is a log-log scatterplot of number of species versus island area, along with the estimated line for the regression of log number of species on log area. The parameter γ in the species–area relation, Median $\{S|A\} = CA^{\gamma}$, is estimated to be 0.250 (a 95% confidence interval is 0.219 to 0.281). It is estimated that the median number of species increases by 19% with each doubling of area.

Scope of Inference

The statistical association from these observational data cannot be used to establish a causal connection. Furthermore, any generalization of these results to islands in other parts of the world or to a wider population, like "islands of rain forest," is purely speculative. Nevertheless, the results from these data are consistent with results for other islands and with small-scale randomized experiments. In

DISPLAY 8.2 Scatterplot and log-log-scatterplot of number of reptile and amphibian species versus area for seven islands in the West Indies

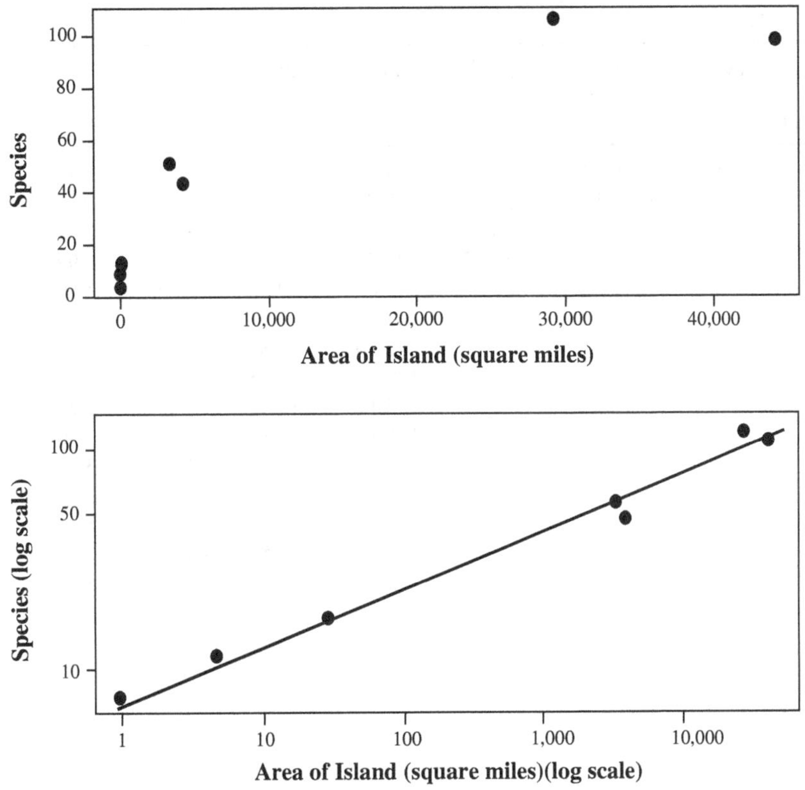

summarizing the studies, Wilson offered a conservatively optimistic guess that the current rate of environmental destruction amounts to 27,000 species vanishing each year and that 20% of all plant and animal species currently on earth will be extinct by the year 2022.

8.1.2 Breakdown Times for Insulating Fluid Under Different Voltages—A Controlled Experiment

In an industrial laboratory, under uniform conditions, batches of electrical insulating fluid were subjected to constant voltages until the insulating property of the fluids broke down. Seven different voltage levels, spaced 2 kilovolts (kV) apart from 26 to 38 kV, were studied. The measured responses were the times, in minutes, until breakdown, as listed in Display 8.3. (Data from W. B. Nelson, Schenectady, N.Y.: GE Co. Technical Report 71-C-011 (1970), as discussed in J. F. Lawless, *Statistical Models and Methods for Lifetime Data*, New York: John Wiley & Sons, 1982, chap. 4.) How does the distribution of breakdown time depend on voltage?

DISPLAY 8.3 Times (in minutes) to breakdown of 76 samples of an insulating fluid subjected to different constant voltages

Group #:	1	2	3	4	5	6	7
Voltage (kV):	26	28	30	32	34	36	38
Sample size:	3	5	11	15	19	15	8
Times (min):	5.79	68.85	7.74	0.27	0.19	0.35	0.09
	1579.52	108.29	17.05	0.40	0.78	0.59	0.39
	2323.70	110.29	20.46	0.69	0.96	0.96	0.47
		426.07	21.02	0.79	1.31	0.99	0.73
		1067.60	22.66	2.75	2.78	1.69	0.74
			43.40	3.91	3.16	1.97	1.13
			47.30	9.88	4.15	2.07	1.40
			139.07	13.95	4.67	2.58	2.38
			144.12	15.93	4.85	2.71	
			175.88	27.80	6.50	2.90	
			194.90	53.24	7.35	3.67	
				82.85	8.01	3.99	
				89.29	8.27	5.35	
				100.59	12.06	13.77	
				215.10	31.75	25.50	
					32.52		
					33.91		
					36.71		
					72.89		

Statistical Conclusion

Display 8.4 shows a scatterplot of the responses on a logarithmic scale plotted against the voltage levels, along with a summary of the parameter estimates in the simple linear regression of log breakdown time on voltage. The median breakdown time decreases by 40% for each 1 kV increase in voltage (95% confidence interval: 32% to 46%).

Scope of Inference

The laboratory setting for this experiment allowed the experimenter to hold all factors constant except the voltage level, which was assigned at different levels to different batches. Therefore it seems reasonable to infer that the different voltage levels must be directly responsible for the observed differences in time to breakdown. It can be inferred that other batches in the same laboratory setting would follow the same pattern so long as the voltage level is within the experimental range. Inference to voltage levels outside the experimental range and to performance under nonlaboratory conditions cannot be made from these data. If such inference is required, further testing or stronger assumptions must be invoked.

DISPLAY 8.4 Scatterplot of breakdown times (natural logarithm scale) versus voltage levels and summary of estimated simple linear regression

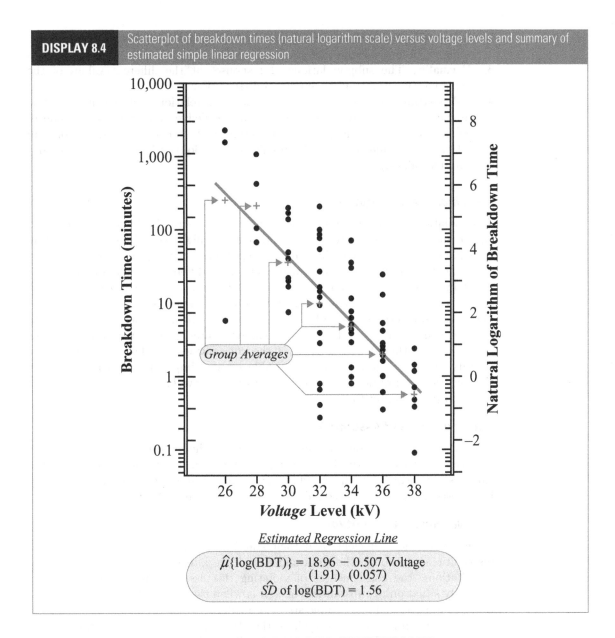

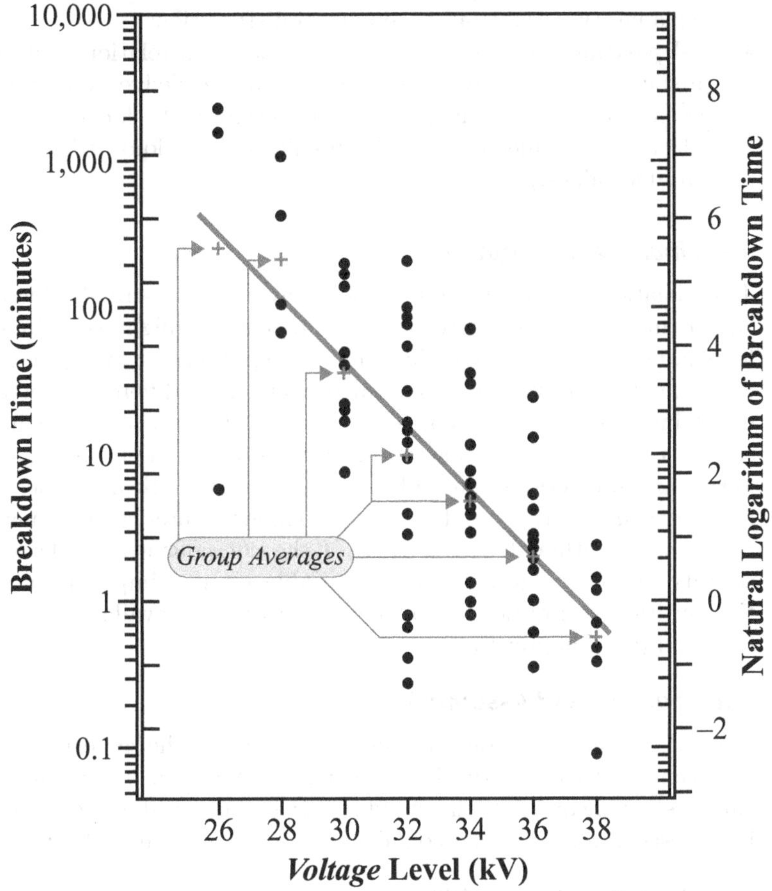

Estimated Regression Line

$\hat{\mu}\{\log(BDT)\} = 18.96 - 0.507$ Voltage
(1.91) (0.057)
$\widehat{SD}$ of log(BDT) = 1.56

8.2 ROBUSTNESS OF LEAST SQUARES INFERENCES

Exact justification of the statistical statements—tests, confidence intervals, and prediction intervals—based on least squares estimates depends on these features of the regression model:

1. **Linearity.** The plot of response means against the explanatory variable is a straight line.

2. **Constant variance.** The spread of the responses around the straight line is the same at all levels of the explanatory variable.
3. **Normality.** The subpopulations of responses at the different values of the explanatory variable all have normal distributions.
4. **Independence.** The location of any response in relation to its mean cannot be predicted, either fully or partially, from knowledge of where other responses are in relation to their means. (Furthermore, the location of any response in relation to its mean cannot be predicted from knowledge of the explanatory variable values.)

The Linearity Assumption

Two violations of the first assumption may occur: A straight line may be an inadequate model for the regression (the regression might contain some curvature, for example); or a straight line may be appropriate for most of the data, but contamination from one or several outliers from different populations may render it inapplicable to the entire set. Both of these violations can cause the least squares estimates to give misleading answers to the questions of interest. Estimated means and predictions can be biased—they systematically under- or overestimate the intended quantity—and tests and confidence intervals may inaccurately reflect uncertainty. Although the severity of the consequences is always related to the severity of the violation, it is undesirable to use simple linear regression when the linearity assumption is not met. Remedies are available for dealing with this situation (see Chapter 9).

The Equal-Spread Assumption

The consequences for violating this assumption are the same as for one-way analysis of variance. Although the least squares estimates are still unbiased even if the variance is nonconstant, the standard errors inaccurately describe the uncertainty in the estimates. Tests and confidence intervals can be misleading.

The Normality Assumption

Estimates of the coefficients and their standard errors are robust to nonnormal distributions. Although the tests and confidence intervals originate from normal distributions, the consequences of violating this assumption are usually minor. The only situation of substantial concern is when the distributions have long tails (outliers are present) and sample sizes are moderate to small.

If prediction intervals are used, on the other hand, departures from normality become important. This is because the prediction intervals are based directly on the normality of the *population distributions* whereas tests and confidence intervals are based on the normality of the *sampling distributions of the estimates* (which may be approximately normal even when the population distributions are not).

The Independence Assumption

Lack of independence causes no bias in least squares estimates of the coefficients, but standard errors are seriously affected. As before, cluster and serial effects, if suspected, should be incorporated with more sophisticated models. See Chapters 13 and 14 for incorporating cluster effects using multiple regression and Chapter 15 for ways to incorporate serial effects into regression models.

8.3 GRAPHICAL TOOLS FOR MODEL ASSESSMENT

The principal tools for model assessment are scatterplots of the response variable versus the explanatory variable and of the residuals versus the fitted values. An initial scatterplot of species number versus island area, on the top of Display 8.2, immediately indicates problems with a straight line model for the means. Similarly, a scatterplot of breakdown time versus voltage, in Display 8.5, shows problems of nonlinearity and nonconstant variance.

DISPLAY 8.5 Scatterplot of breakdown times versus voltage

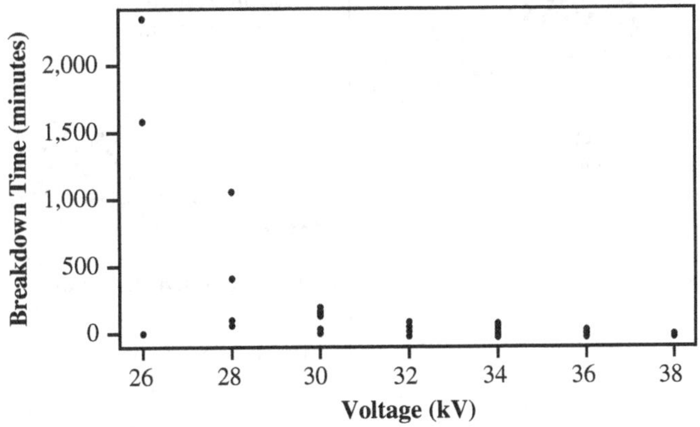

8.3.1 Scatterplot of the Response Variable Versus the Explanatory Variable

Scatterplots with some commonly occurring patterns are shown in Display 8.6. The eye should group points into vertical strips, as shown in (a), to explore the characteristics of subpopulation distributions. Lines connecting strip averages roughly exhibit the relationship between the mean of the responses and the explanatory variable. Vertical lines show the strip standard deviations, which should be examined for patterns in variability. As discussed for each of the plots, the particular patterns for the mean and the variability often suggest a course of action.

(a) This is the ideal situation. The regression is a straight line and the variability is about the same at all locations along the line. Use simple linear regression.

(b) The regression is not a straight line, but it is monotonic (the mean of Y either strictly increases or strictly decreases in X), and the variability is about the

DISPLAY 8.6 Some hypothetical scatterplots of response versus explanatory variable with suggested courses of action; (a) is ideal

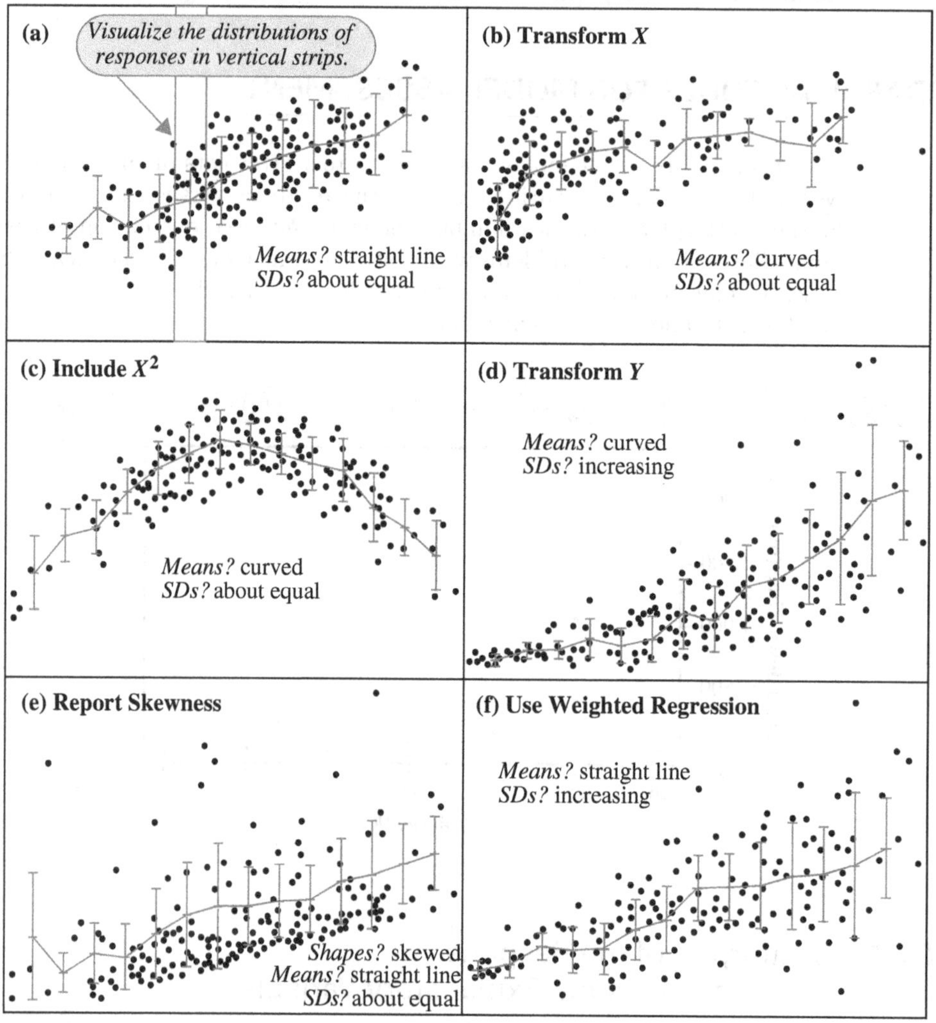

same at all values of X. Try transforming X to a new scale where the straight line assumption appears defensible. Then use simple linear regression.

(c) The regression is not a straight line and is not monotonic, and the variability is about the same at all values of X. No transformation of X can yield a straight line relationship. Try *quadratic regression* ($\mu\{Y|X\} = \beta_0 + \beta_1 X + \beta_2 X^2$), which is discussed in Chapter 9.

(d) The regression is not a straight line, and the variability increases as the mean of Y increases. Try a transformation of Y, such as log, reciprocal, or square root, followed by simple linear regression.

(e) The regression is a straight line, the variability is roughly constant, but the distribution of Y about the regression line is skewed. Remedies are unnecessary, and transformations will create other problems. Use simple linear regression, but report the skewness.

(f) The regression is a straight line but the variability increases as the mean of Y increases. Simple linear regression gives unbiased estimates of the straight line relationship, but better estimates are available using *weighted regression*, as discussed in Section 11.6.1.

8.3.2 Scatterplots of Residuals Versus Fitted Values

Sometimes the patterns in Display 8.6 are difficult to detect because the total variability of the response variable is much larger than the variability around the regression line. Scatterplots of the residuals versus the fitted values are better for finding patterns because the linear component of variation in the responses has been removed, leaving a clearer picture about curvature and spread. The residual plot alerts the user to *nonlinearity, nonconstant variance, and the presence of outliers*.

Example

Display 8.5 is a scatterplot of the breakdown time versus voltage. This scatterplot resembles the pattern in part (d) of Display 8.6, suggesting the need for transformation. Display 8.7 shows the scatterplot and residual plot after applying a square root transformation to the response. The residual plot is more informative than the scatterplot. It has a classic *horn-shaped* pattern, showing a combination of a

DISPLAY 8.7 Scatterplot of the square root of breakdown time versus voltage and a residual plot based on the simple linear regression fit

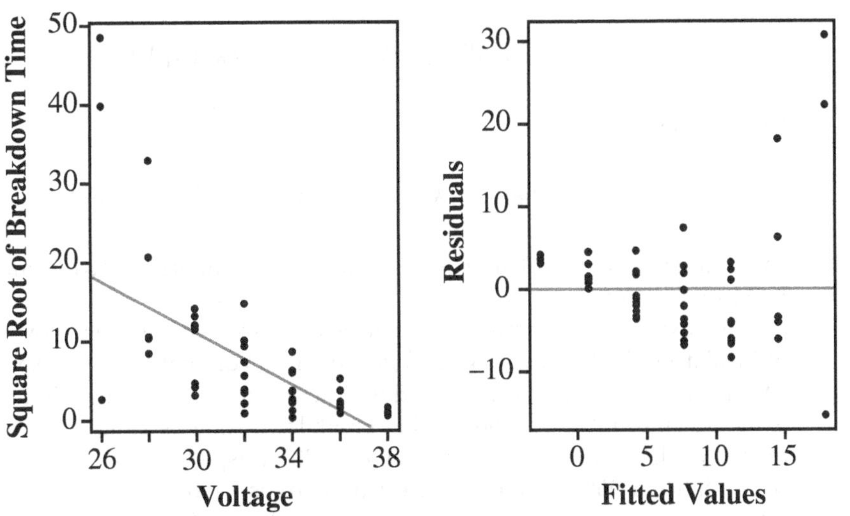

poor fit to the subpopulation averages and increasing variability. It is evident that the square root transformation has not resolved the problem. A log transformation, however, works well (see Display 8.4). (*Note*: The horn-shaped pattern is always a key indicator of the need to transform, whether or not there are replicate samples at specific values of the explanatory variable.)

Transformations Indicated by Horn-Shaped Residual Plots

A horn-shaped pattern in the residual plot suggests a response transformation like the square root, the logarithm, or the reciprocal. The logarithm is the easiest to interpret. The reciprocal, $1/Y$, works better when the nonconstant spread is more severe, and the square root works better when the nonconstant spread is less severe. Judging severity of nonconstant variance in the horn-shaped residual plot is difficult and unnecessary, however. Try one of the transformations, re-fit the regression model on the transformed data, and redraw the residual plot to see if the transformation has worked. If not, then one of the others can be attempted.

8.4 INTERPRETATION AFTER LOG TRANSFORMATIONS

A data analyst must interpret regression results in a way that makes sense to the intended audience. The appropriate wording for inferential statements after logarithmic transformation depends on whether the transformation was applied to the response, to the explanatory variable, or to both.

When the Response Variable Is Logged

If $\mu\{\log(Y)|X\} = \beta_0 + \beta_1 X$, and if the distribution of the transformed responses about the regression is symmetric, then

$$\text{Median}\{Y|X\} = \exp(\beta_0)\exp(\beta_1 X).$$

Consequently,

$$\text{Median}\{Y|(X+1)\}/\text{Median}\{Y|X\} = \exp(\beta_1),$$

so an increase in X of 1 unit is associated with a multiplicative change of $\exp(\beta_1)$ in Median$\{Y|X\}$.

For example, the estimated relationship between the breakdown time (BDT) of insulating fluid and voltage is $\hat{\mu}\{\log(\text{BDT}) \mid \text{Voltage}\} = 19.0 - 0.51\ \text{Voltage}$. A 1 kV increase in voltage is associated with a multiplicative change in median BDT of exp(−0.51), or 0.60. So, the median breakdown time at 28 kV is 60% of what it is at 27 kV; the median breakdown time at 29 kV is 60% of what it is at 28 kV, and so on. Since a 95% confidence interval for β_1 is −0.62 to −0.39, a 95% confidence interval for $\exp(\beta_1)$ is exp(−0.62) to exp(−0.39), or 0.54 to 0.68.

The statement Median $\{Y|(X+1)\} = 0.6 \times \text{Median}\{Y|X\}$ can also be written as Median $\{Y|(X+1)\} - \text{Median}\{Y|X\} = 0.4 \times \text{Median}\{Y|X\}$, which permits the following kind of statement: "It is estimated that the median of Y decreases by 40% for each one unit increase in X (95% confidence interval: 32% to 46%)."

When the Explanatory Variable Is Logged

The relationship $\mu\{Y\,|\,\log(X)\} = \beta_0 + \beta_1 \log(X)$ can be described in terms of multiplicative changes in X, either as a change in the mean of Y for each doubling of X or a change in the mean of Y for each ten-fold increase in X. The chosen multiple should be consistent with the range of X's in the data set.

Notice that

$$\mu\{Y\,|\,\log(2X)\} - \mu\{Y\,|\,\log(X)\} = \beta_1 \log(2),$$

so a doubling of X is associated with a $\beta_1 \log(2)$ change in the mean of Y. Similarly, a ten-fold increase in X is associated with a $\beta_1 \log(10)$ change in the mean of Y.

For the meat processing data of Section 7.1.2, $\hat{\mu}\{\text{pH}\,|\,\log(\text{Time})\} = 6.98 - 0.726 \log(\text{Time})$, so a doubling of time after slaughter is associated with a $\log(2)(-0.726) = -0.503$ unit change in pH. Since a 95% confidence interval for β_1 is from -0.805 to -0.646, a 95% CI for $\log(2)\beta_1$ is from -0.558 to -0.448. In words: It is estimated that the mean pH is reduced by 0.503 for each doubling of time after slaughter (95% confidence interval 0.448 to 0.558).

When Both the Response and Explanatory Variables Are Logged

The interpretation is a combination of the previous two. If $\mu\{\log(Y)|\log(X)\} = \beta_0 + \beta_1 \log(X)$, then $\text{Median}\{Y|X\} = \exp(\beta_0)X^{\beta_1}$. A doubling of X is associated with a multiplicative change of 2^{β_1} in the median of Y. Or, a ten-fold increase in X is associated with a 10^{β_1}-fold change in the median of Y.

For the island size and number of species data, $\hat{\mu}\{\log(\text{species})|\log(\text{area})\} = 1.94 + 0.250 \log(\text{area})$. Thus, an island area of $2A$ is estimated to have a median number of species that is $2^{0.250}$ (or 1.19) times the median number of species for an island of area A. Since a 95% confidence interval for β_1 is 0.219 to 0.281, a 95% confidence interval for the multiplicative factor in the median is $2^{0.219}$ to $2^{0.281}$, or 1.16 to 1.22.

The Need for Interpretation

An important aspect of the log transformation is that straightforward multiplicative statements emerge. Straightforward interpretations after other transformations only follow in certain instances. For example, the square root of the cross-sectional area of a tree may be re-expressed as the diameter; the reciprocal of the time to complete a race may be interpreted as the speed. In general, however, interpretation for other transformations may be awkward.

For two types of questions, interpretation is not critical. First, if the regression is used only to assess whether the distribution of the response is *associated with* the explanatory variable, it is sufficient to test the hypothesis that the slope in the regression of Y on X is zero, where Y or X are transformed variables. If the distribution of the transformed response is associated with X, the distribution of the response is associated with X, even though that association may be difficult to describe. Secondly, if the purpose is prediction, no interpretation of the regression coefficients is needed. It is only necessary to express the prediction on the original scale, regardless of the expression used to make the prediction.

8.5 ASSESSMENT OF FIT USING THE ANALYSIS OF VARIANCE

An analysis of variance can be used to compare several models. When there are replicate response variables at several explanatory variable values, an analysis of variance F-test for comparing the simple linear regression model to the separate-means (one-way analysis of variance) model supplies a formal assessment of the goodness of fit of simple linear regression. This is called the *lack-of-fit F-test.*

8.5.1 Three Models for the Population Means

In the following three model descriptions for means, the subscript i refers to the group number (e.g., different voltage levels).

1. Separate-means model:	$\mu\{Y\|X_i\} = \mu_i$,	for $i = 1, \ldots, I$.
2. Simple linear regression model:	$\mu\{Y\|X_i\} = \beta_0 + \beta_1 X_i$,	for $i = 1, \ldots, I$.
3. Equal-means model:	$\mu\{Y\|X_i\} = \mu$,	for $i = 1, \ldots, I$.

The separate-means model has no restriction on the values of any of the means. It has I different parameters—the individual group means. The simple linear regression model has two parameters, the slope and the intercept. The equal-means model has a single parameter.

These models form a *hierarchical* set. The equal-means model is a special case of the simple linear regression model, which in turn is a special case of the separate-means model. Viewed conversely, the separate-means model is a generalization of the simple linear regression model, which in turn is a generalization of the equal-means model. A generalization of one model is another model that contains the same features but uses additional parameters to describe a more complex structure.

8.5.2 The Analysis of Variance Table Associated with Simple Regression

Display 8.8 contains two different analysis of variance tables. Table (b) comes from a one-way analysis of variance (Chapter 5), showing the details of an F-test that compares the separate-means model to the equal-means model for the insulating fluid example of Section 8.1.2. Table (a) comes from a simple linear regression analysis showing the details of an F-test that compares the simple linear regression model to the equal-means model.

If β_1 is zero, the simple linear regression model reduces to the equal-means model, $\mu\{Y|X\} = \beta_0$. The hypothesis that $\beta_1 = 0$ can therefore be tested with a comparison of the sizes of the residuals from fits to these two models. An analysis of

DISPLAY 8.8 Analysis of variances tables for the insulating fluid data from a simple linear regression analysis and from a separate-means (one-way ANOVA) analysis

(a): Analysis of variance table from a simple linear regression analysis

Source	Sum of squares	d.f.	Mean square	F-statistic	p-value
Regression	190.1514	1	190.1514	78.14	<0.0001
Residual	180.0745	74	2.4334		
Total	370.2258	75			

Residual sum of squares, regression model (180.0745)

$\hat{\sigma}^2$ in regression model (2.4334)

compares regression and equal-means models (p-value)

(b): Analysis of variance table from a one-way analysis of variance

Source	Sum of squares	d.f.	Mean square	F-statistic	p-value
Between groups	196.4774	6	32.7462	13.00	<0.0001
Within groups	173.7484	69	2.5181		
Total	370.2258	75			

Residual sum of squares, separate-means model (173.7484)

$\hat{\sigma}^2$ in separate-means model (2.5181)

compares separate-means and equal-means models (p-value)

variance table associated with simple linear regression displays the components of the F-statistic for this test (Display 8.8(a)). Its two main features are the following: The p-value is the same as the two-sided p-value for the t-test of H_0: $\beta_1 = 0$ (the F-statistic is the square of the t-statistic); and the residual mean square is the estimate of variance, $\hat{\sigma}^2$.

The sums of squares column contains the important working pieces. In Table (a), the *residual sum of squares* is the sum of squares of residuals from the regression model, while the *total sum of squares* is the sum of squares of residuals from the equal-means model. The *regression sum of squares* is their difference; that is, it is the amount by which the residual sum of squares decreases when the model for the mean of Y is generalized by adding $\beta_1 X$. The generalization adds one parameter to the model, and that is the number of degrees of freedom associated with the regression sum of squares.

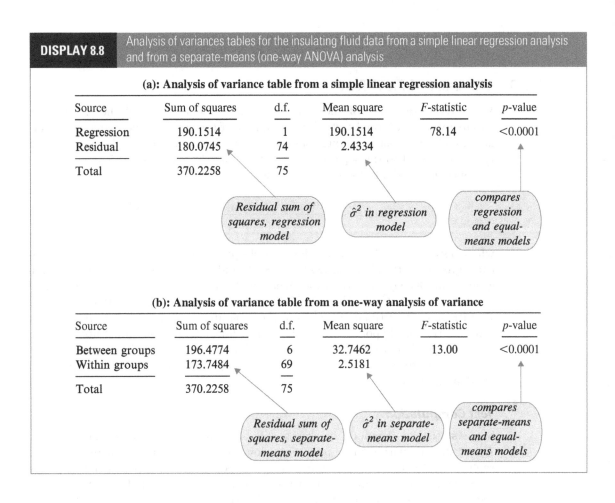

Except for differences in terminology, this is completely analogous to the sums of squares column in Table (b), where the *between sum of squares* is the amount by which the residual sum of squares is reduced by the generalization from the equal-means to the separate-means model, involving the change from one to seven parameters (difference = 6 parameters added).

8.5.3 The Lack-of-Fit *F*-Test

When replicate values of the response occur at some or all of the explanatory variable values, a formal test of the adequacy of the straight-line regression model is available. The test is an extra-sum-of-squares F-test comparing the simple linear (reduced) model to the separate means (full) model. (See Sections 5.3 and 5.4.)

Specifically, the lack-of-fit F-statistic is

$$F\text{stat} = \frac{[\text{SSRes}_{\text{LR}} - \text{SSRes}_{\text{SM}}]/\,[\text{d.f.}_{\text{LR}} - \text{d.f.}_{\text{SM}}]}{\hat{\sigma}^2_{\text{SM}}},$$

where SSRes_{LR} and SSRes_{SM} are the sums of squares of residuals from the simple linear regression ("Residual" in Display 8.8(a)) and separate-means models ("Within groups" in Display 8.8(b)), respectively; and d.f._{LR} and d.f._{SM} are the degrees of freedom associated with these residual sums of squares. The denominator of the F-statistic is the estimate of σ^2 from the separate-means model. The p-value, for a test of the null hypothesis that the simple linear regression model fits, is found as the proportion of values from an F distribution that exceed the F-statistic. The numerator degrees of freedom are $\text{d.f.}_{\text{LR}} - \text{d.f.}_{\text{SM}}$, and the denominator degrees of freedom are d.f._{SM}.

Test Computation

Some statistical computer packages will automatically compute the lack-of-fit F-test statistic within the simple regression procedure, as long as there are indeed replicates. Others might perform this test if requested. If neither of these options is available, the user must obtain the analysis of variance tables from both the regression fit and the one-way analysis of variance, as in Display 8.8, extract the necessary information, and compute the F-statistic manually.

Example—Insulating Fluid

From Display 8.8, $\text{SSRes}_{\text{LR}} = 180.0745$, $\text{SSRes}_{\text{SM}} = 173.7484$, $\text{d.f.}_{\text{LR}} = 74$, and $\text{d.f.}_{\text{SM}} = 69$. So the F-statistic is $[(180.0745 - 173.7484)/(74 - 69)]/2.5181 = 0.502$. The proportion of values from an F-distribution on 5 and 69 degrees of freedom that exceed 0.502 is 0.78. This large p-value provides no evidence of lack-of-fit to the simple linear regression model. A small p-value would have suggested that the variability between group means cannot be explained by a simple linear regression model.

8.5.4 A Composite Analysis of Variance Table

Since the simple linear regression model is intermediate between the equal-means and the separate-means models, the reduction in residual sum of squares (196.4774) associated with generalizing from the equal-means model to the separate-means

DISPLAY 8.9 Reductions in sums of squared residuals in hierarchical models for mean responses in the insulating fluid study

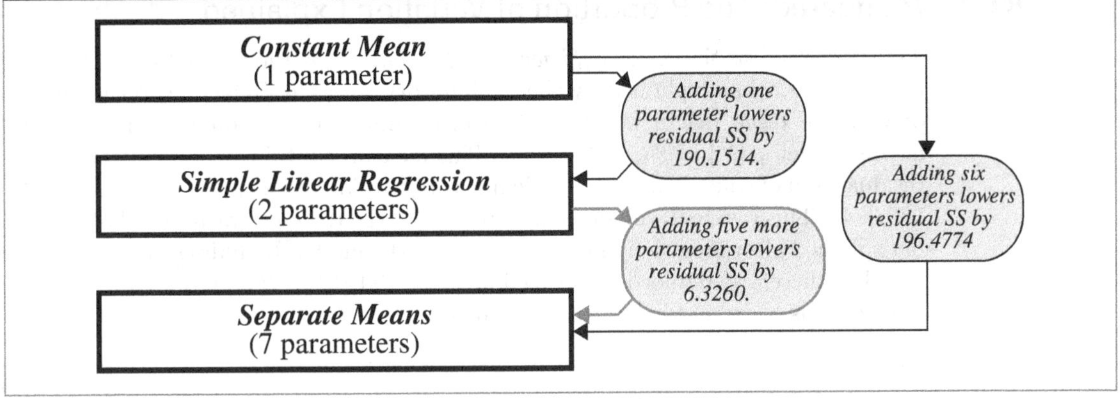

DISPLAY 8.10 Composite analysis of variance table with *F*-test for lack of fit

Source of variation	Sum of squares	d.f.	Mean square	*F*-statistic	*p*-value
Between groups	*196.4774*	*6*	*32.7462*	*13.00*	<0.0001
Regression	190.1514	1	190.1514	75.51	<0.0001
Lack of fit	**6.3260**	**5**	**1.2652**	**0.50**	**0.78**
Within groups	*173.7484*	*69*	*2.5181*		
Total	370.2258	75			

By subtraction

LEGEND

Normal type items come from regression analysis (a).
Italicized items come from separate-means analysis (b).
Boldface items are new and calculated here.

model can be broken up into contributions associated with an initial generalization to the simple linear regression model (190.1514) and with a further generalization from there to the separate-means model. (See Display 8.9.) The amount associated with the latter step is the difference between the two, $196.4774 - 190.1514 = 6.3260$. Therefore, a useful composite of the analysis of variance tables (a) and (b) in Display 8.8 is given in Display 8.10.

Essentially this is the same table as the analysis of variance table for the separate-means model, Display 8.8 (b), except that the "Between group" sum of squares has been broken into two components—one that measures the pattern in the means that follows a straight line, and one that measures any patterns that fail to follow the straight line model. The latter is called the *lack-of-fit* component. The mean squares in each row are the sum of squares divided by degrees of freedom.

8.6 RELATED ISSUES

8.6.1 *R*-Squared: The Proportion of Variation Explained

The *R-squared* statistic, or *coefficient of determination*, is the percentage of the total response variation explained by the explanatory variable. Referring to the analysis of variance table (a) in Display 8.8, the residual sum of squares for the equal-means model, $\mu\{Y\} = \beta_0$, is 370.2258. This measures total response variation. The residual sum of squares for the simple linear regression model $\mu\{Y|X\} = \beta_0 + \beta_1 X$ is 180.0745. This residual sum of squares measures the response variation that remains unexplained after inclusion of the $\beta_1 X$ term in the model. Including the explanatory variable therefore reduced the variability by 190.1513. *R*-squared expresses this reduction as a percentage of total variation:

$$R^2 = 100\left(\frac{\text{Total sum of squares} - \text{Residual sum of squares}}{\text{Total sum of squares}}\right)\%.$$

For the insulating fluids example,

$$R^2 = 100\left(\frac{370.2258 - 180.0745}{370.2258}\right)\% = 51.4\%.$$

This should be read as "Fifty-one percent of the variation in log breakdown times was explained by the linear regression on voltage." The statement is phrased in the past tense because R^2 describes what happened in the analyzed data set.

Notice that if the residuals are all zero (a perfect fit), then R^2 is 100%. At the other extreme, if the best-fitting regression line has slope 0 (and therefore intercept $\overline{Y}$), the residuals will be exactly equal to $Y_i - \overline{Y}$, the residual sum of squares will be exactly equal to the total sum of squares, and R^2 will be zero.

A judgment about what constitutes "good" values for R^2 depends on the context of the study. In precise laboratory work, R^2 values under 90% may be low enough to require refinements in technique or the inclusion of other explanatory information. In some social science contexts, however, where a single variable rarely explains a great deal of the variation in a response, R^2 values of 50% may be considered remarkably good.

For simple linear regression, R^2 is identical to the square of the sample correlation coefficient for the response and the explanatory variable. As with the correlation, R^2 only estimates some population quantity if the (X, Y) pairs are randomly drawn from a population of pairs. It should not be used for inference, and it should never be used to assess the adequacy of the straight line model, because R^2 can be quite large even when the simple linear regression model is inadequate.

DISPLAY 8.11 Estimates of group means and standard errors using different approaches

Means estimated by:
(1) sample averages
(2) regression model

Voltage Level (kV)	n	Average log (BDT)	Internal SE	Pooled SE	Regression Estimate	Regression SE
26	3	5.6240	1.9371	0.9162	5.7640	0.4467
28	5	5.3295	0.5119	0.7907	4.7492	0.3446
30	11	3.8220	0.3350	0.4785	3.7345	0.2536
32	15	2.2285	0.5675	0.4097	2.7198	0.1904
34	19	1.7864	0.3499	0.3640	1.7050	0.1858
36	15	0.9022	0.2866	0.4097	0.6903	0.2432
38	8	−0.4243	0.3506	0.5610	−0.3244	0.3318

Accuracy estimated by:
(i) sample SD/$\sqrt{n}$
(ii) pooled SD/$\sqrt{n}$
(iii) regression method

8.6.2 Simple Linear Regression or One-Way Analysis of Variance?

When the data are arranged in groups that correspond to different levels of an explanatory variable (as in the meat processing and insulating fluid studies), the statistical analysis may be based on either simple linear regression or one-way analysis of variance. The choice between these two techniques is straightforward: *If the simple linear regression model fits* (possibly after transformation), then it is preferred. The regression approach accomplishes four things: It allows for interpolation, it gives more degrees of freedom for error estimation, it gives smaller standard errors for estimates of the mean responses, and it provides a simpler model.

As an illustration of the increased precision from using regression, Display 8.11 shows estimated means in the insulating fluid example from the separate-means model (i.e., the group averages), and the simple linear regression model (i.e., the fitted values). It also shows three standard errors for the estimated means: the internal standard error of the mean, which is the one-sample standard error (equal to the sample SD over the square root of the sample size); the pooled standard error of the mean, which is the one-way analysis of variance standard error (equal to the pooled SD over the square root of the sample size); and the standard error of the estimated mean response from the regression model (Section 7.4.2). The two nonregression standard errors are comparable in size (but, of course, confidence intervals will be narrower in the second case because of the larger degrees of freedom attached to the pooled SD). The standard errors from the regression model are uniformly smaller, however. Since the mean (at $X = 26$, say) is postulated to be $\beta_0 + \beta_1 26$, all 76 observations are used to estimate this, not just those from group 1.

DISPLAY 8.12 Possible patterns in plots of residuals versus time order of data collection

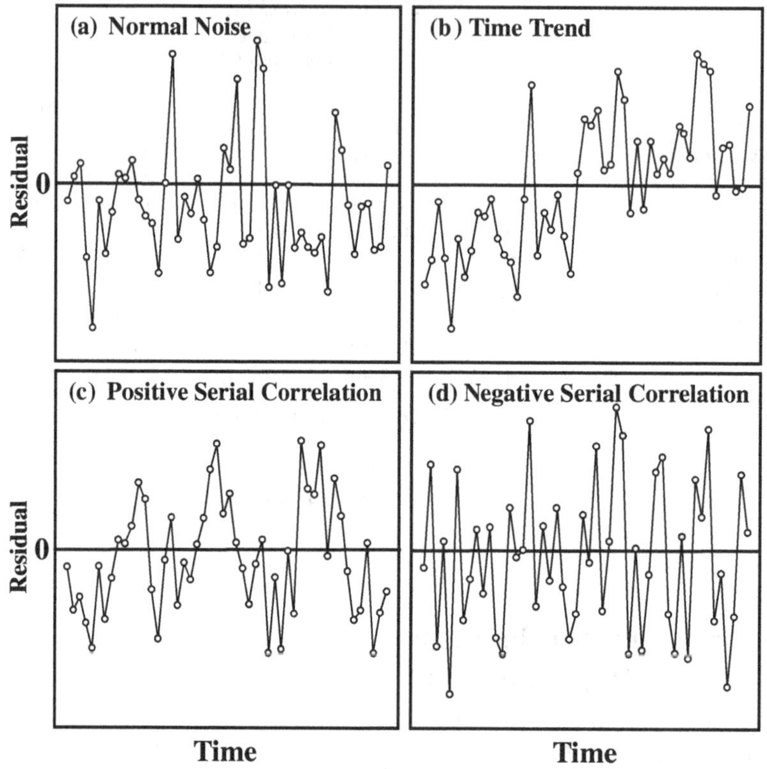

As a general rule, it is appropriate to find the simplest model—the one with the fewest parameters—that adequately fits the data. This ensures both the most precise answers to the questions of interest and the most straightforward interpretation.

8.6.3 Other Residual Plots for Special Situations

Residuals Versus Time Order

If the data are collected over time (or space), serial correlation may occur. A plot of the residuals versus the time order of data collection may be examined for patterns. Four possibilities are shown in Display 8.12. In part (a) no pattern emerges, and no serial correlation is indicated. In part (b) a linear trend over time can be observed. It may be possible to include time as an additional explanatory variable (using multiple linear regression, as in Chapter 9). The pattern in part (c) shows a positive serial correlation, in which residuals tend to be followed in time by residuals of the same sign and of about the same size. The pattern in part (d) shows a negative serial correlation, where residuals of one sign tend to be followed by residuals of the opposite sign. The situations in parts (c) and (d) require time series techniques (Chapter 15).

Normal Probability Plots

The *normal probability plot* is a scatterplot involving the ordered residuals and a set of expected values of an ordered sample of the same size from a standard normal distribution. These expected values are available from statistical theory and are not described further here. Some statistical packages plot the expected values along the y-axis against the ordered residuals along the x-axis (as *The Statistical Sleuth* does), while other packages reverse the axes. Assessing normality visually from a normal plot is easier than from a histogram.

Several normal probability plots for hypothetical data are shown in Display 8.13. When subpopulations have normal distributions, the normal plot is approximately a straight line (a) In a long-tailed distribution (b) residuals in the tails have wider gaps than expected in a normal distribution, with the gaps increasing as one gets further into the tails. A skewed distribution (c) will show wider gaps in one tail and shorter gaps in the other, with a reasonably smooth progression of increase in the longer tail. This is in contrast to a situation with outliers (d) where the bulk of the distribution appears normal with one or a few exceptional cases.

DISPLAY 8.13 Normal probability plots illustrating four distributional patterns

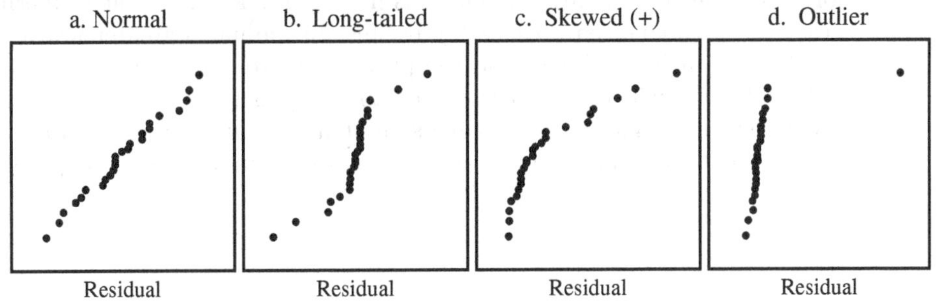

Normal probability plots for the residuals from the meat processing data and the insulating fluid data are shown in Display 8.14. The patterns are close enough to a straight line that the prediction intervals based on the normal assumption should be adequate. (The normal distribution of observations about the regression line is important for the validity of prediction intervals, but not for the validity of estimates, tests, and confidence intervals.)

8.6.4 Planning an Experiment: Balance

Balance means having the same number of experimental units in each treatment group. The insulating fluid data are *unbalanced*, because there are three

DISPLAY 8.14 Normal probability plots of residuals from simple linear regression fits to the meat processing data and the insulating fluid data

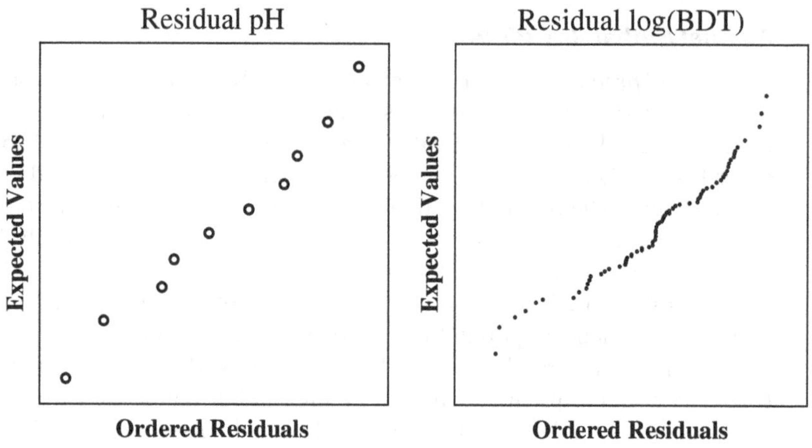

observations in the first treatment group, five in the next, and so on. For the several-treatment experiment, balance is generally desirable in providing equal accuracy for all treatment comparisons, but it is not essential. The voltage experiment was designed as unbalanced presumably because of the much greater waiting time for breakdowns at low voltages and the primary interest in voltages between 30 and 36 kV. Balance will play a more important role when the data are cross-classified according to two factors since some simplifying formulas are only appropriate for balanced data and, more importantly, it allows unambiguous decomposition in the analysis of variance.

8.7 SUMMARY

Exploring statistical relationships begins with viewing scatterplots. Nonlinear regressions, nonconstant spread, and outliers can often be identified at this stage. In cases where problems are less apparent, a simple linear regression can be fit tentatively, and the decision about its appropriateness can be based on the residual plot.

When replicate response variables occur at some of the explanatory variable values, it is possible to conduct a formal lack-of-fit F-test. The test is a special case of the extra-sum-of-squares F-test for comparing two models. The models involved are the simple linear regression (reduced) model and the separate-means (full) model.

Insulating Fluid and Species–Area Studies

Scatterplots and residual plots for the insulating fluid data and for the species–area data reveal nonconstant spread and nonlinear regressions, suggesting transformation of the response variable. In both cases, the spread increases as the mean level increases, indicating a logarithmic, square root, or reciprocal transformation. The scatterplot does not indicate which transformation is best. Several may be tried, with the final choice depending on what is appropriate to the scientific context of the study and to the statistical model assumptions.

After a logarithmic transformation of the times to breakdown, a simple linear regression model fits the insulating fluid data well. No evidence (from a residual plot and a lack-of-fit test) indicates lack of fit or (from the normal plot nonnormality) anything but a normal distribution of the residuals. Mean estimation and prediction can proceed from that model, with results back-transformed to the original scale. Other approaches are possible—another sensible analysis of these data assumes the Weibull distribution on the original scale and gives similar results.

To estimate the parameters in the species–area study, both response and explanatory variables are log-transformed. Here the logarithmic transformations to a simple linear regression model are indicated by theoretical model considerations. Weak evidence remains of increasing variability in the residual plot and of long-tailedness in the normal plot. These data should not, however, be used for predictions. With this small sample size, no further action is required, but confidence limits should be described as approximate.

8.8 EXERCISES

Conceptual Exercises

1. Island Area and Species Count. The estimated regression line for the data of Section 8.1.1 is $\hat{\mu}\{\text{log species} \mid \text{log area}\} = 1.94 + 0.250 \log(\text{area})$. Show how this estimates that islands of area $0.5A$ have a median number of species that is 16% lower than the median number of species for islands of area A.

2. Insulating Fluid. For the insulating fluid data of Section 8.1.2 explain why the regression analysis allows for statements about the distribution of breakdown times at 27 kV while the one-way analysis of variance does not.

3. Big Bang. In the data set of Section 7.1.1 multiple distances were associated with a few recession velocities. Would it be possible to perform the lack-of-fit test for these data?

4. Insulating Fluid. If the sample correlation coefficient between the square root of breakdown time and voltage is −0.648, what is R^2 for the regression of square root of breakdown time on voltage?

5. Why can an R^2 close to 1 not be used as evidence that the simple linear regression model is appropriate?

6. A study is made of the stress response exhibited by a sample of 45 adults to rock music played at nine different volume levels (five adults at each level). What is the difference between using the volume as an explanatory variable in a simple linear regression model and using the volume level as a group designator in a one-way classification model?

7. In a study where four levels of a magnetic resonance imaging (MRI) agent are each given to 3 cancer patients (so there are 12 patients in all), the response is a measure of the degree of seizure activity (an unpleasant side effect). The F-test for lack of fit to the simple linear regression model with X = agent level has a p-value of 0.0082. The t-tools estimate that the effect of increasing the level of the MRI agent by 1 mg/cm^2 is to increase the level of seizure activity by 2.6 units (95% confidence interval from 1.8 to 3.4 units). (a) How should the latter inference be interpreted? (b) How many degrees of freedom are there for (i) the within-group variation? (ii) the lack-of-fit variation?

8. Insulating Fluid. Why would it be of interest to know whether batches of insulating fluid were *randomly* assigned to the different voltage levels?

9. Suppose the (Y, X) pairs are: (5,1), (3,2), (4,3), (2,4), (3,5), and (1,6). Would the least squares fit to these data be much different from the least squares fit to the same data with the first pair replaced by (15,1)?

10. (a) What assumptions are used for exact justification of tests and confidence intervals for the slope and intercept in simple regression? (b) Are any of these assumptions relatively unimportant?

11. Suppose you had data on pairs (Y, X) which gave the scatterplot shown in Display 8.15. How would you approach the analysis?

DISPLAY 8.15 Scatterplot for Exercise 11

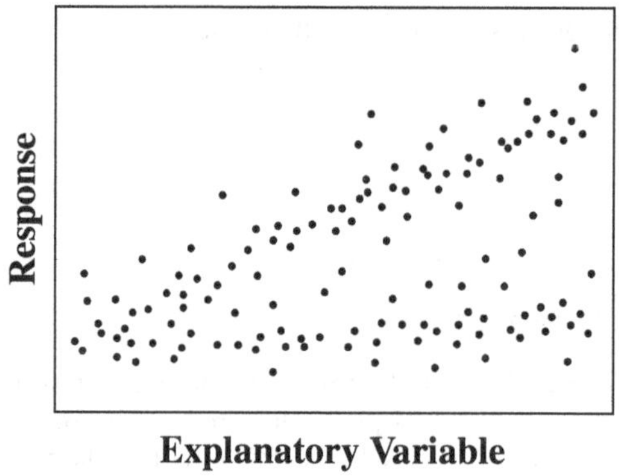

12. What is the technical difficulty with using the separate-means model as a basis for the lack-of-fit F-test when there are no replicate responses?

13. Researchers at a university wish to estimate the effect of class size on course comprehension. An intermediate course in statistics can be taught to classes of any size between 25 and 185 students, and four instructors are available. Suppose the researchers truly believe that the average course comprehension, measured by the average of student scores on a standardized test, is indeed a straight line in class sizes over the range from 25 to 185. What four class sizes should be used in the experiment? Why?

14. Insulating Fluid. Which would you use to predict the log breakdown time for a batch of insulating fluid which is to be put on test at 30 kV: the regression estimate of the mean at 30 kV or the average from the batches that were tested at 30 kV? Why?

Computational Exercises

15. Island Size and Species. (a) Draw a scatterplot of the (untransformed) number of species on the (untransformed) area of the island (Display 8.2, top). (b) Fit the simple linear regression of number of species on area and obtain a residual plot. (c) What features in the two plots indicate a need for transformation?

16. Meat Processing. The data in Display 7.3 are a subset of the complete data on postmortum pH in 12 steer carcasses. (Data from J. R. Schwenke and G. A. Milliken, "On the Calibration Problem Extended to Nonlinear Models," *Biometrics* 47(2) (1991): 563–74). Once again, the purpose is to determine how much time after slaughter is needed to ensure that the pH reaches 6.0. In Chapter 7 the simple linear regression of pH on log(Hour) was fit to the first 10 carcasses only. Refit the model with all 12 carcasses (data given in Display 8.16). (a) Assess lack of fit using a residual plot. (b) Assess lack of fit using the lack-of-fit F-test. (c) The inappropriateness of the simple linear regression model can be remedied by dropping the last two carcasses. Is there any justification for doing so? (*Hint:* In order to answer the question of interest, what range of X's appears to be important?)

DISPLAY 8.16 pH of steer carcasses 1 to 24 hours after slaughter

Animal Number:	1	2	3	4	5	6	7	8	9	10	11	12
Processing Hour:	1	1	2	2	4	4	6	6	8	8	24	24
pH:	7.02	6.93	6.42	6.51	6.07	5.99	5.59	5.80	5.51	5.36	5.30	5.47

17. Biological Pest Control. In a study of the effectiveness of biological control of the exotic weed tansy ragwort, researchers manipulated the exposure to the ragwort flea beetle on 15 plots that had been planted with a high density of ragwort. Harvesting the plots the next season, they measured the average dry mass of ragwort remaining (grams/plant) and the flea beetle load (beetles/gram of ragwort dry mass) to see if the ragwort plants in plots with high flea beetle loads were smaller as a result of herbivory by the beetles. (Data from P. McEvoy and C. Cox, "Successful Biological Control of Ragwort, *Senecio jacobaea*, by Introduced Insects in Oregon," *Ecological Applications* 1(4) (1991): 430–42. The data in Display 8.17 were read from McEvoy and Cox, Figure #2.)

DISPLAY 8.17 Dry mass of ragwort weed on 15 plots exposed to flea beetles

Plot #:	1	2	3	4	5	6	7	8	9	10	11	12	13	14	15
Flea beetle load:	12.2	14.6	15.8	25.3	38.6	76.4	163	182	415	446	628	377	770	1,446	1,012
Ragwort mass:	18.2	17.5	7.22	30.6	6.66	6.14	5.21	0.502	0.611	0.630	0.427	0.011	0.012	0.006	0.002

(a) Use scatterplots of the raw data, along with trial and error, to determine transformations of Y = Ragwort dry mass and of X = Flea beetle load that will produce an approximate linear relationship.

(b) Fit a linear regression model on the transformed scale; calculate residuals and fitted values.

(c) Look at the residual plot. Do you want to try other transformations? What do you suggest?

18. Distance and Order from Sun. Reconsider the planetary distance and order from sun data in Exercise 7.21. Fit a regression model to the data that includes the asteroid belt and fill in the blanks in this conclusion: Aside from some random variation, the distance to the sun increases by __% with each consecutive planet number (95% confidence interval: ____ to ____% increase).

19. Pollen Removal. Reconsider the pollen removal data of Exercise 3.27 and the regression of pollen removed on time spent on flower, for the bumblebee queens only. (a) What problems are evident in the residual plot? (b) Do log transformations of Y or X help any? (c) Try fitting the regression only for those times less than 31 seconds (i.e., excluding the two longest times). Does this fit better? (*Note*: If the linear regression fits for a restricted range of the X's, it is acceptable to fit the model with all the other X's excluded and to report the range of X's for which the model holds.)

20. Quantifying Evidence for Outlierness. In a special election to fill a Pennsylvania State Senate seat in 1993, the Democratic candidate, William Stinson, received 19,127 machine-counted votes and the Republican, Bruce Marks, received 19,691 (i.e., 564 more votes than the Democrat). In addition, however, Stinson received 1,396 absentee ballots and Marks received only 371, so the total tally showed the Democrat, Stinson, winning by 461 votes. The large disparity between the machine-counted and absentee ratios, and the resulting reversal of the outcome due to the absentee ballots, sparked concern about possible illegal influence on the absentee votes. Investigators reviewed data on disparities between machine and absentee votes in prior Pennsylvania State Senate elections to see whether the 1993 disparity was larger than could be explained by historical variation. Display 8.18 shows the data in the form of percentage of absentee and machine-counted ballots cast for the Democratic candidate. The task is to clarify the unusualness of the Democratic absentee percentage in the disputed election. (a) Draw a scatterplot of Democratic percentage of absentee ballots versus Democratic percentage of machine-counted ballots. Use a separate plotting symbol to highlight the disputed election. (b) Fit the simple linear regression of absentee percentage on machine-count percentage, *excluding* the disputed election. Draw this line on the scatterplot. Also include a 95% prediction band. What does this plot reveal about the unusualness of the absentee percentage in the disputed election? (c) Find the prediction and standard error of prediction from this fit if the machine-count percentage is 49.3 (as it is for the disputed election). How many estimated standard deviations is the observed absentee percentage, 79.0, from this predicted value? Compare this answer to a t-distribution (with degrees of freedom equal to the residual degrees of freedom in the regression fit) to obtain a p-value. (d) **Outliers and data snooping.** The p-value in (c) makes sense if the investigation into the 1993 election was prompted by some other reason. Since it was prompted because the absentee percentage seemed too high, however, the p-value in (c) should be adjusted for data snooping. Adjust the p-value with a Bonferroni correction to account for all 22

DISPLAY 8.18 Partial listing of 22 Pennsylvania state senate election results, collected to explore the peculiarity noticed in the 1993 election between William Stinson and Bruce Marks (election number 22 in the data set). The full data set includes the year of the election, the senate district, the number of absentee ballots cast for the Democratic candidate and for the Republican candidate, and the number of machine-counted ballots cast for the Democratic candidate and for the Republican candidate. Also shown are the percentages of absentee and machine ballots cast for the Democratic candidate.

Election	Year	District	DemPctOfAbsenteeVotes	DemPctOfMachineVotes	Disputed
1	82	D2	72.9	69.1	no
2	82	D4	65.6	60.9	no
3	82	D8	74.6	80.8	no
4	84	D1	64.0	60.0	no
5	84	D3	83.3	92.4	no
. . .					
22	93	D2	79.0	49.3	yes

residuals that could have been similarly considered. (Data from Orley Ashenfelter, 1994. Report on Expected Absentee Ballots. Typescript. Department of Economics, Princeton University. See also Simon Jackman (2011). pscl: Classes and Methods for R Developed in the Political Science Computational Laboratory, Stanford University. Department of Political Science, Stanford University. Stanford, California. R package version 1.03.10. URL http://pscl.stanford.edu/)

21. Fish Preferences. Reconsider Case Study 2 in Chapter 6, the study of female preferences among platyfish. Fit the full model in which the mean preference for the yellowtailed male is possibly different for each male pair. Construct a normal probability plot of the residuals and a residual plot. If these suggest a transformation, make it and repeat the analysis including the linear contrast measuring the association of preference with male body size. If they suggest an outlier problem, use the inclusion/exclusion procedure to determine whether the outlying case(s) change(s) the answer to the questions of interest. Also, identify the outlying case(s) and suggest why it might be a true outlier.

Data Problems

22. Ecosystem Decay. As an introduction to their study on the effect of Amazon forest clearing (data from T. E. Lovejoy, J. M. Rankin, R. O. Bierregaard, Jr., K. S. Brown, Jr., L. H. Emmons, and M. E. Van der Woot, "Ecosystem Decay of Amazon Forest Remnants," in M. H. Nitecki, ed., *Extinctions*, Chicago: University of Chicago Press, 1984) the researchers stated: "fragmentation of once continuous wild areas is a major way in which people are altering the landscape and biology of the planet." Their study takes advantage of a Brazilian requirement that 50% of the land in any development project remain in forest and tree cover. As a consequence of this requirement, "islands" of forest of various sizes remain in otherwise cleared areas. The data in Display 8.19 are the number of butterfly species in 16 such islands. Summarize the role of area in the distribution of number of butterfly species. Write a brief statistical report including a summary of statistical findings, a graphical display, and a section detailing the methods used to answer the questions of interest.

DISPLAY 8.19 Forest patch area (hectares) and number of butterfly species found

Reserve	Area	Species	Reserve	Area	Species
1	1	14	9	10	33
2	1	50	10	10	53
3	1	55	11	10	50
4	1	34	12	100	110
5	1	40	13	100	70
6	1	57	14	100	119
7	10	43	15	100	60
8	10	103	16	1,000	145

23. Wine Consumption and Heart Disease. The data in Display 8.20 are the average wine consumption rates (in liters per person) and number of ischemic heart disease deaths (per 1,000 men aged 55 to 64 years old) for 18 industrialized countries. (Data from A. S. St. Leger, A. L. Cochrane, and F. Moore, "Factors Associated with Cardiac Mortality in Developed Countries with Particular Reference to the Consumption of Wine," *Lancet* (June 16, 1979): 1017–20.) Do these data suggest that the heart disease death rate is associated with average wine consumption? If so, how can that

DISPLAY 8.20 First five rows of a data set with wine consumption (liters per person per year) and heart disease mortality rates (deaths per 1,000) in 18 countries

Country	Wine consumption	Heart disease mortality
Norway	2.8	6.2
Scotland	3.2	9.0
England	3.2	7.1
Ireland	3.4	6.8
Finland	4.3	10.2

relationship be described? Do any countries have substantially higher or lower death rates than others with similar wine consumption rates? Analyze the data and write a brief statistical report that includes a summary of statistical findings, a graphical display, and a section detailing the methods used to answer the questions of interest.

24. **Respiratory Rates for Children.** A high respiratory rate is a potential diagnostic indicator of respiratory infection in children. To judge whether a respiratory rate is truly "high," however, a physician must have a clear picture of the distribution of *normal* respiratory rates. To this end, Italian researchers measured the respiratory rates of 618 children between the ages of 15 days and 3 years. Display 8.21 shows a few rows of the data set. Analyze the data and provide a statistical summary. Include a useful plot or chart that a physician could use to assess a normal range of respiratory rate for children of any age between 0 and 3. (Data read from a graph in Rusconi et al., "Reference Values for Respiratory Rate in the First 3 Years of Life," *Pediatrics*, 94 (1994): 350–55.)

DISPLAY 8.21 Partial listing of data on ages (months) and respiratory rates (breaths per minute) for 618 children

Child	1	2	3	4	5	6	...	618
Age:	0.1	0.2	0.3	0.3	0.3	0.4	...	36.0
Rate:	53	38	58	52	42	62	...	31

25. **The Dramatic U.S. Presidential Election of 2000.** The U.S. presidential election of November 7, 2000, was one of the closest in history. As returns were counted on election night it became clear that the outcome in the state of Florida would determine the next president. At one point in the evening, television networks projected that the state was carried by the Democratic nominee, Al Gore, but a retraction of the projection followed a few hours later. Then, early in the morning of November 8, the networks projected that the Republican nominee, George W. Bush, had carried Florida and won the presidency. Gore called Bush to concede. While en route to his concession speech, though, the Florida count changed rapidly in his favor. The networks once again reversed their projection, and Gore called Bush to retract his concession. When the roughly 6 million Florida votes had been counted, Bush was shown to be leading by only 1,738, and the narrow margin triggered an automatic recount. The recount, completed in the evening of November 9, showed Bush's lead to be less than 400.

Meanwhile, angry Democratic voters in Palm Beach County complained that a confusing "butterfly" lay-out ballot caused them to accidentally vote for the Reform Party candidate Pat Buchanan instead of Gore. The ballot, as illustrated in Display 8.22, listed presidential candidates on both a left-hand and a right-hand page. Voters were to register their vote by punching the circle

DISPLAY 8.22 Confusing ballot in Palm Beach County, Florida

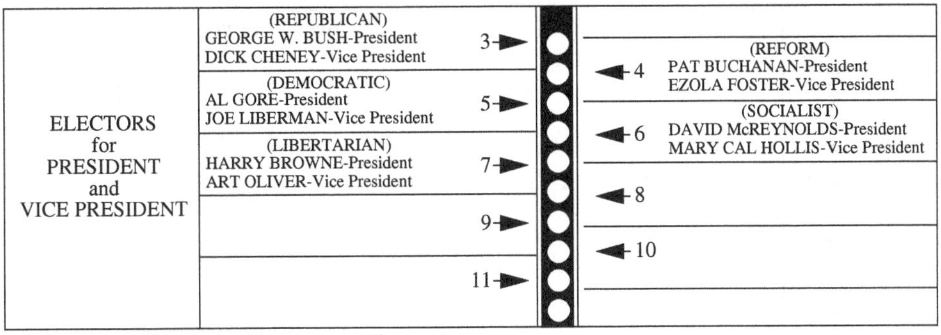

corresponding to their choice, from the column of circles between the pages. It was suspected that since Bush's name was listed first on the left-hand page, Bush voters likely selected the first circle. Since Gore's name was listed second on the left-hand side, many voters—who already knew who they wished to vote for—did not bother examining the right-hand side and consequently selected the second circle in the column; the one actually corresponding to Buchanan. Two pieces of evidence supported this claim: Buchanan had an unusually high percentage of the vote in that county, and an unusually large number of ballots (19,000) were discarded because voters had marked two circles (possibly by inadvertently voting for Buchanan and then trying to correct the mistake by then voting for Gore).

Display 8.23 shows the first few rows of a data set containing the numbers of votes for Buchanan and Bush in all 67 counties in Florida. What evidence is there in the scatterplot of Display 8.24 that Buchanan received more votes than expected in Palm Beach County? Analyze the data without Palm Beach County results to obtain an equation for predicting Buchanan votes from Bush votes. Obtain a 95% prediction interval for the number of Buchanan votes in Palm Beach from this result—assuming the relationship is the same in this county as in the others. If it is assumed that Buchanan's actual count contains a number of votes intended for Gore, what can be said about the likely size of this number from the prediction interval? (Consider transformation.)

DISPLAY 8.23 Votes for Bush and Buchanan in all Florida counties (first 5 of 67 rows)

County	Bush votes	Buchanan votes
Alachua	34,062	262
Baker	5,610	73
Bay	38,637	248
Bradford	5,413	65
Brevard	115,185	570

26. Kleiber's Law. Display 8.25 shows the first five rows of a data set with average mass, metabolic rate, and average lifespan for 95 species of mammals. (From A. T. Atanasov, "The Linear Allometric Relationship Between Total Metabolic Energy per Life Span and Body Mass of Mammals," *Biosystems* 90 (2007): 224–33.) Kleiber's law states that the metabolic rate of an animal species, on

DISPLAY 8.24 Votes for George W. Bush and Pat Buchanan in all Florida counties

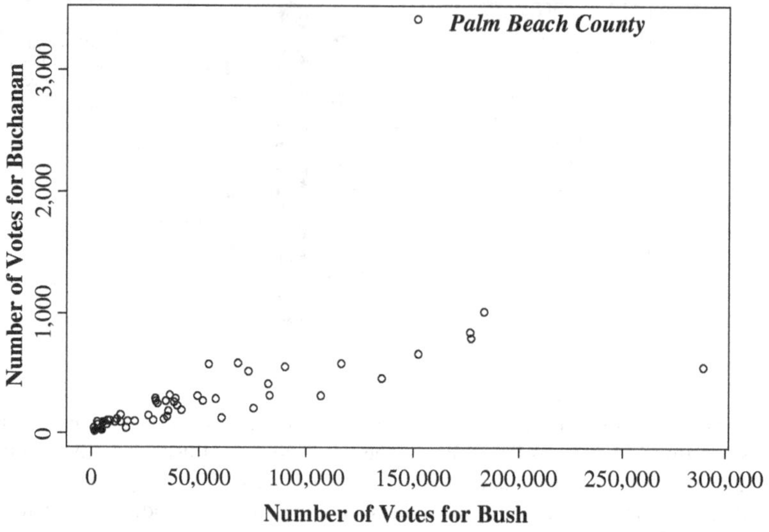

DISPLAY 8.25 Average mass (kg), average basal metabolic rate (kJ per day), and lifespan (years) for 95 mammal species; first 5 of 95 rows

CommonName	Species	Mass	Metab	Life
Echidna	*Tachiglossus aculeatus*	2.50E+00	3.02E+02	14
Long-beaked echidna	*Zaglossus bruijni*	1.03E+01	5.94E+02	20
Platypus	*Ornithorhynchus anatinus*	1.30E+00	2.29E+02	9
Opossum	*Lutreolina crassicaudata*	8.12E−01	1.96E+02	5
South American opossum	*Didelphis marsupialis*	1.33E+00	2.99E+02	6

average, is proportional to its mass raised to the power of 3/4. Judge the adequacy of this theory with these data.

27. Metabolic Rate and Lifespan. It has been suggested that metabolic rate is one of the best single predictors of species lifespan. Analyze the data in Exercise 26 to describe an equation for predicting mammal lifespan from metabolic rate. Also provide a measure of the amount of variation in the distribution of mammal lifespans that can be explained by metabolic rate.

28. IQ, Education, and Future Income. Display 8.26 is a partial listing of a data set with IQ scores in 1981, years of education completed by 2006, and annual income in 2005 for 2,584 Americans who were selected for the National Longitudinal Study of Youth in 1979 (NLSY79), who were re-interviewed in 2006, and who had paying jobs in 2005. (See Exercises 2.22 and 3.30 for a more detailed description of the survey.) (a) Describe the distribution of 2005 income as a function of IQ test score. What percentage of variation in the distribution is explained by the regression? (b) Describe the distribution of 2005 income as a function of years of education. What percentage of variation in the distribution is explained by the regression?

DISPLAY 8.26 IQ test score from 1981 (AFQT—armed forces qualifying test score), number of years of education, and annual income in 2005 (dollars) for 2,584 Americans in the NLSY79 survey; first 5 of 2,584 rows

Subject	AFQT	Educ	Income2005
2	6.841	12	5,500
6	99.393	16	65,000
7	47.412	12	19,000
8	44.022	14	36,000
9	59.683	14	65,000

DISPLAY 8.27 Autism prevalence per 10,000 ten-year olds in each of five years

Year	Prevalence
1992	3.5
1994	5.3
1996	7.8
1998	11.8
2000	18.3

29. Autism Rates. Display 8.27 shows the prevalence of autism per 10,000 ten-year old children in the United States in each of five years. Analyze the data to describe the change in the distribution of autism prevalence per year in this time period. (Data from C. J. Newschaffer, M. D. Falb, and J. G. Gurney, "National Autism Prevalence Trends From United States Special Education Data," *Pediatrics*, 115 (2005): e277–e282.)

Answers to Conceptual Exercises

1. Median {species|area} = exp(1.94)area$^{0.250}$. So Median{species | 0.5 area}/Median{species | area} = $0.5^{.250}$ = 0.84. Thus Median{species | 0.5 area} = 0.84 Median{species | area} and, finally, [Median{species | area} − Median{species | 0.5 area}]/Median{species | area} = 1 − 0.84 = 0.16.

2. The ANOVA model states that there are seven means, one for each voltage level tested, but does not describe any relation between mean and voltage. The regression model establishes a pattern between mean log breakdown time and voltage, for all voltages in the range of 26 to 38 kV.

3. Yes. (*Note*: The multiple observations occurred because of rounding, so they do not represent repeated draws from the same distribution. Nevertheless, they are near replicates, at least, and can be used as such for the lack-of-fit test.)

4. R^2 = square of correlation coefficient = $(-0.648)^2 = 0.420$.

5. Although a high R^2 reflects a strong degree of linear association, this linear association may well be accompanied by curvature (and by nonconstant variance).

6. In the simple linear regression model, the nine mean stress levels lie on a straight line against volume. In the one-way classification (the separate-means model) the mean stress levels may or may not lie on the straight line—their values are not restricted.

7. (a) If the data do not fit the model, then the parameters of the model are not adequate descriptive summaries. No inference should be drawn, at least until a better model is found. (b) (i) 8; (ii) 2.

8. Even in laboratory circumstances, confounding variables are possible. If, for example, batches are spooned from a large container that has density stratification, then assigning consecutive batches to the lowest voltage, then the next lowest, and so on, will confound fluid density with voltage. Randomization never hurts, and usually helps.

9. Yes, very much so. The least squares method is not resistant to the effects of outliers.

10. (a) linearity, constant variance, normality, independence. (b) normality.

11. It may appear that this is a case where the spread of Y increases as the mean of Y does, so a simple transformation may be in order. This will not produce good results. Notice that the average Y at a large X lies in a no-man's land with few observations. Looking at the distributions in strips, you will see two separate groups in each distribution. Look for some important characteristic that separates the data into two groups, then build a separate regression model for each group.

12. In fitting the separate-means model, all residuals are zero. The denominator of the F-statistic is zero, so the F-statistic is not defined.

13. Put two classes at 25 students and two at 185 students. This makes the standard deviation in the sampling distribution of the slope parameter as small as possible, given the constraints of the situation. (But it gives no information on lack of fit; check the degrees of freedom.)

14. The regression estimate. It is more precise (see Display 8.11).

CHAPTER 9

Multiple Regression

Multiple regression analysis is one of the most widely used statistical tools, and for good reason: It is remarkably effective for answering questions involving many variables. Although more difficult to visualize than simple regression, multiple regression is a straightforward extension. It models the mean of a response variable as a function of *several* explanatory variables.

Many issues, tools, and strategies are associated with multiple regression analysis, as discussed in this and the next three chapters. This chapter focuses on the meaning of the regression model and strategies for analysis. The details of estimation and inferential tools are deliberately postponed until the next chapter so that the student may concentrate first on understanding regression coefficients and the types of data structures that may be analyzed with multiple regression analysis.

9.1 CASE STUDIES

9.1.1 Effects of Light on Meadowfoam Flowering—A Randomized Experiment

Meadowfoam (*Limnanthes alba*) is a small plant found growing in moist meadows of the U.S. Pacific Northwest. It has been domesticated at Oregon State University for its seed oil, which is unique among vegetable oils for its long carbon strings. Like the oil from sperm whales, it is nongreasy and highly stable.

Researchers reported the results from one study in a series designed to find out how to elevate meadowfoam production to a profitable crop. In a controlled growth chamber, they focused on the effects of two light-related factors: light intensity, at the six levels of 150, 300, 450, 600, 750, and 900 μmol/m^2/sec; and the timing of the onset of the light treatment, either at photoperiodic floral induction (PFI)—the time at which the photo period was increased from 8 to 16 hours per day to induce flowering—or 24 days before PFI. The experimental design is depicted in Display 9.1. (Data from M. Seddigh and G. D. Jolliff, "Light Intensity Effects on Meadowfoam Growth and Flowering," *Crop Science* 34 (1994): 497–503.)

DISPLAY 9.1 Time line for light variation experiment on meadowfoam

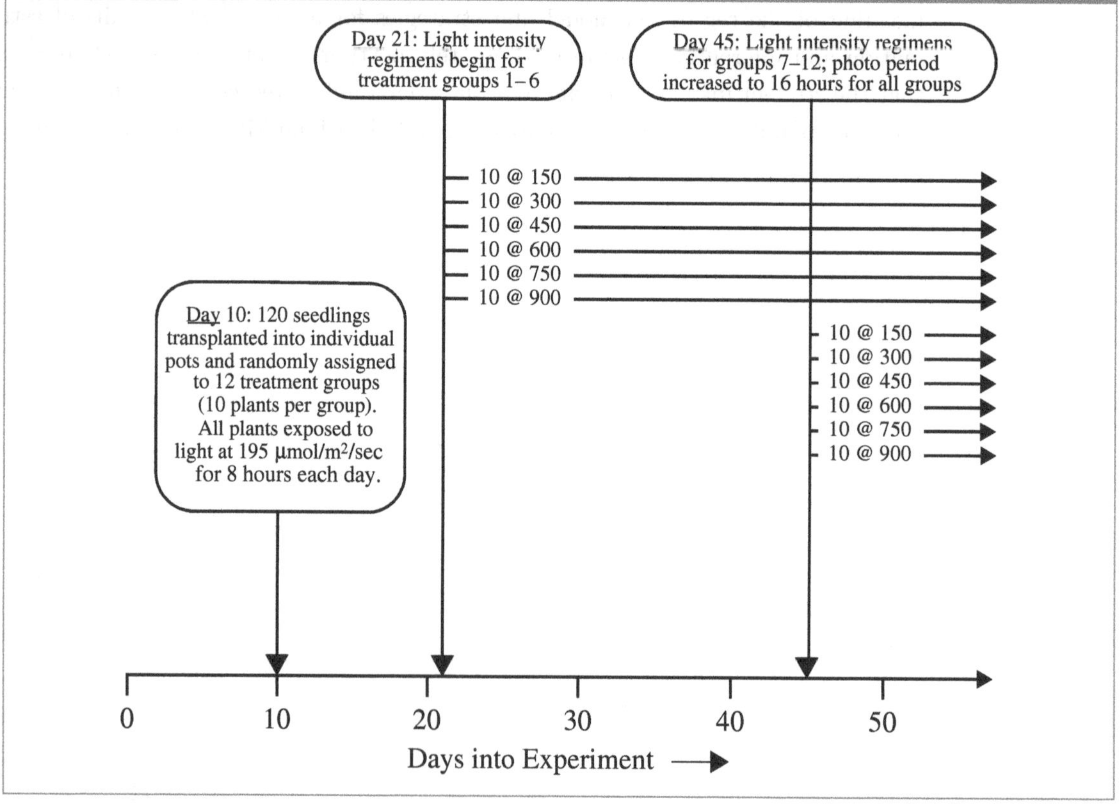

The design consists of 12 treatment groups—the six light intensities at each of the two timing levels. Ten seedlings were randomly assigned to each treatment group. The number of flowers per plant is the primary measure of production, and it was measured by averaging the numbers of flowers produced by the 10 seedlings in each group.

The entire experiment was repeated. No difference was found between the first and second runs of the experiment so, as a suitable starting point for the analysis in this chapter, Display 9.2 presents the two results as replicates. The two observations in each cell of the table are thought of as independent replicates under the specified conditions. What are the effects of differing light intensity levels? What is the effect of the timing? Does the effect of intensity depend on the timing?

DISPLAY 9.2 Average number of flowers per meadowfoam plant, in 12 treatment groups

Timing		Light Intensity (μmol/m^2/sec)					
		150	300	450	600	750	900
Timing	Late (at PFI)	62.3 77.4	55.3 54.2	49.6 61.9	39.4 45.7	31.3 44.9	36.8 41.9
Timing	Early (24 days before PFI)	77.8 75.6	69.1 78.0	57.0 71.1	62.9 52.2	60.3 45.6	52.6 44.4

Statistical Conclusion

Display 9.3 shows the fit of a multiple linear regression model that specifies parallel regression lines for the mean number of flowers as functions of light intensity. Increasing light intensity decreased the mean number of flowers per plant by an estimated 4.0 flowers per plant per 100 μmol/m^2/sec (95% confidence interval from 3.0 to 5.1). Beginning the light treatments 24 days prior to PFI increased the mean numbers of flowers by an estimated 12.2 flowers per plant (95% confidence interval from 6.7 to 17.6). The data provide no evidence that the effect of light intensity depends on the timing of its initiation (two-sided p-value = 0.91, from a t-test for interaction, 20 degrees of freedom).

Scope of Inference

The researchers can infer that the effects above were caused by the light intensity and timing manipulations, because this was a randomized experiment.

9.1.2 Why Do Some Mammals Have Large Brains for Their Size?—An Observational Study

Evolutionary biologists are keenly interested in the characteristics that enable a species to withstand the selective mechanisms of evolution. An interesting variable

DISPLAY 9.3 The average number of flowers per plant versus the applied light intensity for each of the 12 experimental units in the meadowfoam study, with different plotting symbols for units with early (24 days prior to PFI) and late (at PFI) commencement of lighting regimen

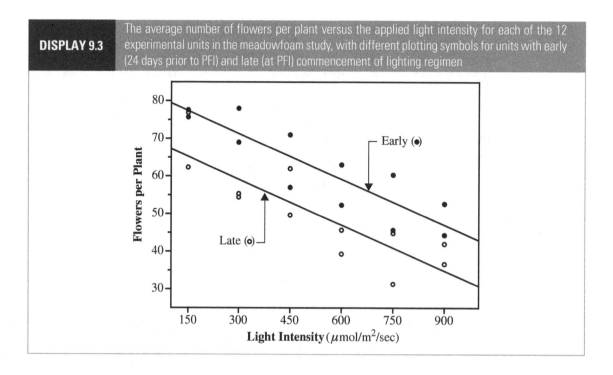

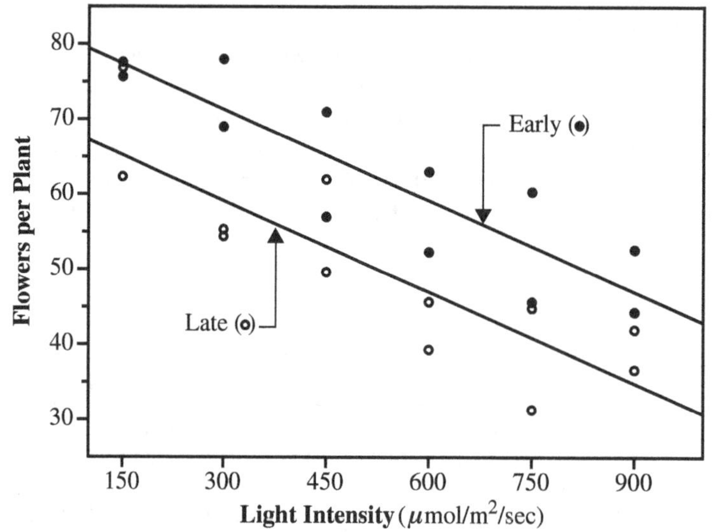

in this respect is brain size. One might expect that bigger brains are better, but certain penalties seem to be associated with large brains, such as the need for longer pregnancies and fewer offspring. Although the individual members of the large-brained species may have more chance of surviving, the benefits for the species must be good enough to compensate for these penalties. To shed some light on this issue, it is helpful to determine exactly which characteristics are associated with large brains, after getting the effect of body size out of the way.

The data in Display 9.4 are the average values of brain weight, body weight, gestation lengths (length of pregnancy), and litter size for 96 species of mammals. (Data from G. A. Sacher and E. F. Staffeldt, "Relation of Gestation Time to Brain Weight for Placental Mammals; Implications for the Theory of Vertebrate Growth," *American Naturalist*, 108 (1974): 593–613. The common names for the species correspond to the Latin names given in the original paper, and those followed by a Roman numeral indicate subspecies.) Since brain size is obviously related to body size, the question of interest is this: Which, if any, variables are associated with brain size, after accounting for body size?

Statistical Conclusion

The data provide convincing evidence that brain weight was associated with either gestation length or litter size, even after accounting for the effect of body weight (p-value < 0.0001; extra sum of squares F-test). There was strong evidence that litter size was associated with brain weight after accounting for body weight and gestation (two-sided p-value $= 0.0089$) and that gestation period was associated

DISPLAY 9.4 Average values of brain weight, body weight, gestation length, and litter size in 96 species of mammal

Species	Brain weight (grams)	Body weight (kilograms)	Gestation period (days)	Litter size
Quokka	17.5	3.5	26	1.0
Hedgehog	3.50	0.93	34	4.6
Tree shrew	3.15	0.15	46	3.0
Elephant shrew I	1.14	0.049	51	1.5
Elephant shrew II	1.37	0.064	46	1.5
Lemur	22	2.1	135	1.0
Slow loris	12.8	1.2	90	1.2
Bush baby	9.9	0.7	135	1.0
Howler monkey	54	7.7	139	1.0
Ring-tail monkey	73	3.7	180	1.0
Spider monkey I	114	9.1	140	1.0
Spider monkey II	109	7.7	140	1.0
Gentle lemur	7.8	0.22	145	2.0
Rhesus monkey I	84.6	6.0	175	1.0
Rhesus monkey II	107	8.7	165	1.1
Hamadryas baboon	183	21	180	1.0
Western baboon	179	32	180	1.0
Vervet guenon	67	4.6	195	1.0
Leaf monkey	65.5	5.8	168	1.0
White-handed gibbon	102	5.5	210	1.0
Orangutan	343	37	270	1.0
Chimpanzee	360	45	230	1.0
Gorilla	406	140	265	1.0
Human being	1,300	65	270	1.0
Long-nosed armadillo	12	3.7	120	4.0
Aardvark	9.6	2.2	31	5.0
Jack rabbit	13.3	2.9	41	2.5
Tree squirrel	6.23	0.33	38	3.0
Flying squirrel	1.89	0.052	40	3.1
Canadian beaver	40	20	128	2.9
Beaver	45	25	128	4.0
Deer mouse I	0.68	0.027	23	3.7
Deer mouse II	0.63	0.026	23	5.0
Deer mouse III	0.52	0.017	24	5.0
Deer mouse IV	0.69	0.024	24	5.0
Hamster I	0.67	0.036	21	4.6
Hamster II	1.12	0.13	16	6.3
Pygmy gerbil	1.04	0.065	21	4.0
Rat I	0.72	0.05	23	7.3
Rat II	2.38	0.34	21	8.0
House mouse	0.45	0.024	19	5.0
Hopping mouse	1.18	0.15	27	5.6
Porcupine I	37	11	112	1.2
Porcupine II	37	14	112	1.2
Porcupine III	24	6.6	113	1.0
Guinea pig	4.28	0.97	67	2.6
Capybara	76	30	123	3.0
Agoutis	20.3	2.8	104	1.3
Acouchis	9.9	0.78	98	1.2
Chinchilla	5.25	0.43	110	2.0
Nutria	23	5.0	132	5.5
Dolphin	1,600	160	360	1.0
Porpoise	537	56	270	1.0
Dog	70.2	8.5	63	4.0
Red fox	48	6.0	52	4.0
Gray fox	37.3	3.8	63	3.7
Bat-eared fox	28.5	3.2	65	4.0
Grizzly bear	400	250	219	2.3
Beaked whale	500	250	240	1.8
Raccoon	41.6	5.3	63	3.5
Kinkajou	31.2	2.0	77	1.1
Badger	53	6.0	60	2.2
Domestic cat	28.4	2.5	63	4.0
Lynx	75	12	60	2.5
Leopard	157	46	92	2.5
Lion	260	180	108	3.0
Tiger	302	210	104	3.0
Fur seal	355	250	254	1.0
Sea lion	363	100	343	1.0
Harp seal	442	110	240	1.0
Weddell seal	550	400	310	1.0
African Elephant	4,480	2,800	655	1.0
Hyrax	20.5	3.8	225	2.4
Horse	712	480	330	1.0
Tapir	250	230	390	1.0
Wild boar	185	150	120	4.0
Domestic pig	180	190	115	8.0
Hippopotamus	590	1,400	240	1.0
Pygmy hippopotamus	260	150	205	1.0
Llama	225	93	330	1.0
Vicuna	198	45	300	1.1
Barking deer	124	16	183	1.1
Fallow deer	223	80	240	1.0
Axis deer	219	89	218	1.0
Red deer	435	200	255	1.0
Elk	365	120	235	1.0
Sambar	383	120	246	1.1
Caribou	288	110	225	1.0
Eland	480	560	255	1.0
Yak	334	250	255	1.0
Cattle	456	520	280	1.0
Duiker	93	13	120	1.0
Blackbuck Antelope	200	39	180	1.0
Barbary sheep	210	66	158	1.2
Domestic sheep	125	49	150	2.4
Domestic goat	106	30	151	2.0

with brain weight after accounting for body weight and litter size (two-sided p-value $= 0.0038$).

Scope of Inference

As suggestive as the findings may be, inferences that go beyond these data are unwise. The data were summarized from available studies and cannot be representative of any wider population. As usual, no causal interpretation can be made from these observational data.

9.2 REGRESSION COEFFICIENTS

9.2.1 The Multiple Linear Regression Model

Regression

The *regression of* Y *on* X_1 *and* X_2 is a rule, such as an equation, that describes the mean of the distribution of a response variable (Y) for particular values of explanatory variables (X_1 and X_2, say). For example, the regression of the response variable Y = *flowers* on X_1 = *light* intensity and X_2 = *time* of light manipulation specifies how the mean number of flowers per plant depends on the levels of light intensity and timing. (The italicized words specify variable names.)

The symbol for the regression is $\mu\{$*flowers* | *light, time*$\}$, which is read as the "mean number of flowers, as a function of light intensity and timing." In terms of a generic response variable, Y, and two explanatory variables, X_1 and X_2:

The regression of Y on X_1 and X_2 is $\mu\{Y|X_1, X_2\}$.

With specific values for X_1 and X_2 inserted, the same expression may be read somewhat differently. For example, $\mu\{$*flowers* | *light* $= 300$, *time* $= 24\}$ is read as "the mean number of flowers when light intensity is 300 μmol/m^2/sec and time is 24 days prior to PFI."

Multiple Regression Models

In *multiple regression* there is a single response variable and *multiple explanatory variables*. The regression rule giving the mean response for each combination of explanatory variables will not be very helpful if it is a huge list or a complicated function. Furthermore, it is usually unwise to think that there is some exact, discoverable regression equation. Many possible *models* are available, however, for describing the regression. Although they should not be thought of as the truth, one or two models may adequately approximate the mean of the response as a function of the explanatory variables, and conveniently allow for the questions of interest to be investigated.

Multiple Linear Regression Model

One family of models is particularly easy to deal with and works for the majority of regression problems—the family of linear regression models. The term *linear* means linear in regression coefficients, as in the following examples:

Examples of Multiple Linear Regression Models

$$\mu\{Y|X_1, X_2\} = \beta_0 + \beta_1 X_1 + \beta_2 X_2,$$
$$\mu\{Y|X_1\} = \beta_0 + \beta_1 X_1 + \beta_2 X_1^2,$$
$$\mu\{Y|X_1, X_2\} = \beta_0 + \beta_1 X_1 + \beta_2 X_2 + \beta_3 X_1 X_2,$$
$$\mu\{Y|X_1, X_2\} = \beta_0 + \beta_1 \log(X_1) + \beta_2 \log(X_2).$$

The extension to more than two explanatory variables is straightforward. Notice that the first term in the examples is β_0 (which multiplies the trivial function, 1). This *constant term* appears in multiple linear regression models unless a specific reason for excluding it exists.

Multiple Regression Model with Constant Variance

The ideal regression model includes an assumption of constant variation. For the meadowfoam example, the assumption states that

$$\text{Var}\{\textit{flowers} \mid \textit{light}, \textit{time}\} = \sigma^2.$$

The notation Var{*flowers* | *light, time*} is read as "the variance of the numbers of flowers, as a function of light and time." The right-hand side of the equation asserts that this variance is constant—the same for all values of *light* and *time*. As in previous chapters, the constant variance assumption is important for two reasons: (1) the regression interpretation is more straightforward when the explanatory variables are only associated with the mean of the response distribution and not other characteristics of it (as described in Section 4.3.2 for the two-sample problem); and (2) the assumption justifies the standard inferential tools, which are presented in the next chapter.

9.2.2 Interpretation of Regression Coefficients

Regression analysis involves finding a good fitting model for the response mean, wording the questions of interest in terms of model parameters, estimating the parameters from the available data, and employing appropriate inferential tools for answering the questions of interest. Before turning to issues of estimation and inference, it is important to discuss the meaning of regression coefficients and what questions can be answered through them.

DISPLAY 9.5 Model for the regression surface of flowers per plant under 12 treatment levels as a regression plane

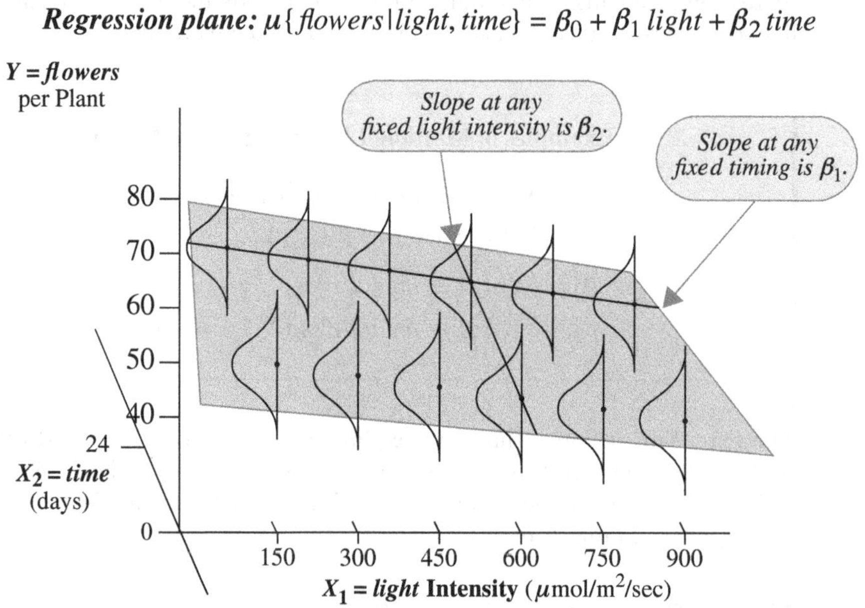

Regression Surfaces

The multiple linear regression model with two explanatory variables, such as

$$\mu\{flowers \mid light, time\} = \beta_0 + \beta_1 light + \beta_2 time,$$

describes the regression surface as a *plane*. The parameter β_0 is the height of the plane when both *light* and *time* equal zero; the parameter β_1 is the slope of the plane as a function of *light* for any fixed value of *time*, and the parameter β_2 is the slope of the plane as a function of *time* for any fixed value of *light*. This model is represented in Display 9.5.

When more than two explanatory variables are present, it is difficult and not generally useful to consider the geometry of regression surfaces. Instead, the regression coefficients are interpreted in terms of the effects that the selected explanatory variables have on the mean of the response when other explanatory variables are also included in the model.

Effects of Explanatory Variables

The *effect* of an explanatory variable is the change in the mean response that is associated with a one-unit increase in that variable while holding all other explanatory variables fixed. In a regression for the meadowfoam study,

$$\begin{aligned} \textit{light} \text{ effect } &= \mu\{\textit{flowers} \mid \textit{light} + 1, \textit{time}\} - \mu\{\textit{flowers} \mid \textit{light}, \textit{time}\}, \\ \textit{time} \text{ effect } &= \mu\{\textit{flowers} \mid \textit{light}, \textit{time} + 1\} - \mu\{\textit{flowers} \mid \textit{light}, \textit{time}\}. \end{aligned}$$

In the planar model, effects are the same at all levels of the explanatory variables. The change in mean *flowers* associated with increasing *light* from 150 to 151 μmol/m^2/sec with *time* at 24 days, for example, is the same as the change associated with increasing *light* from 600 to 601 μmol/m^2/sec with *time* at 0 days. The precise value of the effect is found by performing the subtraction with the specified form of the model:

$$\begin{aligned} \textit{light} \text{ effect } &= \mu\{\textit{flowers} \mid \textit{light} + 1, \textit{time}\} - \mu\{\textit{flowers} \mid \textit{light}, \textit{time}\} \\ &= [\beta_0 + \beta_1(\textit{light} + 1) + \beta_2 \textit{time}] - [\beta_0 + \beta_1 \textit{light} + \beta_2 \textit{time}] = \beta_1. \end{aligned}$$

In this model, the coefficient of one explanatory variable measures the effect of that variable at fixed values of the other.

When regression analysis is performed on the results of a randomized experiment the interpretation of an "effect" of an explanatory variable is straightforward, and causation is implied. For example, "A one-unit increase in light intensity causes the mean number of flowers to increase by β_1."

For observational studies, on the other hand, interpretation is more complicated. First, as usual, we cannot make causal conclusions from statistical association. Second, the X's *cannot* be held fixed independently of one another as they can in a controlled experiment. We must therefore use wording like the following: "For any subpopulation of mammal species with the same body weight and litter size, a one-day increase in the species' gestation length is associated with a β_2 gram increase in mean brain weight." This wording requires the reader to imagine a subpopulation of mammals with fixed values of body weight and litter size, but varying gestation lengths. Whether this interpretation is useful depends on whether such subpopulations exist.

Interpretation Depends on What Other X's Are Included

For the mammal brain weight data, furthermore, the interpretation of β_1 in the model

$$\mu\{\textit{brain} \mid \textit{gestation}\} = \beta_0 + \beta_1 \textit{gestation}$$

differs from the interpretation of β_1 in the model

$$\mu\{\textit{brain} \mid \textit{gestation}, \textit{body}\} = \beta_0 + \beta_1 \textit{gestation} + \beta_2 \textit{body}.$$

In the first model, β_1 measures the rate of change in mean brain weight with changes in gestation length *in the population of all mammal species* ... where body size is variable. In the second model, β_1 measures the rate of change in mean brain weight with changes in gestation length *within subpopulations of fixed body size*. If in the second model there is no effect of gestation length on mean brain weight within subpopulations of fixed body size ($\beta_1 = 0$), and if, as expected, both brain weight and gestation length vary consistently with body size, then β_1 in the first model will not be zero. It will measure the association between brain weight and body size indirectly through their joint associations with gestation length.

9.3 SPECIALLY CONSTRUCTED EXPLANATORY VARIABLES

One can dramatically expand the scope of multiple linear regression by using specially constructed explanatory variables. With them, regression models can exhibit curvature, interactive effects of explanatory variables, and effects of categorical variables.

9.3.1 A Squared Term for Curvature

Display 9.6 is a scatterplot of the corn yield versus rainfall in six U.S. corn-producing states (Iowa, Nebraska, Illinois, Indiana, Missouri, and Ohio), recorded for each year from 1890 to 1927 (see also Exercise 15). A straight-line regression model is not adequate. Although increasing rainfall is associated with higher mean yields for rainfalls up to 12 inches, increasing rainfall at higher levels is associated with no change or perhaps a decrease in mean yield. In short, the rainfall effect depends on the rainfall level.

DISPLAY 9.6 Yearly corn yield versus rainfall (1890–1927) in six U.S. states

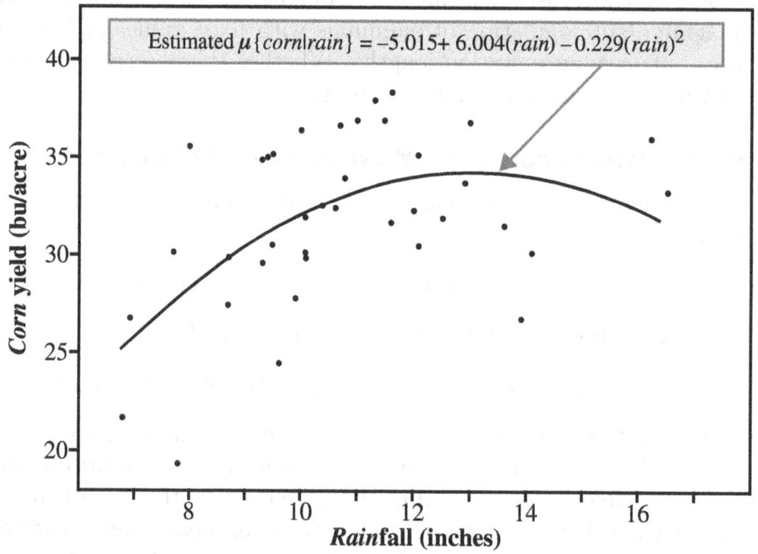

One model for incorporating curvature includes squared rainfall as an additional explanatory variable:

$$\mu\{corn \mid rain\} = \beta_0 + \beta_1 rain + \beta_2 rain^2.$$

It is not necessary that this model corresponds to any natural law (but it could). It is, however, a convenient way to incorporate curvature in the regression of *corn* on *rain*. Remember that *linear* regression means linear in β's, not linear in X, so quadratic regression is a special case of multiple linear regression.

The model incorporates curvature by having the effect of rainfall be different at different levels of rainfall:

$$\begin{aligned}\mu\{corn \mid rain + 1\} - \mu\{corn \mid rain\} &= [\beta_0 + \beta_1(rain + 1) + \beta_2(rain + 1)^2] \\ &\quad - [\beta_0 + \beta_1 rain + \beta_2 rain^2] \\ &= \beta_1 + \beta_2[(2 \times rain) + 1].\end{aligned}$$

From the fitted model shown in Display 9.6, the effect of a unit increase in rainfall from 8 to 9 inches is estimated to be an increase in mean yield of about 2.1 bushels of corn per acre. But the effect of a unit increase in rainfall from 14 to 15 inches is estimated to be a *decrease* in mean yield of about 0.6 bu/acre.

Attempting to interpret the individual coefficients in this example is difficult and unnecessary. The statistical significance of the squared rainfall coefficient is that it highlights the inadequacy of the straight line regression model and suggests that increasing yield is associated with increasing rainfall only up to a point. In many applications squared terms are useful for incorporating slight curvature and the purpose of the analysis does not require that the coefficient of the squared term be interpreted. In specialized situations, higher-order polynomial terms may also be included as explanatory variables.

9.3.2 An Indicator Variable to Distinguish Between Two Groups

Display 9.3 shows a scatterplot of the *flower* numbers versus *light* intensity, with $time = 0$ days coded differently from $time = 24$ days. The lines on this scatterplot do not represent simple linear regression equations for $time = 0$ and $time = 24$ separately. They are the result of a multiple linear regression model that incorporates an *indicator variable* to represent the two levels of the timing variable. The model requires that the lines have equal slopes.

An *indicator variable* (or dummy variable) takes on one of two values: "1" (one) indicates that an attribute is present, and "0" (zero) indicates that the attribute is absent. The variable *early*, for example, is set equal to 1 for units where the timing was 24 days prior to PFI and is set equal to 0 for units where light intensity was not varied prior to PFI. The indicator variable *early* (along with variables to be introduced shortly) appears in Display 9.7.

DISPLAY 9.7 The meadowfoam data (first three columns) and three sets of constructed explanatory variables that are discussed in this chapter

Original variables			Timing indicators		Light level indicators						Interaction
flowers	light	time	day24	day0	L150	L300	L450	L600	L750	L900	light × day24
62.3	150	0	0	1	1	0	0	0	0	0	0
77.4	150	0	0	1	1	0	0	0	0	0	0
77.8	150	24	1	0	1	0	0	0	0	0	150
75.6	150	24	1	0	1	0	0	0	0	0	150
55.3	300	0	0	1	0	1	0	0	0	0	0
54.2	300	0	0	1	0	1	0	0	0	0	0
69.1	300	24	1	0	0	1	0	0	0	0	300
78.0	300	24	1	0	0	1	0	0	0	0	300
49.6	450	0	0	1	0	0	1	0	0	0	0
61.9	450	0	0	1	0	0	1	0	0	0	0
57.0	450	24	1	0	0	0	1	0	0	0	450
71.1	450	24	1	0	0	0	1	0	0	0	450
39.4	600	0	0	1	0	0	0	1	0	0	0
45.7	600	0	0	1	0	0	0	1	0	0	0
62.9	600	24	1	0	0	0	0	1	0	0	600
52.2	600	24	1	0	0	0	0	1	0	0	600
31.3	750	0	0	1	0	0	0	0	1	0	0
44.9	750	0	0	1	0	0	0	0	1	0	0
60.3	750	24	1	0	0	0	0	0	1	0	750
45.6	750	24	1	0	0	0	0	0	1	0	750
36.8	900	0	0	1	0	0	0	0	0	1	0
41.9	900	0	0	1	0	0	0	0	0	1	0
52.6	900	24	1	0	0	0	0	0	0	1	900
44.4	900	24	1	0	0	0	0	0	0	1	900

Consider the regression model

$$\mu\{flowers \mid light, early\} = \beta_0 + \beta_1 light + \beta_2 early.$$

If *time* = 0, then *early* = 0, and the regression line is

$$\mu\{flowers \mid light, early = 0\} = \beta_0 + \beta_1 light.$$

If *time* = 24, then *early* = 1, and the regression line is

$$\mu\{flowers \mid light, early = 1\} = \beta_0 + \beta_1 light + \beta_2.$$

This multiple regression model states that the mean number of flowers is a straight line function of light intensity for both levels of timing. The slope of both lines is β_1; the intercept for units with timing at PFI (*late*) is β_0; and the intercept for units with timing 24 days prior to PFI (*early*) is $\beta_0 + \beta_2$. Because the slopes are the same, the model is called the *parallel lines* regression model. Furthermore, the coefficient of the indicator variable, β_2, is the amount by which the mean number of flowers with prior timing at 24 days exceeds that with no prior timing, after accounting for the effect of light intensity differences. In Display 9.3, the lines showing (estimates for) the mean numbers of flowers versus intensity are separated by a constant vertical difference of β_2 units.

The indicator variable can be defined for either group without affecting the model. If, instead of the indicator variable *early*, the regression had been based on the indicator variable *late*, taking 1 for *time* = 0 and 0 for *time* = 24, then the resulting fit would be exactly the same; it would produce exactly the same lines on Display 9.3. The only difference would be that the intercept for units with timing prior to PFI is β_0 and the intercept for units with timing at PFI is $\beta_0 + \beta_2$. This β_2 is the negative of the β_2 in the previous version of the model.

An indicator variable may be included in a regression model just as any other explanatory variable. The coefficient of the indicator variable is the difference between the mean response for the indicated category (= 1) and the mean response for the other category (= 0), at fixed values of the other explanatory variables.

9.3.3 Sets of Indicator Variables for Categorical Explanatory Variables with More Than Two Categories

A categorical explanatory variable with more than two categories can also be incorporated into a regression model. When a categorical variable is used in regression it is called a *factor* and the individual categories are called the *levels* of the factor. If there are k levels, then $k-1$ indicator variables are needed as explanatory variables.

Light intensity in the meadowfoam study can be viewed as a categorical variable having six levels. For use in regression, indicator variables are created for each level. Let *L150* be equal to 1 for all experimental units that received light intensity of 150 μmol/m^2/sec and equal to 0 for all other units. Let *L300* similarly indicate the units that received light intensity of 300 μmol/m^2/sec; and so on, to *L900*, indicating the units that received light intensity of 900 μmol/m^2/sec. This creates an indicator variable for every level of intensity (Display 9.7). To treat light intensity as a factor, all but one of these are included as explanatory variables in the regression model. The level whose indicator variable is not used in the set is called the *reference level* for that factor.

Intensity has six levels, so five indicator variables will represent all levels. Selecting the first level, 150 μmol/m^2/sec intensity, as the reference level, the multiple linear regression model is

$$\begin{aligned}\mu\{flowers \mid light, day24\} &= \beta_0 + \beta_1 L300 + \beta_2 L450 + \beta_3 L600 \\ &\quad + \beta_4 L750 + \beta_5 L900 + \beta_6 day24.\end{aligned}$$

The coefficients of the indicator variables in this model allow the mean number of flowers to vary arbitrarily with light intensity.

The questions of interest in the meadowfoam study, however, are more easily addressed with the simpler model that uses *light* as a numerical explanatory variable, $\beta_0 + \beta_1 light + \beta_2 day24$. So, an important issue that will arise in later chapters is how best to represent a variable like light intensity—as a factor with several levels or as a numerical variable?

9.3.4 A Product Term for Interaction

Two explanatory variables are said to *interact* if the effect that one of them has on the mean response depends on the value of the other. In multiple regression, an explanatory variable for interaction can be constructed as the product of the two explanatory variables that are thought to interact.

An Interaction Model for the Meadowfoam Data

A secondary question of interest in the meadowfoam study was one of interaction: Does the effect of light intensity on mean number of flowers depend on the timing of the light regime? The product variable *light* × *early* (see Display 9.7) models an interaction between light intensity and timing. Consider the model

$$\mu\{flowers \mid light,\ early\} = \beta_0 + \beta_1 light + \beta_2 early + \beta_3(light \times early).$$

This is a way of expressing the regression of *flowers* on *light*, for different levels of timing. When $early = 0$, the regression equation is a straight-line function of intensity with intercept β_0 and slope β_1. When $early = 1$, the mean number of flowers is also a straight line function of intensity, but with intercept $\beta_0 + \beta_2$ and slope $\beta_1 + \beta_3$. Rearranging the model as

$$\mu\{flowers \mid light,\ early\} = (\beta_0 + \beta_2 early) + (\beta_1 + \beta_3 early) light$$

shows how both the intercept and slope in the regression of *flowers* on *light* depend on the timing. The light intensity effect in the interaction model is $(\beta_1 + \beta_3 early)$; while the timing effect is $(\beta_2 + \beta_3 light)$. The effect of the light intensity depends on the timing; the effect of the timing depends on the light intensity. This representation of the regression leads directly to a graph that shows the mean of flowers as a function of light intensity, with different lines corresponding to different levels of timing, as shown in the top panel of Display 9.8.

As with squared terms in models for curvature, it is often difficult to interpret individual coefficients in an interaction model. The coefficient, β_1, of *light* has changed from being a global slope to being the slope when $time = 0$. The coefficient, β_3, of the product term is the difference between the slope of the regression line on *light* when $time = 24$ and the slope when $time = 0$. If it is only necessary to test whether interaction is present, no lengthy interpretation is needed. The best method of communicating findings about the presence of significant interaction may be to present a table or graph of estimated means at various combinations of the interacting variables.

When to Include Interaction Terms

Interaction terms are not routinely included in regression models. Inclusion is indicated in three situations: when a question of interest pertains to interaction (as in

DISPLAY 9.8 Regression models for separate lines, parallel lines, and equal lines in two groups—meadowfoam study

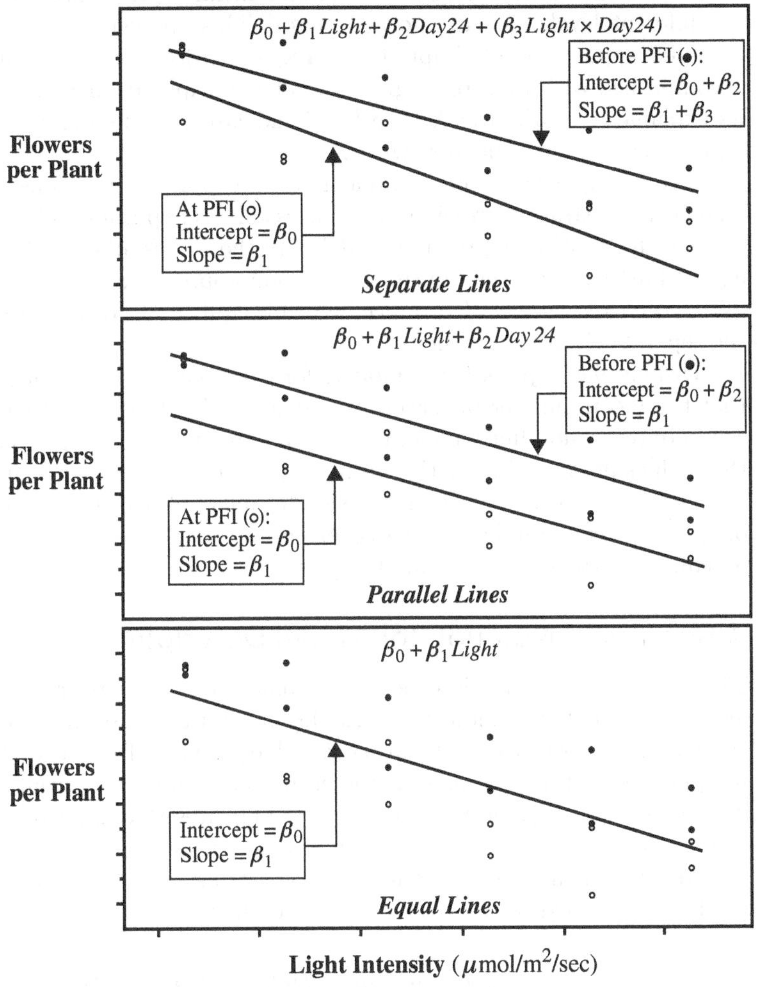

the meadowfoam study); when good reason exists to suspect interaction; or when interactions are proposed as a more general model for the purpose of examining the goodness of fit of a model without interaction.

Except in special circumstances, a model including a product term for interaction between two explanatory variables should also include terms with each of the explanatory variables individually, even though their coefficients may not be significantly different from zero. Following this rule avoids the logical inconsistency of saying that the effect of X_1 depends on the level of X_2 but that there is no effect of X_1.

Completely Separate Regression Lines for Different Levels of a Factor

Sometimes the analyst wishes to fit simple linear regressions of Y on X separately for different levels of a categorical factor. This analysis can be accomplished by repeated application of simple linear regression for each level; but it can also be accomplished with multiple regression. The multiple linear regression model has as its explanatory variables: X, the $k-1$ indicators to distinguish the k levels, and all products of the indicators with X.

The multiple regression approach has advantages. Questions about similarities among the separate regressions are expressible as hypotheses about specific parameters in the multiple regression model: Are the slopes all equal? Are the intercepts equal? Furthermore, the multiple regression approach conveniently provides a single, combined estimate of variance, which is equivalent to the pooled estimate from the separate simple regressions.

Display 9.8 depicts three models for the meadowfoam study, each specifying that the mean number of flowers is a straight line function of light intensity. The separate regression lines model in the top panel is the most general of the three. If the coefficient of the interaction term, β_3, is zero, then the separate regression lines model reduces to the parallel regression lines model, which indicates no interaction of *light* and *time*. Similarly, if β_2 is zero in the parallel regression lines model, then it reduces to the equal lines model.

9.3.5 A Shorthand Notation for Model Description

The notation for regression models is sometimes unnecessarily bulky when there are a large number of indicator variables or interaction terms. An abbreviation for specifying a categorical explanatory variable is to list the categorical variable name in uppercase letters to represent the entire set of indicator variables used to model it. For example, $\mu\{\mathit{flowers, LIGHT, early}\}$ indicates that light intensity is a factor, whose effects are to be modeled by indicator variables.

Additional savings are achieved by omitting the parameters when specifying the terms included in the model. For example,

$$\mu\{\mathit{flowers} \mid \mathit{light}, \mathit{TIME}\} = \mathit{light} + \mathit{TIME}$$

describes the parallel lines model with *light* treated as a numerical explanatory variable and *TIME* as a factor. Similarly,

$$\mu\{\mathit{flowers} \mid \mathit{light}, \mathit{TIME}\} = \mathit{light} + \mathit{TIME} + (\mathit{light} \times \mathit{TIME})$$

specifies the separate lines model with the effects of the numerical explanatory variable *light*, the factor *TIME*, and the interaction terms formed as the products of *light* with the indicator variable(s) for the factor *TIME*. Extensions of this notation conveniently describe models with any number of numerical and categorical variables.

DISPLAY 9.9 A strategy for data analysis using statistical models

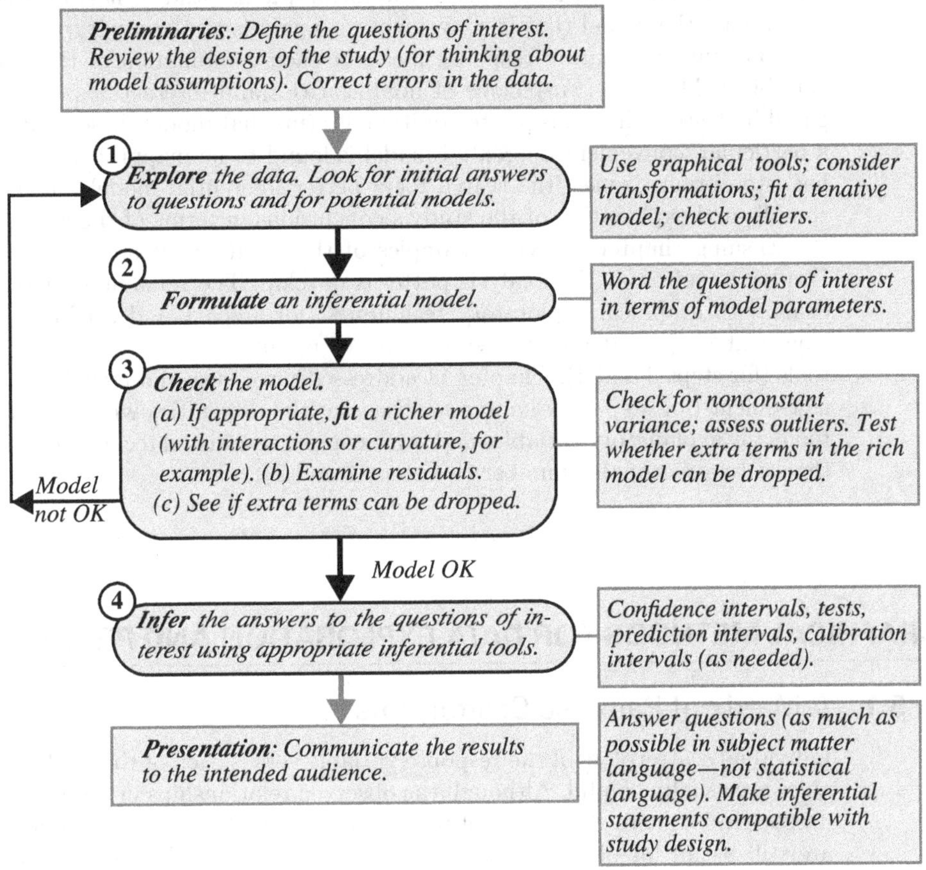

Preliminaries: Define the questions of interest. Review the design of the study (for thinking about model assumptions). Correct errors in the data.
1
Explore the data. Look for initial answers to questions and for potential models.
Use graphical tools; consider transformations; fit a tenative model; check outliers.
2
Formulate an inferential model.
Word the questions of interest in terms of model parameters.
3
Check the model.
(a) If appropriate, fit a richer model (with interactions or curvature, for example). (b) Examine residuals.
(c) See if extra terms can be dropped.
Check for nonconstant variance; assess outliers. Test whether extra terms in the rich model can be dropped.
Model not OK
Model OK
4
Infer the answers to the questions of interest using appropriate inferential tools.
Confidence intervals, tests, prediction intervals, calibration intervals (as needed).
Presentation: Communicate the results to the intended audience.
Answer questions (as much as possible in subject matter language—not statistical language). Make inferential statements compatible with study design.

9.4 A STRATEGY FOR DATA ANALYSIS

Display 9.9 shows a general scheme for using statistical models to analyze data. Data analysis involves more than inserting data into a computer and pressing the right button. Substantial exploratory analysis may be required at the outset, and several dead ends may be encountered in the pursuit of suitable models.

The strategy for data analysis centers on the development of an *inferential model* where answers to the questions of interest can be found by drawing inferences about key parameters. The choice of an inferential model may be guided by the graphical displays. Further checking on model adequacy is accomplished by residual analysis and by informal testing of terms in a richer model that contains additional parameters representing possible curvature, interactions, or more complex features.

The adequacy of a model may be clear from the graphical displays and residual analysis alone. When it is not, informal testing is an important part of finding an adequate model. In particular, if replicates of the response occur at each combination of explanatory variables (as in the meadowfoam study), it is always possible to compare the model that is convenient for analysis to the separate-means model (the one with a separate response mean at each combination of the explanatory variables). Moreover, suspicions of nonlinearities and interactions can be investigated by testing the appropriate additional terms that model these inadequacies. If a particular, convenient inferential model is found to be inappropriate, the analyst has two options: renew the search for a better inferential model, or offer a more complicated explanation of the study's conclusions in terms of the richer model.

Ensuing chapters provide examples of this strategy. Readers should refer to Display 9.9 often as the analysis patterns develop. The remainder of this chapter deals with graphical exploratory techniques (for steps 1 and 2 of this strategy). The next chapter details the estimation of parameters and the main inferential tools (for steps 3 and 4). Chapter 11 addresses new issues and techniques for model assessment (for step 3). Chapter 12 discusses the difficulties with a set containing too many explanatory variables and some computer-assisted techniques for reducing the set to a reasonable number (for steps 1 and 2).

9.5 GRAPHICAL METHODS FOR DATA EXPLORATION AND PRESENTATION

9.5.1 A Matrix of Pairwise Scatterplots

Individual scatterplots of the response variable versus each of the explanatory variables are usually helpful. Although the observed relationships in these plots will not necessarily indicate the effects of interest in a *multiple* regression model, the plots are still useful for observing the marginal (one at a time) relationships, drawing attention to interesting points, and suggesting the need for transformations.

A *matrix of scatterplots* (or *draftsman plot*) is a consolidation of all possible pairwise scatterplots from a set of variables, as shown for the mammal brain weight data in Display 9.10. The bottom row shows the response variable versus each of the explanatory variables and the higher rows show the scatterplots of the explanatory variables versus each other. There is a lot to look at in such a display. Typically, one would investigate the scatterplots of the response versus each of the explanatory variables (the bottom row) first.

Different statistical computer packages have different display formats. Instead of the lower left triangular arrangement shown in Display 9.10, some packages display a corresponding upper right triangle (in addition to the lower left one) that contains the same plots but with axes reversed. This does not contain any more information, although it may provide a different visual perspective. The resolution with which the plots can be drawn may also be an issue. It becomes increasingly difficult to read a matrix of scatterplots as the number of variables is increased. For data sets with many potential explanatory variables, it may be necessary to select subsets for display.

DISPLAY 9.10 Matrix of scatterplots for brain weight data

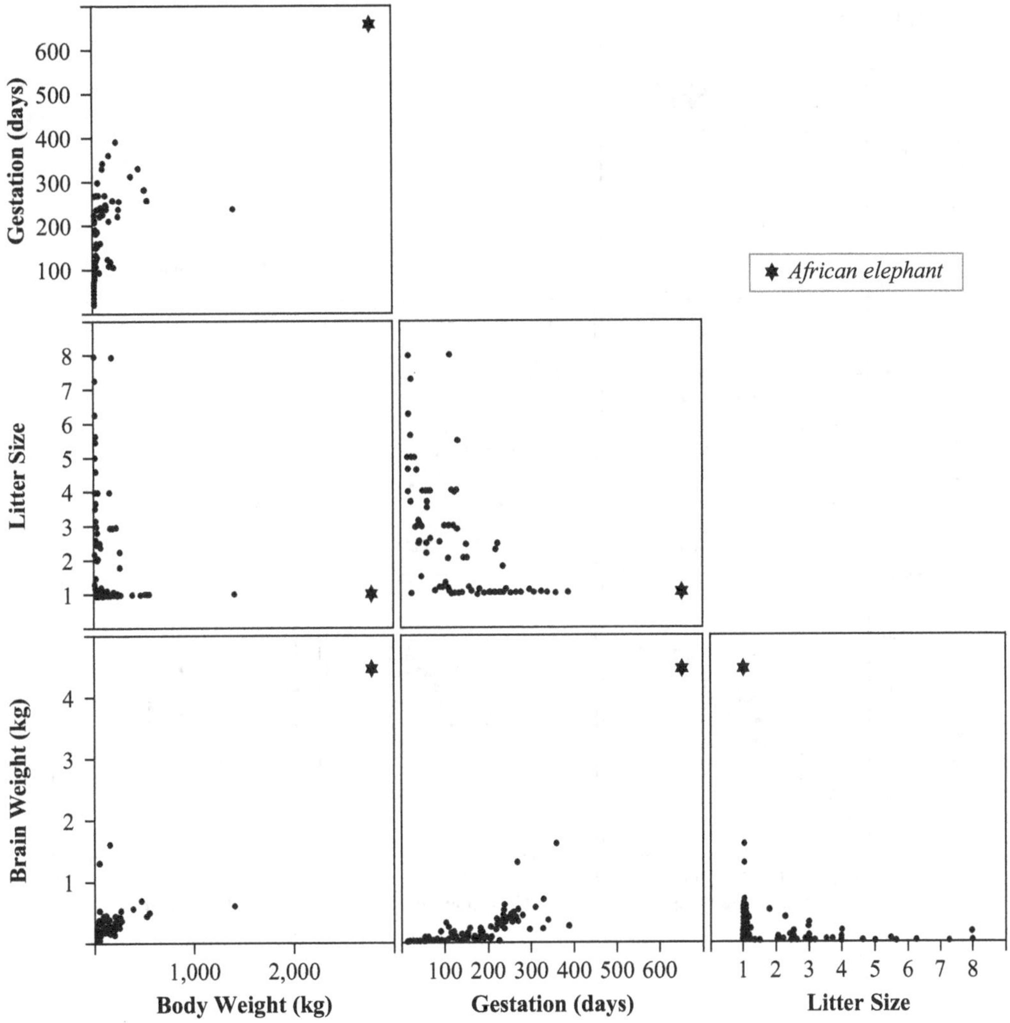

Notes About the Scatterplots for the Brain Weight Data

Consider the bottom row of Display 9.10 first. The scatterplot of brain weight versus body weight is not very helpful since most of the data points are clustered on top of each other in the bottom left-hand corner. The upper bounds for the X and Y axes are determined largely by one point (the African elephant). This scatterplot indicates what should have been evident from the start—that the mammals in this list differ in size by orders of magnitude. The display should be redrawn after brain weight and body weight are transformed to their logarithms.

DISPLAY 9.11 Matrix of scatterplots for brain weight data, after log transformations

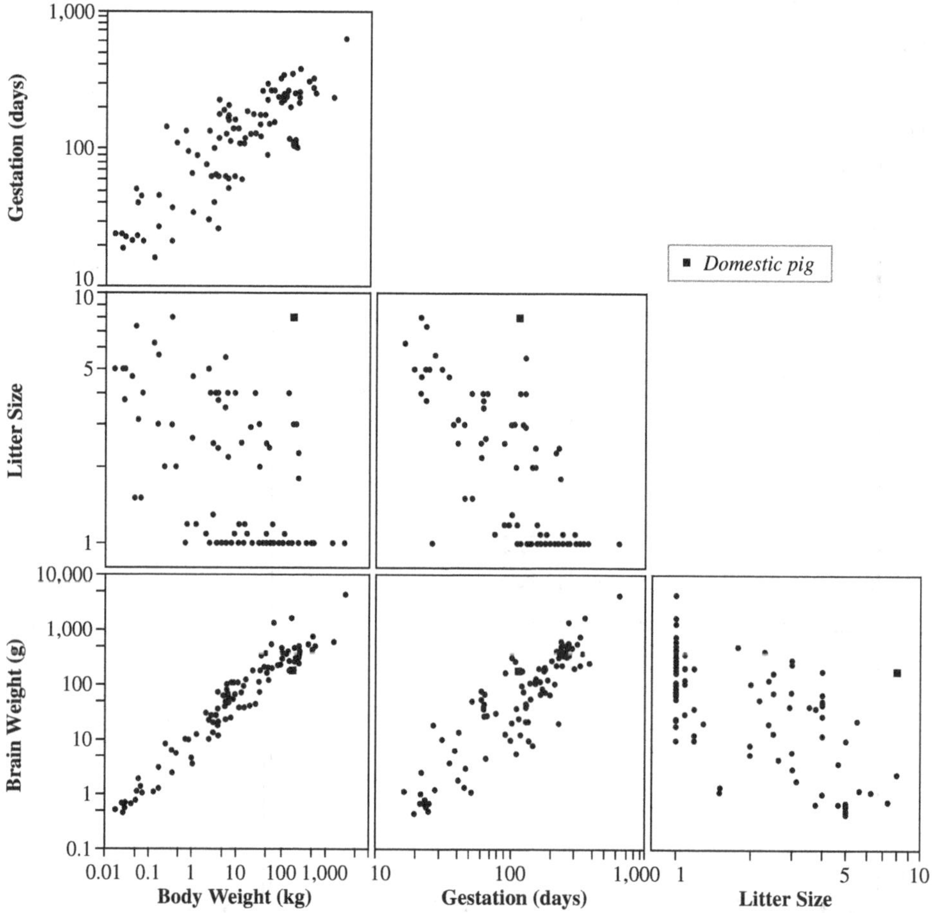

Gestation values also appear to be quite skewed. A trial plot using the log of this variable shows that it too should be transformed. Display 9.11 shows a revised version of the matrix of scatterplots where all variables have been placed on their natural logarithmic scales.

The bottom row of Display 9.11 shows less overlap and a more distinct pattern. Notice the pronounced relationship between log brain weight and each of the explanatory variables. Notice also, from the upper plots in the first column, that gestation and litter size are also related to body weight. Therefore the question of whether there is an association between gestation and brain weight after accounting for the effect of body weight, and the question of whether there is an association between litter size and brain weight after accounting for the effect of body weight, are not resolved by these plots. Nevertheless, they suggest the next course of action, which is to fit a regression model for log brain weight on log body weight, log gestation, and log litter size.

In the scatterplot of log brain weight versus log litter size, there is one data point that stands out—a mammal with litter size eight whose brain size is quite a bit larger than those for other mammals whose litter size is eight. The analyst should determine which mammal this is. It may be influential in the analysis, and it may provide useful information or additional questions for research. Modern statistical computer packages with interactive labeling features make the identification of such points on a graph fairly easy.

9.5.2 Coded Scatterplots

Display 9.3 is an example of a *coded scatterplot*—a scatterplot with different symbols or letters used as plotting marks to distinguish two or more groups. In the example, the groups correspond to different levels of timing. This is a great way of observing the joint effects of one numerical and one categorical explanatory variable. For observing the joint effects of two numerical explanatory variables, it is often very helpful to plot the response versus one of them, and to use a plotting code to represent groups defined by ranges of the other (such as educational level less than 8 years, between 9 and 12 years, or greater than 12 years).

9.5.3 Jittered Scatterplots

The scatterplot of log brain weight versus log litter size (Display 9.11) contains many points that overlap because many species of mammals have the same litter size. This makes it difficult to judge relative frequency. When a variable has only a few distinct values it may be possible to improve the visual information of the plot by *jittering* that variable—adding small computer-generated random numbers to it before plotting. In Display 9.12, log brain weight is plotted against a jittered version of litter size. A jittered variable is only used in the scatterplot; subsequent numerical analyses must be based on the unjittered version. The choice as to the size of the random numbers to add in order to make each point visually distinct requires trial and error.

9.5.4 Trellis Graphs

In Section 9.2 it was noted that β_2 in the model $\mu\{Y|X_1, X_2\} = \beta_0 + \beta_1 X_1 + \beta_2 X_2$ refers to the change in the mean of Y associated with changes in X_2, for subpopulations of fixed X_1. A useful graphical procedure for exploring this relationship is a trellis graph. A simple form is shown in Display 9.13, in which the response variable, log brain weight, is plotted against an explanatory variable of interest—log of gestation length—separately for species in the lowest 25% of body weights, the next lowest, and so on.

While the data in the four panels are not subpopulations with fixed body weights, they are subpopulations with similar values. This display suggests a positive relationship exists between brain weight and gestation length even accounting for different body weights. Furthermore, the slope in this relationship appears to

DISPLAY 9.12 Jittered scatterplot: log brain weight versus litter size (jittered)

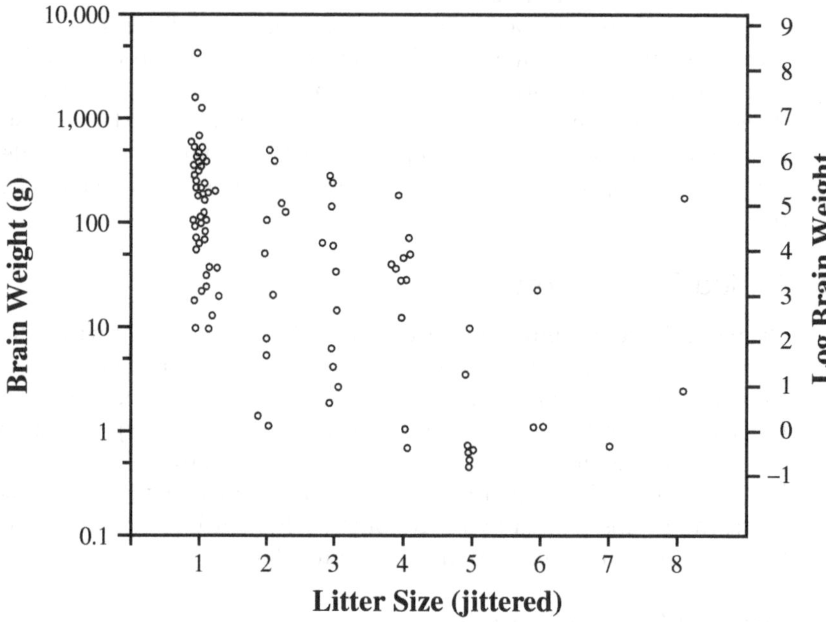

be the same in the different panels of body weight; that is, there is no evidence of an interactive effect of gestation and body weight on brain weight.

9.6 RELATED ISSUES

9.6.1 Computer Output

Regression output from statistical computer packages can be intimidating for the simple reason that it tends to include everything that the user could possibly want and more. One important component, which all packages display, is the list of estimates and their standard errors. Display 9.14 and Display 9.15 show the least squares estimates of the coefficients in specified models for the two examples of this chapter, their standard errors, and two-sided p-values from t-tests for the hypotheses that each β is zero in the specified model.

Although the least squares estimates and the inferential tools are discussed further in the next chapter, it is not difficult to see which questions are addressed by the p-values. The p-value of 0.9096 for the interaction term from Display 9.14 indicates that zero is a plausible value for the coefficient of the interaction explanatory variable; so there is no evidence of an interaction. The p-value of 0.0089 for the coefficient of *llitter* indicates strong evidence that the litter size is associated with brain weight, even after accounting for body weight and gestation.

DISPLAY 9.13 Trellis graph showing scatterplots of log brain weight versus log gestation length, separately for species in each of four categories of body weight

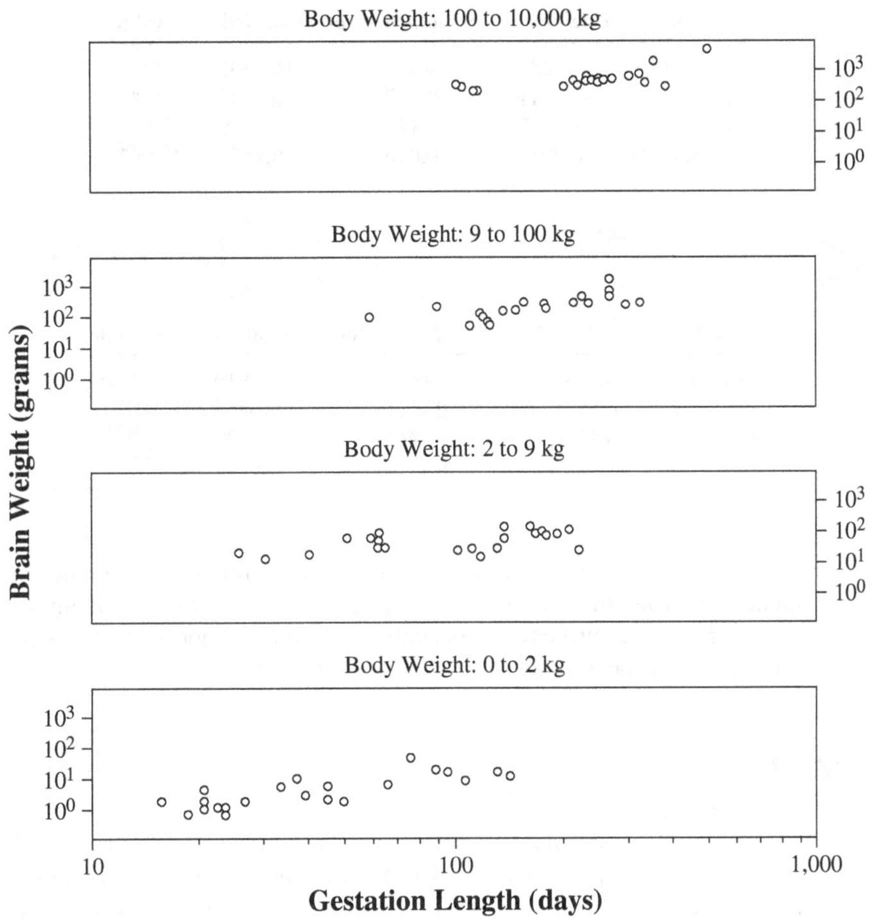

9.6.2 Factorial Treatment Arrangement

The meadowfoam study investigated two treatment factors applied jointly to each experimental unit: the light intensity (at one of six levels) and the timing (at one of two levels). All 12 combinations of these two treatment factors were applied to some experimental units. The term *factorial treatment arrangement* describes the formation of experimental treatments from all combinations of the levels of two or more distinct, individual treatment factors. This study was a 6×2 factorial arrangement, meaning all six levels of one factor appear in combination with both levels of the other.

Three benefits of testing multiple factors all at once rather than through one-at-a-time experiments are the following: Interactive effects can be explored; there is

DISPLAY 9.14 Estimates of regression coefficients in the multiple regression of *flowers* on *light*, *early*, and *light* × *early*—the meadowfoam study

Variable	Coefficient	Standard error	*t*-statistic	*p*-value
Constant	71.6233	4.3433	16.4905	<0.0001
light	−0.0411	0.0074	5.5247	<0.0001
early	11.5233	6.1424	1.8760	0.0753
light × *early*	0.0012	0.0105	0.1150	0.9096

DISPLAY 9.15 Estimates of regression coefficients in the multiple regression of log brain weight on log body weight, log gestation, and log litter size—brain weight data

Variable	Coefficient	Standard error	*t*-statistic	*p*-value
Constant	0.8548	0.6617	1.2919	0.1996
lbody	0.5751	0.0326	17.6468	<0.0001
lgest	0.4179	0.1408	2.9687	0.0038
llitter	−0.3101	0.1159	2.6747	0.0089

more efficient use of the available experimental units with the multifactor arrangement (resulting in smaller standard errors for parameters of interest); and the results for the multifactor experiment are more general, since each treatment is investigated at several levels of the other treatment.

9.7 SUMMARY

Multiple regression analysis refers to a large set of tools associated with developing and using regression models for answering questions of interest. Multiple regression models describe the mean of a single response variable as a function of several explanatory variables. Transformations, indicator variables for grouped data (factors), squared explanatory variables for curvature, and product terms for interaction greatly enhance the usefulness of this model.

Substantial exploratory analysis is recommended for gaining initial insight into what the data have to say in answer to the questions of interest and for suggesting possible regression models for answering them more formally. Some standard graphical procedures are presented here. The data analyst should be prepared to be creative in using the available computer tools to best display the data, while keeping in mind the questions of interest and the statistical tools that might be useful for answering them.

Brain Weight Study

Is brain weight associated with gestation period and/or litter size after accounting for the effect of body weight? This is exactly the type of question for which multiple regression is useful. The regression coefficient of gestation in the regression of brain weight on body weight and gestation describes the effect of gestation for species of roughly the same body weight. With multiple linear regression, the coefficients can be estimated from all the animals without a need for grouping into subsets of similar body weight. Initial scatterplots (or initial inspection of the data) indicate that the regression model should be formed after transforming all the variables to their logarithms.

Meadowfoam Study

A starting point for the analysis is the coded scatterplot of number of flowers per plant versus light intensity, with different codes to represent the two levels of the timing factor (Display 9.3). The plot suggests that the mean number of flowers per plant decreases with increasing light intensity, that the rate of decrease does not depend on timing, and that (for any light intensity value) a larger mean number of flowers is associated with the "before PFI" level of the timing factor. Using an indicator variable for one of the timing levels permits fitting the parallel regression lines model. Further inclusion of the interaction of timing and intensity produces a model that fits separate regression lines for each level of timing. This model permits a check on whether the regression lines are indeed parallel.

9.8 EXERCISES

Conceptual Exercises

1. **Meadowfoam.** (a) Write down a multiple regression model with parallel regression lines of *flowers* on *light* for the two separate levels of *time* (using an indicator variable). (b) Add a term to the model in (a) so that the regression lines are not parallel.

2. **Meadowfoam.** A model (without interaction) for the mean *flowers* is estimated to be $71.3058 - 0.0405 light + 12.1583 early$. For a fixed level of timing, what is the estimated difference between the mean *flowers* at 600 and 300 μmol/m^2/sec of *light* intensity?

3. **Meadowfoam.** (a) Why were the numbers of flowers from 10 plants averaged to make a response, rather than representing them as 10 different responses? (b) What assumption is assisted by averaging the numbers from the 10 plants?

4. **Mammal Brain Weights.** The three-toed sloth has a gestation period of 165 days. The Indian fruit bat has a gestation period of 145 days. From Display 9.14 the estimated model for the mean of log brain weight is $0.8548 + 0.5751 lbody + 0.4179 lgest - 0.3101 llitter$. Since *lgest* for the sloth is 0.1292 more than *lgest* for the fruit bat, does this imply that an estimate of the mean log brain weight for the sloth is (0.4179)(0.1292) more than the mean log brain weight for the bat (i.e., the median is 5.5% higher)? Why? Why not?

5. **Insulating Fluid** (Section 8.1.2). Would it be possible to test for lack of fit to the straight line model for the regression of log breakdown time on voltage by including a voltage-squared term in the model, and testing whether the coefficient of the squared term is zero?

6. **Island Area and Species.** For the island area and number of species data in Section 8.1.1, would it be possible to test for lack of fit to the straight line model for the regression of log number

of species on log island area by including the square of log area in the model and testing whether its coefficient is zero?

7. Which of the following regression models are *linear*?

(a) $\mu\{Y|X\} = \beta_0 + \beta_1 X + \beta_2 X^2 + \beta_3 X^3$
(b) $\mu\{Y|X\} = \beta_0 + \beta_1 10^X$
(c) $\mu\{Y|X\} = (\beta_0 + \beta_1 X)/(\beta_0 + \beta_2 X)$
(d) $\mu\{Y|X\} = \beta_0 \exp(\beta_1 X)$.

8. Describe what σ measures in the meadowfoam problem and in the brain weight problem.

9. **Pollen Removal.** Reconsider the data on proportion of pollen removed and duration of visit to the flower for bumblebee queens and honeybee workers, in Exercise 3.28. (a) Write down a model that describes the mean proportion of pollen removed as a straight-line function of duration of visit, with separate intercepts and separate slopes for bumblebee queens and honeybee workers. (b) How would you test whether the effect of duration of visit on proportion removed is the same for queens as for workers?

10. **Breast Milk and IQ.** In a study, intelligence quotient (IQ) test scores were obtained for 300 8-year-old children who had been part of a study of premature babies in the early 1980s. Because they were premature, all the babies were fed milk by a tube. Some of them received breast milk entirely, some received a prepared formula entirely, and some received some combination of breast milk and formula. The proportion of breast milk in the diet depended on whether the mother elected to provide breast milk and to what extent she was successful in expressing any, or enough, for the baby's diet. The researchers reported the results of the regression of the response variable—IQ at age 8—on social class (ordered from 1, the highest, to 5), mother's education (ordered from 1, the lowest, to 5), an indicator variable taking the value 1 if the child was female and 0 if male, the number of days of ventilation of the baby after birth, and an indicator variable taking the value 1 if there was any breast milk in the baby's diet and 0 if there was none. The estimates are reported in Display 9.16 along with the p-values for the tests that each coefficient is zero. (Data from Lucas et al., "Breast Milk and Subsequent Intelligence Quotient in Children Born Preterm," *Lancet* 339 (1992): 261–64).

DISPLAY 9.16 Breast milk and intelligence data

Explanatory variable	Estimated coefficient	p-value
Social class	−3.5	0.0004
Mother's education	2.0	0.01
Female indicator	4.2	0.01
Days of ventilation	−2.6	0.02
Breast milk indicator	8.3	<0.0001

(a) After accounting for the effects of social class, mother's education, whether the child was a female, and days after birth of ventilation, how much higher is the estimated mean IQ for those children who received breast milk than for those who did not?

(b) Is it appropriate to use the variables "Social class" and "Mother's education" in the regression even though in both instances the numbers 1 to 5 do not correspond to anything real but are merely ordered categories?

(c) Does it seem appropriate for the authors to simply report < 0.0001 for the p-value of the breast milk coefficient rather than the actual p-value?

(d) Previous studies on breast milk and intelligence could not separate out the effects of breast milk and the act of breast feeding (the bonding from which might encourage intellectual development of the child). How is the important confounding variable of whether a child is breast fed dealt with in this study?

(e) Why is it important to have social class and mother's education as explanatory variables?

(f) In a subsidiary analysis the researchers fit the same regression model as above except with the indicator variable for whether the child received breast milk replaced by the percentage of breast milk in the diet (between 0 and 100%). The coefficient of that variable turned out to be 0.09. (i) From this model, how much larger is the estimated mean IQ for children who received 100% breast milk than for those who received 50% breast milk, after accounting for the other explanatory variables? (ii) What is the importance of the percentage of breast milk variable in dealing with confounding variables?

11. Glasgow Graveyards. Do persons of higher socioeconomic standing tend to live longer? This was addressed by George Davey Smith and colleagues through the relationship of the heights of commemoration obelisks and the life lengths of the corresponding grave site occupants. In burial grounds in Glasgow a certain design of obelisk is quite prevalent, but the heights vary greatly. Since the height would influence the cost of the obelisk, it is reasonable to believe that height is related to socioeconomic status. The researchers recorded obelisk height, year of death, age at death, and gender for 1,349 individuals who died prior to 1921. Although they were interested in the relationship between mean life length and obelisk height, it is important that they included year of construction as an explanatory variable since life lengths tended to increase over the years represented (1801 to 1920). For males, they fit the regression of life length on obelisk height (in meters) and year of obelisk construction and found the coefficient of obelisk height to be 1.93. For females they fit the same regression and found the coefficient of obelisk height to be 2.92. (Data from Smith et al., "Socioeconomic Differentials in Mortality: Evidence from Glasgow Graveyards," *British Medical Journal* 305 (1992): 1557–60.)

(a) After accounting for year of obelisk construction, each extra meter in obelisk height is associated with Z extra years in mean lifetime. What is the estimated Z for males? What is the estimated Z for females?

(b) Since the coefficients differ significantly from zero, would it be wise for an individual to build an extremely tall obelisk, to ensure a long life time?

(c) The data were collected from eight different graveyards in Glasgow. Since there is a potential blocking effect due to the different graveyards, it might be appropriate to include a graveyard effect in the model. How can this be done?

Computational Exercises

12. Mammal Brain Weights. (a) Draw a matrix of scatterplots for the mammal brain weight data (Display 9.4) with all variables transformed to their logarithms (to reproduce Display 9.11). (b) Fit the multiple linear regression of log brain weight on log body weight, log gestation, and log litter size, to confirm the estimates in Display 9.15. (c) Draw a matrix of scatterplots as in (a) but with litter size on its natural scale (untransformed). Does the relationship between log brain weight and litter size appear to be any better or any worse (more like a straight line) than the relationship between log brain weight and log litter size?

13. Meat Processing. One way to check on the adequacy of a linear regression is to try to include an X-squared term in the model to see if there is significant curvature. Use this technique on the meat processing data of Section 7.1.2. (a) Fit the multiple regression of pH on hour and hour-squared. Is the coefficient of hour-squared significantly different from zero? What is the p-value? (b) Fit the multiple regression of pH on log(hour) and the square of log(hour). Is the coefficient of the squared-term significantly different from zero? What is the p-value? (c) Does this exercise suggest a potential way of checking the appropriateness of taking the logarithm of X or of leaving it untransformed?

14. Pace of Life and Heart Disease. Some believe that individuals with a constant sense of time urgency (often called type-A behavior) are more susceptible to heart disease than are more relaxed individuals. Although most studies of this issue have focused on individuals, some psychologists have investigated geographical areas. They considered the relationship of city-wide heart disease rates and general measures of the pace of life in the city.

For each region of the United States (Northeast, Midwest, South, and West) they selected three large metropolitan areas, three medium-size cities, and three smaller cities. In each city they measured three indicators of the pace of life. The variable *walk* is the walking speed of pedestrians over a distance of 60 feet during business hours on a clear summer day along a main downtown street. *Bank* is the average time a sample of bank clerks takes to make change for two $20 bills or to give $20 bills for change. The variable *talk* was obtained by recording responses of postal clerks explaining the difference between regular, certified, and insured mail and by dividing the total number of syllables by the time of their response. The researchers also obtained the age-adjusted death rates from ischemic heart disease (a decreased flow of blood to the heart) for each city (*heart*). The data in Display 9.17 were read from a graph in the published paper. (Data from R. V. Levine, "The Pace of Life," *American Scientist* 78 (1990): 450–9.) The variables have been standardized, so there are no units of measurement involved.

DISPLAY 9.17 First five rows of the pace-of-life data set with bank clerk speed, pedestrian walking speed, postal clerk talking speed, and age-adjusted death rates due to heart disease, in 36 cities

City	Bank	Walk	Talk	Heart
Atlanta, GA	25	27	27	19
Bakersfield, CA	29	18	25	11
Boston, MA	31	28	24	24
Buffalo, NY	30	23	23	29
Canton, OH	28	20	18	19

(a) Draw a matrix of scatterplots of the four variables. Construct it so that the bottom row of plots all have *heart* on the vertical axis. If you do not have this facility, draw scatterplots of *heart* versus each of the other variables individually.

(b) Obtain the least squares fit to the linear regression of *heart* on *bank*, *walk*, and *talk*.

(c) Plot the residuals versus the fitted values. Is there evidence that the variance of the residuals increases with increasing fitted values or that there are any outliers?

(d) Report a summary of the least squares fit. Write down the estimated equation with standard errors below each estimated coefficient.

15. Rainfall and Corn Yield. The data on corn yields and rainfall, discussed in Section 9.3.1, appear in Display 9.18. (Data from M. Ezekiel and K. A. Fox, *Methods of Correlation and Regression Analysis*, New York: John Wiley & Sons, 1959; originally from E. G. Misner, "Studies of the Relationship of Weather to the Production and Price of Farm Products, I. Corn" [mimeographed publication, Cornell University, March 1928].)

(a) Plot corn yield versus rainfall.

(b) Fit the multiple regression of corn yield on *rain* and $rain^2$.

DISPLAY 9.18 Partial listing of a data set with average corn yield in a year (bushels) and total rainfall (inches) in six U.S. states (1890–1927).

Year	Yield	Rainfall
1890	24.5	9.6
1891	33.7	12.9
1892	27.9	9.9
1893	27.5	8.7
1894	21.7	6.8
...		
1927	32.6	10.4

(c) Plot the residuals versus year. Is there any pattern evident in this plot? What does it mean? (Anything to do, possibly, with advances in technology?)

(d) Fit the multiple regression of corn yield on *rain*, *rain*^{2}, and *year*. Write the estimated model and report standard errors, in parentheses, below estimated coefficients. How do the coefficients of *rain* and *rain*^{2} differ from those in the estimated model in (b)? How does the estimate of σ differ? (larger or smaller?) How do the standard errors of the coefficients differ? (larger or smaller?) Describe the effect of an increase of one inch of rainfall on the mean yield over the range of rainfalls and years.

(e) Fit the multiple regression of corn yield on *rain*, *rain*^{2}, *year*, and *year* × *rain*. Is the coefficient of the interaction term significantly different from zero? Could this term be used to say something about technological improvements regarding irrigation?

16. Pollen Removal. The data in Exercise 3.27 are the proportions of pollen removed and the duration of visits on a flower for 35 bumblebee queens and 12 honeybee workers. It is of interest to understand the relationship between the proportion removed and duration and the relative pollen removal efficiency of queens and workers. (a) Draw a coded scatterplot of proportion of pollen removed versus duration of visit; use different symbols or letters as the plotting codes for queens and workers. Does it appear that the relationship between proportion removed and duration is a straight line? (b) The logit transformation is often useful for proportions between 0 and 1. If p is the proportion then the logit is $\log[p/(1-p)]$. This is the log of the ratio of the amount of pollen removed to the amount not removed. Draw a coded scatterplot of the logit versus duration. (c) Draw a coded scatterplot of the logit versus log duration. From the three plots, which transformations appear to be worthy of pursuing with a regression model? (d) Fit the multiple linear regression of the logit of the proportion of pollen removed on (i) log duration, (ii) an indicator variable for whether the bee is a queen or a worker, and (iii) a product term for the interaction of the first two explanatory variables. By examining the p-value of the interaction term, determine whether there is any evidence that the proportion of pollen depends on duration of visit differently for queens than for workers. (e) Refit the multiple regression but without the interaction term. Is there evidence that, after accounting for the amount of time on the flower, queens tend to remove a smaller proportion of pollen than workers? Why is the p-value for the significance of the indicator variable so different in this model than in the one with the interaction term?

17. Crab Claw and Force. Using the crab data from Exercise 7.22, (a) draw a scatterplot of claw closing force versus propodus height (both on a log scale), with different plotting symbols to distinguish the three different crab species, and (b) fit the multiple regression of log force on log height and species (as a factor). Provide the estimated model including standard errors of estimated

DISPLAY 9.19 First five rows of a data set with average wing sizes of male and female flies, on logarithmic scale, with standard errors; and average basal length to wing size ratios of the females, at 11 locations in North America and 10 locations in Europe

		Wing size ($10^3\times$log mm)				Basal length to wing size (females)	
Continent	Latitude (N)	Females	SE	Males	SE	Ratio (av)	SE
NA	35.5	901	2.5	797	3.8	0.831	0.010
NA	37.0	896	3.5	806	3.0	0.834	0.014
NA	38.6	906	3.0	812	3.2	0.836	0.012
NA	40.7	907	3.5	807	3.2	0.833	0.013
NA	40.9	898	3.6	818	2.7	0.830	0.012

regression coefficients. (See Exercises 10.9 and 10.10 for analyses that explore a more sophisticated model for these data.)

18. Speed of Evolution. How fast can evolution occur in nature? Are evolutionary trajectories predictable or idiosyncratic? To answer these questions R. B. Huey et al. ("Rapid Evolution of a Geographic Cline in Size in an Introduced Fly," *Science* 287 (2000): 308–9) studied the development of a fly—*Drosophila subobscura*—that had accidentally been introduced from the Old World into North America (NA) around 1980. In Europe (EU), characteristics of the flies' wings follow a "cline"—a steady change with latitude. One decade after introduction, the NA population had spread throughout the continent, but no such cline could be found. After two decades, Huey and his team collected flies from 11 locations in western NA and native flies from 10 locations in EU at latitudes ranging from 35–55 degrees N. They maintained all samples in uniform conditions through several generations to isolate genetic differences from environmental differences. Then they measured about 20 adults from each group. Display 9.19 shows average wing size in millimeters on a logarithmic scale, and average ratios of basal lengths to wing size.

(a) Construct a scatterplot of average wing size against latitude, in which the four groups defined by continent and sex are coded differently. Do these suggest that the wing sizes of the NA flies have evolved toward the same cline as in EU?

(b) Construct a multiple linear regression model with wing size as the response, with latitude as a linear explanatory variable, and with indicator variables to distinguish the sexes and continents. As there are four groups, you will want to have three indicator variables: the continent indicator, the sex indicator, and the product of the two. Construct the model in such a way that one parameter measures the difference between the slopes of the wing size versus latitude regressions of NA and EU for males, one measures the difference between the NA–EU slope difference for females and that for males, one measures the difference between the intercepts of the regressions of NA and EU for males, and one measures the difference between the NA–EU intercepts' difference for females and that for males.

19. Depression and Education. Has homework got you depressed? It could be worse. Depression, like other illnesses, is more prevalent among adults with less education than you have.

R. A. Miech and M. J. Shanahan investigated the association of depression with age and education, based on a 1990 nationwide (U.S.) telephone survey of 2,031 adults aged 18 to 90. Of particular interest was their finding that the association of depression with education strengthens with increasing age—a phenomenon they called the "divergence hypothesis."

DISPLAY 9.20 Kentucky Derby winners, 1896–2011. *Starters* is the number of horses that started the race, *NetToWinner* is the net winnings in dollars; *Time* is the winning time in seconds; *Speed* is the winning average speed, in miles per hour; *Track* is a categorical variable describing track conditions, with seven categories: Fast, Good, Dusty, Slow, Heavy, Muddy, and Sloppy; *Conditions* is a two-category version of *Track*, with categories Fast (which combines categories Fast and Good from *Track*) and Slow (which combines all other categories of *Track*); partial listing.

Year	Winner	Starters	NetToWinner	Time	Speed	Track	Conditions
1896	Ben Brush	8	4,850	127.75	35.23	Dusty	Fast
1897	Typhoon II	6	4,850	132.50	33.96	Heavy	Slow
1898	Plaudit	4	4,850	129.00	34.88	Good	Fast
1899	Manuel	5	4,850	132.00	34.09	Fast	Fast
1900	Lieut. Gibson	7	4,850	126.25	35.64	Fast	Fast
...							
2011	Animal Kingdom	20	2,000,000	122.04	36.87	Fast	Fast

They constructed a depression score from responses to several related questions. Education was categorized as (i) college degree, (ii) high school degree plus some college, or (iii) high school degree only. (See "Socioeconomic Status and Depression over the Life Course," *Journal of Health and Social Behaviour* 41(2) (June, 2000): 162–74.)

(a) Construct a multiple linear regression model in which the mean depression score changes linearly with age in all three education categories, with possibly unequal slopes and intercepts. Identify a single parameter that measures the diverging gap between categories (iii) and (i) with age.

(b) Modify the model to specify that the slopes of the regression lines with age are equal in categories (i) and (ii) but possibly different in category (iii). Again identify a single parameter measuring divergence.

This and other studies found evidence that the mean depression is high in the late teens, declines toward middle age, and then increases towards old age. Construct a multiple linear regression model in which the association has these characteristics, with possibly different structures in the three education categories. Can this be done in such a way that a single parameter characterizes the divergence hypothesis?

Data Problems

20. **Kentucky Derby.** Display 9.20 is a partial listing of data in file ex0920 on Kentucky Derby horse race winners from 1896 to 2011. In all those years the race was 1.25 miles in length so that winning time and speed are exactly inversely related. Nevertheless, a simple regression model for changes over time—such as a straight line model that includes *Year* or a quadratic curve that includes *Year* and $Year^2$—might work better for one of these response variables than the other. (a) Find a model for describing the mean of either winning time or winning speed as a function of year, whichever works better. (b) Quantify the amount (in seconds or miles per hour) by which the mean winning time or speed on fast tracks exceeds the mean on slow tracks (using the two-category variable *Conditions*), after accounting for the effect of year. (c) After accounting for the effects of year and track conditions, is there any evidence that the mean winning time or speed depends on number of horses in the race (*Starters*)? Is there any evidence of an interactive effect of *Starters* and *Conditions*;

DISPLAY 9.21 Dry weight (mg), ingestion rates (mg per day), and percentage of organic matter in the food, for 22 species of aquatic deposit feeders

Species	Weight	Ingestion	Organic
Hydrobia neglecta	0.20	0.57	18.0
Hydrobia ventrosa	0.20	0.86	17.0
Tubifex tubifex	0.27	0.43	29.7
Hyalella azteca	0.32	0.43	50.0
Potamopyrgus jenkinsi	0.46	2.70	14.4
Hydrobia ulvae	0.90	0.67	13.0
Nereis succinea	5.80	20.20	6.8
Pteronarcys scotti	8.40	1.49	93.0
Orchestia grillus	12.40	4.40	88.0
Arenicola grubii	20.40	240.00	2.2
Thoracophelia mucronata	40.00	230.00	1.0
Ilypolax pusilla	53.00	300.00	4.2
Uca pubnax	63.30	19.90	51.0
Scopimera globosa	65.00	50.00	23.6
Pectinaria gouldii	80.00	1,667.00	0.7
Abarenicola pacifica	380.00	3,400.00	1.2
Abarenicola claparedi	380.00	9,400.00	0.4
Arenicola marina	930.00	4,700.00	0.6
Macrophthalmus japonicus	2,050.00	4,680.00	2.1

that is, does the effect of number of horses on the response depend on whether the track was fast or slow? Describe the effect of number of horses on mean winning time or speed. (Data from Kentucky Derby: Kentucky Derby Racing Results, http://www.kentuckyderby.info/kentuckyderby-results.php (July 21, 2011).)

21. Ingestion Rates of Deposit Feeders. Aquatic deposit feeders are organisms that feed on organic material in the bottoms of lakes and oceans. Display 9.21 shows the typical dry weight in mg, the typical ingestion rate (weight of food intake per day for one animal) in mg/day, and the percentage of the food that is composed of organic matter for 19 species of deposit feeders. The organic matter is considered the "quality" part of their food. Zoologist Leon Cammen wondered if, after accounting for the size of the species as measured by the average dry weight, the amount of food intake is associated with the percentage of organic matter in the sediment. If so, that would suggest that either the animals have the ability to adjust their feeding rates according to some perception of food "quality" or that species' ingestion rates have evolved in their particular environments. Analyze the data to see whether the distribution of species' ingestion rates depends on the percentage of organic matter in the food, after accounting for the effect of species weight. Also describe the association. Notice that the values of all three variables differ by orders of magnitude among these species, so that log transformations are probably useful. (Data from L. M. Cammen, "Ingestion Rate: An Empirical Model for Aquatic Deposit Feeders and Detritivores," *Oecologia*, 44 (1980): 303–10.)

22. Mammal Lifespans. The Exercise 8.26 data set contains the mass (in kilograms), average basal metabolic rate (in kilojoules per day), and lifespan (in years) for 95 mammal species. Is metabolic rate a useful predictor of lifespan after accounting for the effect of body mass? What

DISPLAY 9.22 First five rows of a data set with AFQT intelligence test score percentile, years of education achieved by time of interview in 2006, and 2005 annual income in dollars for 1,306 males and 1,278 females in the NLSY survey

Subject	Gender	AFQT	Educ2006	Income2005
2	female	6.841	12	5,500
6	male	99.393	16	65,000
7	male	47.412	12	19,000
8	female	44.022	14	36,000
9	male	59.683	14	65,000

percentage of variation in lifespans (on the log scale) is explained by the regression on log mass alone? What additional percentage of variation is explained by metabolism? Describe the dependence of the distribution of lifespan on metabolic rate for species of similar body mass.

23. Comparing Male and Female Incomes, After Accounting for Education and IQ. Display 9.22 shows the first five rows of a subset of the National Longitudinal Study of Youth data (see Exercise 2.22) with annual incomes in 2005, intelligence test scores (AFQT) measured in 1981, and years of education completed by 2006 for 1,306 males and 1,278 females who were between the ages of 14 and 22 when selected for the survey in 1979, who were available for re-interview in 2006, and who had paying jobs in 2005. Is there any evidence that the mean salary for males exceeds the mean salary for females with the same years of education and AFQT scores? By how many dollars or by what percent is the male mean larger?

Answers to Conceptual Exercises

1. (a) Let *early* = 1 if *time* = 24 and 0 if *time* = 0. Then $\mu\{flowers \mid light, early\} = \beta_0 + \beta_1 light + \beta_2 early$. (b) $\beta_0 + \beta_1 light + \beta_2 early + \beta_3(light \times early)$.

2. The difference is 300 μmol/m^2/sec times the coefficient of *light*, or about −12.15 flowers.

3. (a) The principal reason is that the 10 plants were all treated together and grown together in the same chamber. The experimental unit is always defined as the unit that receives the treatment, here, plants in the same chamber. (b) The assumption of normality is assisted. Averages tend to have normal distributions, so the averaging may alleviate some distributional problems that could arise from looking at separate numbers of flowers.

4. No. The difficulty with interpreting regression coefficients individually, as in a controlled experiment, is that explanatory variables cannot be manipulated individually. In this instance, the sloth and the fruit bat also have different body weights—the sloth weighs 50 times what the fruit bat weighs. (The full model estimates the brain weight of the fruit bat to be only about 35% of the brain weight of the sloth.) One might attempt to envision a fruit bat having the same weight (0.9 kg) as the sloth and the same litter size (1.0), but having a gestation period of 165 instead of 145 days. This approach, however, is generally unsatisfactory because it extrapolates beyond the experience of the data set (resulting in animals like a fish-eating kangaroo with wings).

5. Yes. A common way to explore lack of fit is to introduce curvature and interaction terms to see if measured effects change as the configuration of explanatory variables changes.

6. Yes.

7. Keep your eye on the parameters. If the mean is linear in the parameters, the model is *linear*.

(a) Yes, even though it is not linear in X.
(b) Yes.
(c) No. Both numerator and denominator are linear in parameters and X, but the whole is not.
(d) No. This is a very useful model, however.

8. In both, σ is a measure of the magnitude of the difference between a response and the mean in the population from which the response was drawn. In the meadowfoam problem, σ measures the typical size of differences between seedling flowers (averaged from 10 plants) and the mean seedling flowers (averaged from 10 plants) treated similarly (same intensity and timing potential). In the brain weight problem, it is more difficult to describe what σ measures because the theoretical model invents a hypothetical subpopulation of animal species all having the same body weight, gestation, and litter size.

9. (a) $\mu\{pollen \mid duration, queen\} = \beta_0 + \beta_1 duration + \beta_2 queen + \beta_3(duration \times queen)$, where *queen* is 1 for queens and 0 for workers (or the other way around). (b) H_0: $\beta_3 = 0$.

10. (a) 8.3 points. (b) As long as the effects are indeed linear in these coded variables, it is a useful way to include them in the multiple regression (and more powerful than considering them to be factors). (c) Yes. The evidence is overwhelming that this coefficient is not zero. (d) The use of premature babies who all had to be fed by tube makes it so that some babies received breast milk but all babies were administered their milk in exactly the same way. (e) It is possible that the decision to provide breast milk and the ability to express breast milk are related to social class and mother's education, which are likely to be related to child's IQ score. It is desired to examine the effect of breast milk that is separate from the association with these potentially confounding variables. (f) (i) 4.5 points. (ii) An important confounding variable in the previous result is the mother's decision to provide breast milk, which may be associated with good mothering skills, which may be associated with better intelligence development in the child. Using the proportion of breast milk as an explanatory variable allows the dose–response assessment of breast milk, which indicates that children of mothers who provided breast milk for 100% of the diet tended to score higher on the IQ test than children of mothers who also decided to provide breast milk but were only capable of supplying a smaller proportion of the diet.

11. (a) 1.93 years. 2.92 years. (b) No. No cause-and-effect is implied by the analysis of these observational data. (c) Seven indicator variables can be included to distinguish the eight graveyards.

CHAPTER 10

Inferential Tools for Multiple Regression

Data analysis involves finding a good-fitting model whose parameters relate to the questions of interest. Once the model has been established, the questions of interest can be investigated through the parameter estimates, with uncertainty expressed through p-values, confidence intervals, or prediction intervals, depending on the nature of the questions asked.

The primary inferential tools associated with regression analysis—t-tests and confidence intervals for single coefficients and linear combinations of coefficients, F-tests for several coefficients, and prediction intervals—are described in this chapter. As usual, the numerical calculations for these tools are performed with the help of a computer. The most difficult parts of the task are knowing what model and what inferential tool best suit the need, and knowing how to interpret and communicate the results.

The inferential tools are illustrated in this chapter on models that incorporate special explanatory variables from the previous chapter: indicator variables, quadratic terms, and interaction terms. The tests and confidence statements found in the examples, and their interpretations, are typical of the ones used in many fields and for many different kinds of data.

10.1 CASE STUDIES

10.1.1 Galileo's Data on the Motion of Falling Bodies—A Controlled Experiment

In 1609 Galileo proved mathematically that the trajectory of a body falling with a horizontal velocity component is a parabola. His discovery of this result, which preceded the mathematical proof by a year, was the result of empirical findings in an experiment conducted for another purpose.

Galileo's search for an experimental setting in which horizontal motion was not affected appreciably by friction (to study inertia) led him to construct an apparatus like the one shown in Display 10.1. He placed a grooved, inclined plane on a table, released an ink-covered bronze ball in the groove at one of several heights above the table, and measured the horizontal distance between the table and the resulting ink spot on the floor. The data from one experiment are shown in Display 10.1 in units of *punti* (points). One *punto* is 169/180 millimeters. (Data from S. Drake and J. MacLachlan, "Galileo's Discovery of the Parabolic Trajectory," *Scientific American* 232 (1975): 102–10.)

DISPLAY 10.1 Galileo's experimental results

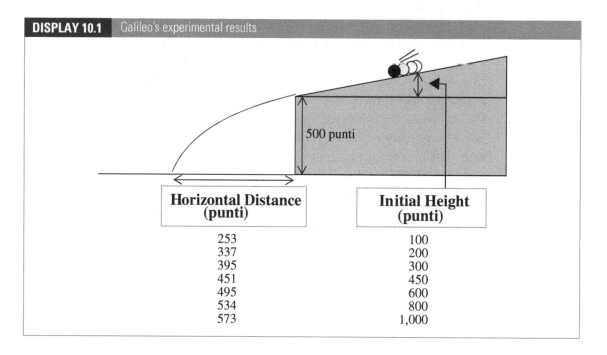

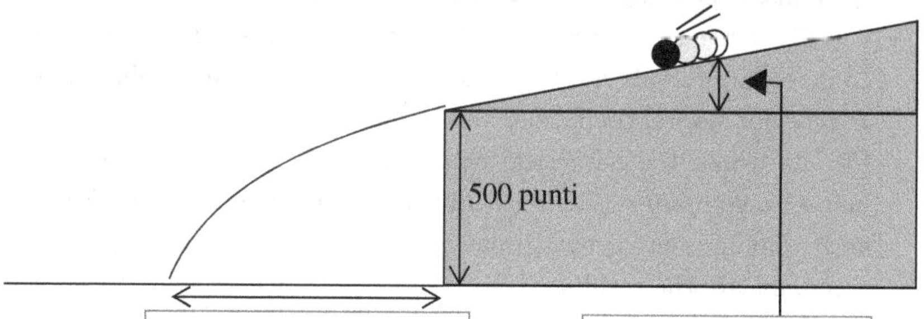

Horizontal Distance (punti)	Initial Height (punti)
253	100
337	200
395	300
451	450
495	600
534	800
573	1,000

Galileo conducted this experiment to determine whether, in the absence of any appreciable resistance, the horizontal velocity of a moving object is constant. While sketching the paths of the trajectories in his notebook, he apparently came to believe that the trajectory was a parabola. Once the idea of a parabola suggested itself to Galileo, he found that proving it mathematically was straightforward.

Although Galileo's experiment preceded Gauss's invention of least squares and Galton's empirical fitting of a regression line by more than 200 years, it is interesting to use regression here to explore the regression of horizontal distance on initial height.

Statistical Conclusion

As shown in Display 10.2, a quadratic curve for the regression of horizontal distance on height fits well for initial heights less than 1,000 punti. The data provide strong evidence that the coefficient of a cubic term differs from zero (two-sided p-value $= 0.007$). Nonetheless, the quadratic model accounts for 99.03% of the variation in measured horizontal distances, and the cubic term explains only an additional 0.91% of the variation. (*Note*: The significance of the cubic term can be explained by the effect of resistance.)

DISPLAY 10.2 Scatterplot of Galileo's horizontal distances versus initial heights, with estimated quadratic regression model (with standard errors in parentheses)

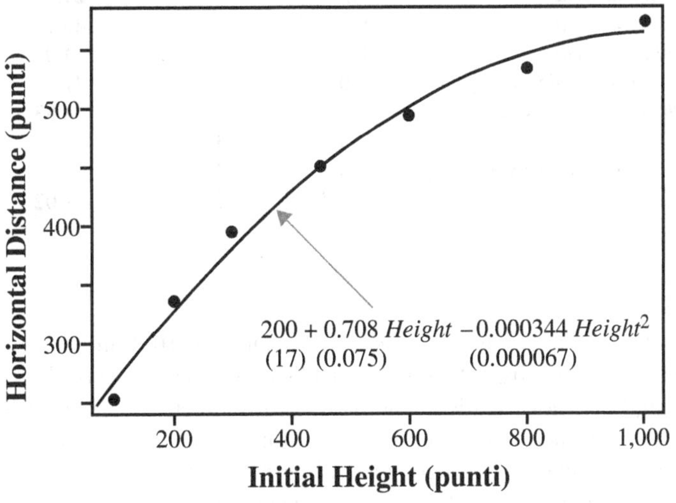

10.1.2 The Energy Costs of Echolocation by Bats—An Observational Study

To orient themselves with respect to their surroundings, some bats use echolocation. They send out pulses and read the echoes that are bounced back from surrounding objects. Such a trait has evolved in very few animal species, perhaps because of the high energy costs involved in producing pulses. Because flight also requires a great deal of energy, zoologists wondered whether the combined energy costs of echolocation and flight in bats was the sum of the flight energy costs and the at-rest echolocation energy costs, or whether the bats had developed a means of echolocation in flight that made the combined energy cost less than the sum.

DISPLAY 10.3 Mass and in-flight energy expenditure for 4 non-echolocating bats (Type = 1), 12 non-echolocating birds (Type = 2), and 4 echolocating bats (Type = 3)

Species	Mass (g)	Type	Flight energy expenditure (W)
Pteropus gouldii	779	1	43.7
Pteropus poliocephalus	628	1	34.8
Hypsignathus monstrosus	258	1	23.3
Eidolon helvum	315	1	22.4
Meliphaga virescens	24.3	2	2.46
Melipsittacus undulatus	35	2	3.93
Sturnus vulgaris	72.8	2	9.15
Falco spaverius	120	2	13.8
Falco tinnunculus	213	2	14.6
Corvus ossifragus	275	2	22.8
Larus atricilla	370	2	26.2
Columba livia	384	2	25.9
Columba livia	442	2	29.5
Columba livia	412	2	43.7
Columba livia	330	2	34.0
Corvus cryptoleucos	480	2	27.8
Phyllostomas hastatus	93	3	8.83
Plecotus auritus	8	3	1.35
Pipistrellus pipistrellus	6.7	3	1.12
Plecotus auritus	7.7	3	1.02

Zoologists considered the data in Display 10.3 on in-flight energy expenditure and body mass from 20 energy studies on three types of flying vertebrates: echolocating bats, non-echolocating bats, and non-echolocating birds. They believed that if the combined energy expenditure for flight and echolocation were additive, the amount of energy expenditure (after accounting for body size) would be greater for echolocating bats than for non-echolocating bats and non-echolocating birds. Display 10.4 shows a log-log scatterplot of in-flight energy expenditure versus body mass for the three types. (Data from J. R. Speakman and P. A. Racey, "No Cost of Echolocation for Bats in Flight," *Nature* 350 (1991): 421–23.)

Statistical Conclusion

The median in-flight energy expenditure for echolocating bats is estimated to be 1.08 times as large as that for non-echolocating bats, after accounting for body mass. A 95% confidence interval for the ratio of the echolocating median to the non-echolocating median, accounting for body mass, is 0.70 to 1.66. The data are consistent with the hypothesis of equal median in-flight energy expenditures in the three groups, after accounting for body size (p-value = 0.66, extra-sum-of-squares F-test).

DISPLAY 10.4 Log-log scatterplot of in-flight energy expenditure versus body mass for 4 non-echolocating bats, 12 non-echolocating birds, and 4 echolocating bats

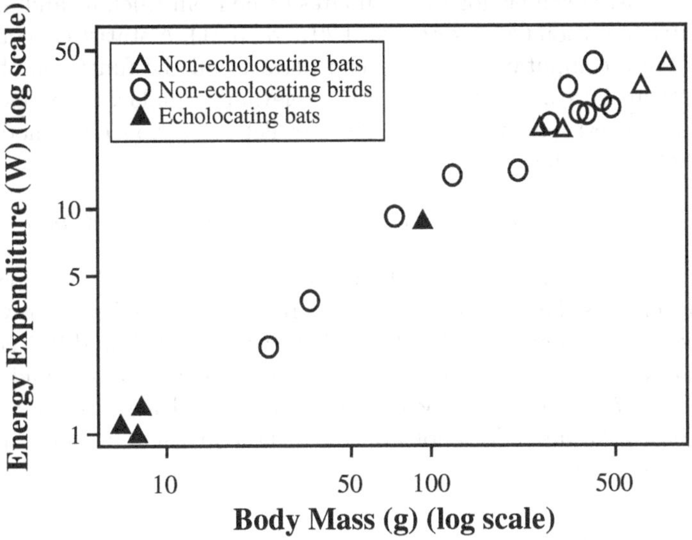

Scope of Inference

The species used in the statistical analysis were those for which relevant data were available; any inference to a larger population of species is speculative. The statistical inferences must also be interpreted in light of the likely violation of the independence assumption, due to treating separate studies on the same species as independent observations.

The scientific results are premised on several facts. Bats emit the echolocation pulses once on each wing beat, starting in the latter phase of the up-stroke, coinciding with the moment that air is expelled from the lungs. The coupling of these three processes apparently accounts for the relative energy economy. The energy used to expel air is also used to send out the pulses, and these events occur just before the greatest demand is made on the wing beat.

10.2 INFERENCES ABOUT REGRESSION COEFFICIENTS

Whether the energy expenditure is the same for echolocating bats as for non-echolocating bats, after accounting for body mass, can be investigated by a test of the coefficient of the indicator variable for echolocating bats in the parallel regression lines model. The simplicity with which the question of interest can be investigated makes this the natural choice for an inferential model. Many regression problems share this feature: questions of interest can be answered through tests or confidence intervals for single coefficients in the model.

10.2.1 Least Squares Estimates and Standard Errors

The least squares estimates are the β values that minimize the sum of squared residuals. Formulas for the estimates come from calculus and are best expressed in matrix notation (see Exercises 10.20 and 10.21). A statistical computer program can do these computations. Users can perform all the functions of regression analysis on their data without knowing the relevant formulas. As an example of computer-provided estimates, consider the parallel regression lines model used for the bat echolocation data:

$$\mu\{lenergy \mid lmass, TYPE\} = \beta_0 + \beta_1 lmass + \beta_2 \, bird + \beta_3 \, ebat,$$

where *lenergy* is the log of the in-flight energy, *lmass* is the log of the body mass, and *TYPE* is the three-level factor represented by the indicator variables *bird* (which takes on a value of 1 if the type of species is a bird, and 0 if not) and *ebat* (which takes on a value of 1 if the species is an echolocating bat, and 0 if not). The third level of *TYPE*, non-echolocating bats, is treated here as the reference level, so its indicator variable does not appear in the regression model. The model is sketched in Display 10.5.

DISPLAY 10.5 The parallel regression lines model for the bat echolocation data

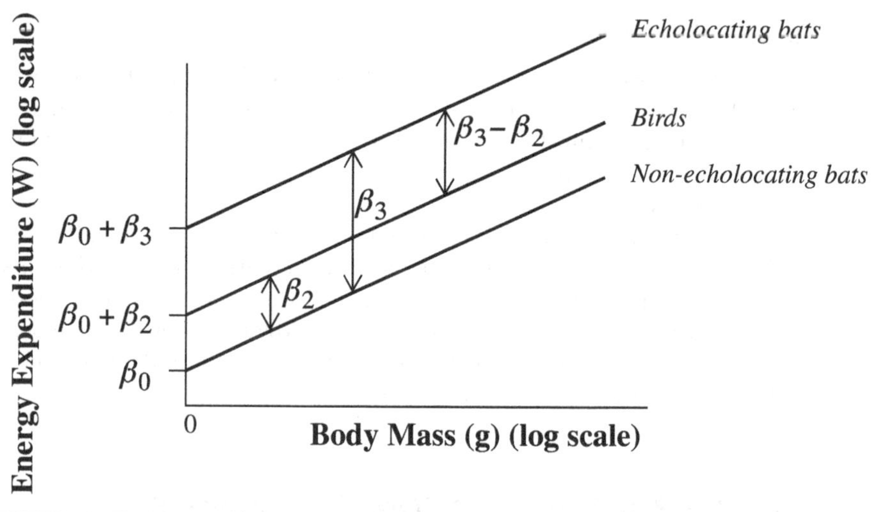

A portion of the output is shown in Display 10.6. The least squares estimates of the β's appear in the column labeled "Coefficient." The estimate of σ^2 is the sum of squared residuals divided by the degrees of freedom associated with the residuals. The usual rule for finding degrees of freedom applies: (Sample size) minus (Number of unknown regression coefficients in the model for the mean). Since the sample size is 20 and there are 4 coefficients here—$\beta_0, \beta_1, \beta_2$, and β_3—the value for degrees of freedom is 16. The square root of the estimate of variance is the

DISPLAY 10.6 Partial summary of the least squares fit to the regression of log energy expenditure on log body mass, an indicator variable for bird, and an indicator variable for echolocating bat

Variable	Coefficient	Standard error	*t*-statistic	Two-sided *p*-value
Constant	−1.5764	0.2872	−5.4880	<0.0001
lmass	0.8150	0.0445	18.2966	<0.0001
bird	0.1023	0.1142	0.8956	0.3837
ebat	0.0787	0.2027	0.3881	0.7030

Estimate of $\sigma = 0.1860$, d.f. = 16

estimated standard deviation about the regression, which is 0.1860. This goes by several names on computer output, including *residual SD*, *residual SE*, and *root mean squared error*.

The column labeled "Standard Error" contains the estimated standard deviations of the sampling distributions of the least squares estimates. The formulas for the standard deviations are once again available from statistical theory. In this instance, the formulas can be conveniently expressed only by using matrix notation, as shown in Exercise 10.22. Again, as with simple regression, they depend on the known values of the explanatory variables in the data set and on the unknown value of σ^2. The standard errors are the values of these standard deviations when σ^2 is replaced by its estimated value. Consequently, the degrees of freedom associated with a standard error are the same as the degrees of freedom associated with the estimate of σ (the sample size minus the number of β's).

Let $\hat{\beta}_j$ ("beta-hat j") represent the estimate of the jth coefficient ($j = 0, 1, 2,$ or 3 in the current example). If the distribution of the response for each combination of the explanatory variables is normal with constant variance, then the t-ratio,

$$t\text{-ratio} = (\hat{\beta}_j - \beta_j)/\text{SE}(\hat{\beta}_j),$$

has a sampling distribution described by the t-distribution with degrees of freedom equal to the degrees of freedom associated with the residuals. This theory leads directly to tests and confidence intervals for individual regression coefficients, in the familiar way.

10.2.2 Tests and Confidence Intervals for Single Coefficients

Do Non-Echolocating Bats Differ from Echolocating Bats? Test for $\beta_3 = 0$

In the parallel regression lines model for log energy on log mass, the mean log energy for non-echolocating bats is $\beta_0 + \beta_1 \, lmass$. The mean log energy for echolocating bats is $\beta_0 + \beta_1 \, lmass + \beta_3$. Thus, for any given mass, the mean log energy expenditure for echolocating bats is β_3 units more than the mean log energy for non-echolocating bats (see Display 10.5). The question of whether the mean log energy expenditure for echolocating bats is the same as the mean log energy expenditure for non-echolocating bats of similar size may be examined through a test of the hypothesis that β_3 equals 0. From Display 10.6, the two-sided p-value is .7030, providing no reason to doubt that β_3 is zero. The p-value was obtained by the computer as the proportion of values in a t-distribution on 16 degrees of freedom that are farther from 0 than 0.3881 (the t-statistic for the hypothesis that β_3 is 0).

Tests of hypotheses $H: \beta_j = c$ for values of c other than zero are occasionally of interest. For these, the user must construct the t-statistic $(\hat{\beta}_j - c)/\mathrm{SE}(\hat{\beta}_j)$ manually and find the p-value by reference to the appropriate t-percentiles.

How Much More Flight Energy Is Expended by Echolocating Bats? Confidence Interval for β_3

Although a test has revealed no reason to doubt that β_3 is zero, neither has it proved that β_3 is zero. Echolocating bats might have a higher energy expenditure, but the available study may not be powerful enough to detect this difference. As is usual for tests of hypotheses, a confidence interval should be reported in addition to the p-value, to emphasize this possibility and to provide an entire set of likely values for β_3.

The estimate of β_3 is 0.0787, with a standard error of 0.2027. The 97.5th percentile of a t-distribution with 16 degrees of freedom is 2.120, so the 95% confidence interval for β_3 is

$$0.0787 - 2.12(0.2027) \quad \text{to} \quad 0.0787 + 2.12(0.2027),$$

or

$$-0.351 \text{ to } +0.508.$$

Coefficients after log transformations are interpreted in the same way as are coefficients described for simple regression in Section 8.4. If the parallel regression lines model is correct, the median in-flight energy expenditure is $\exp(\beta_3)$ times as great for echolocating bats as it is for non-echolocating bats of similar body mass. A 95% confidence interval for $\exp(\beta_3)$ is obtained by taking the antilogs of the endpoints: $\exp(-0.351) = 0.70$ to $\exp(0.508) = 1.66$. The result can be communicated as follows: It is estimated that the median in-flight energy expenditure for echolocating bats is 1.08 times as great as the median expenditure for non-echolocating bats of similar body size. A 95% confidence interval for this multiplicative effect, accounting for body size, is 0.70 to 1.66.

Significance Depends on What Other Explanatory Variables Are Included

The meaning of the coefficient of an explanatory variable depends on what other explanatory variables are included in the regression. The p-value for the test of whether a coefficient is zero must also be interpreted in light of what other variables are included.

Consider three multiple linear regression models for the echolocation study. The first model says that the mean log energy is different for the three types of flying animals, but that it does not involve the body weight: $\mu\{lenergy \mid lmass, TYPE\} = TYPE$. The second model—the parallel lines model—indicates that the regression has the same slope against log body weight in all three groups, but different intercepts: $\mu\{lenergy \mid lmass, TYPE\} = lmass + TYPE$. The third model—the separate lines regression model—allows the three groups to have completely different straight line regressions of log energy on log body weight: $\mu\{lenergy \mid lmass, TYPE\} = lmass + TYPE + lmass \times TYPE$. Least squares fits to these three models yield the following results:

$$(1)\quad \hat{\mu}\{lenergy \mid lmass, TYPE\} = \underset{(0.42)}{3.40} - \underset{(0.60)}{2.74}ebat - \underset{(0.49)}{0.61}bird$$

$$(2)\quad \hat{\mu}\{lenergy \mid lmass, TYPE\} = \underset{(0.29)}{-1.58} + \underset{(0.20)}{0.08}ebat + \underset{(0.11)}{0.10}bird + \underset{(0.045)}{0.815}lmass$$

$$(3)\quad \hat{\mu}\{lenergy \mid lmass, TYPE\} = \underset{(1.26)}{-0.20} - \underset{(1.29)}{1.27}ebat - \underset{(1.30)}{1.38}bird + \underset{(0.21)}{0.59}lmass + \underset{(0.22)}{0.21}(ebat \times lmass) + \underset{(0.21)}{0.25}(bird \times lmass)$$

where the parenthesized numbers beneath the coefficients are their standard errors.

The coefficient of the indicator variable *ebat* is −2.74 in (1), +0.08 in (2), and −1.27 in (3). These might appear to be contradictory findings, since the variable is highly significant in (1) but not significant in (2) and (3), and since the sign of the estimated coefficient differs in the three fits. However, a contradiction arises only if one takes the view that the variable *ebat* plays the same role in each equation, which it does not. In (1), the coefficient of *ebat* measures the difference between mean log energy among echolocating bats and mean log energy among non-echolocating bats, ignoring any explanation of differences based on body size (*lmass* is not in the equation to provide the control). It happened (as indicated in Display 10.3) that the echolocating bats were much smaller than the non-echolocating bats in this study. Without taking body size into account, the statistical model attributes all the difference to group differences.

In (2), however, the coefficient of *ebat* measures the difference between energy expenditure of echolocating bats and of non-echolocating bats after adjusting for body size. Interpreting the coefficient as the difference between energy expenditure of echolocating bats and non-echolocating bats "of about the same size" exceeds the scope of this study, because all the echolocating bats were small and all the non-echolocating bats were large. The echolocating versus non-echolocating difference is thus confounded with the differences based on body size. Because "after adjusting for body size" means that group differences are considered only after the best explanation for body size is taken into account, and because body size explains the differences well, it is easy to understand why the coefficient of *ebat* in (2) can be insignificant even though in (1) it is significant.

In (3) the coefficient of *ebat* measures the difference between the intercept parameter in the regression of echolocating bat log energy versus log body weight and the intercept parameter in the regression of non-echolocating bat log energy versus log body weight. This is a very different interpretation from the one in (2) because here the slopes of the regression equations are allowed to be different. Because the coefficient measures a different quantity in (3) than in (2), there is no reason to expect its statistical significance to be the same.

10.2.3 Tests and Confidence Intervals for Linear Combinations of Coefficients

In the parallel lines model for the echolocation study (represented in Display 10.5), the slope in the regression of log energy on log mass is β_1 for all three groups. The intercept for non-echolocating bats is β_0, the intercept for non-echolocating birds is $\beta_0 + \beta_2$, and the intercept for echolocating bats is $\beta_0 + \beta_3$.

Since the bird line coincides with the non-echolocating bat line if β_2 is 0, a test of the equality of energy distribution in birds and non-echolocating bats, after accounting for body size, is accomplished through a test of whether β_2 equals 0. Similarly, a test of whether the echolocating bat regression line coincides with the non-echolocating bat regression line is accomplished with a test of $\beta_3 = 0$. The bird and echolocating bat regression lines coincide when $\beta_2 = \beta_3$, however, so a test of the equality of these two groups involves the hypothesis that $\beta_3 - \beta_2 = 0$ (or equivalently, that $\beta_2 - \beta_3 = 0$). This is a hypothesis about a linear combination of regression coefficients.

One method of performing the test compares the estimate of $\beta_3 - \beta_2$ with its standard error. Calculating the standard error is a problem here, however, because the estimates are not statistically independent. The correct formula expands the formula for linear combinations encountered in Section 6.2.2 to include the variances and *covariances* in the joint sampling distribution of the regression coefficient estimates. The correct formula appears in Section 10.4.3. For this problem—as for many similar problems—there is an easier solution.

Redefining the Reference Level

Recall that the choice of reference level for any categorical explanatory variable is arbitrary. Thus, any of the three types of flying vertebrates could be used as the reference level. It is a simple matter to refit the model with another choice for the reference level, and either the birds or the echolocating bats will do. The model will remain the same as the one fit in Display 10.6, although the names of the intercepts will change, and the comparison of the non-echolocating birds to the echolocating bats can be accomplished through a single coefficient—the test for which appears in the standard output.

Inference About the Mean at Some Combination of X's

One special linear combination is the mean of Y at some combination of the X's. For Galileo's data and the regression model

$$\mu\{distance \mid height\} = \beta_0 + \beta_1 height + \beta_2 height^2,$$

the mean distance at a height of 250 is

$$\mu\{distance \mid height = 250\} = \beta_0 + (\beta_1 \times 250) + (\beta_2 \times (250)^2),$$

which is estimated by the linear combination of the estimates

$$\hat{\mu}\{distance \mid height = 250\} = (\hat{\beta}_0 \times 1) + (\hat{\beta}_1 \times 250) + (\hat{\beta}_2 \times (250)^2).$$

A standard error for this estimate may be obtained by using the methods of Section 10.4.3, but it may also be extracted from a computer analysis, by redefining the reference level for height. The trick is to create a new explanatory variable by subtracting 250 from each height measurement. Let $ht250 = height - 250$, so $height = 250$ corresponds to $ht250 = 0$, and then fit the model

$$\mu\{distance \mid ht250\} = \beta_0^* + \beta_1^* ht250 + \beta_2^* (ht250)^2.$$

The mean distance at $height = 250$ is $\mu\{distance \mid ht250 = 0\} = \beta_0^*$. Redefining the model terms permits the question of interest to be worded in terms of a single parameter; the standard error for this parameter appears in the usual output. Use the estimated intercept as the estimated mean of interest, and use the reported standard error of the intercept as the standard error for this estimated mean. As with the indicator variable example, the two models are identical, but are parameterized differently. Display 10.7 shows the analysis for both ways of writing the quadratic regression model, emphasizing how to obtain both the estimate of the mean (at $height = 250$ in this example) and its standard error.

DISPLAY 10.7 Estimates of polynomial coefficients with two different reference levels of height, in Galileo's study

(Reference *height* = 0 punti)

Variable	Coefficient	Standard error	t-statistic	Two-sided p-value
Constant	199.91	16.76	11.93	0.0003
height	0.7083	0.0748	9.47	0.0007
height^2	−0.0003437	0.0000668	5.15	0.0068

R-squared = 99.0% Adj. R-squared = 98.6% Estimated SD = 13.6 punti

$\hat{\mu}\{distance|height = 250 \text{ punti}\}$ SE($\hat{\mu}\{distance|height = 250 \text{ punti}\}$)

Reference *height* = 250 punti

Variable	Coefficient	Standard error	t-statistic	Two-sided p-value
Constant	355.51	6.625	53.66	<0.0001
height−250	0.5365	0.0430	12.48	0.0002
(*height*−250)^2	−0.0003437	0.0000668	−5.15	0.0068

R-squared = 99.0% Adj. R-squared = 98.6% Estimated SD = 13.6 punti

DISPLAY 10.8 Estimated median energy expenditures for birds, echolocating bats, and non-echolocating bats as functions of body mass (parallel lines model on log-log scale, with 95% confidence bands)

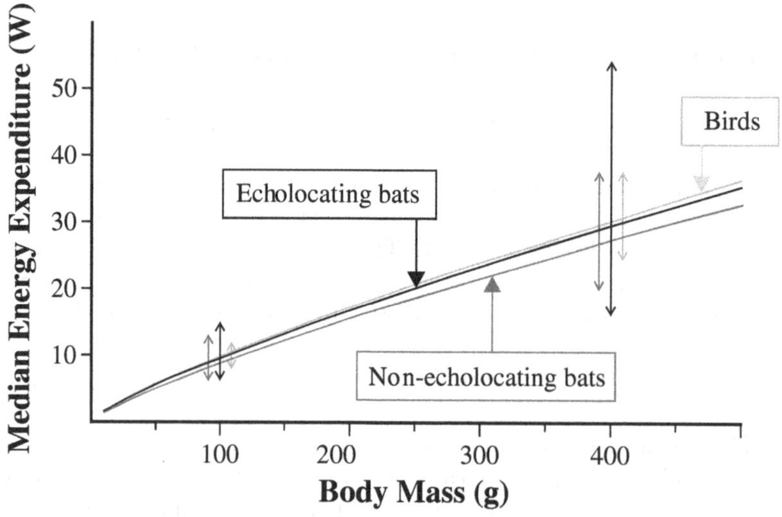

Confidence Bands for Multiple Regression Surfaces

Display 10.8 shows the estimated regressions for the bat data (the parallel regression lines for the log of energy expenditure on the log of body mass) on the original scales of measurement. For non-echolocating bats, for example,

$$\text{Median}\{\mathit{energy} \mid \mathit{mass}\} = \exp(\beta_0) \times \mathit{mass}^{\beta_1}.$$

The curve for this group on the plot was drawn by computing the estimated median at many values of *mass* and connecting the resulting points.

The double arrows in Display 10.8 show the limits of a 95% confidence band for the median energy expenditure at 100 and 400 grams for each of the three groups. These intervals were constructed by finding confidence bands for the regression surface of log(*energy*) and then back-transforming the endpoints. The intervals were constructed by using the computer trick for finding the standard errors of estimated means. This required six different fits to the model, with different reference values, as shown in Display 10.9. The estimated mean for birds of 100-gram body mass, for example, was the estimated intercept in the regression of *lenergy* on (*lmass* – 100) and indicator variables for the two groups other than birds. This turned out to be 2.2789, with a standard error of 0.0604. From these, the confidence interval is calculated and then back-transformed for the light gray double-arrow line at Body mass = 100 in Display 10.8. The calculations appear in the bottom portion of Display 10.9.

Notice that the multiplier used in the confidence band is based on the 95th percentile from an F-distribution. (This example assumes that a confidence band over the entire region—all masses and types—is desired, rather than confidence

DISPLAY 10.9 Construction of the 95% confidence band, using repeated fits of the multiple regression model with different reference points

① *Computer Work*

Reference point		Explanatory variables			
TYPE	Body mass	TYPE indicators	Body mass variable	Intercept estimate	Standard error
birds	100	*nbat, ebat*	*lmass*−log (100)	2.2789	0.0604
	400	"	*lmass*−log (400)	3.4087	0.0635
non-echo bats	100	*ebat, bird*	*lmass*−log (100)	2.1767	0.1144
	400	"	*lmass*−log (400)	3.3064	0.0931
echo bats	100	*nbat, bird*	*lmass*−log (100)	2.2553	0.1277
	400	"	*lmass*−log (400)	3.3851	0.1759

② *Hand Calculations—an Example*

$$\text{Multiplier} = \sqrt{4\,F_{4,16;\,0.95}} = 3.468$$

$$\text{Lower limit} = \exp[2.2789 - (3.468)(0.0604)] = 7.9$$
$$\text{Upper limit} = \exp[2.2789 + (3.468)(0.0604)] = 12.0$$

intervals at a few specific values.) This type of multiplier was first introduced as the Scheffé multiple comparison procedure in Section 6.4.2 and again for confidence bands for simple regression in Section 7.4.2. Here, the numerator and denominator degrees of freedom in F are equal to the number of parameters in model (4) and to the residual degrees of freedom (16), respectively. The multiplier of F inside the square root sign is equal to the number of parameters in the model (4, in this case). Use of the F-based multiplier guarantees at least 95% confidence that the bands contain the regression surface throughout the experimental range. (*Note*: It is dangerous to draw conclusions about the regression outside of the scope of the available data. For example, no non-echolocating bats had body sizes of less than 258 grams. Inference about the distribution of energy expenditure for non-echolocating bats of 100-gram body size requires extrapolation of the model beyond the range within which it was estimated.)

10.2.4 Prediction

Prediction is an important objective in many regression applications. If it is the only objective, there is no need to interpret the coefficients. For example, a researcher may be able to predict individuals' blood pressures from the number of bathrooms in their houses, without having to interpret or understand the relationship between those two variables.

Predicted values are the same as estimated means. They are calculated by evaluating the estimated regression at the desired explanatory variable values, as in the previous section. As in simple linear regression, the error in prediction comes from two sources: the error of estimating the population mean, and the error inherent in the fact that the new observation differs from its mean. Since $\hat{Y} = \text{pred}\{Y|X\} = \hat{\mu}\{Y|X\}$, the prediction error is

$$Y - \hat{Y} = Y - \hat{\mu}\{Y|X\} = [Y - \mu\{Y|X\}] - [\hat{\mu}\{Y|X\} - \mu\{Y|X\}].$$

The variance of a prediction error consists of two corresponding pieces:

$$\text{Prediction variance} = \sigma^2 + [\text{SD}(\hat{\mu}\{Y|X\})]^2.$$

The variance of prediction at some explanatory variable value, therefore, may be estimated by adding the estimate of σ^2 to the square of the standard error of the estimated mean. (See also Section 7.4.3.) The standard error of prediction is the square root of the estimated variance of prediction. Prediction intervals are based on this standard error and on the appropriate percentile from the t-distribution with degrees of freedom equal to those associated with the estimate of σ. As discussed with regard to simple regression, predictions are valid only within the range of values of explanatory variables used in the study. The validity also depends fairly strongly on the assumption of normal distributions in the populations.

Galileo Example

Display 10.7 has all the information needed to predict the horizontal distance of a single ball released at a height of 250 punti. The predicted distance is 355.5 punti. The variance of prediction is $(13.6)^2 + (6.62)^2$, whose square root is the standard error of prediction $=$ 15.1 punti. With $7 - 3 = 4$ degrees of freedom, the t-multiplier for a 95% prediction interval is 2.776, so the interval extends from $355.5 - 42.0 = 313.5$ punti to $355.5 + 42.0 = 397.5$ punti.

10.3 EXTRA-SUMS-OF-SQUARES *F*-TESTS

In multiple regression, data analysts often need to test whether *several* coefficients are all zero. For example, the model

$$\mu\{lenergy \mid lmass, TYPE\} = \beta_0 + \beta_1 lmass + \beta_2 bird + \beta_3 ebat$$

presents the regression as three parallel straight lines. The lines are identical under the hypothesis

$$H: \beta_2 = 0 \text{ and } \beta_3 = 0.$$

The alternative hypothesis is that at least one of the coefficients—β_2 or β_3—is nonzero. t-tests, either individually or in combination, cannot be used to test such a hypothesis involving more than one parameter.

10.3.1 Comparing Sizes of Residuals in Hierarchical Models

The extra-sum-of-squares method, on the other hand, is ideally suited for this purpose. It directly compares a full model (*lmass* + *TYPE*) to a reduced model (*lmass*):

$$\textbf{Full:} \quad \mu\{lenergy \mid lmass, TYPE\} = \beta_0 + \beta_1 lmass + \beta_2 bird + \beta_3 ebat,$$
$$\textbf{Reduced:} \quad \mu\{lenergy \mid lmass, TYPE\} = \beta_0 + \beta_1 lmass.$$

If the coefficients in the last two terms of the full model are zero, their estimates should be close to zero, and fits to the two models should give about the same results. In particular, the residuals should be about the same size. If either β_2 or β_3 is not zero, however, the full model should do a better job of explaining the variation in the response variables, and the residuals should tend to be smaller in magnitude than those from the reduced model. Even if the reduced model is correct, however, the squared residuals in the full model must be somewhat smaller, since the full model has more flexibility to match chance variations in the data. The F-test is used to assess whether the difference between the sums of squared residuals from the full and reduced models is greater than can be explained by chance variation. The sum of squared residuals in the reduced model minus the sum of squared residuals in the full model (as in Section 5.3) is the *extra sum of squares*:

> *Extra sum of squares* =
>
> *Sum of squared residuals from reduced model* −
>
> *Sum of squared residuals from full model.*

The sum of squared residuals is a measure of the response variation that remains unexplained by a model. The extra sum of squares may be interpreted as being the amount by which the unexplained variation decreases when the extra terms are added to the reduced model. Or it may be interpreted as being the amount by which the unexplained variation increases when the extra terms are dropped from the full model.

10.3.2 F-Test for the Joint Significance of Several Terms

The extra-sum-of-squares F-test has been encountered previously as a one-way analysis of variance tool for comparing the separate-means model to the single-mean model (Section 5.3), and as a tool in simple regression for comparing the regression model to the single-mean model or to the separate-means model if there are replicates (Section 8.5). These are all special cases of a general form that is often called a *partial* F*-test*.

The F-statistic, based on the extra sum of squared residuals in a reduced model over a full model is defined as follows:

$$F\text{-statistic} = \frac{\left[\dfrac{\textit{Extra sum of squares}}{\textit{Number of betas being tested}}\right]}{\textit{Estimate of } \sigma^2 \textit{ from full model}}.$$

The part in brackets is the average size of the extra-sum-of-squares per coefficient being tested. If the reduced model is correct, then this per-coefficient variation should be roughly equal to the per observation variation, σ^2. The F-statistic, therefore, should be close to 1.

When the reduced model is correct and the rest of the model assumptions (including normality) hold, the sampling distribution of this F-statistic is an F-distribution. (Although exact justification is based on normality, the F-test and the t-tests are robust against departures from normality.) The numerator degrees of freedom are the number of β's being tested. This is either the number of β's in the full model minus the number of β's in the reduced model or, equivalently, the residual degrees of freedom in the reduced model minus the residual degrees of freedom in the full model. The denominator degrees of freedom are those associated with the estimate of σ^2 in the full model. If the F-statistic is larger than expected from this F-distribution, this is interpreted as evidence that the reduced model is incorrect. The test's p-value—the chance that an F-variable exceeds the calculated value of the F-statistic—measures the strength of that evidence.

Example—Bat Echolocation Data

Computations in the extra-sum-of-squares F-test are detailed in Display 10.10, leading to a conclusion that there is no evidence of a group difference.

Special Case: Testing the Significance of a Single Coefficient

One special hypothesis is that a single coefficient is zero; for example, $H\colon \beta_3 = 0$. Since the t-test is available for this hypothesis, one might wonder whether the F-test leads to the same result. It does. The F-statistic is the square of the t-statistic, and the p-value from the F-test is the same as the two-sided p-value from the t-test. As a practical matter, the t-test results are more convenient to obtain from computer output.

Special Case: Testing the "Overall Significance" of the Regression

Another special hypothesis is that all regression coefficients except β_0 are zero. This hypothesis proposes that none of the considered explanatory variables are useful in explaining the mean response. Although this hypothesis only occasionally corresponds to a question of interest, the extra-sums-of-squares F-test for this hypothesis—often called the *F-test for overall significance of the regression*—is routine output in most statistical computer programs. As discussed in the following section,

DISPLAY 10.10 The extra-sum-of-squares *F*-test for testing equality of intercepts in the parallel regression lines model (bat echolocation data)

(1) *Fit the FULL model:* $\mu\{lenergy|\ lmass,\ TYPE\} = \beta_0 + \beta_1 lmass + \beta_2 bird + \beta_3 ebat.$

Sum of squared residuals = 0.55332 **d.f. = 16** $\hat{\sigma}^2 = 0.03458$

(2) *Fit the REDUCED model:* $\mu\{lenergy|\ lmass,\ TYPE\} = \beta_0 + \beta_1 lmass.$

Sum of squared residuals = 0.58289 **d.f. = 18**

(3) *The extra sum of squares is the difference between the two residual sums of squares.* → **Extra SS = 0.58289 − 0.55332 = 0.02957**

(4) *Numerator degrees of freedom are the number of β's in the full model minus the number of β's in the reduced model.* → **Numerator d.f. = 4 − 2 = 2**

(5) *Calculate the F-statistic.* →

$$F\text{-statistic} = \frac{\left[\dfrac{0.02957}{2}\right]}{0.03458} = \frac{0.014785}{0.03458} = 0.428$$

(6) *Find* $Pr(F_{2,16} > 0.428)$ *from table, computer, or calculator.* → ***p*-value = 0.66**

Conclusion: There is no evidence that mean log energy differs for birds, echolocating bats, and non-echolocating bats, after accounting for body mass.

the component calculations of this *F*-test are routinely available in an analysis of variance table.

10.3.3 The Analysis of Variance Table

The reduced model for the mean contains only the constant term β_0. Hence the least squares estimate of β_0 is the average response ($\overline{Y}$), and the sum of squared residuals is $(n-1)s_Y^2$. This is the "Total sum of squares" as it measures the total variation of the responses about their average. The sum of squared residuals from the model under investigation is labeled the "Residual sum of squares" (or sometimes the "Error sum of squares") in the analysis of variance table. The difference between these two—the extra-sum-of-squares—is the amount of total variation in the response variable that can be explained by the regression on the explanatory variables. It is called the "Regression sum of squares" or the "Model sum of squares." An analysis of variance table for the quadratic fit to Galileo's data appears in Display 10.11.

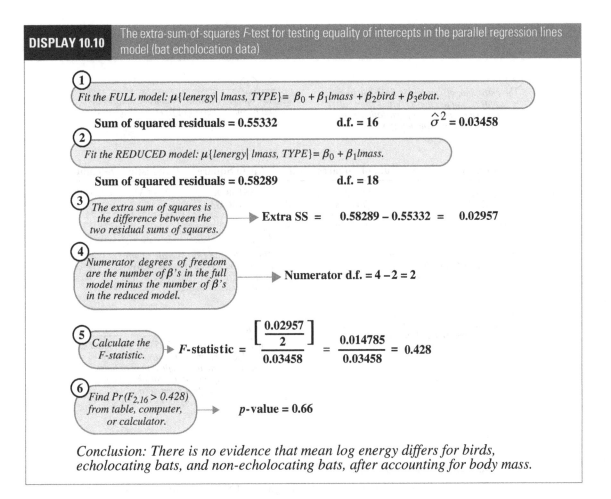

DISPLAY 10.11 Analysis of variance table for Galileo's data: fit of the data to the quadratic model: $\mu\{distance \mid height\} = \beta_0 + \beta_1\,height + \beta_2\,height^2$ (based on seven observations)

Source of Variation	Sum of Squares	d.f.	Mean Square	F-Statistic	p-Value
Regression	76277.92	2	38138.96	205.03	<0.0001
Residual	744.08	4	186.02		
Total	77022.00	6			

(1) *Residual sum of squares from reduced model: Mean{distance} = β_0.*

(2) *Residual sum of squares from full model: Mean{distance} = $\beta_0 + \beta_1 height + \beta_2 (height)^2$.*

(3) *The extra sum of squares is the total sum of squares minus the residual sum of squares (the amount of variation explained by the two explanatory variables).*

(4) *Mean squares are always sums of squares divided by their degrees of freedom. The residual mean square is the estimate of σ^2.*

(5) *The F-statistic (for overall significance of regression) is the regression mean square divided by the residual mean square.*

(6) *p-value = $Pr(F_{2,4} > 205.03)$. The small p-value here indicates overwhelming evidence that the coefficient of at least one of the explanatory variables height and height^2 differs from zero.*

Using Analysis of Variance Tables for Extra-Sums-of-Squares Tests

The calculations for extra-sums-of-squares tests can be carried out manually, as shown in Display 10.10. The sums of squared residuals in steps 1 and 2 are obtained by fitting the full and reduced models, obtaining the analysis of variance tables for each, and reading the sum of squared residuals from the tables. Some statistical computing packages will perform the calculations with an analysis of variance command that includes the full and reduced models as user-specified arguments.

Analysis of the Bat Echolocation Data

The analysis of the bat echolocation data involves several extra-sums-of-squares F-tests. The main question of interest is, "Is there a difference in the in-flight energy expenditures of echolocating and non-echolocating bats after body size is accounted for?" This question does not involve non-echolocating birds, but including them in the analysis is important for obtaining the best possible estimate of σ^2. A secondary question compares birds to the two bat groups.

Inspection of the data and initial graphical exploration indicated the need to analyze both energy and body mass on the log scale. The statistical analysis starts coming into focus with a graphical display like the coded scatterplot in Display 10.4.

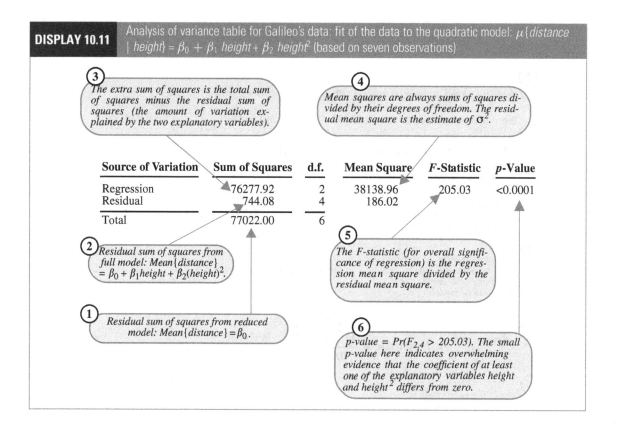

It is apparent that a straight line regression of log energy on log body mass is probably appropriate for each of the three types, and there is no obvious evidence of a difference in the slopes of these lines for the three types of flying vertebrates. These results are fortunate, because the question of interest is most easily addressed if the group differences are the same for all body weights. Hence, the parallel lines model offers the most convenient inferential model.

To buttress the argument, the analyst must make sure that the regression lines are, indeed, parallel before testing whether the intercepts are equal. This involves first fitting a rich model that incorporates different intercepts and different slopes. The separate lines model is

$$\mu\{lenergy \mid lmass, TYPE\} = \\ \beta_0 + \beta_1 lmass + \beta_2 bird + \beta_3 ebat + \beta_4(lmass \times bird) + \beta_5(lmass \times ebat).$$

The next step is to fit the rich model and examine a residual plot (plot of residuals versus fitted values). This plot (not shown) indicates no problems with the model assumptions, so it is appropriate to perform the F-test for the hypothesis that the interaction terms can be dropped (namely, $H\colon \beta_4 = \beta_5 = 0$). This F-test, based on the analysis of variance tables, is shown in Display 10.12.

Since the p-value is 0.53, it is safe to drop the interaction terms and address the comparison of interest in terms of the parallel lines model. It is then appropriate to test whether the intercepts in the parallel regression lines model are the same for the three groups. This is accomplished by conducting another extra-sum-of-squares F-test, as previously illustrated in Display 10.10. The p-value of 0.66 implies that the data are consistent with the hypothesis that no difference in energy expenditure exists among the three groups, after the effect of body mass is accounted for. Nevertheless, confidence intervals for the group differences should still be presented, as in the summary of statistical findings in Section 10.1.2, to reveal other possibilities that are not excluded by the data.

10.4 RELATED ISSUES

10.4.1 Further Notes on the *R*-Squared Statistic

Example—Galileo's Data

From the analysis of variance table in Display 10.11, it is evident that the quadratic regression of *distance* on *height* explains 76277.92/77022.00, or 99.03% of the total variation in the observed distances. Only 0.97% of the variation in distances remains unexplained.

Display 10.13 shows the regression output for another model for these data: the regression of distance on *height*, *height-squared*, and *height-cubed*. The p-value for the coefficient of *height-cubed* (0.0072) provides strong evidence that the cubic term is significant even when the linear and quadratic terms are in the model. On the other hand, the percentage of variation explained by the cubic model is 99.94%, only 0.91% more than the percentage of variation explained by the quadratic model.

DISPLAY 10.12 The extra sum of squares F-test comparing the separate regression lines model to the parallel regression lines model—bat echolocation data

① FIT FULL MODEL: $\mu\{lenergy|lmass, TYPE\} = \beta_0 + \beta_1 lmass + \beta_2 bird + \beta_3 ebat + \beta_4(lmass \times bird) + \beta_5(lmass \times ebat).$

Source of Variation	Sum of Squares	d.f.	Mean Square	F-Statistic	p-Value
Regression	29.46993	5	5.89399	163.4	<0.0001
Residual	0.50487	14	0.03606		
Total	29.97480	19			

Residual SS ← *Estimate of σ^2* · *d.f.*

② FIT REDUCED MODEL: $\mu\{lenergy|lmass, TYPE\} = \beta_0 + \beta_1 lmass + \beta_2 bird + \beta_3 ebat.$

Source of Variation	Sum of Squares	d.f.	Mean Square	F-Statistic	p-Value
Regression	29.42148	3	9.80716	283.6	<0.0001
Residual	0.55332	16	0.03458		
Total	29.97480	19			

Residual SS

③ *The extra sum of squares is the difference between residual sums of squares.*

Extra SS = 0.55332 − 0.50487 = 0.04845

Numerator d.f. = Number of β's in full model minus Number of β's in reduced model.

⑤ *Calculate the F-statistic.*

$$F\text{-statistic} = \frac{\left[\dfrac{0.04845}{2}\right]}{0.03606} = 0.672$$

⑥ Look up $Pr(F_{2,14} > 0.672).$ → p-value = 0.53

Conclusion: *There is no evidence that the association between energy expenditure and body size differs among the three types of flying vertebrates (p-value = 0.53).*

Of course that 0.91% accounts for 94% of the remaining variability from the quadratic model, which helps explain why the cubic term is statistically significant.

Two very useful features of R^2 have been revealed with Galileo's data. First, the summary measure—for instance, 99.03% of the variation explained—provides a valuable image of the tightness of the fit. Second, the proportion of additional

DISPLAY 10.13 Partial output from the regression of distance on *height*, *height-squared*, and *height-cubed*, for Galileo's data

Variable	Coefficient	Standard error	*t*-statistic	*p*-value
Constant	155.78	8.33	18.71	0.0003
height	1.1153	0.0657	16.98	0.0004
height-squared	−0.001245	0.000138	−8.99	0.0029
height-cubed	5.477×10^{-7}	0.838×10^{-7}	6.58	0.0072

Estimate of standard deviation about the regression: 4.011 on 3 degrees of freedom
$R^2 = 99.94\%$.

variation explained by one newly introduced variable after the others have been accounted for can be used in assessing its practical significance. Despite these useful summarizing applications, however, R^2 suffers considerable abuse. It is rarely an appropriate statistic to use for model checking, model comparison, or inference.

R^2 Can Always Be Made 100% by Adding Explanatory Variables

Display 10.14 shows the scatterplot of distance versus height for Galileo's seven measurements. Drawn through the points is a sixth-order polynomial regression line:

$$\mu\{distance \mid height\} = \beta_0 + \beta_1 height + \beta_2(height)^2 + \beta_3(height)^3 + \beta_4(height)^4 + \beta_5(height)^5 + \beta_6(height)^6.$$

This fit produces residuals that are all zero and, consequently, an R^2 of 100%. In fact, for n data points with distinct explanatory variable values, an $n-1$ polynomial regression will always fit exactly. That does not imply that the equation has any usefulness, though. While the data at hand have been modeled exactly, the particular equation is unlikely to fit as well on future data.

This same warning applies to multiple regression in general. R-squared can always be made 100% by adding enough explanatory variables. The question is whether the model is appropriate for the population or instead is only appropriate for the particular data on hand. If inference to some population (real or hypothetical) is desired, it is important not to craft a model that is overly customized to the data on hand. Tests of significance and residual plots are more appropriate tools for model building and checking.

The Adjusted R^2 Statistic

The adjusted R^2 is a version of R^2 that includes a penalty for unnecessary explanatory variables. It measures the proportion of the observed spread in the responses that is explained by the regression model, where spread is measured by residual mean squares rather than by residual sums of squares. This approach provides a

DISPLAY 10.14 Scatterplot of Galileo's horizontal distances versus initial heights, with estimated sixth-order polynomial regression curve ($R^2 = 100\%$)

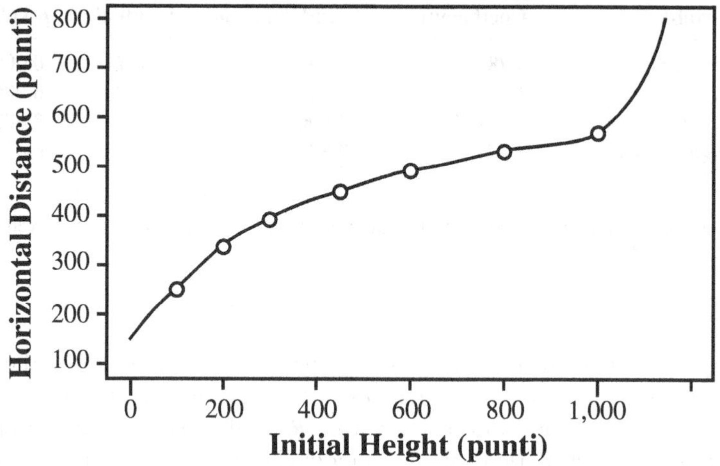

better basis for judging the improvement in a fit due to adding an explanatory variable, but it does not have the simple summarizing interpretation that R^2 has:

$$\text{Adjusted } R^2 = 100\frac{(\text{Total mean square}) - (\text{Residual mean square})}{\text{Total mean square}}\%.$$

Thus adjusted R^2 is useful for casual assessment of improvement of fit, but R^2 is better for description, as illustrated in the summary of statistical findings in Section 10.1.1.

10.4.2 Improving Galileo's Design with Replication

Since Galileo's study contained no replication, it is impossible to separate the repeatable part of the relationship (if any) from the real variability in replicate measurements of horizontal distances of balls started at the same height. In other words, all estimates of σ are necessarily based on a particular form of the regression model and may be biased as a result of model inadequacies.

Suppose that Galileo instead had selected four evenly spaced heights—say 100, 400, 700, and 1,000 punti—and had replicated the experiment at each height. Using four heights would have allowed Galileo to fit a cubic polynomial, giving him a model on which to judge the proposed parabola (through a test of the cubic term). The revised design also yields 4 degrees of freedom for estimating residual variability free from possible model complications. The cubic model can be judged by comparing it to the separate-means model for the four groups.

A natural, *design-based* estimate of variance is obtained by pooling sample variances from groups of data at identical values of the explanatory variables. It can only be obtained if there are replicates. Without replication, the only available estimate of variance is the one obtained from the residuals about the fit to the presumed model for the mean. There is always a possibility of bias in this *model-based* estimate of variance, due to inadequacies of the regression model. With a design-based estimate no such bias can occur, because the size of the residuals does not hinge on the accuracy of any presumed regression model. This is a reason for incorporating some replication into an experimental design.

10.4.3 Variance Formulas for Linear Combinations of Regression Coefficients

On some occasions it may seem more convenient to estimate a linear combination of regression parameters directly than to instruct the computer to change the reference level (as in Section 10.2.3). In general, a linear combination can be written as

$$\gamma = C_0\beta_0 + C_1\beta_1 + C_2\beta_2 + \cdots + C_p\beta_p,$$

where the C's are known coefficients, some of which may be zeros. The estimate of this combination is

$$g = C_0\hat{\beta}_0 + C_1\hat{\beta}_1 + C_2\hat{\beta}_2 + \cdots + C_p\hat{\beta}_p.$$

The formula for the variance of this linear combination differs from the one for variance of a linear combination of independent averages in Section 6.2.2. Since the estimated coefficients are not independent of one another, the formula here involves their covariances.

For a population of pairs U and V, the *covariance* of U and V is the mean of $(U - \mu_U)(V - \mu_V)$, and it describes how the two variables covary. The covariance of $\hat{\beta}_1$ and $\hat{\beta}_2$ is the mean of $(\hat{\beta}_1 - \beta_1)(\hat{\beta}_2 - \beta_2)$ in their sampling distribution. A formula for the covariances of least squares estimates is available, and estimates are available from least squares computer routines. If Cov represents covariance, the variance of the sampling distribution of the estimated linear combination is

$$\text{Var}\{g\} = C_0^2\text{SE}(\hat{\beta}_0)^2 + C_1^2\text{SE}(\hat{\beta}_1)^2 + \cdots + C_p^2\text{SE}(\hat{\beta}_p)^2 + 2C_0C_1\text{Cov}(\hat{\beta}_0, \hat{\beta}_1)$$
$$+ 2C_0C_2\text{Cov}(\hat{\beta}_0, \hat{\beta}_2) + \cdots + 2C_{p-1}C_p\text{Cov}(\hat{\beta}_{p-1}, \hat{\beta}_p).$$

To see why this is so consider the linear combination $\gamma = C_1\beta_1 + C_2\beta_2$ and its estimate $g = C_1\hat{\beta}_1 + C_2\hat{\beta}_2$. Then

$$(g - \gamma) = C_1(\hat{\beta}_1 - \beta_1) + C_2(\hat{\beta}_2 - \beta_2)$$

and

$$(g - \gamma)^2 = C_1^2(\hat{\beta}_1 - \beta_1)^2 + C_2^2(\hat{\beta}_2 - \beta_2)^2 + 2C_1C_2(\hat{\beta}_1 - \beta_1)(\hat{\beta}_2 - \beta_2).$$

DISPLAY 10.15 Inference about $\beta_2 - \beta_3$, the coefficient of the indicator variable for birds minus the coefficient of the indicator variable for echolocating bats

(1) *Estimate the linear combination of coefficients as the same linear combination of estimated coefficients.*

Estimate of $\beta_2 - \beta_3$, from Display 10.6: 0.1023 – 0.0787 = 0.0236

(2) *Obtain the estimated variance–covariance matrix of the estimated regression coefficients.*

Estimated variance–covariance matrix (from computer):

	(Constant)	*lmass*	*bird*	*ebat*
Constant	0.08250	–0.01211	–0.01921	–0.05056
lmass	–0.01211	0.00198	0.00173	0.00687
bird	–0.01921	0.00173	0.01304	0.01464
ebat	–0.05056	0.00687	0.01464	0.04108

Notes: *The matrix is symmetric. The square roots of the diagonal elements are the standard errors of the estimated coefficients (as reported in Display 10.6).*

(3) *The estimated variance of $\hat{\beta}_2 - \hat{\beta}_3$ is*

$1^2\text{Var}(\hat{\beta}_2) + (-1)^2\text{Var}(\hat{\beta}_3) + 2(1)(-1)\text{Cov}(\hat{\beta}_2, \hat{\beta}_3)$.

Estimated variance of $\hat{\beta}_2 - \hat{\beta}_3$: 0.01304 + 0.04108 – (2 × 0.01464) = 0.02484

$\text{SE}(\hat{\beta}_2 - \hat{\beta}_3) = (0.02484)^{1/2} = 0.1576$ (d.f.=16)

The variance of g is Mean$(g - \gamma)^2$, which can be expressed as

$$C_1^2\,\text{Mean}(\hat{\beta}_1 - \beta_1)^2 + C_2^2\,\text{Mean}(\hat{\beta}_2 - \beta_2)^2 + 2C_1C_2\,\text{Mean}(\hat{\beta}_1 - \beta_1)(\hat{\beta}_2 - \beta_2)$$
$$= C_1^2\,\text{Var}(\hat{\beta}_1) + C_2^2\,\text{Var}(\hat{\beta}_2) + 2C_1C_2\,\text{Cov}(\hat{\beta}_1, \hat{\beta}_2).$$

The estimate of this expression is obtained by substituting the estimates of the variances and covariance.

The Estimated Variance–Covariance Matrix for the Coefficients

The estimated covariances of the coefficients' sampling distributions are calculated by most statistical computer programs. They are stored as an array, or *matrix*, for use in formulas such as the one just discussed. (See also Exercise 10.22.) The easy computer trick for finding the standard error of $\hat{\beta}_2 - \hat{\beta}_3$ for the parallel regression lines model fit to the echolocation data was shown in Section 10.2.3. Display 10.15 illustrates the alternative, direct approach, using the estimated coefficient covariances supplied by a statistical computer program.

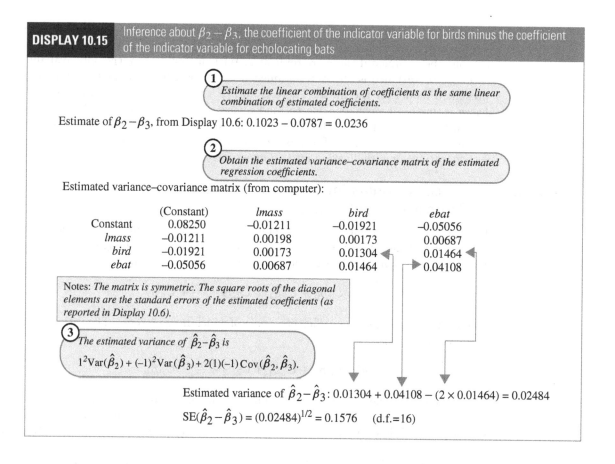

10.4.4 Further Notes About Polynomial Regression

A Second-Order Model in Two Explanatory Variables

A full, *second-order model* in explanatory variables X_1 and X_2 is

$$\mu\{Y|X_1, X_2\} = \beta_0 + \beta_1 X_1 + \beta_2 X_2 + \beta_3 X_1^2 + \beta_4 X_2^2 + \beta_5 X_1 X_2.$$

This is a useful model in *response surface studies*, where the goal is to model a nonlinear response surface in terms of a polynomial. It is rarely useful to attempt cubic and higher-order terms.

When Should Squared Terms Be Included?

As with interaction terms, quadratic terms should not routinely be included. They are useful to consider in four situations: when the analyst has good reason to suspect that the response is nonlinear in some explanatory variable (through knowledge of the process or by graphical examination); when the question of interest calls for finding the values that maximize or minimize the mean response; when careful modeling of the regression is called for by the questions of interest (and presumably this is only the case if there are just a few explanatory variables); or when inclusion is used to produce a rich model for assessing the fit of an inferential model.

10.4.5 Finding Where the Mean Response Is at Its Maximum in Quadratic Regression

The value of X that maximizes (or minimizes) the quadratic expression

$$\mu\{Y|X\} = \beta_0 + \beta_1 X + \beta_2 X^2$$

is $X_{max} = -\beta_1/(2\beta_2)$ (a result from calculus). It is a maximum or minimum according to whether β_2 is negative or positive. This is a *nonlinear* combination of coefficients, so the usual method for obtaining standard errors for linear combinations is of no use.

Fieller's Method

Fieller's method tests and constructs confidence intervals for ratios of regression coefficients. The method is illustrated here for X_{max}.

For a specific value of M, the hypothesis $H_0\colon X_{max} = M$ can be re-expressed as $H_0\colon \beta_1 = -2M\beta_2$. If this hypothesis is correct, the mean for Y above becomes

$$\mu\{Y|X\} = \beta_0 - 2M\beta_2 X + \beta_2 X^2 = \beta_0 + \beta_2 X^*,$$

where $X^* = X^2 - 2MX$ is a new variable, in a simple linear regression model. Applying the extra-sum-of-squares F-test of this as the reduced model against the full model above provides a test of H_0. A 95% confidence interval is constructed, by trial and error, as the range of M-values that result in test p-values larger than or equal to 0.05. (See Section 4.2.4.)

10.4.6 The Principle of Occam's Razor

The principle of Occam's Razor is that simple models are to be preferred over complicated ones. Named after the 14th-century English philosopher, William of Occam, this principle has guided scientific research ever since its formulation. It has no underlying theoretical or logical basis; rather, it is founded in common sense and successful experience. It is often called the Principle of Parsimony.

In statistical applications, the idea translates into a preference for the more simple of two models that fit data equally well. One should seek a parsimonious model that is as simple as possible and yet adequately explains all that can be explained. Methods for paring down large sets of explanatory variables are discussed in Chapter 12.

10.4.7 Informal Tests in Model Fitting

Tests for hypotheses about regression coefficients—t-tests and extra-sum-of-squares F-tests—are valuable for two purposes: for formally providing evidence regarding questions of interest in a final model and for exploring models by testing potential terms at the exploratory stage. The attitude toward the p-values is somewhat different for these two purposes.

In answering questions of interest, p-values are given their formal interpretation. A small p-value provides evidence against the null hypothesis and in favor of the alternative. The null hypothesis may be true and the particular sample may have happened by chance to be an unusual one; the p-value provides a measure of just how unusual it would be. A large p-value means either that the null hypothesis is correct or that the study was not powerful enough to detect a departure from it.

For example, the adequacy of a straight line regression model may be examined through casual testing of an additional quadratic term. The statistical significance of the quadratic term helps to clarify possible curvature. In this usage of testing, some decision—in this case about a suitable model—is made. It should be realized that this decision is sample size dependent; one is more likely to find evidence of curvature in a large data set than in a small one. Nevertheless, the device of informal testing is useful in conjunction with other exploratory tools.

For answering questions of interest, tests should not be overemphasized. Even if the question of interest calls for a test, reporting a confidence interval to indicate the possible sizes of the effects of interest remains important. This is true whether the p-value for the test is small or large.

10.5 SUMMARY

Galileo's Study

Galileo's data are used to find a polynomial describing the mean distance as a function of height. The scatterplot shows that the relationship is not a straight line. The coefficient of a height-squared term, when added to the simple linear regression model, significantly differs from zero. R^2 is a useful summary here: the quadratic regression model explains 99.03% of the variation in the horizontal distances. The large R^2 does not mean that all other terms are insignificant. A test of hypothesis is used to resolve that matter. In fact, when a height-cubed term is added to the model, the p-value for testing whether its coefficient is zero is 0.007, and the value of R^2 increases to 99.94%. This example demonstrates the benefit of R^2 for summarizing a fit and for indicating the degree of relevance of a significant term.

Echolocation by Bats

The goal is to see whether the in-flight energy expended by echolocating bats differs from the energy used by non-echolocating bats of similar body mass. There is no major question involving the birds, but their inclusion helps to clarify a common relationship between energy and body mass. A useful starting point in the analysis is a coded scatterplot of energy versus body mass. It is apparent by inspection that both energy and body mass should be transformed to their logarithms. A plot on this scale (Display 10.4) reveals a straight line relationship but very little additional difference in energy expenditure among the three types of flying vertebrates. Comparing the in-flight energy expenditures of echolocating bats and non-echolocating bats is difficult because the former are all small bats and the latter all big bats. Multiple linear regression permits a comparison after accounting for body mass, but the comparison must be made cautiously, since the ranges of body mass for the two types do not overlap. In light of this limitation, the comparison is made on the basis of an indicator variable in a parallel regression lines model. The data are consistent with the hypothesis that echolocating bats pay no extra energy price for their echolocating skills.

10.6 EXERCISES

Conceptual Exercises

1. **Galileo's Data.** Why is horizontal distance, rather than height, the response variable?

2. **Brain Weight.** Display 10.16 shows two possible *influence diagrams* relating the variables in the brain weight study of Section 9.1.2. If brain weight is directly associated with gestation period and litter size, as in part (a) of the figure, then animals that have approximately the same body size but different gestation period and different litter size should have different brain sizes, and scatterplots of brain size versus gestation and litter size individually should show some association. But the scatterplots can also show indirect associations, as in part (b) of the figure. If brain weight, gestation period, and litter size are all driven by body size, they should show mutual associations, whether or not a direct association exists. Can a statistical analysis distinguish between direct and indirect association? Explain how or, if not, why not.

3. **Brain Weight.** Consider the mammal brain weight data from Section 9.1.2, the model

$$\mu\{lbrain \mid lbody, lgest, llitter\} = \beta_0 + \beta_1 lbody + \beta_2 lgest + \beta_3 llitter,$$

and the hypothesis $H\colon \beta_2 = 0$ and $\beta_3 = 0$. (a) Why can this not be tested by the two t-tests reported in the standard output? (b) Why can this not be tested by the two t-tests along with an adjustment for multiple comparisons?

DISPLAY 10.16 Associations of brain weight with gestation period and litter size: direct or indirect?

(a) DIRECT ASSOCIATION

Animals with about the same body size will have different brain sizes when their litter sizes and gestation periods are different.

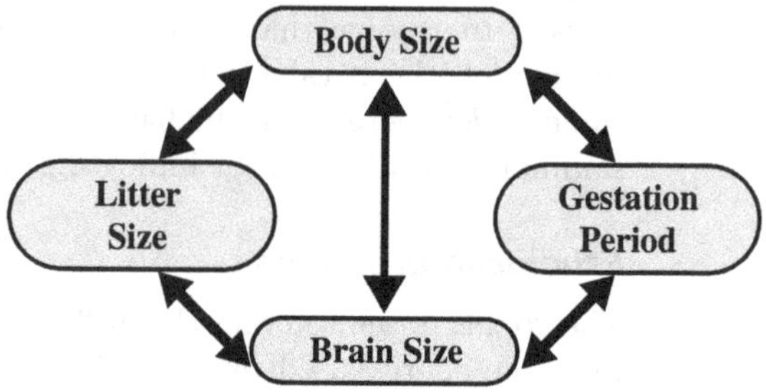

(b) INDIRECT ASSOCIATION

Animals with about the same body size will have about the same brain size even if their gestation periods and litter sizes are different. Any apparent association between brain size and litter size, say, is a result of their both being related to body size.

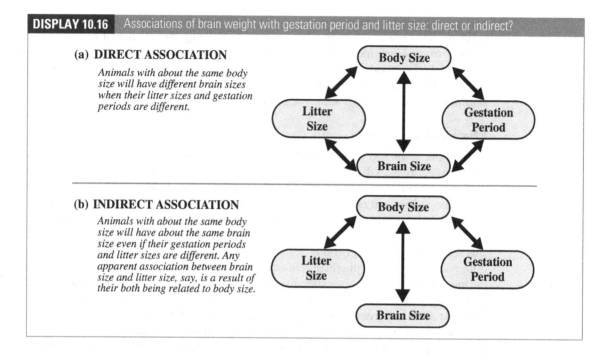

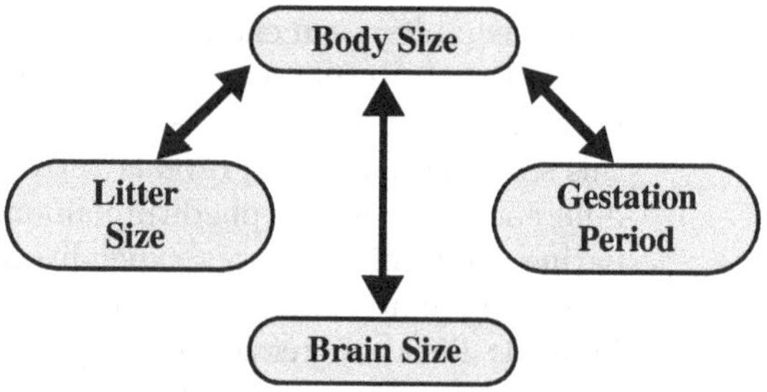

4. **Stocks and Jocks.** Based on data from nine days in June 1994, a multiple regression equation was fit to the Dow Jones Index on the following seven explanatory variables: the high temperature in New York City on the previous day; the low temperature on the previous day; an indicator variable taking on the value 1 if the forecast for the day was sunny and 0 otherwise; an indicator variable taking on the value 1 if the New York Yankees won their baseball game of the previous day and 0 if not; the number of runs the Yankees scored; an indicator variable taking on the value of 1 if the New York Mets won their baseball game of the previous day and 0 if not; and the number of runs the Mets scored. As the chart in Display 10.17 shows, the predicted values of the stock market index were strikingly close to the actual values. R^2 was 89.6%. Why is this unremarkable?

5. **Bat Echolocation.** Consider these three models:

$$\mu\{lenergy \mid lmass, TYPE\} = \beta_0 + \beta_1 lmass$$

$$\mu\{lenergy \mid lmass, TYPE\} = \beta_0 + \beta_1 lmass + \beta_2 bird + \beta_3 ebat$$

$$\mu\{lenergy \mid lmass, TYPE\} = \beta_0 + \beta_1 lmass + \beta_2 bird + \beta_3 ebat + \beta_4 (lmass) \times (bird) + \beta_5 (lmass \times (ebat)).$$

(a) Explain why they can be described as representing the single line, parallel lines, and separate lines models, respectively. (b) Explain why the second model can be the "reduced model" in one F-test but the "full model" in another.

6. **Bat Echolocation.** A possible statement in conclusion of the analysis is this: "It is estimated that the median in-flight energy expenditure for echolocating bats is 1.08 times as large as the median in-flight energy expenditure for non-echolocating bats of similar body size." Referring to Display 10.4, explain why including the phrase, "of similar body size" in this statement is suspect. What alternative wording is available?

DISPLAY 10.17 Actual values and predicted values of Dow Jones Index

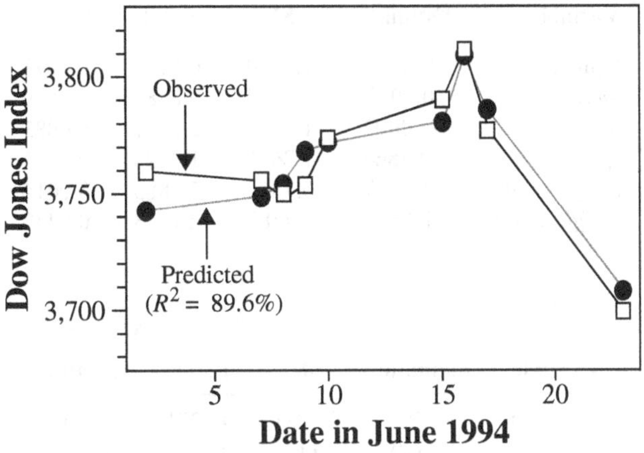

7. Life Is a Rocky Road. A regression of the number of crimes committed in a day on volume of ice cream sales in the same day showed that the coefficient of ice cream sales was positive and significantly differed from zero. Which of the following is the most likely explanation? (a) The content of ice cream (probably the sugar) encourages people to commit crimes. (b) Successful criminals celebrate by eating ice cream. (c) A pathological desire for ice cream is triggered in a certain percentage of individuals by certain environmental conditions (such as warm days), and these individuals will stop at nothing to satisfy their craving. (d) Another variable, such as temperature, is associated with both crime and ice cream sales.

8. In the terminology of extra-sums-of-squares F-tests, does the reduced model correspond to the case where the null hypothesis is true? Is the full model the one that corresponds to the alternative hypothesis's being true?

Computational Exercises

9. Crab Claws. Reconsider the data on claw closing force and claw size for three species of crabs, shown in Exercise 7.22. Display 10.18 shows output from the least squares fit to the separate lines model for the regression of log force on log height. The regression model for log force was

$$\mu\{lforce \mid lheight, SPECIES\} = \beta_0 + \beta_1 lheight + \beta_2 lb + \beta_3 cp + \beta_4(lheight \times lb) + \beta_5(lheight \times cp),$$

where *lheight* represents log height, *lb* represents an indicator variable for the second species, and *cp* represents an indicator variable for the third species. The sample size was 38.

(a) How many degrees of freedom are there in the estimate of σ?
(b) What is the p-value for the test of the hypothesis that the slope in the regression of log force on log height is the same for species 2 as it is for species 1?
(c) What is a 95% confidence interval for the amount by which the slope for species 3 exceeds the slope for species 1?

10. Crab Claws. The sum of squared residuals from the fit described in Exercise 9 is 5.99713, based on 32 degrees of freedom. The sum of squared residuals from the fit without the last two terms

DISPLAY 10.18 Least squares output for Exercise 9

Variable	Estimate	SE	*t*-stat	*p*-value
Constant	0.5191	1.0000	0.5191	0.6073
lheight	0.4083	0.4868	0.8387	0.4079
lb	−4.2992	1.5283	2.8131	0.0083
cp	−2.4864	1.7606	1.4123	0.1675
lheight × *lb*	2.5653	0.7354	3.4885	0.0014
lheight × *cp*	1.6601	0.7889	2.1043	0.0433

DISPLAY 10.19 Regression output data for Exercise 11

Variable	Estimate	SE	*t*-stat	*p*-value
Constant	3.775	0.3881	9.7321	<0.0000
lsize	0.0809	0.1131	0.7139	0.2443
days	0.0774	0.1447	0.5346	0.5104

Estimated SD about the regression is 0.8234 on 13 degrees of freedom; $R^2 = 11.41\%$.

is 8.38155, based on 34 degrees of freedom. Form an F-statistic and find the p-value for the test that the slopes are the same for the three species.

11. Butterfly Occurrences. Display 10.19 summarizes results from the regression of the log of the number of butterfly species observed on the log of the size of the reserve and the number of days of observations, from 16 reserves in the Amazon River Basin.

- **(a)** What is the two-sided p-value for the test of whether size of reserve has any effect on number of species, after accounting for the days of observation? What is the one-sided p-value if the alternative is that size has a positive effect? Does this imply that there is no evidence that the median number of species is related to reserve size? The researchers tended to spend more days searching for butterflies in the larger reserves. How might this affect the interpretation of the results?
- **(b)** What is a two-sided p-value for the test that the coefficient of *lsize* is 1? (This is simply a computational exercise; there is no obvious reason to conduct this test with these data.)
- **(c)** What is a 95% confidence interval for the coefficient of *lsize*?
- **(d)** What proportion of the variation in log number of species remains unexplained by log size and days of observations?

12. Brain Weights. With the data described in Section 9.1.2, construct an extra sum of squares F-test for determining whether gestation period and litter size are associated with brain weight after body weight is accounted for.

13. Bat Echolocation. (a) Fit the parallel regression lines model to duplicate the results in Display 10.6. (b) From these results, what are the estimated intercept and estimated slope for the regression of log energy on log mass for (i) non-echolocating bats, (ii) non-echolocating birds, and (iii) echolocating bats? (c) Refit the model using, instead, the indicator variables *bird* and *nbat*, where *nbat* takes on the value 1 for species of non-echolocating bats and 0 for other species. (d) Based on the results in part (c), what are the estimated intercept and estimated slope for the regression of log energy on log mass for (i) non-echolocating bats, (ii) non-echolocating birds, and (iii) echolocating bats? How

DISPLAY 10.20 Protein in minnow larvae exposed to copper and zinc

Copper (ppm)	Zinc (ppm)	Protein (μg/larva)	Copper (ppm)	Zinc (ppm)	Protein (μg/larva)
0	0	201	112.5	0	188
0	375	186	112.5	375	172
0	750	173	112.5	750	157
0	1,125	110	112.5	1,125	115
0	1,500	115	112.5	1,500	108
37.5	0	202	150	0	133
37.5	375	161	150	375	125
37.5	750	172	150	750	184
37.5	1,125	138	150	1,125	135
37.5	1,500	133	150	1,500	114
75	0	204			
75	375	165			
75	750	148			
75	1,125	143			
75	1,500	123			

do these compare to the estimates obtained in part (b)? (e) With the results of (c), test whether the lines for the echolocating bats and the non-echolocating birds coincide.

14. Toxic Effects of Copper and Zinc. In a study of the joint toxicity of copper and zinc, researchers randomly allocated 25 beakers containing minnow larvae to receive one of 25 treatment combinations. The treatment levels were all combinations of 5 levels of zinc and 5 levels of copper added to a beaker. Following a four-day exposure, a sample of the minnow larvae were homogenized and analyzed for protein. The results are shown in Display 10.20. (Data from D. A. J. Ryan, J. J. Hubert, J. B. Sprague, and J. Parrott, "A Reduced-Rank Multivariate Regression Approach to Aquatic Joint Toxicity Experiments," *Biometrics* 48 (1992): 155–62.) Fit a full second-order model for the regression of protein on copper and zinc, and examine the plot of residuals versus fitted values. Repeat after taking the log of protein. Which model is preferable?

15. Kentucky Derby. Reconsider the Kentucky Derby winning times and speeds from Exercise 9.20. Test whether there is any effect of the categorical factor "Track" (with seven categories) on winning speed, after accounting for year. The full model will have *Year* and the categorical factor *Track*; the reduced model will have only *Year*.

16. Galileo's Data. Use Galileo's data in Display 10.1 (data file: case1001) to perform the following operations.

- **(a)** Fit the regression of distance on height and height-squared. Obtain the estimates, their standard errors, the estimate of σ^2, and the variance–covariance matrix of the estimated coefficients.
- **(b)** Verify that the square roots of the diagonal elements are equal to the standard errors reported with the estimated coefficients.
- **(c)** Compute the estimated mean distance when the initial height is 500 punti.
- **(d)** Calculate the standard error for the estimated mean in part (c).
- **(e)** Use the answer to parts (a) and (d) and the relationship between the variance of the estimated mean and the variance of prediction to obtain the standard error of prediction at an initial height of 500 punti.

17. Galileo's Data. Use Galileo's data in Display 10.1 (data file case1001) to fit the regression of distance on (a) height; (b) height and height2; (c) height, height2, and height3; (d) height, height2, height3, and height4; (e) height, height2, height3, height4, and height5; (f) height, height2, height3, height4, height5, and height6. For each part, find R^2 and R^2_{adj}.

18. Corn Yield and Rainfall. Reconsider the corn yield and rainfall data (Display 9.18; data file ex0915). Fit the regression of yield on rainfall, rainfall-squared, and year. Use the approach of Section 10.4.5 to find the rainfall that maximizes mean yield.

19. Meadowfoam. Carry out a *lack-of-fit* F-test for the regression of number of flowers on light intensity and an indicator variable for time, using the data in Display 9.2 (data file case0901): (a) Fit the regression of *flowers* on *light* and an indicator variable for *time* $= 24$, and obtain the analysis of variance table. (b) Fit the same regression except with *light* treated as a factor (using 5 indicator variables to distinguish the 6 groups), and with the interaction of these two factors, and obtain the analysis of variance table. (c) Perform an extra-sum-of-squares F-test comparing the full model in part (b) to the reduced model in part (a). (*Note:* The full model contains 12 parameters, which is equivalent to the model in which a separate mean exists for each of the 12 groups. No pattern is implied in this model. See Displays 9.7 and 9.8 for help.)

20. Calculus Problem. The least squares problem in multiple linear regression is to find the parameter values that minimize the sum of squared differences between responses and fitted values,

$$SS(\beta_0, \beta_1, \dots, \beta_p) - \sum_{i=1}^{n}(Y_i - \beta_0 - \beta_1 X_{1i} - \dots - \beta_p X_{pi})^2.$$

Set the partial derivatives of SS with respect to each of the unknowns equal to zero. Show that the solutions must satisfy the set of *normal equations*, as follows:

$$\beta_0 n + \beta_1 \Sigma X_{1i} + \beta_2 \Sigma X_{2i} + \dots + \beta_p \Sigma X_{pi} = \Sigma Y_i$$
$$\beta_0 \Sigma X_{1i} + \beta_1 \Sigma X_{1i}^2 + \beta_2 \Sigma X_{1i} X_{2i} + \dots + \beta_p \Sigma X_{1i} X_{pi} = \Sigma X_{1i} Y_i$$
$$\beta_0 \Sigma X_{2i} + \beta_1 \Sigma X_{2i} X_{1i} + \beta_2 \Sigma X_{2i}^2 + \dots + \beta_p \Sigma X_{2i} X_{pi} = \Sigma X_{2i} Y_i$$

$$\vdots \qquad \vdots \qquad \vdots \qquad \vdots \qquad \vdots$$

$$\beta_0 \Sigma X_{pi} + \beta_1 \Sigma X_{pi} X_{1i} + \beta_2 \Sigma X_{pi} X_{2i} + \dots + \beta_p \Sigma X_{pi}^2 = \Sigma X_{pi} Y_i,$$

where each Σ indicates summation over all cases ($i = 1, 2, \dots, n$). Show, too, that solutions to the normal equations minimize SS.

21. Matrix Algebra Problem. Let **Y** be the $n \times 1$ column vector containing the responses, let **X** be the $n \times (p + 1)$ array whose first column consists entirely of ones and whose other columns are the explanatory variable values, and let **b** be the $(p + 1) \times 1$ column containing the resulting parameter estimates. Show that the normal equations in Exercise 20 can be written in the form

$$(\mathbf{X}^T \mathbf{X})\mathbf{b} = \mathbf{X}^T \mathbf{Y}.$$

Therefore, as long as the matrix inversion is possible, the least squares solution is

$$\mathbf{b} = (\mathbf{X}^T \mathbf{X})^{-1} \mathbf{X}^T \mathbf{Y}.$$

When is the inversion possible?

22. Continuing Exercise 21, statistical theory says that the means of the estimates in the vector **AY**, where **A** is a matrix, are the elements of the vector $\mathbf{A}\mu\{\mathbf{Y}\}$; and the matrix of covariances of these estimates is $\mathbf{A}\text{Cov}(\mathbf{Y})\mathbf{A}^T$. Use the theory and the model $\mu\{\mathbf{Y}\} = \mathbf{X}\beta, \text{Cov}(\mathbf{Y}) = \sigma^2\mathbf{I}$ (where **I** is an $n \times n$ identity matrix) to show that the mean in the sampling distributions of the least squares estimate **b** is β. Then show that the matrix of covariances is $\text{Cov}\{\mathbf{b}\} = \sigma^2(\mathbf{X}^T\mathbf{X})^{-1}$.

23. Speed of Evolution. Refer back to Exercise 9.18. The authors of that study concluded that although the wing size of North American flies was converging rapidly to the same cline as exhibited by the European flies, the means by which the cline is achieved is different in the North American population.

(a) As evidence that the means of convergence is different, the authors concluded that there was a marked difference between the NA and the EU patterns of the basal length-to-wing size ratios versus latitude (in females). Fit a multiple linear regression that allows for different slopes and different intercepts. In a single F-test, evaluate the evidence against there being a single straight line that describes the cline on both continents. If you conclude there is a difference, is the difference one of slope alone? of intercept alone? or of both?

(b) Return to the basic question of whether the wing sizes in NA flies have established a cline similar to their EU ancestors. Using the model developed in Exercise 9.18, answer these questions: (i) Is there a nonzero slope to the cline of NA females? (ii) Is there a nonzero slope to the cline of NA males? (iii) Is there a difference between the clines of NA and EU females, and if so, what is its nature? (iv) Repeat (iii) for males.

24. Speed of Evolution. (Refer again to Exercise 9.18 and also to Exercise 10.23.) Many software systems allow the user to perform weighted regression, in which different squared residuals from regression receive different weights in deciding which set of parameter estimates provide the smallest sum of squared residuals. If each individual response has an independent estimate of its likely error, the weight given to each residual is usually taken to be the reciprocal of the square of that likely error. The standard error of wing sizes are standard errors of the averages around 2 individual (log) wing sizes. If your software allows for weights, construct a weight variable as the inverse square of the standard errors. Then repeat both parts of Exercise 10.23 using weighted regression. Do the results differ? Why is this preferable to using each fly as a separate case?

25. Potato Yields. Nitrogen and water are important factors influencing potato production. One study of their roles was conducted at sites in the St. John River Valley of New Brunswick. (G. Belanger et al., "Yield Response of Two Potato Cultivars to Supplemental Irrigation and N fertilization in New Brunswick," *American Journal of Potato Research* 77 (2000): 11–21.) Nitrogen fertilizer was applied at six different levels in combination with two water conditions: irrigated or nonirrigated. This design was repeated at four different sites in 1996, with the resulting yields depicted in Display 10.21. Notice that the patterns of responses against nitrogen level are fit reasonably well by quadratic curves.

Each quadratic requires 3 parameters, so a model that would allow for separate quadratic curves for each site-by-irrigation combination would have 24 parameters. (a) Using indicator functions for sites and for irrigation, construct a multiple linear regression model with 23 variables that will allow for completely different quadratic curves. Interpret the parameters in this model, if possible. (b) Describe how you would answer the following questions: (i) Is there evidence that the manner in which the quadratic terms differ by water condition changes from site to site (or is the difference the same at all four sites)? (ii) If the quadratic term differences are the same at all sites, is there strong evidence of a difference by water condition? (iii) If there is no difference between quadratic terms by water or by site, is there evidence of any quadratic term at all? (iv), (v), and (vi): Repeat (i), (ii), and (iii) for the linear terms, if there is no evidence of any quadratic terms. (c) Why are the questions in (b) ordered as they are?

DISPLAY 10.21 Potato yield versus nitrogen at four sites

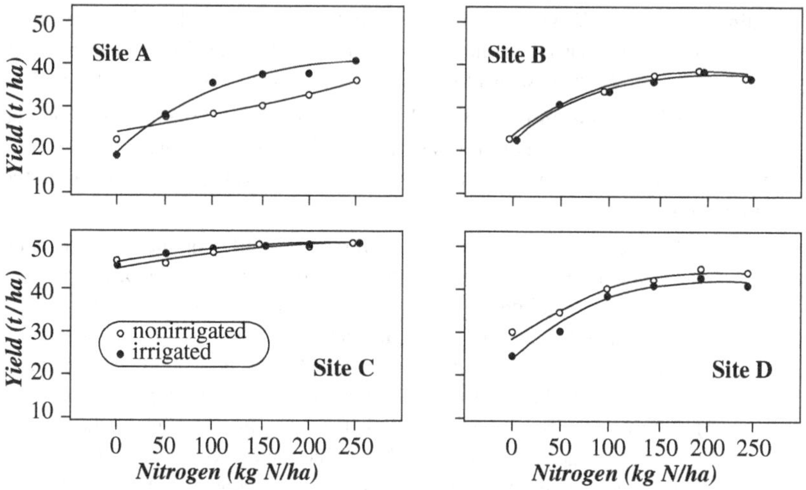

Data Problems

26. Thinning of Ozone Layer. Thinning of the protective layer of ozone surrounding the earth may have catastrophic consequences. A team of University of California scientists estimated that increased solar radiation through the hole in the ozone layer over Antarctica altered processes to such an extent that primary production of phytoplankton was reduced 6 to 12%.

Depletion of the ozone layer allows the most damaging ultraviolet radiation—UVB (280–320 nm)—to reach the earth's surface. An important consequence is the degree to which oceanic phytoplankton production is inhibited by exposure to UVB, both near the ocean surface (where the effect should be slight) and below the surface (where the effect could be considerable).

To measure this relationship, the researchers sampled from the ocean column at various depths at 17 locations around Antarctica during the austral spring of 1990. To account for shifting of the ozone hole's positioning, they constructed a measure of UVB exposure integrated over exposure time. The exposure measurements and the percentages of inhibition of normal phytoplankton production were extracted from their graph to produce Display 10.22. (Data from R. C. Smith et al., "Ozone Depletion: Ultraviolet Radiation and Phytoplankton Biology in Antarctic Waters," *Science* 255 (1992): 952–57.) Does the effect of UVB exposure on the distribution of percentage inhibition differ at the surface and in the deep? How much difference is there? Analyze the data, and write a summary of statistical findings and a section of details documenting those findings. (*Suggestion*: Fit the model with different intercepts and different slopes, even if some terms are not significantly different from zero.)

27. Factors Affecting Extinction. The data in Display 10.23 are measurements on breeding pairs of land-bird species collected from 16 islands around Britain over the course of several decades. For each species, the data set contains an average time of extinction on those islands where it appeared (this is actually the reciprocal of the average of $1/T$, where T is the length of time the species remained on the island, and $1/T$ is taken to be zero if the species did not become extinct on the island); the average number of nesting pairs (the average, over all islands where the birds appeared, of the number of nesting pairs per year); the size of the species (categorized as large or small); and the migratory status of the species (migrant or resident). (Data from S. L. Pimm, H. L. Jones, and

DISPLAY 10.22 First 5 of 17 rows of a data set with exposure to ultraviolet B radiation and percentage inhibition of primary phytoplankton production in Antarctic water

Location	Percent inhibition	UVB exposure	Surface (S) or Deep (D)
1	0.0	0.0000	D
2	1.0	0.0000	D
3	6.0	0.0100	D
4	7.0	0.0150	S
5	7.0	0.0185	S

J. Diamond, "On the Risk of Extinction," *American Naturalist* 132 (1988): 757–85.) It is expected that species with larger numbers of nesting pairs will tend to remain longer before becoming extinct. Of interest is whether, after accounting for number of nesting pairs, size or migratory status has any effect. There is also some interest in whether the effect of size differs depending on the number of nesting pairs. If any species have unusually small or large extinction times compared to other species with similar values of the explanatory variables, it would be useful to point them out. Analyze the data. Write a summary of statistical findings and a section of details documenting the findings.

28. El Niño and Hurricanes. Shown in Display 10.24 are the first few rows of a data set with the numbers of Atlantic Basin tropical storms and hurricanes for each year from 1950 to 1997. The variable *storm index* is an index of overall intensity of the hurricane season. (It is the average of number of tropical storms, number of hurricanes, the number of days of tropical storms, the number of days of hurricanes, the total number of intense hurricanes, and the number of days they last—when each of these is expressed as a percentage of the average value for that variable. A *storm index* score of 100, therefore, represents, essentially, an average hurricane year.) Also listed are whether the year was a cold, warm, or neutral El Niño year, a constructed numerical variable *temperature* that takes on the values −1, 0, and 1 according to whether the El Niño temperature is cold, neutral, or warm; and a variable indicating whether West Africa was wet or dry that year. It is thought that the warm phase of El Niño suppresses hurricanes while a cold phase encourages them. It is also thought that wet years in West Africa often bring more hurricanes. Analyze the data to describe the effect of El Niño on (a) the number of tropical storms, (b) the number of hurricanes, and (c) the storm index after accounting for the effects of West African wetness and for any time trends, if appropriate. (These data were gathered by William Gray of Colorado State University, and reported on the *USA Today* weather page: www.usatoday.com/weather/whurnum.htm)

29. Wage and Race 1987. Shown in Display 10.25 are the first few rows of a data set from the 1988 March U.S. Current Population Survey. The set contains weekly wages in 1987 (in 1992 dollars) for a sample of 25,437 males between the age of 18 and 70 who worked full-time, their years of education, years of experience, whether they were black, whether they worked in a standard metropolitan statistical area (i.e., in or near a city), and a code for the region in the U.S. where they worked (Northeast, Midwest, South, and West). Analyze the data and write a brief statistical report to see whether and to what extent black males were paid less than nonblack males in the same region and with the same levels of education and experience. Realize that the extent to which blacks were paid differently than nonblacks may depend on region. (Suggestion: Refrain from looking at interactive effects, except for the one implied by the previous sentence.) (These data, from the Current Population Survey (CPS), were discussed in the paper by H. J. Bierens and D. K. Ginther, "Integrated Conditional Moment Testing of Quantile Regression Models," *Empirical Economics* 26 (2001): 307–24; and made available at the Web site http://econ.la.psu.edu/~hbierens/MEDIAN.HTM (April, 2008).)

DISPLAY 10.23 Bird extinction data

Species	Ave. extinction time (years)	Ave. number of nesting pairs	Size (large or small)	Migratory status (resident or migrant)
Sparrowhawk	3.030	1.000	L	R
Buzzard	5.464	2.000	L	R
Kestrel	4.098	1.210	L	R
Peregrine	1.681	1.125	L	R
Grey partridge	8.850	5.167	L	R
Quail	1.493	1.000	L	M
Red-legged partridge	7.692	2.750	L	R
Pheasant	3.846	5.630	L	R
Water rail	16.667	3.000	L	R
Corncrake	4.219	4.670	L	M
Moorhen	8.130	4.056	L	R
Coot	5.000	1.000	L	R
Lapwing	7.299	6.960	L	M
Golden plover	1.000	1.670	L	M
Ringed plover	27.027	5.560	L	R
Curlew	3.106	2.830	L	M
Redshank	4.000	4.375	L	M
Snipe	16.129	4.125	L	M
Stock dove	3.484	3.670	L	R
Rock dove	37.037	8.330	L	R
Wood pigeon	7.299	2.750	L	R
Cuckoo	2.525	1.430	L	M
Short-eared owl	4.132	2.000	L	R
Little owl	2.000	2.750	L	R
Magpie	10.000	4.500	L	R
Jackdaw	2.667	7.120	L	R
Carrion crow	4.587	4.580	L	R
Raven	58.824	2.350	L	R
Skylark	32.258	6.870	S	R
Swallow	2.571	3.830	S	M
House martin	2.160	5.000	S	M
Yellow wagtail	1.000	1.250	S	M
Pied wagtail	2.967	2.270	S	R
Meadow pipit	9.524	5.350	S	R
Wren	11.111	8.700	S	R
Dunnock	7.299	6.100	S	R
Robin	4.000	3.330	S	R
Stonechat	2.381	3.640	S	R
Wheatear	2.611	4.830	S	M
Blackbird	3.257	4.670	S	R
Song thrush	1.701	1.700	S	R
Mistle thrush	1.795	1.330	S	R
Grasshopper warbler	1.198	1.000	S	M
Sedge warbler	3.185	1.900	S	M
Whitethroat	2.273	4.420	S	M
Willow warbler	1.111	1.250	S	M
Chiffchaff	1.000	1.000	S	M
Goldcrest	1.000	1.000	S	R
Spotted flycatcher	1.230	1.000	S	M
Great tit	6.061	2.500	S	R
Blue tit	3.175	1.500	S	R
Yellowhammer	2.000	2.500	S	R
Reed bunting	5.076	5.630	S	R
Chaffinch	1.934	2.370	S	R
Goldfinch	1.493	1.500	S	R
Redpoll	1.000	1.000	S	R
Linnet	5.102	6.500	S	R
House sparrow	3.003	4.500	S	R
Tree sparrow	1.898	2.170	S	R
Starling	41.667	11.620	S	R
Pied flycatcher	1.000	1.000	S	M
Siskin	1.000	1.000	S	R

30. Wages and Race 2011. Display 10.26 is a partial listing of a data set with weekly earnings for 4,952 males between the age of 18 and 70 sampled in the March 2011 Current Population Survey (CPS). These males are a subset who had reported earnings and who responded as having race as either "Only White" or "Only Black." Also recorded are the region of the country (with four categories: Northeast, Midwest, South, and West), the metropolitan status of the men's employment (with three categories: Metropolitan, Not Metropolitan, and Not Identified), age, education category (with 16 categories ranging from "Less than first grade" to "Doctorate Degree"), and education code, which is a numerical value that corresponds roughly to increasing levels of education (and so may be useful for plotting). What evidence do the data provide that the distributions of weekly earnings differ in the populations of white and black workers after accounting for the other variables? By how many dollars or by what percent does the White population mean (or median) exceed the Black population mean (or median)? (Data from U.S. Bureau of Labor Statistics and U.S. Bureau of the Census: Current Population Survey, March 2011 http://www.bls.census.gov/cps_ftp.html#cpsbasic; accessed July 25, 2011.)

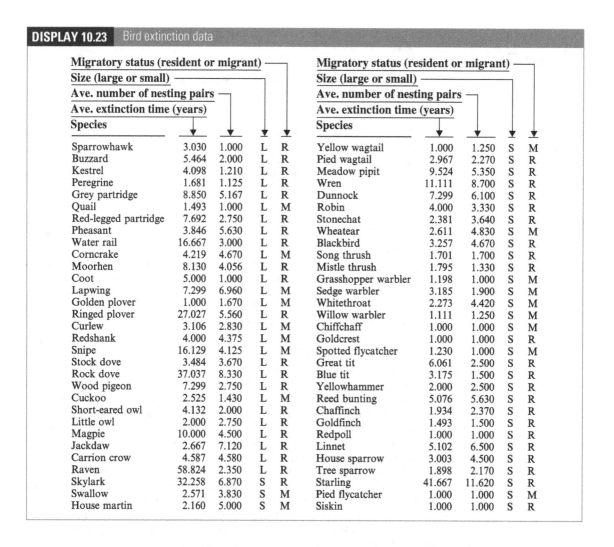

DISPLAY 10.24 Atlantic Basin hurricane and El Niño data for 1950–1997; partial listing

Year	El Niño	Temperature	West Africa	Storms	Hurricanes	Storm index
1950	cold	−1	1	13	11	243
1951	warm	1	0	10	8	121
1952	neutral	0	1	7	6	97
1953	warm	1	1	14	6	121
...						

DISPLAY 10.25 Data on the first 5 individuals (out of 25,437) in the 1987 wage and race data set

Region	MetropolitanStatus	Exper	Educ	Race	WeeklyEarnings
South	NotMetropolitanArea	8	12	NotBlack	859.71
Midwest	MetropolitanArea	30	12	NotBlack	786.73
West	MetropolitanArea	31	14	NotBlack	1424.5
West	MetropolitanArea	17	16	NotBlack	959.16
West	MetropolitanArea	6	12	NotBlack	154.32

DISPLAY 10.26 First five rows of a data set with weekly earnings (in U.S. dollars) in 2011 for 4,952 males who specified either "White Only" or "Black Only" as their race

Region	MetropolitanStatus	Age	EducCat	EducCode	Race	WeeklyEarnings
West	Not Metropolitan	64	SomeCollegeButNoDegree	40	White	1418.84
Midwest	Metropolitan	51	AssocDegAcadem	42	White	1000.00
South	Metropolitan	25	NinthGrade	35	White	420.00
West	Not Metropolitan	46	MastersDegree	44	White	1980.00
Northeast	Metropolitan	31	BachelorsDegree	43	White	1750.00

31. **Who Looks After the Kids?** Different bird species have different strategies: Maternal, Paternal, and BiParental care. In 1984 J. Van Rhijn argued in the *Netherlands Journal of Zoology* that parental care was the ancestral condition, going back to the dinosaur predecessors of birds. D. J. Varricchio et al. (*Science*, Vol. 322, Dec. 19, 2008, pp. 1826–28) tested that argument by comparing the relationships between Clutch Volume and adult Body Mass in six different groups: modern maternal-care bird species ($Mat; n = 171$), modern paternal-care bird species ($Pat; n = 40$), modern biparental-care bird species ($BiP; n = 204$), modern maternal-care crocodiles ($Croc; n = 19$), non-avian maniraptoran dinosaurs thought to be ancestors of modern birds ($Mani; n = 3$), and other non-avian dinosaurs ($Othr; n = 6$). The question of interest was which group of modern creatures most closely matches the relationship in the maniraptoran dinosaurs. A partial listing of the data appears in Display 10.27.

Fit a single model, with clutch volume (possibly transformed) as the response, that allows for separate straight lines in each group, using indicator variables for the different groups. For example, by selecting *Mani* as the reference group, you can determine how close any other group is to *Mani* by dropping out the indicator for that group and its product with the body mass variable. Do this for all other groups. (a) Are there differences among the relationships in the groups other than the *Mani*

DISPLAY 10.27 Body mass (in kilograms) and clutch volume (in cubic milimeters) for 443 species of animals in 6 groups (*Mat*: modern maternal-care birds, *Pat*: modern paternal-care birds, *BiP*: modern biparental-care birds, *Croc*: modern maternal-care crocodiles, *Mani*: non-avian maniraptoran dinosaurs, and *Othr*: other non-avian dinosaurs; first 5 of 443 rows

CommonName	Genus	Species	Group	BodyMass	ClutchVolume
Mallard	*Anas*	*platyrhynchos*	Mat	1.14E+00	5.42E+05
Greylag Goose	*Anser*	*anser*	Mat	3.31E+00	7.97E+05
Mute Swan	*Cygnus*	*olor*	Mat	1.07E+01	1.75E+06
Common Eider	*Somateria*	*mollissima*	Mat	2.07E+00	4.75E+05
Southern Screamer	*Chauna*	*torquata*	BiP	4.40E+00	5.13E+05

DISPLAY 10.28 First five rows of a data set with component test scores in Arithmetic Reasoning, Word Knowledge, Paragraph Comprehension, and Mathematics Knowledge taken in 1981; the AFQT score, which is a linear combination of them; and annual income in 2005 for 12,139 Americans who were selected in the NLSY79 sample, who were available for re-interview in 2006 and who had complete values of variables used in this and related analyses

Subject	Arith	Word	Parag	Math	AFQT	Income2005
2	8	15	6	6	6.841	5,500
6	30	35	15	23	99.393	65,000
7	14	27	8	11	47.412	19,000
8	13	35	12	4	44.022	36,000
9	21	28	10	13	59.683	65,000

group? What if there were none? (b) What do you conclude about which other group is nearest the maniraptoran dinosaur group? (c) Is the model defensible? (d) Comment on the study design.

32. Galton's Height Data. Reconsider the data in Exercise 7.26 on heights of adult children and their parents. Ignore the possible dependence of observations on children from the same family for now to answer the following: (a) What is an equation for predicting a child's adult height from their mother's height, their father's height, and their gender? (b) By how much does the male mean height exceed the female mean height for children whose parents' heights are the same? (c) Find a 95% prediction interval for a female whose father's height is 72 inches and whose mother's height is 64 inches.

33. IQ Score and Income. Display 10.28 is a partial listing of the National Longitudinal Study of Youth (NLSY79) subset (see Exercise 2.22) with annual incomes in 2005 (in U.S. dollars, as recorded in a 2006 interview) and scores on the Word Knowledge, Paragraph Comprehension, Arithmetic Reasoning, and Mathematics Knowledge portions of the Armed Forces Vocational Aptitude Battery (ASVAB) of tests taken in 1981. The AFQT is the Armed Forces Qualifying Test percentile, which is based on a linear combination of these four components and which is sometimes used as a general intelligence test score. A previous exercise had to do with the possible dependence of 2005 income on AFQT score. An interesting question is whether there might be some better linear combination of the four components than AFQT for predicting income. Investigate this by seeing whether the component scores are useful predictors of Income2006 in addition to AFQT. Also see whether

AFQT is a useful predictor in addition to the four test scores. Which test scores seem to be the most important predictors of 2006 income? (Answer with a single p-value conclusion.)

Answers to Conceptual Exercises

1. The initial heights were controlled by Galileo. Distance is the only random quantity.

2. Yes. The extra-sum-of-squares F-test, which compares the model with all three explanatory variables to the model with *lbody* only, addresses precisely this issue.

3. (a) The test where β_2 and β_3 both equal zero can be cast in terms of comparing the full model ($\beta_0 + \beta_1 lbody + \beta_2 lgest + \beta_3 llitter$) to the reduced model ($\beta_0 + \beta_1 lbody$). The t-test where β_2 is zero, on the other hand, implies a reduced model of $\beta_0 + \beta_1 lbody + \beta_3 llitter$; and the t-test where β_3 is zero implies a reduced model of $\beta_0 + \beta_1 lbody + \beta_2 lgest$. Neither of these reduced models is the same as the one sought, nor can the results from them be combined in any way to give some answer. (b) Same reason. The t-tests consider models where only one parameter is zero, but the model with both β_1 and β_2 equal to zero does not enter the picture. Multiple comparison adjustment does nothing to resolve the fact that these tests are different.

4. The model has nearly as many free parameters (8) as it has observations (9). One should expect a good fit to the data at hand, even if the explanatory variables have little relationship to the response.

5. (a) Each of the models represents the relationship between *lenergy* and *lmass* as a straight line, within the groups. The first model says that the intercept and slope—and hence the full line—is the same in all groups. The second model says that the slope is the same in each model while the intercepts are different. The third model allows the slopes and the intercepts to differ among all groups. (b) The second model is a reduced model in a test for equal slopes (with possibly differing intercepts); it is the full model for a test of equal intercepts (given equal slopes).

6. No bats of comparable size could be found in both groups. A more correct wording here might be "after adjustment for body size," but the underlying difficulty is not avoided.

7. (d).

8. The reduced model takes the null hypothesis to be true. The full model, however, encompasses both the null hypothesis and the alternative hypothesis. Thus, the full model is also correct when the reduced model is correct. Put another way, the full model is thought to be adequate from the start; the reduced model is obtained by imposing the constraints of the null hypothesis on the full model.

CHAPTER 11

Model Checking and Refinement

Multiple regression analysis takes time and care. Dead ends in the pursuit of models are expected and common, especially when many explanatory variables are involved. Frustration and wasted effort can be avoided, however, by going about the analysis in a proper order. In particular, transformations and outliers must be dealt with early on. Although each analysis will be guided by the peculiarities of the particular data and the questions of interest, initial assessment and graphical analysis are usually followed by the fitting of a rich, tentative model and by examination of residual plots. This part of the analysis should suggest whether more investigation into transformation or whether examination of outliers is needed. Some special tools for the latter are provided in this chapter. Once the data analyst has checked that the inferential tools are valid and not seriously influenced by one or two observations, the structure of the model itself can be refined by testing terms to see which should be included.

11.1 CASE STUDIES

11.1.1 Alcohol Metabolism in Men and Women—An Observational Study

Women exhibit a lower tolerance for alcohol and develop alcohol-related liver disease more readily than men. When men and women of the same size and drinking history consume equal amounts of alcohol, the women on average carry a higher concentration of alcohol in their bloodstream. According to a team of Italian researchers, this occurs because alcohol-degrading enzymes in the stomach (where alcohol is partially metabolized before it enters the bloodstream and is eventually metabolized by the liver) are more active in men than in women. The researchers studied the extent to which the activity of the enzyme explained the first-pass alcohol metabolism and the extent to which it explained the differences in first-pass metabolism between women and men. Their data (read from a graph) are listed in Display 11.1. (Data from M. Frezza et al., "High Blood Alcohol Levels in Women," *New England Journal of Medicine* 322 (1990): 95–99.)

The subjects were 18 women and 14 men, all volunteers living in Trieste. Three of the women and five of the men were categorized as alcoholic. All subjects received ethanol, at a dose of 0.3 grams per kilogram of body weight, orally one day and intravenously another, in randomly determined order. Since the intravenous administration bypasses the stomach, the difference in blood alcohol concentration—the

DISPLAY 11.1 First-pass metabolism of alcohol in the stomach (mmol/liter-hour) and gastric alcohol dehydrogenase activity in the stomach (μmol/min/g of tissue) for 18 women and 14 men

Subject	Metabolism	Gastric activity	Female (1 = F, 0 = M)	Alcoholic (1 = A, 0 = N)	Subject	Metabolism	Gastric activity	Female (1 = F, 0 = M)	Alcoholic (1 = A, 0 = N)
1	0.6	1.0	1	1	17	2.5	3.0	1	0
2	0.6	1.6	1	1	18	2.9	2.2	1	0
3	1.5	1.5	1	1	19	1.5	1.3	0	1
4	0.4	2.2	1	0	20	1.9	1.2	0	1
5	0.1	1.1	1	0	21	2.7	1.4	0	1
6	0.2	1.2	1	0	22	3.0	1.3	0	1
7	0.3	0.9	1	0	23	3.7	2.7	0	1
8	0.3	0.8	1	0	24	0.3	1.1	0	0
9	0.4	1.5	1	0	25	2.5	2.3	0	0
10	1.0	0.9	1	0	26	2.7	2.7	0	0
11	1.1	1.6	1	0	27	3.0	1.4	0	0
12	1.2	1.7	1	0	28	4.0	2.2	0	0
13	1.3	1.7	1	0	29	4.5	2.0	0	0
14	1.6	2.2	1	0	30	6.1	2.8	0	0
15	1.8	0.8	1	0	31	9.5	5.2	0	0
16	2.0	2.0	1	0	32	12.3	4.1	0	0

DISPLAY 11.2 First-pass metabolism and gastric alcohol dehydrogenase activity in alcoholic and nonalcoholic men and women

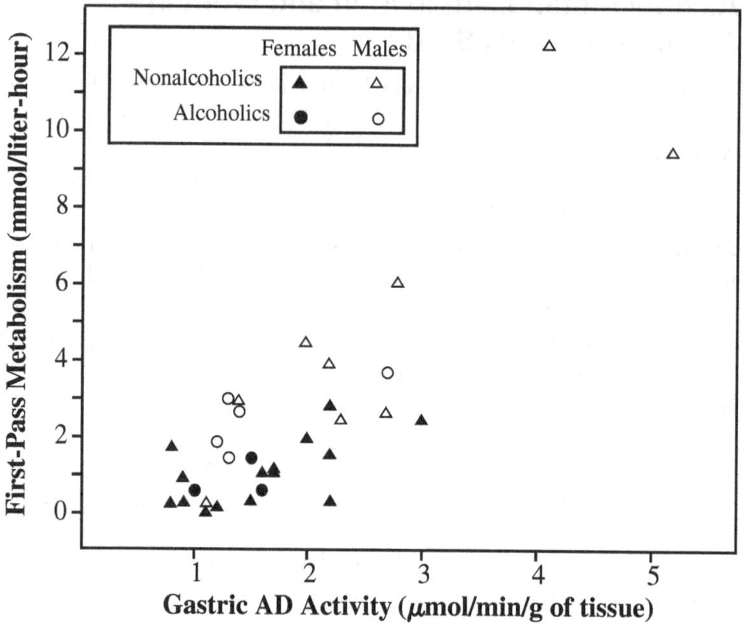

concentration after intravenous administration minus the concentration after oral administration—provides a measure of the "first-pass metabolism" in the stomach. In addition, gastric alcohol dehydrogenase (AD) activity (activity of the key enzyme) was measured in mucus samples taken from the stomach linings. The data are plotted in Display 11.2.

Several questions arise. Do levels of first-pass metabolism differ between men and women? Can the differences be explained by postulating that men have more dehydrogenase activity in their stomachs? Are the answers to these questions complicated by an alcoholism effect?

Statistical Conclusion

The following inferences pertain only to individuals with gastric AD activity levels between 0.8 and 3.0 μmol/min/g. No reliable model could be determined for values greater than 3.0. There was no evidence from these data that alcoholism was related to first-pass metabolism in any way (p-value $= 0.93$, from an F-test for significance of alcoholism and its interaction with gastric activity and sex.) Convincing evidence exists that first-pass metabolism was larger for males than for females overall (two-sided p-value $= 0.0002$, from a rank-sum test) and that gastric AD activity was larger for males than for females (two-sided p-value $= 0.07$ from a rank-sum test). Males had higher first-pass metabolism than females even after accounting for differences in gastric AD activity (two-sided p-value $= 0.0003$ from a t-test for equality of male and female slopes when both intercepts are zero). For a given level of gastric dehydrogenase activity, the mean first-pass alcohol metabolism for men is estimated to be 2.20 times as large as the mean first-pass alcohol metabolism for women (approximate 95% confidence interval from 1.37 to 3.04).

Scope of Inference

Because the subjects were volunteers, no inference to a larger population is justified. The inference that men and women do have different first-pass metabolism is greatly strengthened, however, by the existence of a physical explanation for the difference. The conclusions about the relationship between first-pass metabolism, gastric AD dehydrogenase activity, and sex are restricted to individuals whose gastric AD activity is less than 3. The sparseness of data for individuals with greater gastric AD activity levels prevents any resolution of the answers in the wider range.

11.1.2 The Blood–Brain Barrier—A Controlled Experiment

The human brain is protected from bacteria and toxins, which course through the bloodstream, by a single layer of cells called the *blood–brain barrier*. This barrier normally allows only a few substances, including some medications, to reach the brain. Because chemicals used to treat brain cancer have such large molecular size, they cannot pass through the barrier to attack tumor cells. At the Oregon Health Sciences University, Dr. E. A. Neuwelt developed a method of disrupting the barrier by infusing a solution of concentrated sugars.

As a test of the disruption mechanism, researchers conducted a study on rats, which possess a similar barrier. (Data from P. Barnett et al., "Differential Permeability and Quantitative MR Imaging of a Human Lung Carcinoma Brain Xenograft in the Nude Rat," *American Journal of Pathology* 146(2) (1995): 436–49.) The rats were inoculated with human lung cancer cells to induce brain tumors. After 9 to 11 days they were infused with either the barrier disruption (BD) solution or, as a control, a normal saline (NS) solution. Fifteen minutes later, the rats received a standard dose of the therapeutic antibody L6-F$(ab')_2$. After a set time they were sacrificed, and the amounts of antibody in the brain tumor and in normal tissue were measured. The time line for the experiment is shown in Display 11.3. Measurements for the 34 rats are listed in Display 11.4.

DISPLAY 11.3 Time line for blood–brain barrier disruption experiment

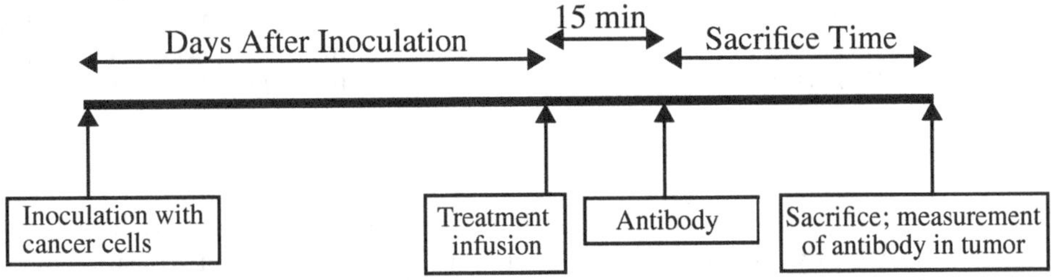

DISPLAY 11.4 Response variable, design variables (explanatory variables associated with the assignment of experimental units to groups), and several covariates (explanatory variables not associated with the assignment) for 34 rats in the blood–brain barrier disruption experiment

	Response variable	Design variables		Covariates				
Case	Brain tumor count (per gm)/ Liver count (per gm)	Sacrifice time (hours)	Treatment	Days post inoculation	Sex	Initial weight (grams)	Weight loss (grams)	Tumor weight (10^{-4} grams)
1	41081/ 1456164	0.5	BD	10	F	239	5.9	221
2	44286/ 1602171	0.5	BD	10	F	225	4.0	246
3	102926/ 1601936	0.5	BD	10	F	224	−4.9	61
4	25927/ 1776411	0.5	BD	10	F	184	9.8	168
5	42643/ 1351184	0.5	BD	10	F	250	6.0	164
6	31342/ 1790863	0.5	NS	10	F	196	7.7	260
7	22815/ 1633386	0.5	NS	10	F	200	0.5	27
8	16629/ 1618757	0.5	NS	10	F	273	4.0	308
9	22315/ 1567602	0.5	NS	10	F	216	2.8	93
10	77961/ 1060057	3	BD	10	F	267	2.6	73
11	73178/ 715581	3	BD	10	F	263	1.1	25
12	76167/ 620145	3	BD	10	F	228	0.0	133
13	123730/ 1068423	3	BD	9	F	261	3.4	203
14	25569/ 721436	3	NS	9	F	253	5.9	159
15	33803/ 1019352	3	NS	10	F	234	0.1	264
16	24512/ 667785	3	NS	10	F	238	0.8	34
17	50545/ 961097	3	NS	9	F	230	7.0	146
18	50690/ 1220677	3	NS	10	F	207	1.5	212
19	84616/ 48815	24	BD	10	F	254	3.9	155
20	55153/ 16885	24	BD	10	M	256	−4.7	190
21	48829/ 22395	24	BD	10	M	247	−2.8	101
22	89454/ 83504	24	BD	11	F	198	4.2	214
23	37928/ 20323	24	NS	10	F	237	2.5	224
24	12816/ 15985	24	NS	10	M	293	3.1	151
25	23734/ 25895	24	NS	10	M	288	9.7	285
26	31097/ 33224	24	NS	11	F	236	5.9	380
27	35395/ 4142	72	BD	11	F	251	4.1	39
28	18270/ 2364	72	BD	10	F	223	4.0	153
29	5625/ 1979	72	BD	10	M	298	12.8	164
30	7497/ 1659	72	BD	10	M	260	7.3	364
31	6250/ 928	72	NS	10	M	272	11.0	484
32	11519/ 2423	72	NS	11	F	226	2.2	168
33	3184/ 1608	72	NS	10	M	249	−4.4	191
34	1334/ 3242	72	NS	10	F	240	6.7	159

Since the amount of the antibody in normal tissue indicates how much of it the rat actually received, a key measure of the effectiveness of transmission across the blood–brain barrier is the ratio of the antibody concentration in the brain tumor to the antibody concentration in normal tissue outside of the brain. The brain tumor concentration divided by the liver concentration is a measure of the amount of the antibody that reached the brain relative to the amount of it that reached other parts of the body. This is the response variable: both the numerator and denominator of this ratio are listed in Display 11.4. The explanatory variables in the table comprise two categories: ***design variables*** are those that describe manipulation by the researcher; ***covariates*** are those measuring characteristics of the subjects that were not controllable by the researcher.

Was the antibody concentration in the tumor increased by the use of the blood–brain barrier disruption infusion? If so, by how much? Do the answers to these two questions depend on the length of time after the infusion (from 1/2 to 72 hours)? What is the effect of treatment on antibody concentration after weight loss, total tumor weight, and other covariates are accounted for? A coded scatterplot relating to the major questions is shown in Display 11.5.

DISPLAY 11.5 Log-log scatterplot of the ratio of antibody concentration in brain tumor to antibody concentration in liver versus sacrifice time, for 17 rats given the barrier disruption infusion and for 17 rats given a saline (control) infusion

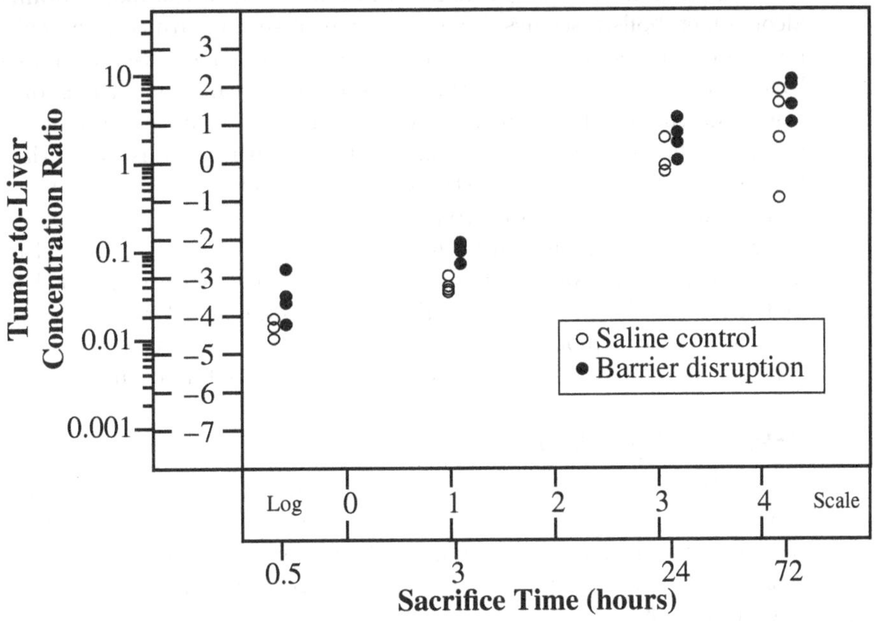

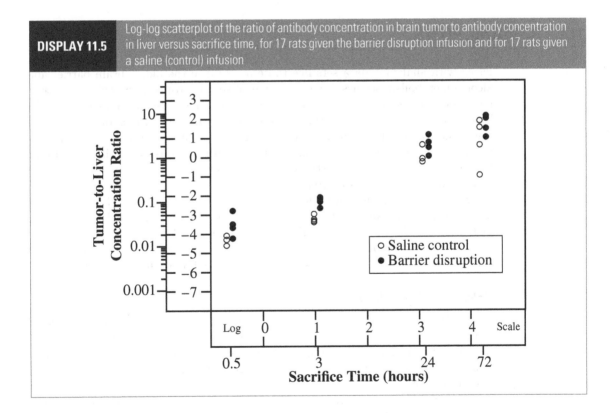

Statistical Conclusion

The median antibody concentration in the tumor (relative to that in the liver) was estimated to be 2.22 times as much for rats receiving the barrier disruption infusion than for those receiving the control infusion (95% confidence interval, from 1.56 to

3.15 times as much). This multiplicative effect appears to be constant between 1/2 and 72 hours after the infusion (the p-value for a test of interaction between treatment and sacrifice time is 0.92, from an F-test on 3 and 26 degrees of freedom).

Scope of Inference

One hitch in this study is that randomization was not used to assign rats to treatment groups. This oversight raises the possibility that the estimated relationships might be related to confounding variables over which the experimenter exercised no control. Including the measured covariates in the model helps alleviate some concern, and the results appear not to have been affected by these potential confounding variables. Nevertheless, causal implications can only be justified on the tenuous assumption that the assignment method used was as effect-neutral as a random assignment would have been.

11.2 RESIDUAL PLOTS

Faced with analyzing data sets like those involved in the blood–brain barrier and alcohol metabolism studies, a researcher must seek good-fitting models for answering the questions of interest, bearing in mind the model assumptions required for least squares tools, the robustness of the tools against violations of the assumptions, and the sensitivity of these tools to outliers. Since model-building efforts are wasted if the analyst fails to detect problems with nonconstant variance and outliers early on, it is wise to postpone detailed model fitting until after outliers and transformation have been thoroughly considered.

Much can be resolved from initial scatterplots and inspection of the data, but it is almost always worthwhile to obtain the finer picture provided by a residual plot. Creating this plot involves fitting some model in order to get residuals. On the basis of the scatterplots, the analyst can choose some tentative model or models and conduct residual analysis on these, recognizing that further modeling will follow.

Selecting a Tentative Model

A tentative model is selected with three general objectives in mind: The model should contain parameters whose values answer the questions of interest in a straightforward manner; it should include potentially confounding variables; and it should include features that capture important relationships found in the initial graphical analysis.

It is disadvantageous to start with either too many or too few explanatory variables in the tentative model. With too few, outliers may appear simply because of omitted relationships. With too many (lots of interactions and quadratic terms, for example), the analyst risks overfitting the data—causing real outliers to be explained away by complex, but meaningless, structural relationships. Overfitting becomes less of a problem when the sample sizes are substantially larger than the number of model parameters.

For large sample sizes, therefore, the initial tentative model for residual analysis can err on the side of being rich, including potential model terms that may not be retained in the end. For small sample sizes, several tentative models may be needed for residual analysis; and the data analyst must guard against including terms whose significance hinges on one or two observations. As evident in the strategy for data analysis laid out in Display 9.9, the process of trying a model and plotting residuals is often repeated until a suitable inferential model is determined.

Example—Preliminary Steps in the Analysis of the Blood–Brain Barrier Data

The coded scatterplot in Display 11.5 is a good starting point for the analysis. Apparently, the disruption solution does allow more antibody to reach the brain than the control solution does; this effect is about the same for all sacrifice times (time between antibody treatment and sacrifice); an increasing proportion of antibody reaches the brain with increasing time after infusion; and this increasing relationship appears to be slightly nonlinear. A matrix of scatterplots and a correlation matrix (an array showing the sample correlation coefficients for all possible pairs of variables), which are not shown here, indicate further that the covariates—days after inoculation, initial weight, and sex of the rat—are associated with the response. These covariates are also related to the treatment given. (Recall that randomization was not used.) In particular, rats treated at longer days after inoculation were also assigned to the longer sacrifice times. Furthermore, all male rats were assigned to the longer sacrifice times.

This initial investigation suggests the following tentative regression model (using the shorthand model specification of Section 9.3.5):

$$\mu\{antibody \mid SAC, TREAT, DAYS, FEM, weight, loss, tumor\}$$
$$= SAC + TREAT + (SAC \times TREAT) + DAYS + FEM + weight + loss + tumor,$$

where *antibody* is the logarithm of the ratio of antibody in the brain tumor to that in the liver. *SAC* is the sacrifice time factor with four levels; *TREAT* is treatment, with two levels; *DAYS* is days after inoculation, with three levels; and *FEM* is sex, with two levels. *Weight*, *loss*, and *tumor* are the initial weight, weight loss, and tumor weight variables. Display 11.5 shows a strong linear effect of log sacrifice time on the response, but some additional curvature may be present as well. To avoid mismodeling the effect of sacrifice time at the start, it is treated as a factor with four levels. Similarly, the coded scatterplot suggests that the difference between the two treatments may be greater for the shorter sacrifice times than for the longer ones. Consequently, the sacrifice time by treatment interaction terms are included in the tentative model. Although more terms may be added to this model later, it captures the most prominent features of the scatterplot. Display 11.6 shows the plot of residuals versus the fitted values from the regression model. (*Note:* Even if prior experience or initial inspection had not led the researchers to consider the logarithms of the response, the coded scatterplot and residual plot would have revealed that the variability increases with increasing response, leading them to the same consideration.)

DISPLAY 11.6 Scatterplot of residuals versus fitted values from the fit of the logged response on a rich model for explanatory variables—brain–barrier data

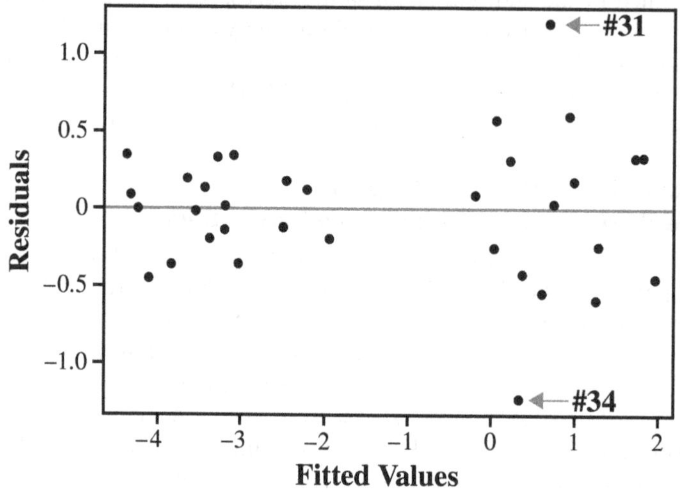

The residual plot in Display 11.6 exemplifies the ambiguity that can arise with small data sets. Is there a funnel-shaped pattern, or is the apparent funnel only due to a few outliers? The usual course of action consists of three steps:

1. Examine the outliers for recording error or contamination.
2. Check whether a standard transformation resolves the problem.
3. If neither of these steps works, examine the outliers more carefully to see whether they influence the conclusions (following the strategy suggested in Section 11.3).

The residual plot is based on a response that has already been transformed into its logarithm. A reciprocal transformation corrects more pronounced funnel-shaped patterns than does the log. Here, however, it does not help, and there is no suggestion of a recording error. Consequently, the analyst must proceed with further model fitting, paying careful attention to the roles of observations 31 and 34.

Example—Preliminary Steps in the Analysis of the Alcohol Metabolism Data

Refer to the coded scatterplot in Display 11.2. The next step is to examine a residual plot for outliers and to assess the need for transformation. The plot in Display 11.7 is a residual plot from the regression of first-pass metabolism on gastric AD activity (*gast*), an indicator variable for females (*fem*), an indicator for alcoholics (*alco*), and the interaction terms *gast* × *fem*, *fem* × *alco*, *gast* × *alco*, and *gast* × *fem* × *alco*. (The last term is a *three-factor interaction* term, formed as the product of three explanatory variables.)

DISPLAY 11.7 Residual plot from the regression of first-pass metabolism on gastric activity, sex indicator, alcoholism indicator, and all two- and three-factor interactions

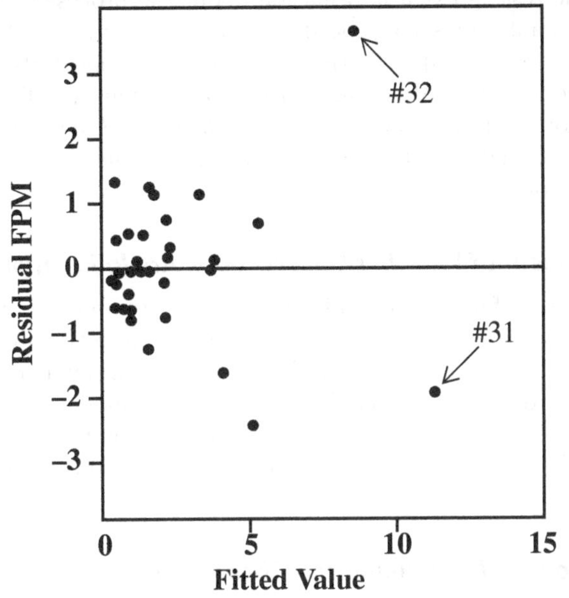

The plot draws attention to two observations: one that has a considerably larger residual than the rest and one that has a fitted value quite a bit larger than the rest. These are cases 31 and 32, and they appear in the coded scatterplot of Display 11.2 in the upper right-hand corner, separated from the rest of the points. There appears to be a downward trend in the residual plot, excluding cases 31 and 32. This could reflect a model that is heavily influenced by one or two observations and consequently does not fit the bulk of the observations well.

11.3 A STRATEGY FOR DEALING WITH INFLUENTIAL OBSERVATIONS

Least squares regression analysis is not resistant to outliers. One or two observations can strongly influence the analysis, to the point where the answers to the questions of interest change when these isolated cases are excluded. Although any influential observation that comes from a population other than the one under investigation should be removed, removing an observation simply because it is influential is not justified. In any circumstance, it is unwise to state conclusions that hinge on one or two data points. Such a statistical study should be considered extremely fragile.

There are two approaches for dealing with excessively influential observations in regression analysis. One is to use a robust and resistant regression procedure. The other is to use least squares but to examine outliers and influence closely to see whether the suspect observations are indeed influential, why they are influential (this can help dictate the subsequent course of action), and whether they provide some interesting extra information about the process under study. Using a robust and resistant regression procedure is particularly useful if past experience indicates that the response distribution tends to have long tails and that outliers are an expected nuisance. When influential observations are discovered in the course of a regression analysis, however, *The Statistical Sleuth* recommends closer examination. This often clarifies the problems and sometimes reveals additional, unexpected information.

Assessment of Whether Observations Are Influential

The strategy for assessing influence involves temporarily removing suspected influential observations to see whether the answers to the questions of interest change. Does the evidence from a test change from slight evidence to convincing evidence? Does the decision to include a term in the model change? Does an important estimate change by a practically relevant amount? If not, the observation is not influential and the analysis can proceed as usual. If so, further action must be taken.

What to Do About Influential Observations

If an observation is influential and it substantially differs from the remaining data in its explanatory variable values, perhaps it is being accorded too much weight in fitting a model over a sparsely represented region of the explanatory variables. For example, given 30 observations for X between 0 and 5 and a single value of X at 15, it is unrealistic to try to model the regression of Y on X over the entire range of 0 to 15. The observations in the sparsely represented region can be removed, the model fit on the remaining data points, and conclusions stated only for the restricted range of explanatory variables. This restriction should be stated explicitly, with the added comment that more data would be needed to model the regression accurately over the broader range of explanatory variables.

If an influential observation is not particularly unusual in its explanatory variable values, and if no definitive explanation for its unique behavior can be found, omitting it cannot be justified. More data are needed to answer the questions of interest. As a last resort, the results can be reported with and without the influential observation.

A complete strategy is summarized in Display 11.8. Although this is not stated explicitly in the strategy, the analyst should first check any unusual observation for recording accuracy and for alternative explanations of its unusualness before proceeding to this statistical approach.

Example—Alcohol Metabolism Study

The tentative model is fit with and without cases 31 and 32, to examine which aspects of the fit change and by how much. The resulting changes in estimates, standard errors, and p-values are shown in Display 11.9.

DISPLAY 11.8 A strategy for dealing with suspected influential cases

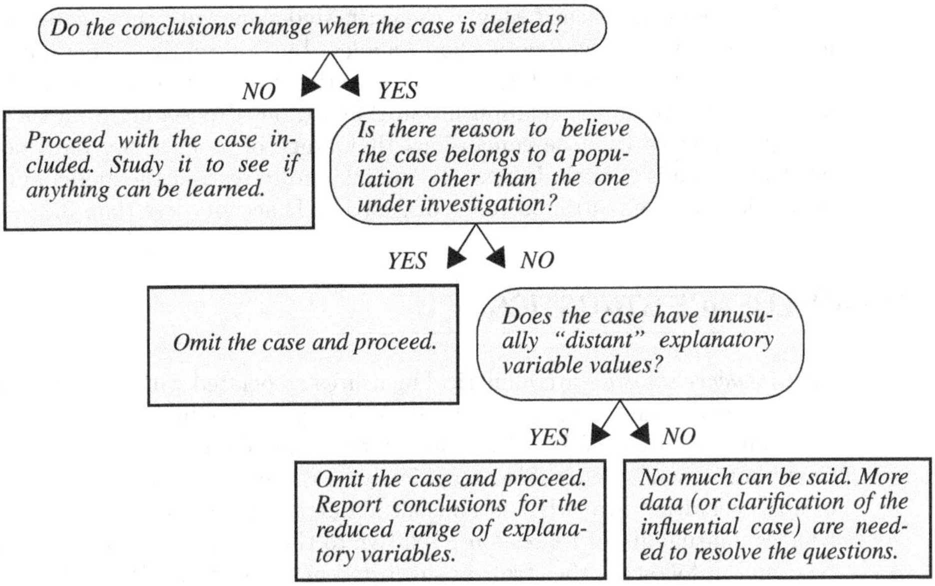

DISPLAY 11.9 Regression parameter estimates, standard errors, and *p*-values from the regression of first-pass metabolism on gastric activity, an indicator for female, an indicator for alcoholic, and all second- and third-order interactions: first with all observations and then with all observations except 31 and 32

	All 32 observations			**Cases 31 and 32 removed**		
Variable	Estimate	Standard error	Two-sided *p*-value	Estimate	Standard error	Two-sided *p*-value
Constant	−1.660	1.000	0.11	−0.680	1.309	0.61
Gastric activity (G)	2.514	0.343	<0.0001	1.921	0.608	0.0045
Female (F)	1.466	1.333	0.28	0.486	1.467	0.74
Alcoholic (A)	2.552	1.946	0.20	1.572	1.812	0.40
G×F	−1.673	0.620	0.013	−1.081	0.721	0.15
F×A	−2.252	4.394	0.61	−1.272	3.467	0.72
G×A	−1.459	1.053	0.18	−0.866	0.963	0.38
G×F×A	1.199	2.998	0.69	0.606	2.316	0.80

A striking consequence of the exclusion of cases 31 and 32 is the drop in significance of the interaction of gastric activity and sex from a *p*-value of 0.013 to one of 0.15. The reason for this change is evident from the coded scatterplot in Display 11.2. Ignoring the effect of alcoholism, imagine the slope in the regression of first-pass metabolism on gastric AD activity for males and females separately.

The slope for males is substantially greater with cases 31 and 32 included, than with them excluded.

What does this mean? Maybe the estimated slope with those cases is accurate and males have a larger slope than females; but alternatively, perhaps the relationship is not a straight line for gastric activity greater than 3. It is difficult to know how to model the relationship in that region. Proceeding from the guideline that it is unwise to state conclusions that hinge on one or two data points, the prudent action is to exclude cases 31 and 32, and to restrict the model building and conclusions to the restricted range of gastric AD activity less than 3.

11.4 CASE-INFLUENCE STATISTICS

Case-influence statistics are numerical measures associated with the individual influence of each observation (each case). When provided by a statistical computer program, they are useful for two reasons: they can help identify influential observations that may not be revealed graphically; and they partition the overall influence of an observation into what is unusual about its explanatory variable values and what is unusual about its response relative to the fitted model. This partition may be useful in following the strategy suggested in Display 11.8 for dealing with cases of suspect influence.

11.4.1 Leverages for Flagging Cases with Unusual Explanatory Variable Values

The ***leverage*** of a case is a measure of the distance between its explanatory variable values and the average of the explanatory variable values in the entire data set. As illustrated in the alcohol metabolism study, cases with high leverage may exert strong influence on the results of model fitting.

When only a single explanatory variable X is involved, the leverage of the ith case is

$$h_i = \frac{1}{(n-1)}\left[\frac{X_i - \overline{X}}{s_X}\right]^2 + \frac{1}{n} \quad \text{or} \quad \frac{(X_i - \overline{X})^2}{\sum(X - \overline{X})^2} + \frac{1}{n},$$

where s_X is the sample standard deviation of X. Aside from the parts involving n, the first formulation shows that the leverage is a distance of X_i from $\overline{X}$, in units of standard deviations; and the second formulation shows that the leverage is the proportion of the total sum of squares of the explanatory variable contributed by the ith case.

When two or more explanatory variables are involved, the leverage can only be expressed in matrix notation. The interpretation is a straightforward extension of the one-variable case, however. The leverage for case i is a measure of the distance of case i from the average (in a multivariable sense). The one aspect of leverage in the multiple regression setting that is not an obvious extension of the preceding formulas is that the distance from the average is relative to the joint spread of the explanatory variables. Display 11.10 illustrates a problem in which one case is

DISPLAY 11.10 An illustration of what is meant by "far from the average" of multiple explanatory variables when they are correlated

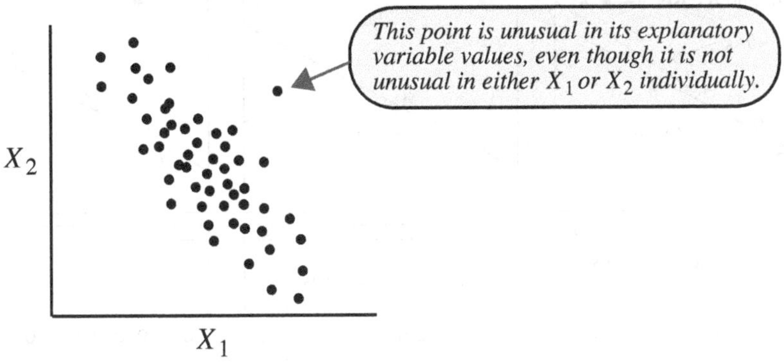

relatively far from the two-dimensional scatter of the other cases, although it is not particularly unusual in either of the single dimensions.

It is important to identify this case, since it stands alone. Because it would be largely responsible for dictating the location of the estimated regression surface in the surrounding region, the point has a high potential for influence. The multiple regression version of leverage would be large for such an observation.

Cases with Large Leverage

Leverages are greater than $1/n$ and less than 1 for each observation, and the average of all the h_i's in a data set is always p/n, where p is the number of regression coefficients. Leverages also depend on what model is being entertained. The standard deviation of the residual for the ith case is related to its leverage:

$$\text{SD}(\text{Residual}_i) = \sigma\sqrt{(1 - h_i)}.$$

A case with large leverage has a residual with low variability. Because its explanatory variable values are so unusual, it dictates the location of the estimated regression over the whole region in its vicinity; no other points in the region share the responsibility. Because its residual must be small, this case acts like a magnet on the estimated regression surface. If, however, its response falls close to the regression surface (as determined by the remaining observations alone), it is not necessarily influential. Therefore, while a large leverage does not necessarily indicate that the case is influential, it does imply that the case has a high *potential* for influence.

Cases with large leverage can easily be overlooked in scatterplots and other graphical displays. Display 11.10 shows how a case with high leverage can be visible in a two-dimensional scatterplot, even though it would not be exceptional in either of the component one-dimensional histograms. Similarly, a case with high leverage may be visible in a three-dimensional plot but not in any of the component two-dimensional scatterplots. Therefore, examining the numerical measures of leverage

DISPLAY 11.11 Case influence statistics for three sample situations

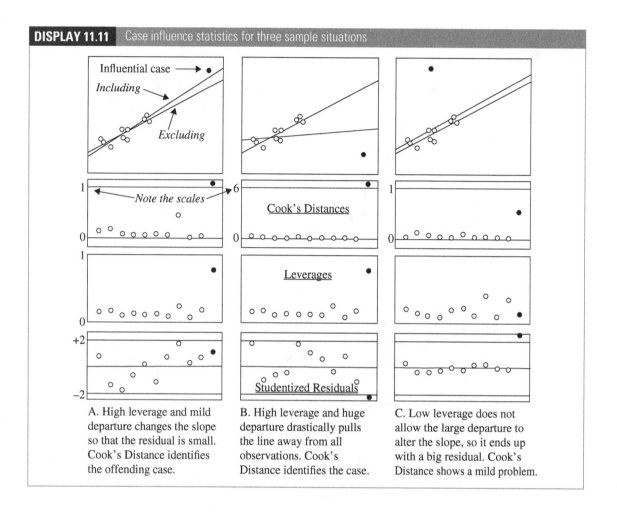

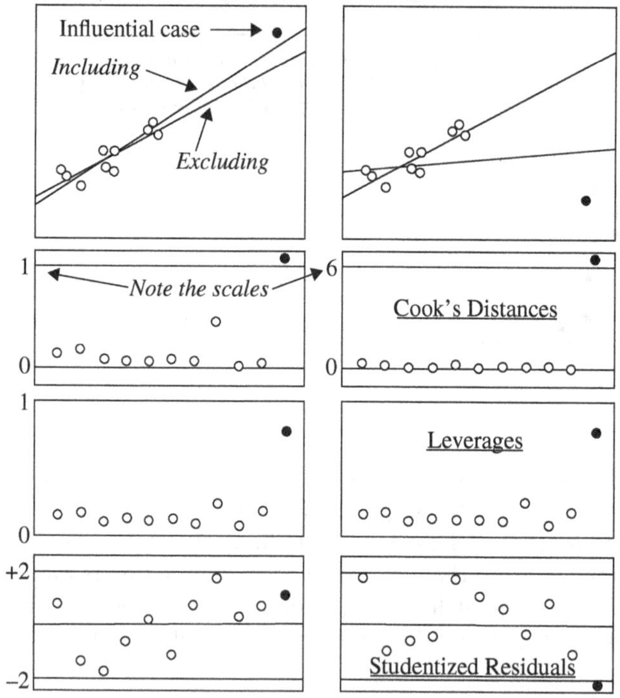

A. High leverage and mild departure changes the slope so that the residual is small. Cook's Distance identifies the offending case.

B. High leverage and huge departure drastically pulls the line away from all observations. Cook's Distance identifies the case.

C. Low leverage does not allow the large departure to alter the slope, so it ends up with a big residual. Cook's Distance shows a mild problem.

by themselves is important. The leverage measure h_i would identify the case in Display 11.10 because its distance takes into account the correlations among the variables.

It is difficult to say how large a value of h_i is sufficiently large to warrant further attention. Since the average of the h_i's is p/n, some statisticians (and statistical computer programs) use twice the average—$2p/n$—as a lower cutoff for flagging cases that have a high potential for excessive influence. This formulation is somewhat arbitrary. The main point is to use the leverage measure in conjunction with the other case influence statistics to get some overall assessment of influence. Display 11.11 will show one way to display the leverages in conjunction with other case influence statistics.

Leverage Computations

Most statistical computer programs will carry out the calculations for leverage as part of their regression routine. If not, the calculations may be made by using the formula

$$h_i = \left[\frac{\text{SE}(\text{fit}_i)}{\hat{\sigma}}\right]^2.$$

The values for $\text{SE}(\text{fit}_i)$ are often available. If not, they may be found with the method in Section 10.2.3.

11.4.2 Studentized Residuals for Flagging Outliers

A *studentized residual* is a residual divided by its estimated standard deviation. Some residuals naturally have less variation than others because of their leverages. Therefore, the usual residual plot may not direct attention to cases whose residuals lie much farther from zero than expected. The studentized residuals,

$$studres_i = \frac{res_i}{\hat{\sigma}\sqrt{1-h_i}},$$

put all residuals on a common scale: numbers of standard deviations. Since roughly 95% of normally distributed values fall within two standard deviations of their mean, it is common to investigate observations whose studentized residuals are smaller than −2 or larger than 2. Of course, it is not unusual to find roughly 5% of observations outside this range, and 5% can be a sizeable number if the sample size is large. (Technically, this is called the *internally studentized residual*; see Note 1 at the end of Section 11.4.4.) As with leverages, studentized residual calculations are typically performed by the statistical computer program.

Example—Observation #31 from the First-Pass Metabolism Study

For the fit of the regression of metabolism on gastric activity, sex indicator, and their interaction, the observed value for the 31st observation is 9.5 and its fitted value is 11.0024. This leaves the residual of $res_{31} = (9.5 - 11.0024) = -1.5024$. From computer calculations, its leverage is $h_{31} = 0.5355$. Notice that the rough lower cutoff for flagging large influence, $2p/n$, is $2 \times 4/32 = 0.25$; according to that rule, case 31 has a high potential for influence. The estimated standard deviation of the residual is $1.20730(1 - 0.5355)^{1/2} = 0.8228$. The studentized residual is −1.8260. This is moderately far from zero, but not alarmingly so.

11.4.3 Cook's Distances for Flagging Influential Cases

Cook's Distance is a case statistic that measures *overall* influence—the effect that omitting a case has on the estimated regression coefficients. For case i, Cook's Distance can be represented as

$$D_i = \sum_{j=1}^{n} \frac{(\hat{Y}_{j(i)} - \hat{Y}_j)^2}{p\hat{\sigma}^2},$$

where $\hat{Y}_j$ is the jth fitted value in a fit using all the cases; $\hat{Y}_{j(i)}$ is the jth fitted value in a fit that excludes case i from the data set; p is the number of regression coefficients; and $\hat{\sigma}^2$ is the estimated variance from the fit, based on all observations. The numerator measures how much the fitted values (and, therefore, the parameter estimates) change when the ith case is deleted. The denominator scales this difference in a useful way. A case that is influential, in terms of the least squares estimated coefficients changing when it is deleted, will have a large value of Cook's Distance.

An equivalent expression for Cook's Distance is

$$D_i = \frac{1}{p}(studres_i)^2\left(\frac{h_i}{1-h_i}\right).$$

This alternate expression is useful for computing, since it does not require that the ith case actually be deleted. More importantly, it shows that an influential case—with a large Cook's Distance—is influential because it has a large studentized residual, a large leverage, or both.

Cook's Distance for case 31 of the alcohol metabolism data is 0.9610. Some statisticians use a rough guideline that a value of D_i close to or larger than 1 indicates a large influence. Realize, however, that Cook's Distance measures the influence on all the regression coefficients. By identifying potentially problematic cases, the data analyst can directly refit the model with and without the indicated observations to see whether the answers to the important questions of interest change. The effect of the case removal on a single coefficient of interest may be much more or much less dramatic than what is indicated by the general measure D_i. Nevertheless, Cook's Distance is very useful for calling attention to potentially problematic cases.

11.4.4 A Strategy for Using Case Influence Statistics

Inspecting graphical displays and scatterplots may alert the analyst to the need for case influence investigations. In addition, the presence of statistically significant, complex effects that make little sense may indicate problems with influence. But when the residual plot from fitting a good inferential model fails to suggest any problems, there is generally no need to examine case influence statistics at all.

The suggested strategy in Display 11.8 requires some assessment of influence, of the unusualness of a case's explanatory variable values, and of the degree to which a case is an outlier. Therefore, the trio of case influence measures—D_i, h_i, and $studres_i$ (or similar trios described later)—are examined jointly. Since values of these measures exist for every observation in the data set, the examination of a full list may be difficult. Some computer programs use rough guidelines (such as $D_i > 1$, $h_i > 2p/n$, or $|studres_i| > 2$) to flag the relevant cases. Ideally, a graphical display of the case statistics, like Display 11.12 for the alcohol metabolism data, makes case influence analysis convenient.

In the top plot, the values of Cook's Distance for cases 31 and 32 are obviously substantially larger than the rest, indicating what was previously observed: these two observations are influential. In addition, they both have high leverages, as one would expect from their gastric AD activity values' being so much larger than the rest. One other case—number 17—has a large leverage. This is the female with

DISPLAY 11.12 Case-influence statistics for the fit of first-pass metabolism on gastric activity, sex indicator, and their interaction

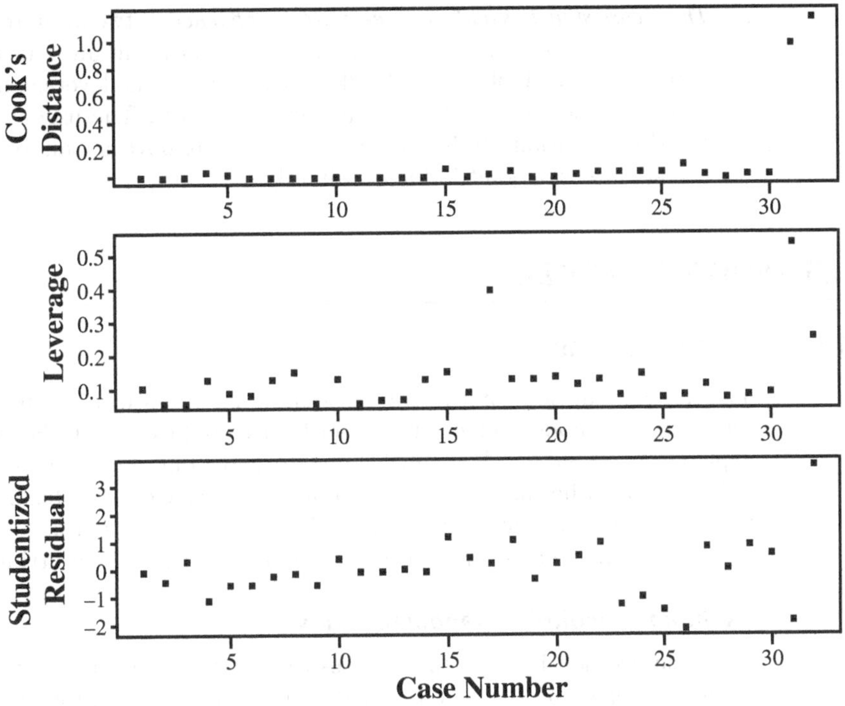

the largest gastric AD activity value. No evidence indicates that this observation is influential, however, so it need not be studied further.

Notes About Case Influence Statistics

1. *Externally studentized residuals.* A potential problem with the internally studentized residual is that the estimate of σ from the fit with all observations may be tainted by case i if i is indeed an outlier. The *externally studentized residual* is

$$studres_i^* = \frac{res_i}{\hat{\sigma}_{(i)}\sqrt{1-h_i}},$$

where $\hat{\sigma}_{(i)}$ is the estimate of the standard deviation about the regression line from the fit that excludes case i.

2. *External studentization in measures of influence.* If external studentization makes sense for residuals, it also makes sense for the measure of influence. Although it is not widely used, an externally studentized version of Cook's Distance, D_i^*, can be obtained by using $\hat{\sigma}_{(i)}$ in place of $\hat{\sigma}$ in the definition—or, equivalently, by using $studres_i^*$ in place of $studres_i$ in the computing version. A measure that *is* widely used is DFFITS_i, which is $(pD_i^*)^{1/2}$. It measures influence in the

same way as the externally studentized version of Cook's Distance, but on a slightly different scale. The decision about which of these measures to use is based largely on which is available in the statistical computer program.

3. *These case statistics address one-at-a-time influence.* If two or more cases are jointly influential and, in particular, if they are together in an unusual region of the explanatory variables, then Cook's Distance and the leverage measure may fail to detect their joint influence and joint leverage. The user must be on guard for this situation. It is always possible to remove a pair of observations to investigate their joint influence directly.

11.5 REFINING THE MODEL

11.5.1 Testing Terms

The key assumption for the correctness of the statistical conclusions is that the terms in the model sufficiently describe the regression of the response on the explanatory variables. It is misleading to leave out important explanatory variables. On the other hand it is also important to use Occam's Razor (Section 10.4.6) to trim away nonessential terms. Therefore, after transformations and outliers have been resolved, the analysis focuses on finding a simple, good-fitting model.

Example—Alcohol Metabolism Study

Once the decision is made to set cases 31 and 32 aside, the analyst must refit the model and examine the new residual plot. No additional problem is indicated, so model refinement may begin.

Since alcoholism is not of primary concern, and since so few alcoholics are included in the data set, it would be helpful if this variable could be ignored. One approach in the investigation is to test whether all the terms involving alcoholism in the tentative model can be dropped, using an extra-sum-of-squares F-test. There are four such terms: *alco*, *gast*×*alco*, *fem*×*alco*, and *gast*×*fem*×*alco*. The F-statistic is 0.21. By comparison to an F-distribution on 4 and 22 degrees of freedom, the p-value is 0.93, so the data are consistent with there being no alcoholism effect.

With alcoholism left out, the resulting model for consideration is

$$\mu\{metabolism \mid gast, fem\} = \beta_0 + \beta_1 gast + \beta_2 fem + \beta_3 (gast \times fem),$$

the separate regression lines model. The estimates of regression parameters, their standard errors, and their p-values appear in Display 11.13.

It may appear that sex plays no significant role, because the parameters β_2 (the amount by which the women's intercept exceeds the men's intercept) and β_3 (the amount by which the women's slope exceeds the men's slope), have nonsignificant p-values individually. This is deceiving. A sex difference is distinctly noticeable on the scatterplot (Display 11.2). The p-values suggest, however, that there is no need for different slopes if different intercepts are included in the model, and no need for different intercepts if different slopes are included.

DISPLAY 11.13 Least squares estimates for the regression of first-pass metabolism on gastric AD activity, sex, and their interaction (excluding cases 31 and 32)

Variable	Estimate	Standard error	*t*-statistic	Two-sided *p*-value
Constant	0.070	0.802	0.087	0.93
gast	1.565	0.407	3.843	0.0007
fem	−0.267	0.993	−0.269	0.79
gast × *fem*	−0.728	0.539	−1.351	0.19

In Display 11.2, the intercepts for both men and women are near zero. Since, in addition, a zero intercept makes sense in this application (because there is no first-pass metabolism if there is no activity of the enzyme), it is useful to force both lines to go through the origin, by dropping the constant term and the female indicator variable:

$$\mu\{metabolism \mid gast, fem\} = \beta_1 gast + \beta_2(gast \times fem).$$

In this model, first-pass metabolism is directly proportional to gastric activity, but the constant of proportionality differs for men and for women. The F-statistic comparing this model to the one whose fit is summarized in Display 11.13 is 0.06 with 2 and 26 degrees of freedom, so the smaller model is adequate.

The fit to the new model appears in Display 11.14. In this model, both terms are essential, so this is accepted as the final version for inference. The conclusions stated in the summary of statistical findings are based on this fit. Notice in particular that, for any level of gastric AD activity (in the range of 0.8 to 3.0), the mean first-pass metabolism for males divided by the mean first-pass metabolism for females is $\beta_1/(\beta_1 + \beta_2)$, which is estimated as 2.20. Thus, the mean for males is estimated to be 2.20 times the mean for females, even after accounting for gastric dehydrogenase activity.

DISPLAY 11.14 Results for mean metabolism being proportional to gastric activity

Variable	Estimate	Standard error	*t*-statistic	Two-sided *p*-value
gast	1.5989	0.1249	12.800	<0.0001
gast × *fem*	−0.8732	0.1740	−5.019	<0.0001

Residual SD = 0.8518; d.f. = 28

11.5.2 Partial Residual Plots

A scatterplot exhibits only the marginal association of two variables, which may depend heavily on their mutual association with a third variable. The question

DISPLAY 11.15 Scatterplot of log brain weight versus log gestation length, and scatterplot of the partial residuals of log brain weight (adjusted for log body weight) versus log gestation length—mammal brain weight data from Section 9.1.2

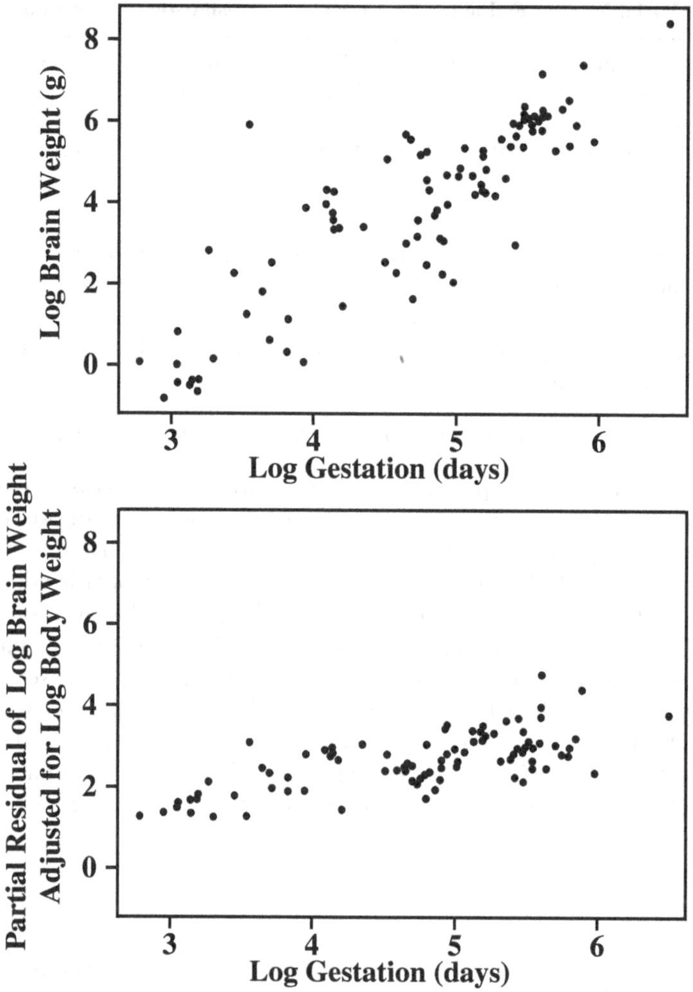

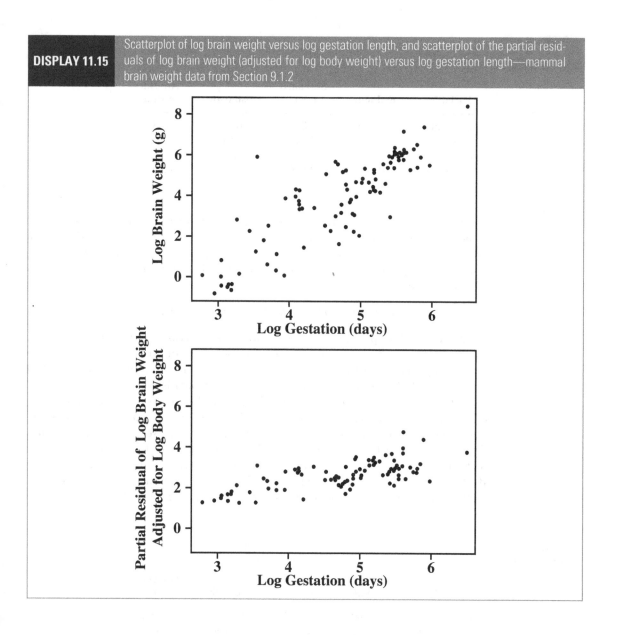

of interest may be better addressed by a plot showing the association of the two variables, after getting the effect of the third variable out of the way.

The top scatterplot in Display 11.15 shows log brain weight versus log gestation length for the 96 species of mammals examined in the study discussed in Section 9.1.2. An important question of interest is whether an association exists between brain weight and gestation, after body weight is accounted for. The multiple regression coefficient addresses this directly, but a graphical display in conjunction with it would be helpful.

Suppose that

$$\mu\{lbrain \mid lbody, lgest\} = \beta_0 + \beta_1 lbody + f(lgest),$$

where *lbrain*, *lbody*, and *lgest* represent the logarithms of brain weight, body weight, and gestation length, and where the last term is some unspecified function of *lgest*. The purpose of a *partial residual plot* is to explore the nature of the function $f(lgest)$, including whether it can be adequately approximated by a single linear term $\beta_2 lgest$ and whether the estimated relationship might be affected by one or several influential cases.

In this setup,

$$f(lgest) = \mu\{lbrain \mid lbody, lgest\} - (\beta_0 + \beta_1 lbody),$$

so it would be useful to plot $lbrain - (\beta_0 + \beta_1 lbody)$ versus *lgest* to visually explore the function $f(lgest)$. Since the β's are unknown, however, this is impossible. They can be replaced by estimates, but not just any estimates will work. It would not be appropriate to estimate them by the regression of *lbrain* on *lbody* alone, since the necessary coefficients β_0 and β_1 are those in the regression of *lbrain* on *lbody* and *lgest*. On the other hand, since $f(lgest)$ is unspecified, it is unclear how to include it in the model.

Partial Residuals

The idea behind a partial residual plot is to approximate $f(lgest)$ with the linear function $\beta_2 lgest$. This may be a crude approximation, but it is often good enough to estimate the correct β_0 and β_1 for plotting purposes. The following steps are used to draw a *partial residual plot* of *lbrain* versus *lgest*, adjusting for *lbody*:

1. Obtain the estimated coefficients in the linear regression of *lbrain* on *lbody* and *lgest*: $\mu\{lbrain \mid lbody, lgest\} = \beta_0 + \beta_1 lbody + \beta_2 lgest$.
2. Compute the *partial residuals* as $pres = lbrain - \hat{\beta}_0 - \hat{\beta}_1 lbody$.
3. Plot the partial residuals versus *lgest*.

A partial residual plot is shown in the lower scatterplot of Display 11.15. After getting the effect of log body weight out of the way, less of an association remains between log brain weight and log gestation, but some association does persist and a linear term should be adequate to model it.

When the effects of many other explanatory variables need to be "subtracted," an easier calculation formula is available. Steps 1 and 2 of the method just described can be replaced with these:

1. Obtain the residuals, *res*, from the fit to the linear regression of *lbrain* on *lbody* and *lgest* (include all explanatory variables—those whose effects are to be subtracted, and the one that is to be plotted).
2. Compute the partial residuals as $pres = res + \hat{\beta}_2 lgest$ (the residuals from the fit with all explanatory variables plus the estimated component associated with the effect of the explanatory variable under question).

The meaning of the partial residuals is clearer in the first version, but the calculation is often more straightforward with the second. Because of this calculating formula, partial residual plots are sometimes referred to as *component plus residuals plots*.

Notes About Partial Residuals

When Should Partial Residual Plots Be Used? Partial residuals are primarily useful when analytical interest centers on one explanatory variable whose effect is expected to be small relative to the effects of others. They are also useful when uncertainty exists about a particular explanatory variable that needs to be modeled carefully or when the underlying explanation for why an observation is influential on the estimate of a single coefficient needs to be understood.

Augmented Partial Residuals. Rather than using $\beta_2 lgest$ as an approximation to $f(lgest)$, some statisticians prefer to use $\beta_2 lgest + \beta_3 lgest^2$. Here, partial residuals are obtained just as in the preceding algorithms, except that $lgest^2$ is also included as an explanatory variable in step 1. (In step 2 of the component-plus-residual version, $pres = res + \hat{\beta}_2 lgest + \hat{\beta}_3 lgest^2$.) If they are equally convenient to use, the augmented partial residuals are preferred. In many cases, however, the difference between the partial residual and the augmented partial residual is slight.

Example—Blood–Brain Barrier

The residual plot in Display 11.6 indicated some potential outliers, but further investigation does not show that these points are influential in determining the structure of the model or in answering the questions of interest (see Exercise 11.18).

The key explanatory variables are the indicator variable for whether the rat received the disruption infusion or the control infusion, the length of time after infusion that the rat was sacrificed, and the interaction of these. The additional covariates should be given a chance to be included in the model, for two reasons. First (and most importantly), since randomization was not used, it behooves the researchers to demonstrate that the differences in treatment effects cannot be explained by differences in the types of rats that received the various treatments. Second, even if randomization had been used, including important covariates can yield higher resolution. If the covariates have some additional association with the response, smaller standard errors and more powerful tests should result from their inclusion.

Among the covariates, sex and days after inoculation are associated with both the response and the design variables. To some extent the effects of these variables are confounded, since their effects on the response cannot be separated. On the other hand, the effects of the design variables can be examined after the covariates are accounted for, and the effects of the covariates can be examined after the design variables are accounted for. This is shown graphically in the partial residual plots of Display 11.16.

The top scatterplot indicates that the relationship between the response and the design variables (sacrifice time and treatment) is much the same when the

DISPLAY 11.16 Some partial residual plots for the blood–brain barrier data, with log of antibody concentration ratio (brain tumor-to-liver) as response

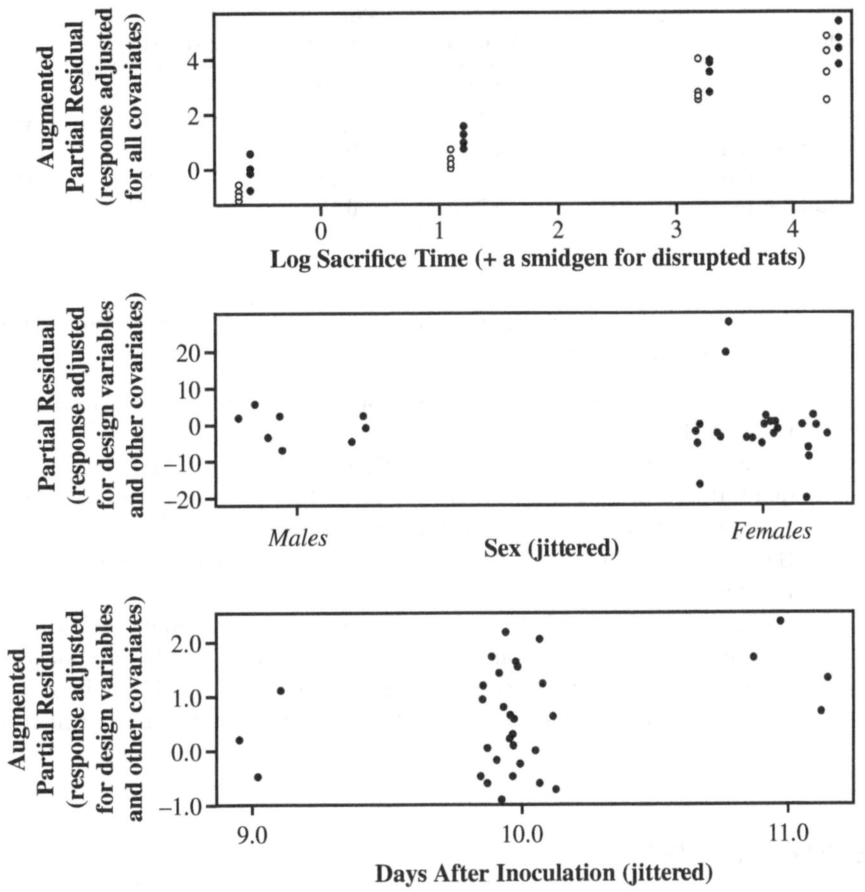

effects of the covariates are included as when they are ignored (Display 11.5). The lower two plots show that, after the effects of the design variables are accounted for, little evidence exists of a sex effect, although slight visual evidence exists of a days-after-inoculation effect.

This conclusion is further investigated through model fitting. A search through possible models that contain covariates shows that sex and days after inoculation (treated as a factor) are the only ones associated with the response. When the design variables are included as well, three conclusions are supported:

1. The covariates are not significant when the design variables are also included in the model.
2. The design variables *are* significant when the covariates are also included in the model.

DISPLAY 11.17 Results from the regression of log ratio of antibody concentration (brain tumor-to-liver) on sacrifice time (treated as a factor) and treatment

Variable	Estimate	Standard error	*t*-statistic	Two-sided *p*-value
Constant	−4.302	0.205	−21.01	<0.0001
Indicator for time = 3	1.134	0.252	4.50	0.0001
Indicator for time = 24	4.257	0.259	16.43	<0.0001
Indicator for time = 72	5.154	0.259	19.89	<0.0001
Indicator for treatment = BD	0.797	0.183	4.35	0.0002

3. The conclusions regarding the design variables depend very little on whether the covariates are in the model.

These results suggest that the conclusions can be based satisfactorily on the model without the covariates.

Since the effect of log sacrifice time is not linear (and since the addition of a quadratic term does not remedy the lack-of-fit), sacrifice time is treated as a factor with four levels. Therefore, the final model used to estimate the treatment effect has the following terms: *TIME*+*TREAT*. The estimates and standard errors are shown in Display 11.17. The coefficient of the indicator variable for the blood–brain barrier disruption treatment is 0.797. So, expressed in accordance with the interpretation for log-transformed responses, the median ratio of antibody concentration in the brain tumor to antibody concentration in the liver is estimated to be exp(0.797) = 2.22 times greater for the blood–brain barrier diffusion treatment than for the control.

11.6 RELATED ISSUES

11.6.1 Weighted Regression for Certain Types of Nonconstant Variance

Although nonconstant variance can sometimes be corrected by a transformation of the response, in many situations it cannot. If enough information is known about the form of the nonconstant variance, the method of *weighted least squares* may be used.

The *weighted regression* model, written here with two explanatory variables, is

$$\mu\{Y_i \mid X_{1i}, X_{2i}\} = \beta_0 + \beta_1 X_{1i} + \beta_2 X_{2i}$$

$$\mathrm{Var}\{Y_i \mid X_{1i}, X_{2i}\} = \sigma^2 / w_i,$$

where the w_i's are known constants called *weights* (because cases with larger w_i's have smaller variances and should be weighted more in the analysis).

This model arises in at least three practical situations:

1. *Responses are estimates; SEs are available.* Sometimes the response values are measurements whose estimated standard deviations, $\text{SE}(Y_i)$, are available. In the preceding model, the w_i's are taken to be $1/[\text{SE}(Y_i)]^2$; that is, the responses with smaller standard errors should receive more weight.
2. *Responses are averages; only the sample sizes are known.* If the responses are averages from samples of different sizes and if the ordinary regression model applies for the individual observations (the ones going into the average), then the weighted regression model applies to the averages, with weights equal to the sample sizes. The averages based on larger samples are given more weight.
3. *Variance is proportional to X.* Sometimes, while the regression of a response on an explanatory variable is a straight line, the variance increases with increases in the explanatory variable. Although a log transformation of the response might correct the nonconstant variance, it would induce a nonlinear relationship. A weighted regression model, with $w_i = 1/X_i$ (or possibly $w_i = 1/X_i^2$) may be preferable.

The weighted regression model can be estimated by *weighted least squares* within the standard regression procedure in most statistical computing programs. The estimated regression coefficients are chosen to minimize the weighted sum of squared residuals (see Exercise 21 for the calculus). It is necessary for the user to specify the response, the explanatory variables, and the weights.

11.6.2 The Delta Method

When, as in the alcohol metabolism study, there is a quantity of interest that is a nonlinear function of model parameters, calculating a standard error for the estimate of the quantity requires advanced methods. One such method—the *delta method*—requires some calculus and is therefore presented as an optional topic.

Taking the alcohol metabolism study's example, suppose interest centers on the parameter $\theta = \beta_1/(\beta_1 + \beta_2)$, where estimates of β_1 and β_2 are available. Substituting the estimates into the equation for θ produces an estimate for θ. Two inputs are required for calculating its standard error: (1) the variance–covariance matrix of the β-estimates, which should be available from the computer, and (2) the partial derivatives of θ with respect to each of the β's. Display 11.18 illustrates how these pieces combine to produce the standard error for this θ.

11.6.3 Measurement Errors in Explanatory Variables

Sometimes a theoretical model specifies that the mean response depends on certain explanatory variables that cannot be measured directly. This is called the *errors-in-variables problem.* If, for example, a study is examining the relationship between blood cholesterol (Y) and the dietary intake of polyunsaturated fat (X), and if the intake of polyunsaturated fat is estimated from a questionnaire individuals supply on what they eat in a typical week, then the questionnaire results will not measure X precisely.

DISPLAY 11.18 The delta method for calculating the standard error of a nonlinear function of parameter estimates, applied to the alcohol metabolism study

① *Express the parameter of interest, θ, as a function of parameters estimated by computer analysis.*

$$\theta = \frac{\beta_1}{\beta_1 + \beta_2}$$

② *Obtain β-estimates and their variance–covariance matrix from the computer.*

	Estimate	***Variance–Covariance Matrix***	
β_1	1.5989	0.01561	−0.01561
β_2	−0.8732	−0.01561	0.03027

$\hat{\text{Var}}(\hat{\beta}_1)$ $\hat{\text{Cov}}(\hat{\beta}_1, \hat{\beta}_2)$ $\hat{\text{Var}}(\hat{\beta}_2)$

③ *Caclulate and estimate the partial derivatives of θ with respect to the β's.*

$$\theta_1 = \frac{\partial\theta}{\partial\beta_1} = \frac{\beta_2}{(\beta_1+\beta_2)^2} \cong \frac{-0.8732}{(1.5989-0.8732)^2} = -1.6581$$

$$\theta_2 = \frac{\partial\theta}{\partial\beta_2} = \frac{-\beta_1}{(\beta_1+\beta_2)^2} \cong \frac{-1.5989}{(1.5989-0.8732)^2} = -3.036$$

④ *The estimate of θ has approximate variance*

$$\text{Var}(\hat{\theta}) = \sum_i \theta_i^2 \text{Var}(\hat{\beta}_i) + 2\sum_{i<j} \theta_i\theta_j \text{Cov}(\hat{\beta}_i, \hat{\beta}_j)$$

$$\cong (-1.6581)^2(0.01561) + (-3.0360)^2(0.03027) + 2(-1.6581)(-3.0362)(-0.01561) = 0.1648$$

⑤ *The (approximate) standard error of the estimate is the square root* ... $= \sqrt{0.1648} = 0.4060$

If the purpose of the regression is prediction, and if the prediction of future responses is to be based on measured explanatory variables that have the same kind of measurement errors as the ones used to estimate the regression, then measurement errors in explanatory variables do not present a problem. The model of interest in this case is the one for the regression of the response on the measured explanatory variables. The regression on the exact explanatory variables is not relevant.

For other purposes, however, measurement errors in explanatory variables present a problem. The least squares estimates based on the imprecisely measured explanatory variables are biased estimates of the coefficients in the regression of the response on the precisely measured explanatory variables. If there is a single explanatory variable, the slope estimate tends to be closer to zero than it should be. If there are many explanatory variables, some of which are measured with error, the individual coefficients (including the coefficients of variables that are free of measurement error) may be over- or underestimated, depending on the correlation structure of the explanatory variables.

Alternatives to least squares that account for measurement errors in explanatory variables do exist, but they are sophisticated and require extra information about the measurement errors (for example, that replicate measurements are available on a subset of subjects). The advice of a professional statistician may be required for this problem.

11.7 SUMMARY

This chapter focuses on the first three steps—exploring the data, fitting a model, and checking the model—of the strategy for data analysis using statistical models (Display 9.9). After initial exploration, the analyst fits a tentative model and uses the residuals to check on the need for transformation or outlier action. The exploration continues by entertaining various models and testing terms within the model. There is not necessarily any "best" or "final" model. Often different questions of interest are addressed through different models. In the blood–brain barrier data set, for example, one of the questions of interest is investigated via interaction terms. When these are found to be insignificant, they are dropped in favor of a different model for making inferences about the treatment effect.

Alcohol Metabolism Study

Several questions are asked of these data. For two-sample questions like "Does the first-pass metabolism of alcohol differ between males and females?" the rank-sum test is used to avoid considering the outliers. To investigate whether metabolism differs between males and females once the effect of gastric AD activity has been accounted for, a regression model for metabolism as a function of gastric AD activity and an indicator variable for sex should be used. Finding a model is made difficult by the need to consider quite a few terms on the basis of a small data set, as well as by the influence of two observations with atypical values of the explanatory variable. Since the data are not capable of resolving a regression relationship for gastric AD activities larger than 3, the two extreme cases are set aside and the subsequent analysis and conclusions are restricted to the reduced range. Within that reduced range (and ignoring the effect of alcoholism), either of two models can be used: the parallel regression lines model or the regression model forced through the origin with different slopes for males and females. The latter is selected for inference primarily because it seems somewhat more likely to apply in the broader range (and because it results in a smaller estimate of residual variation).

Blood–Brain Barrier Study

The lack of random assignment of the rats to the treatment groups is a major concern. In partial compensation for this design flaw, tentative regression models include potentially confounding covariates—measured but uncontrolled characteristics of the experimental units. Although some covariate differences are detected among the treatment groups, the regression analysis is able to clarify the treatment effects after accounting for the covariates. Without randomization, however, the causal conclusions remain tied to the adequacy of the models and to the speculation that no additional unmeasured covariates that might explain the treatment effects differ between the groups.

11.8 EXERCISES

Conceptual Exercises

1. Alcohol Metabolism. The subjects in the study were given alcohol on two consecutive days. On one of the days it was administered orally and on the other it was administered intravenously. The type of administration given on the first day was decided by a random mechanism. Why was this precaution taken?

2. Alcohol Metabolism. Here are two models for explaining the mean first-pass metabolism:

$$\text{Model 1: } \beta_0 + \beta_1 gast + \beta_2 fem$$
$$\text{Model 2: } \beta_0 + \beta_1 gast + \beta_2 gast \times fem.$$

(a) Why are there no formal tools for comparing these two models? (b) For a given value of gastric activity, what is the mean first-pass metabolism for men minus the mean first-pass metabolism for women (i) from Model 1? (ii) from Model 2?

3. Alcohol Metabolism. What would be the meaning of a third-order interactive effect of gastric activity, sex, and alcoholism on the mean first-pass metabolism?

4. Blood–Brain Barrier. (a) How should rats have been randomly assigned to treatment groups? How many treatment groups were there? What is the name of this type of experimental design and this type of treatment structure? (b) How should rats have been randomly assigned to treatments if the researchers suspected that the sex of the rat might be associated with the response? What is the name of this type of experimental design?

5. Blood–Brain Barrier. The residual plot in Display 11.6 contains two distinct groups of points: a cluster on the left and a cluster on the right. (a) Why is this? (b) Does it imply any problems with the model?

6. Robustness and resistance are different properties. (a) What is the difference? (b) Why are they both relevant when the response distribution is "long-tailed"?

7. Display 11.19 shows a hypothetical scatterplot of salary versus experience, with different codes for males and females. The male and female slopes differ significantly if the male with the most experience is included, but not if he is excluded. What course of action should be taken, and why?

DISPLAY 11.19 Hypothetical scatterplot of salary versus experience for males and females

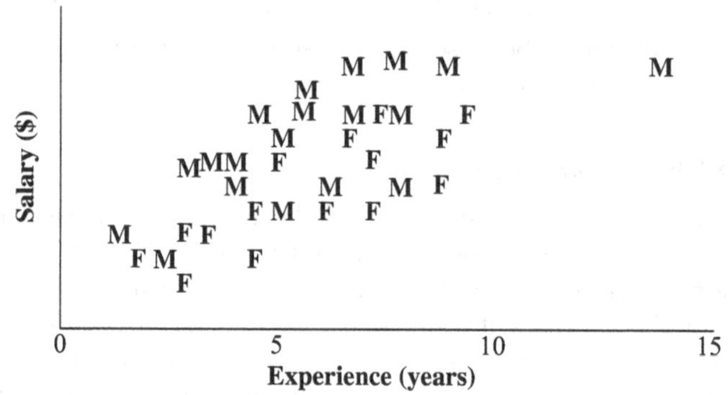

8. (a) Why does a case with large leverage have the *potential* to be influential? Why is it not necessarily influential? (b) Draw a hypothetical scatterplot of Y versus a single X, where one observation

has a high leverage but is not influential. (c) Draw a hypothetical scatterplot of Y versus a single X, where one observation has a high leverage and is influential.

9. Suppose it is desired to obtain partial residuals in order to plot Y versus X_2 after getting the effect of X_1 out of the way. The first task is to fit the regression of Y on X_1 and X_2. Let the estimated coefficients be $\hat{\beta}_0$, $\hat{\beta}_1$, and $\hat{\beta}_2$, and let the residuals from this fit be represented by res_i. The definition of the ith partial residual is $pres_i = Y_i - \hat{\beta}_0 - \hat{\beta}_1 X_{1i}$. The alternative computational formula is $res_i + \hat{\beta}_2 X_{2i}$. Why are these formulas equivalent?

Computational Exercises

10. Pollen Removal. Reconsider the pollen removal data in Exercise 3.27. (a) Draw a coded scatterplot of the log of the proportion of pollen removed relative to the proportion unremoved (if p is the proportion removed, take $Y = \log[p/(1-p)]$) versus the log of the duration of visit, with a code to distinguish queens from workers. (b) Fit the regression of Y on the two explanatory variables and their interaction, and obtain a residual plot. Does the residual plot indicate any problems? (c) Obtain a set of case influence statistics. Are there any problem observations? What is the most advisable course of action? (d) Does a significant interaction appear to exist, or can the simpler parallel regression lines model be used?

11. Chernobyl Fallout. The data in Display 11.20 are the cesium (^{134}Cs + ^{137}Cs) concentrations (in Bq/kg) in soil and in mushrooms at 17 wooded locations in Umbria, Central Italy, from August 1986 to November 1989. Researchers wished to investigate the cesium transfer from contaminated soil to plants—after the Chernobyl nuclear power plant accident in April 1986—by describing the distribution of the mushroom concentration as a function of soil concentration. (a) Obtain a set of case influence statistics from the simple linear regression fit of mushroom concentration on soil concentration. What do these indicate about case number 17? (b) Repeat part (a) after taking the logarithms of both variables. (Data from R. Borio et al., "Uptake of Radiocesium by the Mushrooms," *Science of the Total Environment* 106 (1991): 183–90.)

DISPLAY 11.20 Cesium (^{134}Cs + ^{137}Cs) concentrations (in Bq/kg) in soil and in mushrooms at 17 locations in Italy, after the Chernobyl nuclear power plant accident; first 5 of 17 rows

Mushroom	Soil
1	33
9	55
14	138
17	319
20	415

12. Brain Weights. Reconsider the brain weight data of Display 9.4. (a) Fit the regression of brain weight on body weight, gestation, and log litter size, using no transformations. Obtain a set of case-influence statistics. Is any mammal influential in this fit? (b) Refit the regression without the influential observation, and obtain the new set of case influence statistics. Are there any influential observations from this fit? (c) What lessons about the connection between the need for a log transformation and influence can be discerned?

13. Brain Weights. Identifying which mammals have larger brain weights than were predicted by the regression model might point the way to further variables that can be examined. Fit the regression of log brain weight on log body weight, log gestation, and log litter size, and compute the studentized residuals. Which mammals have substantially larger brain weights than were predicted by the model? Do any mammals have substantially smaller brain weights than were predicted by the model?

14. Corn Yield and Rainfall. Reconsider the data in Exercise 9.15. Fit the regression of corn yield on rainfall, rainfall-squared, and year. (a) Obtain the partial residuals of corn yield, adjusted for rainfall, and plot them versus year. (b) Obtain the augmented partial residual of corn yield, adjusted for year, and plot these values versus rainfall. (c) In your opinion, do these plots provide any clarification over the ordinary scatterplots?

15. Election Fraud. Reconsider the disputed election data from Exercise 8.20. (a) Draw a scatterplot of the Democratic percentage of absentee votes versus the Democratic percentage of machine votes for the 22 elections. Fit the simple linear regression of Democratic percentage of absentee votes on the Democratic percentage of machine votes and include the line on the plot. (b) Find the *internally* studentized residual for case number 22 from this fit. (c) Find the *externally* studentized residual for case number 22 from this fit. (d) Are the externally and internally studentized residuals for case 22 very different? If so, what can explain the difference? (e) What do the studentized residuals for case 22 indicate about the unusualness of the absentee ballot percentage for election 22 relative to the pattern of absentee and machine percentages established from the other 21 elections?

16. First-Pass Metabolism. Calculate the leverage, the studentized residual, and Cook's Distance for the 32nd case. Use the model with gastric activity, a sex indicator variable, and the interaction of these two.

17. Blood–Brain Barrier. (a) Using the data in Display 11.4, compute "jittered" versions of treatment, days after inoculation, and an indicator variable for females by adding small random numbers to each (uniform random numbers between −0.15 and 0.15 work well). (b) Obtain a matrix of the correlation coefficients among the same five variables (not jittered!). (c) In pencil, write the relevant correlation (two digits is enough) in a corner of each of the scatterplots in the matrix of scatterplots. (d) On the basis of this, what can be said about the relationship between the covariates (sex and days after inoculation), the response, and the design variables (treatment and sacrifice time)?

18. Blood–Brain Barrier. Using the data in Display 11.4, fit the regression of the log response (brain tumor-to-liver antibody ratio) on all covariates, the treatment indicator, and sacrifice time, treated as a factor with four levels (include three indicator variables, for sacrifice time = 3, 24, and 72 hours). (a) Obtain a set of case influence statistics, including a measure of influence, the leverage, and the studentized residual. (b) Discuss whether any influential observations or outliers occur with respect to this fit.

19. Blood–Brain Barrier. (a) Using the data in Display 11.4, fit the regression of the log response (brain tumor-to-liver antibody ratio) on an indicator variable for treatment and on sacrifice time treated as a factor with four levels (include three indicator variables, for sacrifice time = 3, 24, and 72 hours). Use the model to find the estimated mean of the log response at each of the eight treatment combinations (all combinations of the two infusions and the four sacrifice times). (b) Let X represent log of sacrifice time. Fit the regression of the log response on an indicator variable for treatment, X, X^2, and X^3. Use the estimated model to find the estimated mean of the log response at each of the eight treatment combinations. (c) Why are the answers to parts (a) and (b) the same?

20. Warm-Blooded *T. Rex*? The data in Display 11.21 are the isotopic composition of structural bone carbonate (X) and the isotopic composition of the coexisting calcite cements (Y) in 18 bone samples from a specimen of the dinosaur *Tyrannosaurus rex*. Evidence that the mean of Y is positively associated with X was used in an argument that the metabolic rate of this dinosaur resembled warm-blooded more than cold-blooded animals. (Data from R. E. Barrick and W. J. Showers, "Thermophysiology of *Tyrannosaurus rex*: Evidence from Oxygen Isotopes," *Science* 265 (1994): 222–24.) (a) Examine the effects on the p-value for significance of regression and on

R-squared of deleting (i) the case with the smallest value of X, and (ii) the two cases with the smallest values of X. (b) Why does R-squared change so much? (c) Compute the case influence statistics, and discuss interesting cases. (d) Recompute the case statistics when the case with the smallest X is deleted. (e) Comment on the differences in the two sets of case statistics. Why may pairs of influential observations not be found with the usual case influence statistics? (f) What might one conclude about the influence of the two unusual observations in this data set?

DISPLAY 11.21 Isotopic composition of carbonate and of calcite cements in 18 samples of bone from a *Tyrannosaurus rex* specimen

Carbonate	Calcite
21.3	20.5
23.6	23.2
26.7	27.0
26.8	28.7
27.0	27.9
27.1	27.9
27.2	29.0
27.3	27.7
27.5	27.6
27.6	29.6
28.0	28.0
28.0	29.2
28.2	27.5
29.3	28.4
29.4	28.5
29.4	29.3
29.4	30.0
29.5	31.0

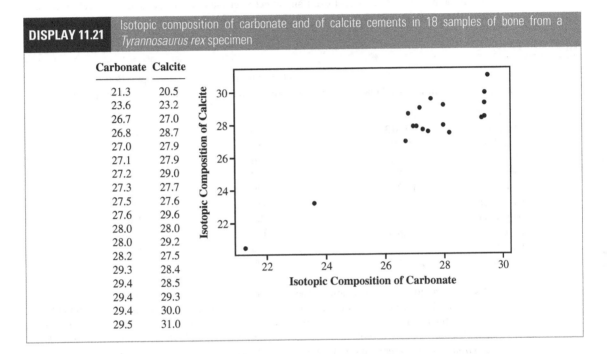

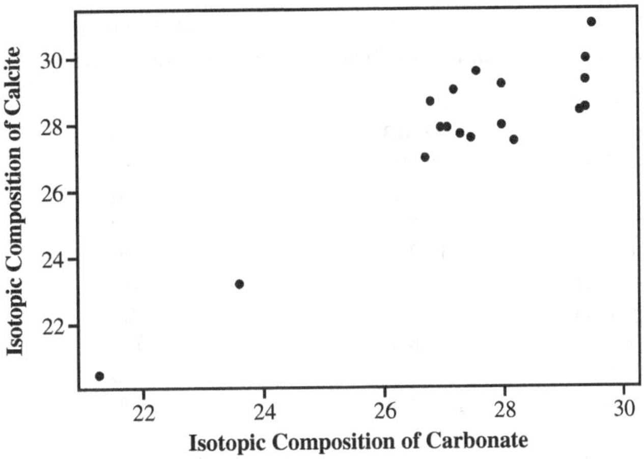

21. **Calculus Problem.** The weighted least squares problem in multiple linear regression is to find the parameter values that minimize the weighted sum of squares,

$$SS_w(\beta_0, \beta_1, \ldots, \beta_p) = \sum_{i=1}^{n} w_i (Y_i - \beta_0 - \beta_1 X_{1i} - \cdots - \beta_p X_{pi})^2,$$

with all $w_i > 0$. (a) Setting the partial derivatives of SS_w with respect to each of the parameters equal to zero, show that the solutions must satisfy this set of normal equations:

$$\begin{array}{llllll}
\beta_0 \sum w_i & + \ \beta_1 \sum w_i X_{1i} & + \ \beta_2 \sum w_i X_{2i} & + \cdots + & \beta_p \sum w_i X_{pi} & = \sum w_i Y_i \\
\beta_0 \sum w_i X_{1i} + & \beta_1 \sum w_i X_{1i}^2 & + \beta_2 \sum w_i X_{1i} X_{2i} & + \cdots + & \beta_p \sum w_i X_{1i} X_{pi} & = \sum w_i X_{1i} Y_i \\
\beta_0 \sum w_i X_{2i} & + \beta_1 \sum w_i X_{2i} X_{1i} + & \beta_2 \sum w_i X_{2i}^2 & + \cdots + & \beta_p \sum w_i X_{2i} X_{pi} & = \sum w_i X_{2i} Y_i \\
\vdots & \vdots & \vdots & & \vdots & \vdots \\
\beta_0 \sum w_i X_{pi} & + \beta_1 \sum w_i X_{pi} X_{1i} & + \beta_2 \sum w_i X_{pi} X_{2i} & + \cdots + & \beta_p \sum w_i X_{pi}^2 & = \sum w_i X_{pi} Y_i .
\end{array}$$

(b) Show that solutions to the normal equations minimize SS.

Data Problems

22. **Deforestation and Debt.** It has been theorized that developing countries cut down their forests to pay off foreign debt. Two researchers examined this belief using data from 11 Latin American nations. (Data from R. T. Gullison and E. C. Losos, "The Role of Foreign Debt in Deforestation in Latin America," *Conservation Biology* 7(1) (1993): 140–7.) The data on debt, deforestation, and population appear in Display 11.22. Does the evidence significantly support the theory that debt causes deforestation? Does debt exert any effect after the effect of population on deforestation is accounted for? Describe the effect of debt, after accounting for population.

DISPLAY 11.22 Foreign debt, annual deforestation area, and population for 11 Latin American countries

Country	Debt (millions of dollars)	Deforestation (thousands of hectares)	Population (thousands of people)
Brazil	86,396	12,150	128,425.0
Mexico	79,613	2,680	74,194.5
Ecuador	6,990	1,557	8,750.5
Colombia	10,101	1,500	27,254.0
Venezuela	24,870	1,430	16,170.5
Peru	10,707	1,250	18,496.5
Nicaragua	3,985	550	3,021.5
Argentina	36,664	400	29,400.5
Bolivia	3,810	300	5,970.5
Paraguay	1,479	250	3,424.5
Costa Rica	3,413	90	2,439.5

23. **Air Pollution and Mortality.** Does pollution kill people? Data in one early study designed to explore this issue came from five Standard Metropolitan Statistical Areas (SMSA) in the United States, obtained for the years 1959–1961. (Data from G. C. McDonald and J. A. Ayers, "Some Applications of the 'Chernoff Faces': A Technique for Graphically Representing Multivariate Data," in *Graphical Representation of Multivariate Data*, New York: Academic Press, 1978.) Total age-adjusted mortality from all causes, in deaths per 100,000 population, is the response variable. The explanatory variables listed in Display 11.23 include mean annual precipitation (in inches); median number of school years completed, for persons of age 25 years or older; percentage of 1960 population that is nonwhite; relative pollution potential of oxides of nitrogen, NO_X; and relative pollution potential of sulfur dioxide, SO_2. "Relative pollution potential" is the product of the tons emitted per day per square kilometer and a factor correcting for SMSA dimension and exposure. The first three explanatory variables are a subset of climate and socioeconomic variables in the original data set. (*Note*: Two cities—Lancaster and York—are heavily populated by members of the Amish religion,

DISPLAY 11.23 Air pollution and mortality data for 5 U.S. cities, 1959–1961; first 5 of 60 rows

City	Mortality	Precipitation	Education	Nonwhite	NO_x	SO_2
San Jose, CA	790.73	13	12.2	3.0	32	3
Wichita, KS	823.76	28	12.1	7.5	2	1
San Diego, CA	839.71	10	12.1	5.9	66	20
Lancaster, PA	844.05	43	9.5	2.9	7	32
Minneapolis, MN	857.62	25	12.1	2.0	11	26

who prefer to teach their children at home. The lower years of education for these two cities do not indicate a social climate similar to other cities with similar years of education.) Is there evidence that mortality is associated with either of the pollution variables, after the effects of the climate and socioeconomic variables are accounted for? Analyze the data and write a report of the findings, including any important limitations of this study. (*Hint*: Consider looking at case-influence statistics.)

24. Natal Dispersal Distances of Mammals. Natal dispersal distances are the distances that juvenile animals travel from their birthplace to their adult home. An assessment of the factors affecting dispersal distances is important for understanding population spread, recolonization, and gene flow—which are central issues for conservation of many vertebrate species. For example, an understanding of dispersal distances will help to identify which species in a community are vulnerable to the loss of connectedness of habitat. To further the understanding of determinants of natal dispersal distances, researchers gathered data on body weight, diet type, and maximum natal dispersal distance for various animals. Shown in Display 11.24 are the first 6 of 64 rows of data on mammals. (Data from G. D. Sutherland et al., "Scaling of Natal Dispersal Distances in Terrestrial Birds and Mammals," *Conservation Ecology* 4(1) (2000): 16.) Analyze the data to describe the distribution of maximum dispersal distance as a function of body mass and diet type. Write a summary of statistical findings.

DISPLAY 11.24 Natal dispersal distances and explanatory variables for 64 mammals; first 6 of 64 rows

Species	Body mass (kg)	Diet type	Maximum dispersal distance (km)
1. *Didellphis virginianus*	2.41	Omnivore	5.15
2. *Phascogale tapotafa*	0.17	Carnivore	6.80
3. *Trichosurus vulpecula*	2.93	Carnivore	12.80
4. *Sorex araneus*	0.004	Carnivore	0.87
5. *Scapanus townsendii*	0.15	Omnivore	0.86
6. *Ursus americanus*	104.45	Omnivore	225.00

25. Ingestion Rates of Deposit Feeders. The ingestion rates and organic consumption percentages of deposit feeders were considered in Exercise 9.21. The data set ex1125 repeats these data, but this time with three additional bivalve species included (the last three). The researcher wished to see if ingestion rate is associated with the percentage of organic matter in food, after accounting for animal weight, but was unsure about whether bivalves should be included in the analysis. Analyze the data to address this question of interest. (Data from L. M. Cammen, "Ingestion Rate: An Empirical Model for Aquatic Deposit Feeders and Detritivores," *Oecologia* 44 (1980): 303–10.)

26. Metabolism and Lifespan of Mammals. Use the data from Exercise 8.26 to describe the distribution of mammal lifespan as a function of metabolism rate, after accounting for the effect of body mass. One theory holds that for a given body size, animals that expend less energy per day (lower metabolic rate) will tend to live longer.

Answers to Conceptual Exercises

1. By randomly determining order, the researchers avoid bias in determining first-pass metabolism that would occur if an order effect existed.

2. (a) Neither of the models is a subset of the others, so it is impossible to test a term in a "full model" to see whether the "reduced model" does just as well. (Incidentally, the models explain the data about equally well.) (b) (i) β_2 (ii) $\beta_2 gast$.

3. It would mean that the effect of gastric AD activity on first-pass metabolism differed between males and females, and that the amount of the sex difference differed between alcoholics and nonalcoholics. (Two-factor interactions are hard enough to describe in words. Three- and higher-factor interactions typically involve very long and confusing sentences. A theory without interactions is obviously simpler than one with interactions.)

4. (a) Rats should have been randomly assigned to one of eight groups, corresponding to the eight combinations of treatment and sacrifice time. This is a completely randomized design with factorial (2 × 4) treatment structure. (b) If the response is suspected to be related to the sex of the rat, a randomized block experiment should be performed. The procedure in part (a) can be followed separately for male and female rats.

5. (a) The fitted values are noticeably larger for the rats in the groups with longer sacrifice times. (b) No.

6. (a) *Robustness* describes the extent to which inferential statements are correct when specific assumptions are violated. *Resistance* describes the extent to which the results remain unchanged when a small portion of the data is changed, perhaps drastically (and therefore describes the extent to which individual observations can be influential). (b) When the distribution is long-tailed, the robustness against departures from normality cannot be guaranteed. Put another way, there are likely to be outliers, which can have undue influence on the results, since the least squares method is not resistant.

7. In the absence of further knowledge, it is safest to exclude the very experienced male from the data set and restrict conclusions about the differences between male and female salaries to individuals with 10 years of experience or fewer. It may be that males and females have different slopes over the wider range of experience, or it may be that the straight line is not an adequate model over the wider range of experiences. There is certainly insufficient data in the more-than-10-years range to resolve this issue.

8. (a) A large leverage indicates that a case occupies a position in the "X-space" that is not densely populated. It therefore plays a large role in shaping the estimated regression model in that region. Since it does not share its role (much) with other nearby points, it must draw the regression surface close to it. For this reason it has a high potential for influence. If, however, the fit of the model without that point is about the same as the fit of the model with it, it is not influential.

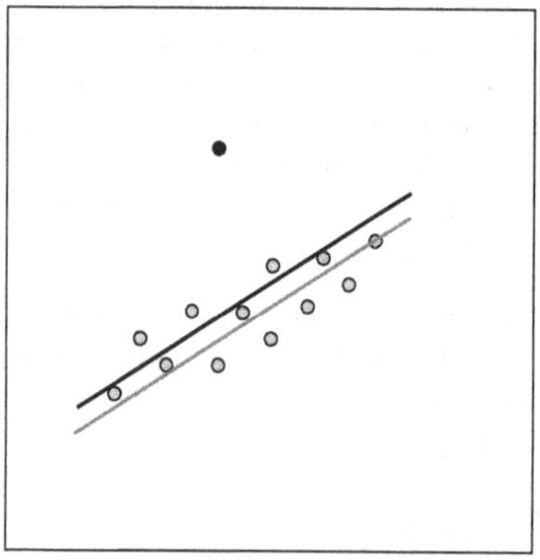

(b) The case with high leverage (•) exerts no influence on the slope. The line without the case (–) only has its intercept changed when the case is included (–).

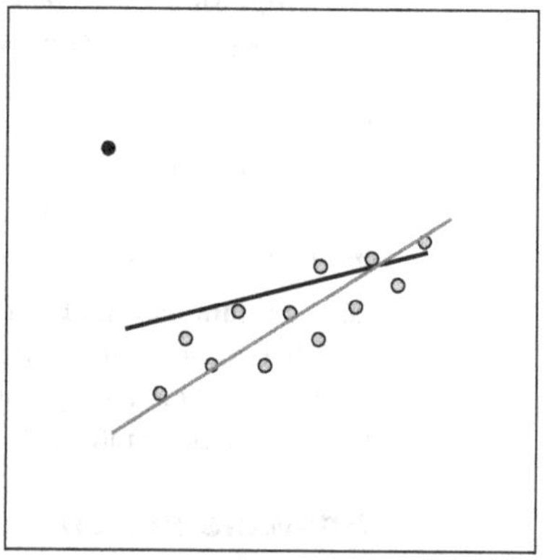

(c) When the case with high leverage has its explanatory variable value outside those of the other cases, the slope of the regression can be changed dramatically.

9. Since $res_i = Y_i - \hat{\beta}_0 - \hat{\beta}_1 X_{1i} - \hat{\beta}_2 X_{2i}$, the alternative calculating formula is $Y_i - \hat{\beta}_0 - \hat{\beta}_1 X_{1i} - \hat{\beta}_2 X_{2i} + \hat{\beta}_2 X_{2i}$. The last two terms cancel, leaving the original definition.

CHAPTER 12

Strategies for Variable Selection

There are two good reasons for paring down a large number of explanatory variables to a smaller set. The first reason is somewhat philosophical: Simplicity is preferable to complexity. Thus, redundant and unnecessary explanatory variables should be excluded on principle. The second reason is more concrete: Unnecessary terms in the model yield less precise inferences.

Various statistical tools are available for choosing a good subset from a large pool of explanatory variables. Discussed in this chapter are sequential variable-selection techniques, and comparisons among all possible subsets through examination of Cp and the Bayesian Information Criterion. The most important practical lessons are that the variable selection process should be sensitive to the objectives of the study and that the particular subset chosen is relatively unimportant.

12.1 CASE STUDIES

12.1.1 State Average SAT Scores—An Observational Study

When, in 1982, average Scholastic Achievement Test (SAT) scores were first published on a state-by-state basis in the United States, the huge variation in the scores was a source of great pride for some states and of consternation for others. Average scores ranged from a low of 790 (out of a possible 1,600) in South Carolina to a high of 1,088 in Iowa. This 298-point spread dwarfed the 20-year national decline of 80 points. Two researchers set out to "assess the extent to which the compositional/demographic and school-structural characteristics are implicated in SAT differences." (Data from B. Powell and L. C. Steelman, "Variations in State SAT Performance: Meaningful or Misleading?" *Harvard Educational Review* 54(4) (1984): 389–412.)

Display 12.1 contains state averages of the total SAT (verbal + quantitative) scores, along with six variables that may be associated with the SAT differences among states. Some explanatory variables come from the Powell and Steelman article, while others were obtained from the College Entrance Examination Board (by Robert Powers). The variables are the following: *Takers* is the percentage of the total eligible students (high school seniors) in the state who took the exam; *income* is the median income of families of test-takers, in hundreds of dollars; *years* is the average number of years that the test-takers had formal studies in social sciences, natural sciences, and humanities; *public* is the percentage of the test-takers who attended public secondary schools; *expend* is the total state expenditure on secondary schools, expressed in hundreds of dollars per student; and *rank* is the median percentile ranking of the test-takers within their secondary school classes.

Notice that the states with high average SATs had low percentages of takers. One reason is that these are mostly midwestern states that administer other tests to students bound for college in-state. Only their best students planning to attend college out of state take the SAT exams. As the percentage of takers increases for other states, so does the likelihood that the takers include lower-qualified students. After accounting for the percentage of students who took the test and the median class rank of the test-takers (to adjust, somewhat, for the selection bias in the samples from each state), which variables are associated with state SAT scores? After accounting for the percentage of takers and the median class rank of the takers, how do the states rank? Which states perform best for the amount of money they spend?

Statistical Conclusion

The percentage of eligible students taking the test and the median class rank of the students taking the test explain 81.5% of the variation in state average test scores. This confirms the expectation that much of the between-states variation in SAT scores can be explained by the relative quality of the students within the state who decide to take the test. After the percentage of students in a state who take the test and the median class rank of these students are accounted for, convincing

DISPLAY 12.1 Average SAT scores by U.S. state in 1982, and possible associated factors

Rank	State	SAT	Takers	Income	Years	Public	Expend	Rank
1	Iowa	1,088	3	326	16.79	87.8	25.60	89.7
2	South Dakota	1,075	2	264	16.07	86.2	19.95	90.6
3	North Dakota	1,068	3	317	16.57	88.3	20.62	89.8
4	Kansas	1,045	5	338	16.30	83.9	27.14	86.3
5	Nebraska	1,045	5	293	17.25	83.6	21.05	88.5
6	Montana	1,033	8	263	15.91	93.7	29.48	86.4
7	Minnesota	1,028	7	343	17.41	78.3	24.84	83.4
8	Utah	1,022	4	333	16.57	75.2	17.42	85.9
9	Wyoming	1,017	5	328	16.01	97.0	25.96	87.5
10	Wisconsin	1,011	10	304	16.85	77.3	27.69	84.2
11	Oklahoma	1,001	5	358	15.95	74.2	20.07	85.6
12	Arkansas	999	4	295	15.49	86.4	15.71	89.2
13	Tennessee	999	9	330	15.72	61.2	14.58	83.4
14	New Mexico	997	8	316	15.92	79.5	22.19	83.7
15	Idaho	995	7	285	16.18	92.1	17.80	85.9
16	Mississippi	988	3	315	16.76	67.9	15.36	90.1
17	Kentucky	985	6	330	16.61	71.4	15.69	86.4
18	Colorado	983	16	333	16.83	88.3	26.56	81.8
19	Washington	982	19	309	16.23	87.5	26.53	83.2
20	Arizona	981	11	314	15.98	80.9	19.14	84.3
21	Illinois	977	14	347	15.80	74.6	24.41	78.7
22	Louisiana	975	5	394	16.85	44.8	19.72	82.9
23	Missouri	975	10	322	16.42	67.7	20.79	80.6
24	Michigan	973	10	335	16.50	80.7	24.61	81.8
25	West Virginia	968	7	292	17.08	90.6	18.16	86.2
26	Alabama	964	6	313	16.37	69.6	13.84	83.9
27	Ohio	958	16	306	16.52	71.5	21.43	79.5
28	New Hampshire	925	56	248	16.35	78.1	20.33	73.6
29	Alaska	923	31	401	15.32	96.5	50.10	79.6
30	Nevada	917	18	288	14.73	89.1	21.79	81.1
31	Oregon	908	40	261	14.48	92.1	30.49	79.3
32	Vermont	904	54	225	16.50	84.2	20.17	75.8
33	California	899	36	293	15.52	83.0	25.94	77.5
34	Delaware	897	42	277	16.95	67.9	27.81	71.4
35	Connecticut	896	69	287	16.75	76.8	26.97	69.8
36	New York	896	59	236	16.86	80.4	33.58	70.5
37	Maine	890	46	208	16.05	85.7	20.55	74.6
38	Florida	889	39	255	15.91	80.5	22.62	74.6
39	Maryland	889	50	312	16.90	80.4	25.41	71.5
40	Virginia	888	52	295	16.08	88.8	22.23	72.4
41	Massachusetts	888	65	246	16.79	80.7	31.74	69.9
42	Pennsylvania	885	50	241	17.27	78.6	27.98	73.4
43	Rhode Island	877	59	228	16.67	79.7	25.59	71.4
44	New Jersey	869	64	269	16.37	80.6	27.91	69.8
45	Texas	868	32	303	14.95	91.7	19.55	76.4
46	Indiana	860	48	258	14.39	90.2	17.93	74.1
47	Hawaii	857	47	277	16.40	67.6	21.21	69.9
48	North Carolina	827	47	224	15.31	92.8	19.92	75.3
49	Georgia	823	51	250	15.55	86.5	16.52	74.0
50	South Carolina	790	48	214	15.42	88.1	15.60	74.0

evidence exists that both state expenditures (one-sided p-value < 0.0001) and years of formal study in social sciences, natural sciences, and humanities (one-sided p-value $= 0.0005$) are associated with SAT averages. Alaska had a substantially higher expenditure than other states and was excluded from the analysis.

The ranking of states after accounting for the relative quality of students in the state who take the exam (as represented by the percentage of takers and the median class rank of the students taking the exam) is shown on the left side of Display 12.2. Here, the states are ordered according to the size of their residuals in the regression of SAT scores on percentage of takers and median class rank. For example, the SAT average for New Hampshire is 49 points higher than the estimated mean SAT for states with the same percentage of students taking the exam and median class rank. This ranking and the ranking based on raw averages show some dramatic differences. South Dakota ranks second in its raw SAT scores, for example (see Display 12.1), but only 2% of its eligible students took the exam. After percentage of takers and median class rank are accounted for, South Dakota ranks 24th.

The ranking of states after additionally accounting for state expenditure is shown on the right side of Display 12.2. This ranking provides a means of gauging how states perform for the amount of money they spend (but see the cautionary notes in the next subsection regarding scope of inference). For example, Oregon ranks 31st in its unadjusted average SAT. But after accounting for the percentage of takers, their median class rankings, and the state's expenditure per student, it ranks 46th, indicating that its students performed below what might have been expected given the amount of money the state spends on education.

Scope of Inference

The participating students were *self-selected*; that is, each student included in a state's sample decided to take the SAT, as opposed to being a randomly selected student. Therefore, a state's average does not represent all its eligible students. Including the percentage of takers and the median class rank of takers in the regression models adjusts for this difference, but only to the extent that these variables adequately account for the relative quality of students taking the test. No check on this assumption is possible with the existing data. In addition, these are observational data, for which many possible confounding factors exist, so the statistical statements of significance (as for the effect of expenditure, for example) do not imply causation.

12.1.2 Sex Discrimination in Employment—An Observational Study

Display 12.3 lists data on employees from one job category (skilled, entry-level clerical) of a bank that was sued for sex discrimination. These are the same 32 male and 61 female employees, hired between 1965 and 1975, who were considered in Section 1.1.2. The measurements are of annual salary at time of hire, salary as of March 1977, sex (1 for females and 0 for males), seniority (months since first hired), age (months), education (years), and work experience prior to employment with the bank (months).

DISPLAY 12.2 State SAT scores after adjustment (in points above or below average)

SATs adjusted for % taking exam and their median class rank			SATs adjusted for % taking exam, rank, and expenditure		
Rank	State	Adjusted SAT average	Rank	State	Adjusted SAT average
1	New Hampshire	49	1	New Hampshire	67
2	Iowa	41	2	Tennessee	49
3	Montana	38	3	Vermont	39
4	Connecticut	38	4	Connecticut	28
5	Washington	34	5	Nebraska	22
6	Minnesota	34	6	Virginia	20
7	Wisconsin	31	7	Maine	18
8	Colorado	31	8	Arizona	18
9	New York	29	9	Minnesota	17
10	Kansas	29	10	North Dakota	17
11	Massachusetts	27	11	Illinois	16
12	Illinois	25	12	Ohio	15
13	Nebraska	24	13	Washington	15
14	Vermont	21	14	Iowa	14
15	North Dakota	20	15	Colorado	11
16	Tennessee	16	16	Idaho	10
17	Delaware	13	17	Utah	10
18	Maryland	12	18	Missouri	8
19	Virginia	11	19	Maryland	7
20	Ohio	11	20	New Mexico	5
21	New Mexico	8	21	South Dakota	4
22	Rhode Island	8	22	Indiana	4
23	New Jersey	8	23	Wisconsin	3
24	South Dakota	7	24	Rhode Island	3
25	Arizona	6	25	Kentucky	1
26	Pennsylvania	4	26	Montana	0
27	Missouri	4	27	Florida	−1
28	Oregon	3	28	Kansas	−1
29	Maine	2	29	Hawaii	−5
30	Michigan	−1	30	Delaware	−5
31	Wyoming	−2	31	Alabama	−5
32	Utah	−3	32	Massachusetts	−6
33	Idaho	−5	33	New Jersey	−8
34	Florida	−6	34	Oklahoma	−11
35	California	−6	35	New York	−13
36	Oklahoma	−14	36	Arkansas	−13
37	Hawaii	−19	37	Pennsylvania	−14
38	Kentucky	−23	38	Michigan	−14
39	Indiana	−24	39	California	−18
40	Nevada	−28	40	West Virginia	−19
41	West Virginia	−32	41	Texas	−22
42	Louisiana	−33	42	Georgia	−23
43	Arkansas	−34	43	Nevada	−25
44	Alabama	−38	44	Wyoming	−27
45	Texas	−40	45	Louisiana	−28
46	Georgia	−58	46	Oregon	−29
47	Mississippi	−60	47	Mississippi	−40
48	North Carolina	−61	48	North Carolina	−42
49	South Carolina	−94	49	South Carolina	−55

DISPLAY 12.3 Sex discrimination data

Beginning salary	1977 salary	Fsex (1 = F)	Seniority	Age	Education	Experience
5,040	12,420	0	96	329	15	14
6,300	12,060	0	82	357	15	72
6,000	15,120	0	67	315	15	35.5
6,000	16,320	0	97	354	12	24
6,000	12,300	0	66	351	12	56
6,840	10,380	0	92	374	15	41.5
8,100	13,980	0	66	369	16	54.5
6,000	10,140	0	82	363	12	32
6,000	12,360	0	88	555	12	252
6,900	10,920	0	75	416	15	132
6,900	10,920	0	89	481	12	175
5,400	12,660	0	91	331	15	17.5
6,000	12,960	0	66	355	15	64
6,000	12,360	0	86	348	15	25
5,100	8,940	1	95	640	15	165
4,800	8,580	1	98	774	12	381
5,280	8,760	1	98	557	8	190
5,280	8,040	1	88	745	8	90
4,800	9,000	1	77	505	12	63
4,800	8,820	1	76	482	12	6
5,400	13,320	1	86	329	15	24
5,520	9,600	1	82	558	12	97
5,400	8,940	1	88	338	12	26
5,700	9,000	1	76	667	12	90
3,900	8,760	1	98	327	12	0
4,800	9,780	1	75	619	12	144
6,120	9,360	1	78	624	12	208.5
5,220	7,860	1	70	671	8	102
5,100	9,660	1	66	554	8	96
4,380	9,600	1	92	305	8	6.25
4,290	9,180	1	69	280	12	5
5,400	9,540	1	66	534	15	122
4,380	10,380	1	92	305	12	0
5,400	8,640	1	65	603	8	173
5,400	11,880	1	66	302	12	26
4,500	12,540	1	96	366	8	52
5,400	8,400	1	70	628	12	82
5,520	8,880	1	67	694	12	196
5,640	10,080	1	90	368	12	55
4,800	9,240	1	73	590	12	228
5,400	8,640	1	66	771	8	228
4,500	7,980	1	80	298	12	8
5,400	11,940	1	77	325	12	38
5,400	9,420	1	72	589	15	49
6,300	9,780	1	66	394	12	86.5
5,160	10,680	1	87	320	12	18
5,100	11,160	1	98	571	15	115
4,800	8,340	1	79	602	8	70

DISPLAY 12.3 Sex discrimination data—continued

Beginning salary	1977 salary	Fsex (1 = F)	Seniority	Age	Education	Experience
5,400	9,600	1	98	568	12	244
4,020	9,840	1	92	528	10	44
4,980	8,700	1	74	718	8	318
5,280	9,780	1	88	653	12	107
5,700	8,280	1	65	714	15	241
4,800	8,340	1	87	647	12	163
4,800	13,560	1	82	338	12	11
5,700	10,260	1	82	362	15	51
4,380	9,720	1	93	303	12	4.5
4,380	10,500	1	89	310	12	0
5,400	10,680	0	88	359	12	38
5,400	11,640	0	96	474	12	113
5,100	7,860	0	84	535	12	180
6,600	11,220	0	66	369	15	84
5,100	8,700	0	97	637	12	315
6,600	12,240	0	83	536	15	215.5
5,700	11,220	0	94	392	15	36
6,000	12,180	0	91	364	12	49
6,000	11,580	0	83	521	15	108
6,000	8,940	0	80	686	12	272
6,000	10,680	0	87	364	15	56
4,620	11,100	0	77	293	12	11.5
5,220	10,080	0	85	344	12	29
6,600	15,360	0	83	340	15	64
5,400	12,600	0	78	305	12	7
6,000	8,940	0	78	659	8	320
5,400	9,480	0	88	690	15	359
6,000	14,400	0	96	402	16	45.5
5,700	10,620	1	88	410	15	61
5,400	10,320	1	78	584	15	51
4,440	9,600	1	97	341	15	75
6,300	10,860	1	84	662	15	231
6,000	9,720	1	69	488	12	121
5,100	9,600	1	85	406	12	59
4,800	11,100	1	87	349	12	11
5,100	10,020	1	87	508	16	123
5,700	9,780	1	74	542	12	116.5
5,400	10,440	1	72	604	12	169
5,100	10,560	1	84	458	12	36
4,800	9,240	1	84	571	16	214
6,000	11,940	1	86	486	15	78.5
4,380	10,020	1	93	313	8	7.5
5,580	7,860	1	69	600	12	132.5
4,620	9,420	1	96	385	12	52
5,220	8,340	1	70	468	12	127

Did the females receive lower starting salaries than similarly qualified and similarly experienced males? After accounting for measures of performance, did females receive smaller pay increases than males?

Statistical Conclusion

The data provide convincing evidence that the median starting salary for females was lower than the median starting salary for males, even after the effects of age, education, previous experience, and time at which the job began are taken into account (one-sided p-value <0.0001). The median beginning salary for females was estimated to be only 89% of the median salary for males, after accounting for the variables mentioned above (a 95% confidence interval for the ratio of adjusted medians is 85% to 93%). There is little evidence that the pay increases for females differed from those for males (one-sided p-value $= 0.27$ from a t-test for the difference in mean log of average annual raise; one-sided p-value $= 0.72$ after further adjustment for other variables except beginning salary; one-sided p-value $= 0.033$ after further adjustment for other variables including beginning salary).

Inferences

Since these are observational data, the usual dictum applies: No cause-and-effect inference of discrimination can be drawn. Of course that argument disallows nearly all claims of discrimination, because supporting data are always observational. To evaluate such evidence, U.S. courts have adopted a method of burden-shifting. Plaintiffs present statistical analyses as *prima facie* evidence of discrimination. If the evidence is substantial, the burden shifts to the defendants to show either that these analyses are flawed or that alternative, nondiscriminatory factors explain the differences. If the defendants' arguments are accepted, the burden shifts back to the plaintiffs to show that the defendants' analyses are wrong or that their explanations were pretextual. If plaintiffs succeed in this, the inference of discrimination stands.

12.2 SPECIFIC ISSUES RELATING TO MANY EXPLANATORY VARIABLES

12.2.1 Objectives

Several convenient tools are available for paring down large sets of explanatory variables. But without understanding what they offer, one may incorrectly suppose that the computer is uncovering some law of nature. Like any other statistical tool, these paring tools are most helpful when used to address well-defined questions of interest. Although it may be tempting to think that finding "the important set" of explanatory variables is a question of interest, the actual set chosen is usually one of several (or many) equally good sets.

Objective I: Adjusting for a Large Set of Explanatory Variables

In the sex discrimination example the objective is to examine the effect of sex after accounting for other legitimate determinants of salary. A game plan for the analysis is to begin by finding a subset of the other explanatory variables with a variable-selection technique, and then to add the indicator variable for sex into the model. The variable-selection techniques are entirely appropriate for this purpose. The set of explanatory variables chosen is evidently one of several (or perhaps many) equally useful sets. Although a particular explanatory variable may not be included in the final model, it has still been adjusted for, since it was given a chance to be in the model (prior to inclusion of the sex indicator). No interpretation is made of the particular set chosen, nor is any interpretation made of the coefficients. The coefficient of the sex indicator variable is interpreted, but this is straightforward: It represents the association between sex and salary *after* accounting for the effects of the other explanatory variables.

Objective II: Fishing for Explanation

In many studies, unfortunately, no such well-defined question has been posed. One question likely to be asked of the SAT data is vague: "Which variables are important in explaining state average SAT scores, after accounting for percentage of takers and median class rank?" Variable-selection techniques can be used to identify a set of explanatory variables, but the analyst may be tempted to attach meaning to the particular variables selected and to interpret coefficients—temptations not encountered in Objective I.

There are several reasons to exercise great caution when using regression for this purpose:

1. The explanatory variables chosen are not necessarily special. Inclusion or exclusion of individual explanatory variables is affected strongly by the correlations between them.
2. Interpreting the coefficients in a multiple regression with correlated explanatory variables is extremely difficult. For example, an estimated coefficient may have the opposite sign of the one suggested by a scatterplot, since a coefficient shows the effect of an explanatory variable on a response after accounting for the other variables in the model. It is often difficult to recognize what "after accounting for the other variables in the model" means when many variables are included.
3. A regression coefficient continues to be interpreted (technically) as a measure of the effect of one variable while holding all other variables fixed. Unfortunately, increases in one variable are almost always accompanied by changes in many others, so the situation described by the interpretation lies outside the experience provided by the data.
4. Finally, of course, causal interpretations from observational studies are always suspect. The selected variables may be related to confounding variables that are more directly responsible for the response.

Objective III: Prediction

If the purpose of the model is prediction, no interpretation of the particular set of explanatory variables chosen or their coefficients is needed, so there is little room for abuse. The variable-selection techniques are useful for showing a few models that work well. From these, the researcher selects one with explanatory variables that can be obtained conveniently for future prediction.

Different Questions Require Different Attitudes Toward the Explanatory Variables

With regard to the SAT score example, a business firm looking for a place to build a new facility may have as its objective in analyzing the data a ranking of states in terms of their ability to train students. The only relevant question to this firm is whether the raw SAT averages accurately reflect the educational training. The problem of selection bias is critical, so inclusion of the percentage of SAT takers or of the median rank is essential. All other variables reflect factors that will not affect the firm's decision, so they are not relevant. This firm could fit a model using the percentage of takers as an explanatory variable and use the residuals as a way of ranking the states.

A legislative watchdog committee, on the other hand, may have the objective of determining the impact of state expenditures on state SAT scores. It might include all variables that could affect SAT scores before including expenditures. Under this arrangement, the committee would claim an effect for expenditures only when the effect could not be attributed to some other related factor.

If prediction is the purpose, all explanatory variables that can be used in making the prediction should be considered eligible for inclusion. (Prediction is not reasonable in the SAT score example, because one would not know the explanatory variables until the SAT scores themselves were available.)

12.2.2 Loss of Precision

The precision in estimating an important regression coefficient or in predicting a future response may be decreased if too many explanatory variables are included in the model. The standard error of the estimator for a multiple regression coefficient of an explanatory variable, X, can be expressed as the standard error in the simple linear regression multiplied by the square root of the product of two factors, a *mean square ratio* and a *variance inflation factor (VIF)*. The *mean square ratio* is the ratio of estimate of residual variance from the multiple and the simple regression models. If useful predictors are added in the multiple regression model then this ratio should be less than one.

Often more important, however, is the variance inflation factor,

$$VIF = 1/(1 - R_X^2)$$

where R_X^2 is the proportion of the variation in X that is explained by its linear relationship to other explanatory variables in the model. When X can be explained well by the additional variables—a condition called *multicollinearity*—the inflation factor can be very high.

Adding additional explanatory terms, therefore, may either decrease or increase precision depending on the relative effects on the *mean square ratio* and the *VIF*. These conflicting effects highlight a fundamental tension in statistical modeling. If the purpose is to estimate the coefficient of X (or to predict a future Y) as precisely as possible, additional explanatory terms may help if those terms reduce the unexplained variation in the response, but may hurt if this reduction is insufficient to overcome the increase in the variance inflation factor. Sections 12.3 and 12.4 return to this issue, introducing tools for finding sets of explanatory variables that include "enough" but "not too many" terms.

12.2.3 A Strategy for Dealing with Many Explanatory Variables

A strategy for dealing with many explanatory variables should include the following elements:

1. Identify the key objectives.
2. Screen the available variables, deciding on a list that is sensitive to the objectives and excludes obvious redundancies.
3. Perform exploratory analysis, examining graphical displays and correlation coefficients.
4. Perform transformations as necessary.
5. Examine a residual plot after fitting a rich model, performing further transformations and considering outliers.
6. Use a computer-assisted technique for finding a suitable subset of explanatory variables, exerting enough control over the process to be sensitive to the questions of interest.
7. Proceed with the analysis, using the selected explanatory variables.

Example—Preliminary Analysis of the State SAT Data

The objectives of the SAT study were set out earlier. The variables in Display 12.1 were screened from the original source variables. (Several irrelevant variables were not retained.)

Display 12.4 shows the matrix of scatterplots for the SAT data. Examination of the bottom row indicates some nonlinearity in the relationship between SAT and percentage of takers and shows some potential outliers with respect to the variables public school percentage and state expenditure. Using the logarithm of the percentage of takers responds to the first problem, yielding a relationship that looks like a straight line. The outliers were identified: Louisiana is the state with the unusually low percentage of students attending public schools, and Alaska is the state with the very high state expenditure. These require further examination after a model has been fit.

The questions of interest call for adjustment with respect to percentage of SAT takers and median class rank of those takers in each state. When a preliminary regression of SAT was fit on both of these variables (a fit unaffected by Louisiana and Alaska), the variables were found to explain 81.5% of the variation between state SAT averages.

Partial residual plots at this stage help the researcher evaluate the effects of some of the other explanatory variables, after the first two are accounted for. Display 12.5 shows a partial residual plot of the relationship between SAT and expenditure, after the effects of percentage of takers and class rank have been

DISPLAY 12.4 Matrix of scatterplots for SAT scores and six explanatory variables

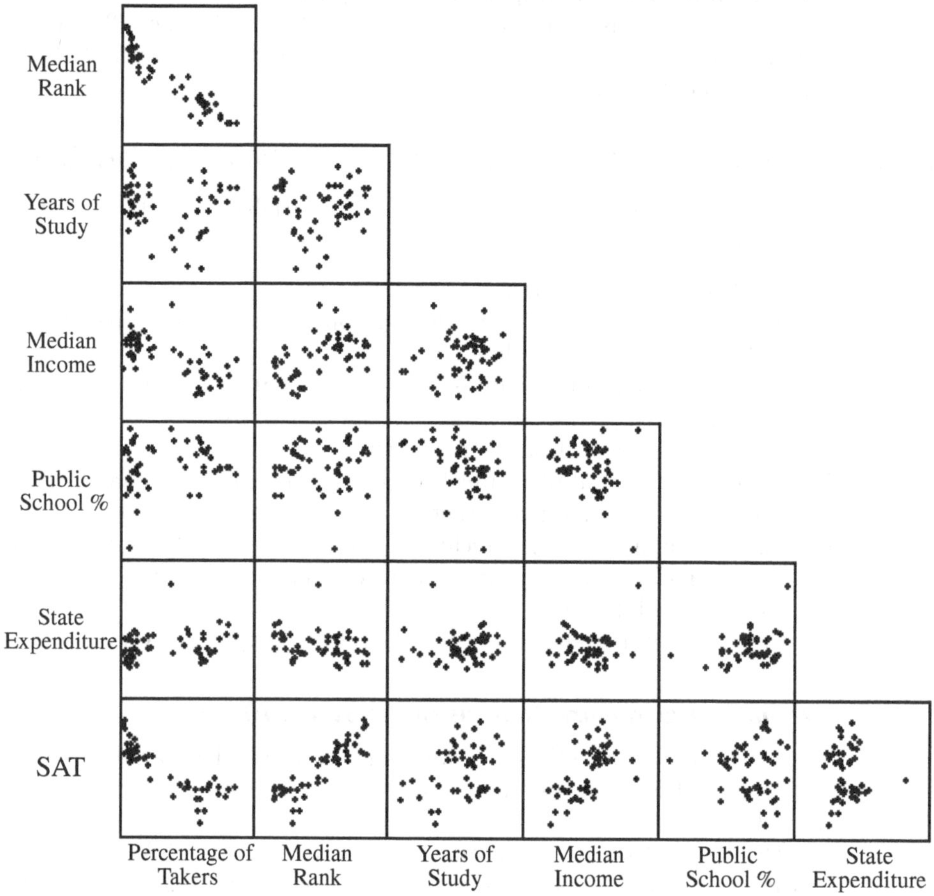

cleared out of the way. This plot has two striking features: It reveals the noticeable effect of expenditure, after the two adjustment variables are accounted for; and it demonstrates that Alaska is very influential in any fit involving expenditure. To clarify this, the diagram shows two possible straight line regression fits—one with and one without Alaska. The slopes of these two lines differ so markedly that neither value would lie within a confidence interval based on the other. Case influence statistics confirm that Alaska is influential. It has a large leverage ($h_i = 0.458$, from a statistical computer program) and does not fit well to the model that fits the other states (studentized residual = −3.28).

The prudent step is to set Alaska aside for the remainder of the analysis. It can be argued that the role of expenditure is somewhat different in that state, since the costs of heating school buildings and transporting teachers and students long distances are far greater in Alaska, and these expenses do not contribute directly to educational quality. Furthermore, too few states have similar expenditures to

DISPLAY 12.5 Partial residual plot of state average SAT scores (adjusted for percentage of students in the state who took the test and for median class rank of the students who took the test) versus state expenditure on secondary education

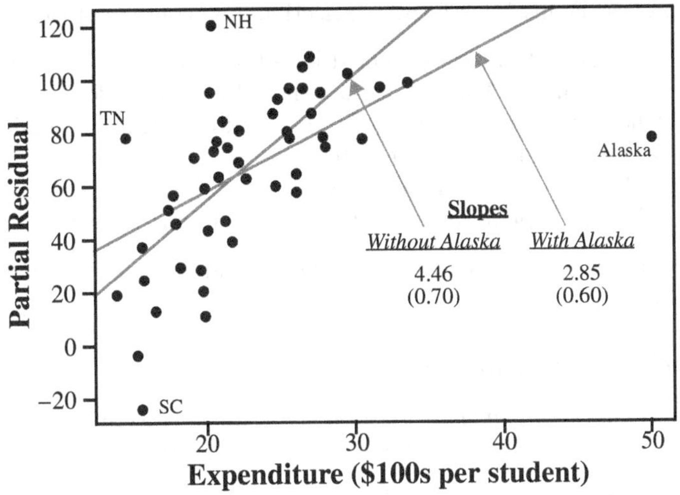

permit an accurate determination of the proper form of the model for expenditures greater than $3,500 per student. A similar plot of partial residuals versus percentage of students in public schools provides no reason to be concerned about Louisiana. The effect of the percentage of students who attend public schools in a state appears to be insignificant whether Louisiana is included or not.

12.3 SEQUENTIAL VARIABLE-SELECTION TECHNIQUES

The search for a suitable subset of explanatory variables may encompass a large array of possible models. Computers facilitate the process immensely, since hundreds or even thousands of models can be analyzed in a short period of time. Yet in some instances the search may be so wide as to tie up even the best computer for an intolerable amount of time. Sequential variable selection procedures offer the option of exploring some (but not all) of the possible models.

A step in any sequential (or *stepwise*) procedure begins with a tentative *current model*. Several other models in the neighborhood of the current model (but differing from the current model in having one of its variables excluded or having one additional variable included) are examined to determine whether they are in some sense superior. If so, the best is chosen as the tentative model for the next step. Procedures may differ in their definitions of neighborhood, in their criteria for superiority, and in their initial tentative model.

12.3.1 Forward Selection

The *forward selection* procedure starts with a constant mean as its current model and adds explanatory variables one at a time until no further addition significantly improves the fit. Each forward selection step consists of two tasks:

1. Consider all models obtained by adding one more explanatory variable to the current model. For each variable not already included, calculate its "F-to-enter" (the extra-sum-of-squares F-statistic for testing its significance). Identify the variable with the largest F-to-enter.
2. If the largest F-to-enter is greater than 4 (or some other user-specified number), add that explanatory variable to form a new current model.

Tasks 1 and 2 are repeated until no additional explanatory variables can be added.

12.3.2 Backward Elimination

In the *backward elimination* method, the initial current model contains *all* possible explanatory variables. Each step in this procedure consists of two tasks:

1. For each variable in the current model, calculate the F-to-remove (the extra-sum-of-squares F-statistic for testing its significance). Identify the variable with the smallest F-to-remove.
2. If the smallest F-to-remove is 4 (or some other user-specified number) or less, then remove that explanatory variable to arrive at a new current model.

Tasks 1 and 2 are repeated until none of the remaining explanatory variables can be removed.

12.3.3 Stepwise Regression

The constant mean model with no explanatory variables is the starting current model. Each step consists of the following tasks.

1. Do one step of forward selection.
2. Do one step of backward elimination.

Tasks 1 and 2 are repeated until no explanatory variables can be added or removed.

Inclusion of Categorical Factors

Each categorical factor is represented in a regression model by a set of indicator variables. In the absence of a good reason for doing otherwise, this set should be added or removed as a single unit. Some computer packages allow this automatically.

No Universal Best Model

Forward selection, backward elimination, and stepwise regression can lead to different final models. This fact should not be alarming, since no single best model can be expected. It does, however, emphasize a disadvantage of these methods: Only one model is presented as an answer, which may give the false impression that this model alone explains the data well. As a result, unwarrantedly narrow attention may be accorded to the particular explanatory variables in that single model.

12.3.4 Sequential Variable-Selection with the SAT Data

Display 12.6 is a skeleton of all possible models in the SAT example (with Alaska excluded). Such a display would not be used in practice, but it illustrates how sequential techniques find a good-fitting model even though they search through only a handful of all possible models.

The models on the display are labeled with one-character symbols associated with each explanatory variable. The model labeled 1 has only the intercept. The model labeled T has the intercept and t (the log of the percentage of takers) only. The model labeled *ET* has the intercept, the log of percentage of takers (t), and state expenditure (e). And so on. All models in the same row contain the same number of explanatory variables. Models are coded so that the variables whose F-statistics are greater than 4 are listed in uppercase letters, and those with F-statistics less than or equal to 4 are listed in lowercase letters. So, for example, in the fit to the four-variable model with income, years, public, and expenditure, listed as *IYPe*, expenditure is not significant, but the other variables are. The lines connecting the boxes in the display represent the paths for getting from one model to another that has one fewer or one more explanatory variable.

Forward Selection

Model 1 is taken as the initial current model, and all single-variable models are considered. In the second row of Display 12.6, models I, Y, R, and T all have coefficients whose F-statistics are larger than 4. The one of these with the largest F-to-enter is T, so it is taken as the next current model. Next, all models that include T and one other variable are examined. Only Y and E add significantly when added to the model with T. Of these, E has the larger F-to-enter, so the next current model is ET. Next, all three-variable models that include E and T are considered, but in no case is the additional variable significant. Therefore, the forward selection process stops, and the final model is ET. The entire procedure required the fitting of only 16 of the 64 possible models.

Backward Elimination

The model with all variables, *iYpERT*, is taken as the starting point. Of the variables in the model, p (public) has the smallest F-statistic, and p is dropped because its F is less than 4. The resulting model is *iYERT*. The least significant coefficient in the new current model is the one for i (income), and its F-statistic is less than 4, so it is dropped. The resulting model is *YERT*. All of the variables in this model have F-statistics greater than 4, so this is the final model. This procedure required the fitting of only 3 of the 64 possible models, and none of the 3 was in the set considered by forward selection.

DISPLAY 12.6 Anatomy of sequential variable selection—SAT example

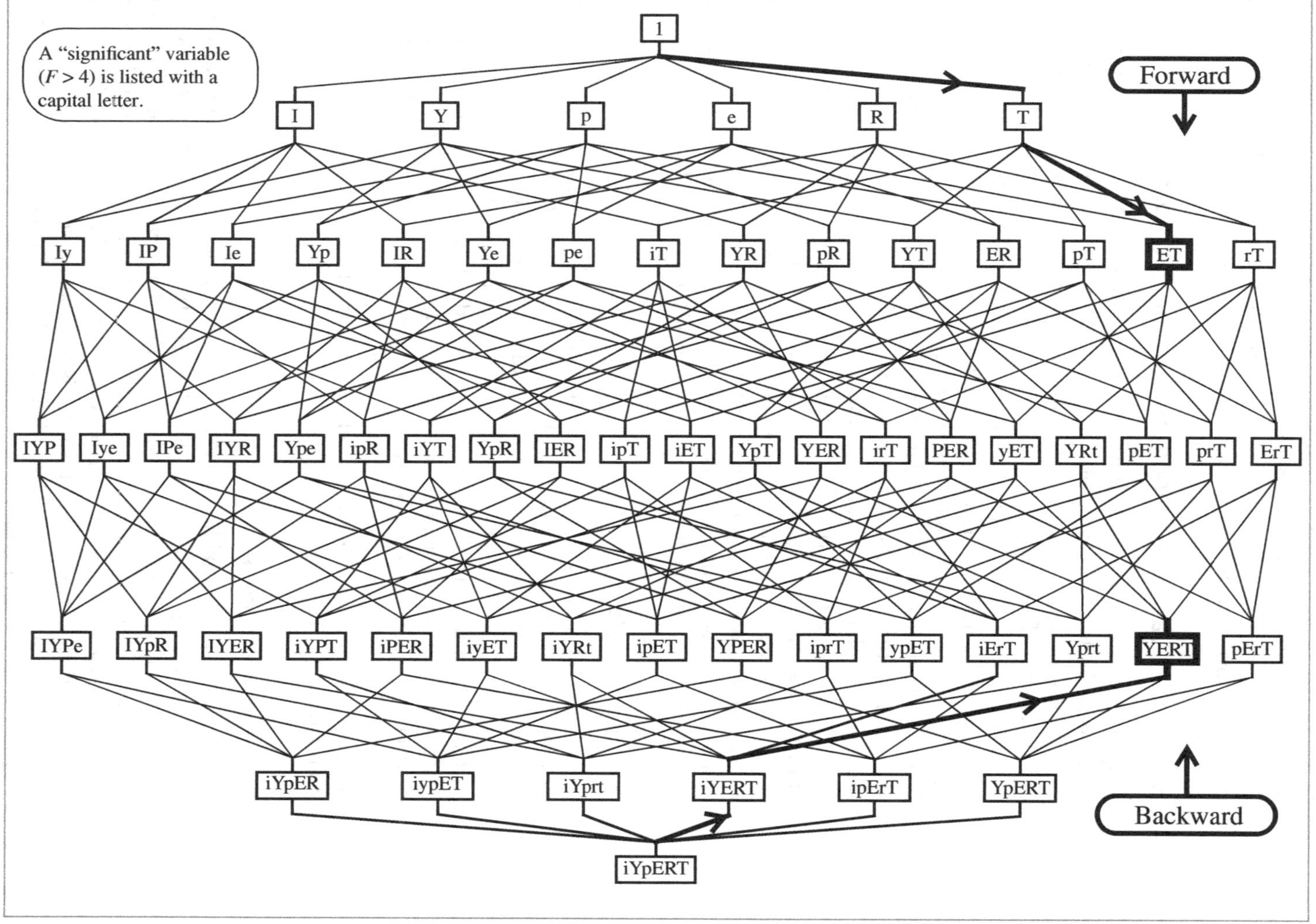

Stepwise Regression

Stepwise regression starts with one step of forward selection to arrive at the model T. No variables can be dropped, so it tries to add another. It arrives at ET. It then checks whether any of the variables in this model can be dropped. None can, so it seeks to add another variable, but no others add significantly, so the procedure stops at ET.

12.3.5 Compounded Uncertainty in Stepwise Procedures

The cutoff value of 4 for the F-statistic (or 2 for the magnitude of the t-statistic) corresponds roughly to a two-sided p-value of less than 0.05. The notion of "significance" cannot be taken seriously, however, because sequential variable selection is a form of data snooping.

At step 1 of a forward selection, the cutoff of $F = 4$ corresponds to a hypothesis test for a single coefficient. But the actual statistic considered is the largest of several F-statistics, whose sampling distribution under the null hypothesis differs sharply from an F-distribution.

To demonstrate this, suppose that a model contained ten explanatory variables and a single response, with a sample size of $n = 100$. The F-statistic for a single variable at step 1 would be compared to an F-distribution with 1 and 98 degrees of freedom, where only 4.8% of the F-ratios exceed 4. But suppose further that all 11 variables were generated completely at random (and independently of each other), from a standard normal distribution. What should be expected of the largest F-to-enter?

This random generation process was simulated 500 times on a computer. Display 12.7 shows a histogram of the largest among ten F-to-enter values, along with the theoretical F-distribution. The two distributions are very different. At least one F-to-enter was larger than 4 in 38% of the simulated trials, even though none of the explanatory variables was associated with the response.

This and related arguments imply that sequential procedures tend to select models that have too many variables if the set contains unimportant ones. This may not be a serious problem for some data problems, but better techniques are available. In part, variable-selection routines are important for historical reasons, since they were the only tools available when computing time was substantially slower. They are also important because they extend in an obvious way to other types of modeling. Now that computers can look at every possible regression model, however, criteria not based on sequential significance address more pertinent features in desired models.

12.4 MODEL SELECTION AMONG ALL SUBSETS

A second approach to model selection involves fitting all possible subset models and identifying the ones that best satisfy some model-fitting criterion. Since sequential selection is avoided, the criteria can be based directly on the key issues: including

DISPLAY 12.7 Simulated distribution of the largest of ten *F*-statistics

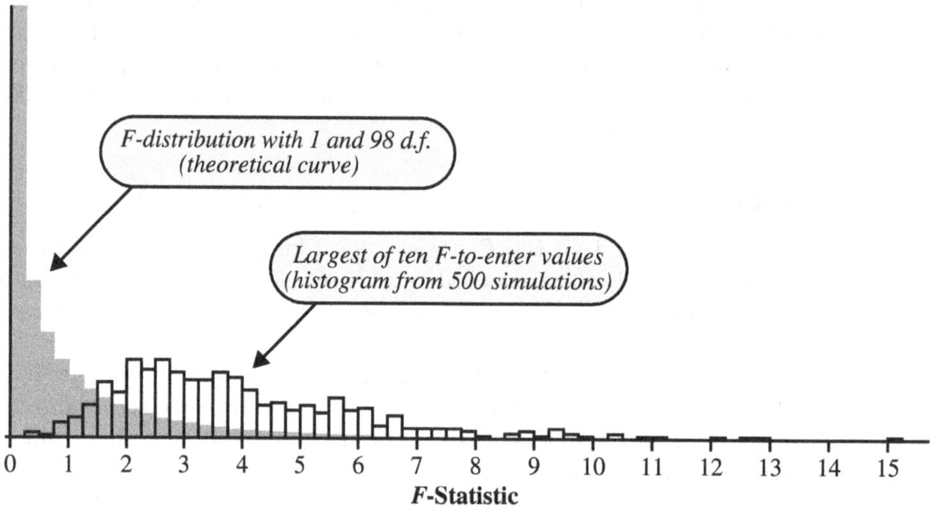

enough explanatory variables to model the response accurately without the loss of precision that occurs when essentially redundant and unnecessary terms are included.

The two criteria presented in this section are the Cp statistic and Schwarz's Bayesian Information Criterion. These are calculated for the fit to a model, so each of the possible regression models receives a numerical score. Before these are examined, some simpler choices should be mentioned.

R^2 and Adjusted R^2

Comparing two models with the same number of parameters is relatively easy: The one with the smaller residual mean square $\hat{\sigma}^2$ is preferred. When applied to comparisons of models that have different numbers of parameters, R^2 (not adjusted) always leads to selecting the model with all variables. Because R^2 is incapable of making a sensible selection, some researchers use the adjusted R^2. This is equivalent to selecting the model with the smallest $\hat{\sigma}^2$—a criterion that generally favors models with too many variables.

12.4.1 The Cp Statistic and Cp Plot

The Cp statistic is a criterion that focuses directly on the trade-off between bias due to excluding important explanatory variables and extra variance due to including too many. The bias in the ith fitted value is

$$\text{Bias}\{\hat{Y}_i\} = \mu\{\hat{Y}_i\} - \mu\{Y_i\}.$$

This is the amount by which the mean in the sampling distribution of the ith fitted value differs from the mean it is attempting to estimate. The mean squared error is the squared bias plus the variance:

$$\text{MSE}\{\hat{Y}_i\} = [\text{Bias}\{\hat{Y}_i\}]^2 + \text{Var}\{\hat{Y}_i\}.$$

The *total mean squared error (TMSE)* for a given model is the sum of these over all observations:

$$\text{TMSE} = \sum_{i=1}^{n} \text{MSE}\{\hat{Y}_i\}.$$

It is desired to find a subset model with a small value of TMSE. The squared bias terms are small if no important explanatory variables are left out, and the variance terms are small if no unnecessary explanatory variables are included.

The Cp statistic is an estimate of TMSE/σ^2, based on assuming that the model with all available explanatory variables has no bias. For a model that has p regression coefficients (including β_0), the Cp statistic is computed as

$$\text{C}_\text{P} = p + (n - p)\frac{(\hat{\sigma}^2 - \hat{\sigma}^2_{\text{full}})}{\hat{\sigma}^2_{\text{full}}},$$

where $\hat{\sigma}^2$ is the estimate of σ^2 from the tentative model, and $\hat{\sigma}^2_{\text{full}}$ is the estimate of σ^2 from the fit with all possible explanatory variables. One Cp statistic can be calculated for each possible model.

Models with small Cp statistics are looked on more favorably. If a model lacks important explanatory variables, it will show greater residual variability than the full model with all explanatory variables will show. Thus, $\hat{\sigma}^2 - \hat{\sigma}^2_{\text{full}}$ will be large. On the other hand, if the difference in estimates of σ^2 is close to zero, including p in the formula will add a penalty in the Cp statistic for having more explanatory variables than necessary.

The Cp Plot

A Cp plot is a scatterplot with one point for each subset model. The y-coordinate is the Cp statistic, and the x-coordinate is p—the number of coefficients in the model. Each point in the plot is labeled to show the model to which it corresponds, often by including one-letter codes corresponding to each explanatory variable that appears in the model.

The Cp plot, which is available in many statistical computer packages, can be scanned to identify several models with low Cp statistics. The line at $\text{Cp} = p$ is often included on the plot, since a model without bias should have Cp approximately equal to p. Some users treat this line as a reference guide for visual assessment of the breakdown of the Cp statistic into bias and variance components; but the Cp statistic itself is the criterion. Thus, the models with smallest Cp statistics are considered. There is nothing magical about the model with the smallest Cp, since the actual ordering is likely to be affected by sampling variability; thus, a different set of data may result in a different ordering. Picking a single model from among all

DISPLAY 12.8 Three criteria for comparing regression models with different subsets of X's. Subsets that produce small values of these are thought to strike a good balance between small sum of squared residuals (SSRes) and not too many regression coefficients (p).

Criterion to make small	=	SS Res. part	+	Penalty for number of terms
Cp	=	$\frac{SSRes}{\hat{\sigma}^2_{full}} - n$	+	$2p$
BIC	=	$n \times \log\left(\frac{SSRes}{n}\right)$	+	$\log(n) \times (p + 1)$
AIC	=	$n \times \log\left(\frac{SSRes}{n}\right)$	+	$2 \times (p + 1)$

those with small Cp statistics is usually a matter of selecting the most convenient one whose coefficients all differ significantly from zero.

12.4.2 Akaike and Bayesian Information Criteria

The Cp statistic takes a measure of the lack of fit of a model and adds a penalty for the number of terms in the model. As more terms are included in the model, lack of fit decreases, but the penalties increase. Models with small Cp statistics balance the lack of fit and the penalty.

Two alternative model selection statistics incorporating similar penalties are Akaike's Information Criterion (AIC) and Schwarz's Bayesian Information Criterion (BIC). Display 12.8 shows the formulas for these statistics.

Some computational versions of BIC and AIC differ from the formulas in Display 12.8 by a constant. Since the constant is the same for all models, this will not affect the ordering of models indicated by the model-fitting criteria.

The BIC penalty for number of betas is larger than the AIC penalty for all sample sizes larger than 7, and the difference increases with increasing sample size. Consequently, BIC will tend to favor models with fewer X's than AIC, especially from large samples. There is no theory for clarifying the relative benefits of these two criteria; they simply reflect different approaches for balancing under- and over-fitting. In general, AIC seems more appropriate if there are not too many redundant and unnecessary X's in the starting set; BIC seems more appropriate if there are many redundant and unnecessary. The distinction between these two situations is rarely clear in practice, but BIC is particularly useful when the researcher places a larger burden of model determination on the computer, such as when a set of measured X's, their quadratic terms, and all possible two-term interactions are included in the starting set.

In any case, it is important to realize that no criterion can be superior in all situations. It is also important to realize that variable-selection routines will not discover a scientific law of nature. There are usually many possible subsets—formed from highly interconnected explanatory variables—that explain the response

DISPLAY 12.9 Cp plot for state SAT averages (excluding Alaska and showing only those models with Cp < 10); t = log percentage of takers, i = income, y = years, p = public, e = expend, and r = rank

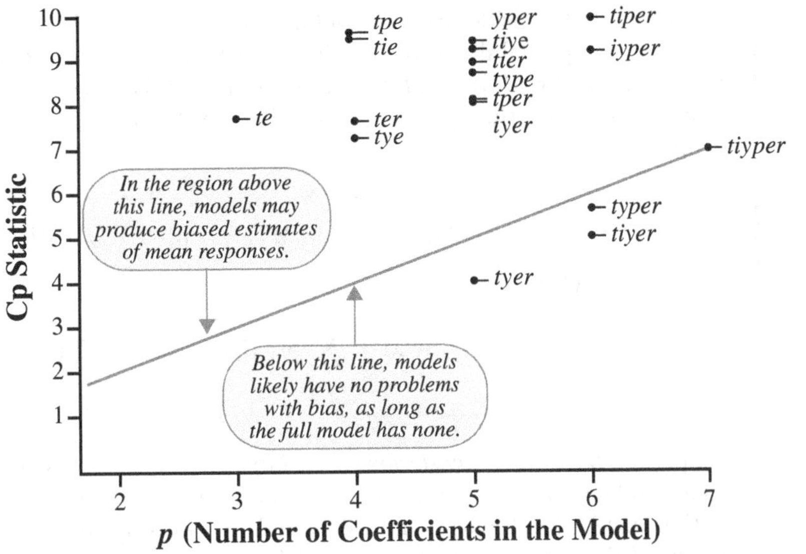

variation equally well. As the famous English statistician George E. P. Box said, "All models are wrong, but some are useful."

SAT Example

Display 12.9 shows a Cp plot for the SAT data (without Alaska), including only models with Cp less than 10. The models are identified on the plot by the variables included, with t = log percentage of takers, i = income, and so on. Observe that the Cp statistic for the general model that includes all variables—*tiyper*—equals the maximum number of terms, 7 (including the constant). The model *tyer*, which was identified by backward elimination, has the smallest Cp statistic.

The model *tyer* has the smallest BIC, at 322.4. It is followed by *te* with BIC = 322.7, *tye* with BIC = 324.2, and *ter* with BIC = 324.6. The models with smallest AIC values are *tyer* with AIC = 311.1, *tiyer* with AIC = 311.9, *typer* with AIC = 312.7, and *tiyper* with AIC = 313.9.

12.5 POSTERIOR BELIEFS ABOUT DIFFERENT MODELS

One branch of statistical reasoning, called ***Bayesian statistics***, assumes that beliefs in various hypotheses can be expressed on a probability scale. The elegant theorem of English mathematician Thomas Bayes (1702–1761), known as Bayes' Theorem, specifies precisely how beliefs held prior to seeing the data should be altered by statistical evidence to arrive at a set of posterior beliefs. In the present setting, Bayesian statistics can be used to update the probability that each subset model is correct, given the available data.

Suppose that consideration is confined to models $M_1, M_2, \ldots, M_K$, and suppose that prior probabilities (probabilities of belief prior to seeing the data) on these models are $\text{pr}\{M_1\}$, $\text{pr}\{M_2\}, \ldots,$ $\text{pr}\{M_K\}$. Applying a Bayesian formulation, Schwarz showed that the updated probabilities for the models, given the data (D), were approximately

$$\text{pr}\{M_j|D\} = \text{pr}\{M_j\} \times \exp\{-\text{BIC}_j\}/\text{SUM}$$

where

$$\text{SUM} = \sum_{i=1}^{K} \text{pr}\{M_i\} \times \exp\{-\text{BIC}_i\},$$

where $\text{pr}\{M_j|D\}$, called a ***posterior probability***, is the probability of M_j after seeing the data. When no prior reasons exist for believing that one model is better than any other, $\text{pr}\{M_j\}$ is the same for all j, and the posterior probabilities are simple functions of the BIC values for each model.

For the SAT example, the values of BIC have been calculated for each model and then used in this formula to identify the posterior probabilities. The posterior probability that the model is *te* is 0.764, followed by *tyer* with 0.115, *tie* with 0.060, and *tpe* with 0.041. The posterior probability that the model is any of the other 60 models is 0.020, and no other single model has a posterior probability greater than 0.006.

12.6 ANALYSIS OF THE SEX DISCRIMINATION DATA

To investigate male/female differences in beginning salaries requires a strategy that convincingly accounts for variables other than the sex indicator before adding it to the equation.

Initial investigation indicates that beginning salary should be examined on the log scale. Scatterplots of log beginning salary versus other variables show that the effect of experience is not linear. Beginning salaries increase with increasing experience up to a certain point; then they level off and even drop down for individuals with more experience. A quadratic term could reproduce this behavior well. A similar effect is seen with age: a correlation between age and experience may be responsible for that effect, but it is wise initially to consider a squared term for both variables. It may seem that seniority (number of months working with the company) is an inappropriate term to include for modeling beginning salary, but its inclusion accounts for increasing beginning salaries over time. Rather than as seniority, it might be thought of as time prior to March 1977 that the individual was hired.

Such considerations point to the necessity of including in the investigation a model rich enough to be extremely unlikely to miss an important relationship between log beginning salary and all explanatory variables other than the sex indicator. A saturated second-order model that includes all quadratic terms and all interaction terms satisfies that criterion. (See Section 12.7.3 for further discussion of second-order models.)

Notation

Fourteen variables appear in the saturated second-order model, so naming variables with single letters utilizes space well, even though it sacrifices clarity. The full complement of explanatory variables is given in Display 12.10. A word represents a model whose explanatory variables consist of the letters that make up the word. For example, the word *saebc* represents the model with seniority, age, education, the square of age, and the product of age with education—a model with six parameters in all (including the constant).

DISPLAY 12.10 Explanatory variables for the sex discrimination data

Main effect variables	Quadratic variables	Interaction variables	
s = Seniority	$t=s^2$	$m=s\times a$	$c=a\times e$
a = Age	$b=a^2$	$n=s\times e$	$k=a\times x$
e = Education	$f=e^2$	$v=s\times x$	$q=e\times x$
x = Experience	$y=x^2$		

A total of 16,384($=2^{14}$) models are formed by all possible combinations of the 14 terms in the preceding box. The goal is to find a good subset of these terms for explaining log salary, to which the sex indicator variable may be added. Although quadratic and interaction terms should be considered, it is not good practice to include quadratic terms without the corresponding linear term; neither is it good practice to include interaction terms without the two corresponding main effects. The preferred approach, therefore, is to identify the better models (of the 16,384) and then to identify the best subset of these that meets the narrower criteria restrictions of "good practice."

Identifying Good Subset Models

Display 12.11 shows a Cp plot. The Cp statistics were based on the full model containing all 14 terms. To reduce clutter, the plot displays Cp statistics only for models that meet the criteria discussed in the previous paragraph and, in addition, have small Cp values. Only a few good-fitting models with more than seven parameters are shown on this plot.

Models *saexck* and *saexnck* have lower Cp statistics than any others, and they appear to have no bias relative to the saturated model (since they fall near or below the line at Cp $= p$). The BIC clarifies the issue further: among 382 models examined, the BIC assigns a posterior probability of 0.7709 to *saexck*, with *saexyc*

DISPLAY 12.11 Cp plot for the sex discrimination study

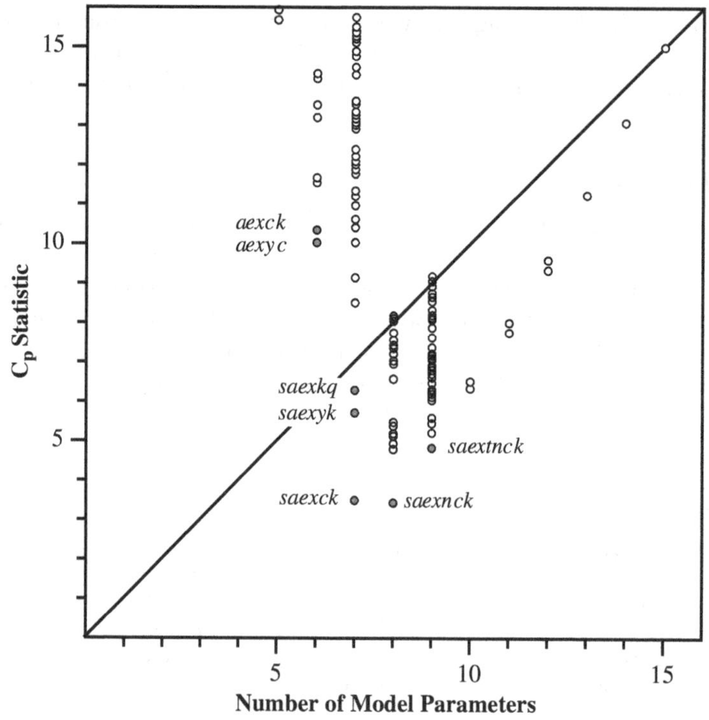

receiving the second highest posterior probability of 0.0625. Display 12.12 shows the 20 models that recorded the highest posterior probabilities.

Evaluating the Sex Effect

When the sex indicator variable is added into model *saexck*, its estimated coefficient is −0.1196, with a standard error of 0.0229. The t-statistic is 5.22, and the test that the coefficient is zero against the alternative that it is negative has a one-sided p-value $= 6.3 \times 10^{-7}$. Thus, conclusive evidence exists that sex is associated with beginning salary, even after seniority, age, education, and experience are accounted for in this model. The median female salary is estimated to be $\exp(-0.1196) = 0.887$ times as large as (that is, about 11% less than) the median male salary, adjusted for the other variables.

Accounting for Model Selection Uncertainty Using the BIC

The preceding inferential statements pertain specifically to the single model *saexck*, selected to represent the influence of the potentially confounding variables. One criticism of this result is that it fails to account for the uncertainty involved in the model selection process. A second criticism is that the addition of the sex indicator

DISPLAY 12.12 Bayesian posterior analysis of the difference between male and female log beginning salaries

Model	BIC	Posterior probability	Coefficient of sex	SE	One-sided *p*-value
saexck	−404.15	0.6909	−0.1196	0.0229	6.16E-07
saexnck	−401.91	0.0738	−0.1173	0.0229	9.31E-07
saexyc	−401.63	0.0558	−0.1287	0.0226	8.21E-08
saexkq	−401.03	0.0306	−0.1244	0.0221	1.16E-07
saexbck	−400.32	0.0150	−0.1195	0.0229	6.57E-07
saextck	−400.21	0.0134	−0.1189	0.0232	8.97E-07
saexbkq	−400.19	0.0132	−0.1206	0.0221	2.36E-07
saexync	−400.16	0.0129	−0.1258	0.0225	1.34E-07
saexfck	−399.95	0.0104	−0.1196	0.0230	6.80E-07
saexckq	−399.89	0.0097	−0.1208	0.0230	5.43E-07
saexyck	−399.74	0.0084	−0.1257	0.0232	2.75E-07
saexmck	−399.68	0.0079	−0.1195	0.0231	7.32E-07
saexvck	−399.63	0.0076	−0.1196	0.0231	7.17E-07
aexyc	−399.49	0.0065	−0.1247	0.0238	5.48E-07
aexck	−399.18	0.0048	−0.1135	0.0246	6.86E-06
saexyq	−398.64	0.0028	−0.1328	0.0218	1.46E-08
saextyc	−398.42	0.0023	−0.1271	0.0228	1.41E-07
saectnck	−398.14	0.0017	−0.1163	0.0232	1.41E-06
saexbc	−397.95	0.0014	−0.1230	0.0237	6.83E-07
saexyvc	−397.93	0.0014	−0.1281	0.0226	9.97E-08
all others combined		0.0294			

s = Seniority	$t = s^2$	$m = s \times a$	$c = a \times e$
a = Age	$b = a^2$	$n = s \times e$	$k = a \times x$
e = Education	$f = e^2$	$v = s \times x$	$q = e \times x$
x = Experience	$y = x^2$		

variable to other good-fitting models might produce less convincing evidence of a sex effect.

The posterior probabilities for various models allow the analyst to address both criticisms. Suppose that δ represents the coefficient of the sex indicator variable and that $f(\delta|M_j)$ represents some function $f(\delta)$ based on the assumption that model M_j is correct. The Bayesian analysis permits an average estimate of $f(\delta)$ over different models. Furthermore, the average is weighted to represent the strength of belief in each of the various models. The estimate of $f(\delta)$ in this way incorporates the uncertainty in model selection. The weighted average is

$$f\{\delta|D\} = \sum_{i=1}^{K} f(\delta|M_j) \times \text{pr}\{M_j|D\}.$$

Two choices for $f(\delta)$ are δ itself (the estimate of the sex effect) and the p-value for testing its significance. The sex indicator variable was added to each of the 382 models and combined as shown in the preceding equation to produce the single (posterior) estimate of -0.1206 and a single (posterior) p-value of 6.7×10^{-7} (this requires some programming). A listing of the top 20 candidate models, together with their contributions to the posterior information, is shown in Display 12.12.

The use of Bayesian statistics requires that somewhat different interpretations be given to the estimate and to the p-value. The average estimate is termed a *Bayesian posterior-mean estimate*, while the average p-value is called the *posterior probability* that δ is greater than zero. The probability refers to a measure of belief about the hypothesis that the sex coefficient is zero. This is a different philosophical attitude than any previously considered, but two issues argue in its favor for this application. First the formula itself gives the p-value for every model a chance to contribute in the final estimate, with a weight appropriate to the support given the model by the data. Second, no known statistical procedures based on other philosophies incorporate model selection uncertainty in this way.

12.7 RELATED ISSUES

12.7.1 The Trouble with Interpreting Significance when Explanatory Variables Are Correlated

When explanatory variables are correlated, interpretation of individual regression coefficients and their significance is quite difficult and perhaps too convoluted to be of use. The central difficulty is that interpreting a single coefficient as the amount by which the mean response changes as the single variable changes by one unit, holding all other variables fixed, does not apply; this is because the data generally lack experience with situations where one variable can be changed in isolation.

Consider the variable *expend*—the per student state expenditure—in the SAT example. This variable can be added to any model that does not already contain it. This was done for all models not containing *expend*, with the results shown in Display 12.13.

The first three columns of Display 12.13 are self-explanatory, the key point being that the p-values vary widely. The last column is the addition to R^2 resulting from the addition of *expend* to the model. The variation in these percentages can be traced to the correlation between *expend* and the other variables used in the model. The last column can be interpreted as the extra sum of squares due to *expend*, expressed as a percentage of the total sum of squares. The fourth column is the extra sum of squares due to *expend*, expressed as a percentage of the sum of squared residuals. The huge variations in that column arise because of the differing amounts of variability explained by the different models not containing *expend*. This is most directly related to the variation in p-values for *expend*, but the correlational aspect also plays a role.

Ultimately, the answer to the often-asked question, "Which variables are significant?" has to be "It depends." Similarly, there is usually no definitive answer

DISPLAY 12.13 Statistical measures for the contribution of *expend* to different models (Alaska removed)

			% Variation explained by *expend*	
Model	*t*-statistic	Two-sided *p*-value	Previous residual	Total variation
E	0.27	0.79	0.2	0.2
I+*E*	0.40	0.69	0.3	0.2
Y+*E*	0.86	0.39	1.6	1.4
P+*E*	0.18	0.86	0.1	0.1
R+*E*	4.78	0.00002	33.2	7.4
T+*E*	6.04	0.0000002	44.3	8.4
IY+*E*	0.07	0.94	0.0	0.0
IP+*E*	0.11	0.92	0.0	0.0
IR+*E*	4.88	0.00001	24.6	6.7
IT+*E*	5.88	0.0000005	43.4	8.1
YP+*E*	1.05	0.30	2.4	2.1
YR+*E*	4.20	0.0001	28.2	4.3
YT+*E*	5.31	0.000003	38.5	6.3
PR+*E*	6.02	0.0000003	44.6	9.4
PT+*E*	5.91	0.0000004	43.7	8.2
RT+*E*	6.13	0.0000002	45.5	8.4
IYP+*E*	0.92	0.36	1.9	0.9
IYR+*E*	4.35	0.00008	30.1	4.2
IYT+*E*	5.20	0.000005	38.0	6.2
IPR+*E*	5.57	0.000001	41.4	8.0
IPT+*E*	5.59	0.000001	41.5	7.5
IRT+*E*	5.92	0.0000004	44.4	7.9
YPR+*E*	4.80	0.00002	34.4	5.3
YPT+*E*	4.78	0.00002	34.2	5.1
YRT+*E*	5.23	0.000005	38.3	5.6
PRT+*E*	6.27	0.0000001	47.2	8.8
IYPR+*E*	4.27	0.0001	29.8	4.1
IYPT+*E*	4.33	0.00009	30.4	4.3
IYRT+*E*	5.00	0.00001	36.8	5.1
IPRT+*E*	5.94	0.0000004	45.1	8.0
YPRT+*E*	5.13	0.000007	38.0	5.4
IYPRT+*E*	4.64	0.00003	33.9	4.5

to the question, "Which variable is most significant?" It is best to try to focus attention on questions that ask specifically for the assessment of significance after accounting for other specified variables or groups of variables.

12.7.2 Regression for Adjustment and Ranking

One valuable use of regression is for adjustment. Consider the regression of state SAT averages on percentage of eligible students in the state who decide to take the test and on the median class rank of the students who take the exam. A residual plot from the fit (excluding Alaska) is shown in Display 12.14. The residuals are the

DISPLAY 12.14 Scatterplot of residuals versus fitted values from the regression of state SAT average on percentage of takers and median class rank of takers

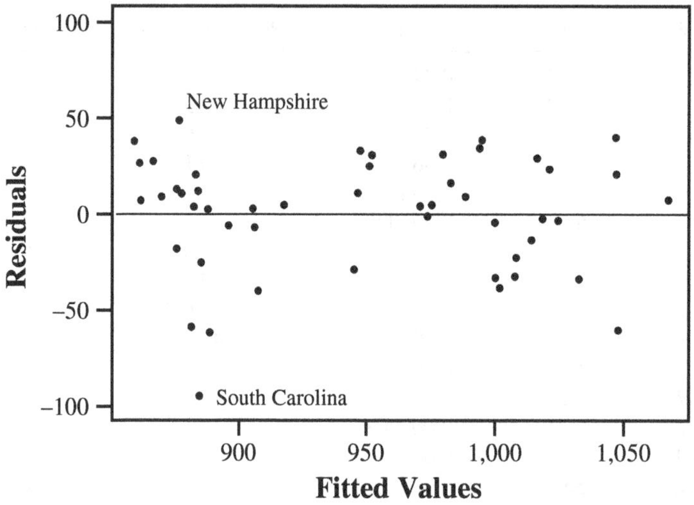

state SATs with the estimated linear effects of log percentage of takers and median class rank removed. Thus, they serve as the SATs adjusted for these two variables (and are scaled so that the average is zero). New Hampshire, for example, whose "raw" SAT average is 925 (which is 23 points less than the overall average of all states) has a very high percentage of eligible students who take the exam and a fairly low median class rank among the takers. The estimated mean SAT score for a state with these values is 876. Since New Hampshire's actual SAT score was 925, its value exceeded the prediction (according to the model with percentage of takers and median class rank) by 49 points. It is therefore appropriate to say that New Hampshire's SAT adjusted for percentage of takers and median class rank of takers is 49 points above average.

The ranking on the left side of Display 12.2 is based directly on the residuals shown in Display 12.14. New Hampshire has the largest residual, and therefore the largest SAT adjusted for percentage of takers and median class rank. South Carolina has the smallest.

The ranking on the right side of Display 12.2 is based on the residuals from a model that also includes expenditure. In this case, New Hampshire's SAT, adjusted for percentage of takers, median class rank, and expenditure, is 67 points above the average predicted for a state with the same values of percentage of takers, median rank, and expenditure.

12.7.3 Saturated Second-Order Models

A saturated second-order model (SSOM) includes the squares of all explanatory variables and all cross products of pairs of explanatory variables. It describes $\mu(Y)$

as a completely arbitrary parabolic surface. The SSOM should contain a subset model that describes the regression surface well.

Numbers of Subset Models

Given K original variables, the SSOM itself contains $[K(K+3)/2]+1$ parameters. When all possibilities are taken into account, the total number of p-parameter hierarchical models (meaning ones where, for example, X_1^2 appears only if X_1 also appears, and X_1X_2 appears only if both X_1 and X_2 appear) available from K original variables is

$$\sum_{j=0}^{K} C_{K,j} \times C_{(C_{j+1,2}),(p-1-j)},$$

where $C_{n,m}$, recall, is $n!/[m!(n-m)!]$ and $C_{n,m}=0$ whenever $n<m$. Holding out the sex indicator in the sex discrimination problem leaves $K=4$ explanatory variables: *seniority*, *age*, *education*, and *experience*. The full SSOM has 15 parameters, and subset models have $1, 2, 3, \ldots,$ 14 parameters. The number of distinct hierarchical models that can be considered is the number of hierarchical modes with one parameter plus the number with two parameters and so on up to the number of hierarchical models with fourteen parameters. According to the formula, a total of 1,337 distinct models can be considered.

Strategies for Exploring Subsets of the SSOM

The SAT study featured $K=6$ explanatory variables. The SSOM for that problem contains 28 parameters. The number of hierarchical models with 17 parameters, for example, is 352,716, and the total of all hierarchical models to consider is 2,104,489. Things rapidly get out of hand, and it becomes necessary to plan a strategy for sorting through some—but not all—of the models for promising candidates.

The strategy employed in the sex discrimination study was to examine all subset models up to some level ($p=7$) first. Forward selection from the best models with $p \le 7$ came next, followed by backward elimination from the saturated second-order model. Finally, models in the neighborhood of promising models were included. A *neighborhood* of one model consists of all models that can be obtained by adding or dropping one variable.

Another strategy involves grouping variables into sets that have a common theme and exploring each set separately to determine its best subset. When these are combined into an overall model, subsets consisting of products of retained variables from one set with those from another may also be examined.

A third strategy is to identify which of the original variables are important, by examining all subsets with main effects only (i.e., excluding squared and interaction terms), and then to build the SSOM from those variables alone and explore subsets.

12.7.4 Cross Validation

Inference after the use of a variable selection technique is tainted by the data snooping involved in the process. The selected model is likely to fit much better to the data that gave it birth than to fresh data. Thus, p-values, confidence intervals, and prediction intervals should be used cautiously.

If the data set is very large, the analyst may benefit from dividing it at random into separate model construction and validation sets. A variable selection technique can be used on the model construction set to determine a set of explanatory variables. The selected model can then be refit on the validation set, without any further exploration into suitable explanatory variables, and inferential questions can be investigated on this fit, ignoring the construction data. When the purpose of the regression analysis is prediction, it is recommended that the validation data set be about 25% of the entire set. Although the saving of the 25% of the data for validation seems reasonable for other purposes, the actual benefits are not very well understood.

12.8 SUMMARY

Model selection requires an overall strategy for analyzing the data with regression tools (see Display 9.9). After giving initial thought to a game plan for investigating the questions of interest, the preliminary analysis consists of a combination of exploration, model fitting, and model checking. Once some useful models have been identified, the answers to the questions of interest can be addressed through inferences about their parameters.

Tools for initial exploration include graphical methods (such as a matrix of scatterplots, coded scatterplots, jittered scatterplots, and interactive labeling) and correlation coefficients between various variables. Certain tricks for modeling are used extensively in the case studies—indicator variables, quadratic terms, and interaction terms. For model checking and model building, a number of other tools are suggested: residual plots, partial residual plots, informal tests of coefficients, case influence statistics, the Cp plot, the BIC, and sequential variable selection techniques. Finally, some inferential tools are presented: t-tests and confidence intervals for individual coefficients and linear combinations of coefficients, extra sums-of-squares F-tests, prediction intervals, and calibration intervals.

SAT Study

Although this example is used to demonstrate variable-selection techniques, the actual analysis is guided by the objectives, and variable selection played only a minor role. These data have been used to rank the states on their success in secondary education, but in this regard selection bias poses a serious problem. In some states, for example, a high SAT average reflects the fact that only a small proportion of students—the very best ones—took the test, rendering the self-selected sample far from representative of high school students in the state overall. One goal of the regression analysis is to establish a ranking that accounts for this selection bias. Although overcoming the limitations of a self-selected sample is impossible, it is possible to rank the states after subtracting out the effects of the different proportions of students taking the test and their different median class rankings. Accomplishing this involves fitting the regression of SAT scores on these two explanatory variables, and ranking the states according to the sizes of their residuals.

A further question is exploratory in nature: Are any other variables associated with SAT score? For example, is the amount of money spent on secondary education related to it? Variable selection techniques may be useful for sorting through various models to identify promising predictors. Whatever models are suggested, though, only the ones that include percentage of takers and median class rank should be chosen, since a question does not make sense unless it addresses the effect on SAT after these two variables associated with the selection bias are accounted for. After performing a transformation of the percentage takers and setting Alaska aside, the Cp plot selects the model *tyer* with log takers, years (studying natural science, social science, and humanities), expenditure, and median rank as explanatory variables.

Sex Discrimination Study

Model selection in this example is motivated by the need to account fully for the nondiscriminatory explanatory variables before the sex indicator variable is introduced. Initial scatterplots indicate the important explanatory variables, which permit fitting a tentative model on which residual analysis can be conducted. After transformation, the analysis consists of using a variable selection technique to identify a good set of variables for explaining beginning salary. An inferential model is formed by this set plus the sex indicator variable, making precise the objective of seeing whether sex constitutes an important explanatory factor *after* everything else is accounted for. It may be disconcerting to realize that the different variable selection routines indicate different subset models; but it is reassuring that the estimated sex effect is about the same in all cases.

12.9 EXERCISES

Conceptual Exercises

1. State SATs. True or false? If the coefficient of the public school percentage is not significant (p-value > 0.05) in one model, it cannot be significant in any model found by adding new variables.

2. State SATs. True or false? If the coefficient of income is significant (p-value < 0.05) in one model, it will also be significant in any model that is found by adding new variables.

3. State SATs. Why are partial residual plots useful for this particular data problem?

4. Sex Discrimination. Suppose that another explanatory variable is length of hair at time of hire. (a) Will the estimated sex difference in beginning salaries be greater, less, or unchanged when length of hair is included in the set of explanatory variables for adjustment? (b) Why should this variable not be used?

5. Suppose that the variance of the estimated slope in the simple regression of Y on X_1 is 10. Suppose that X_2 is added to the model, and that X_2 is uncorrelated with X_1. Will the variance of the coefficient of X_1 still be 10?

6. In a study of mortality rates in major cities, researchers collected four weather-related variables, eight socioeconomic variables, and three air-pollution variables. Their primary question concerned the effects, if any, of air pollution on mortality, after accounting for weather and socioeconomic differences among the cities. (a) What strategy for approaching this problem should be adapted? (b) What potential difficulties are involved in applying a model-selection strategy similar to the one used in the sex-discrimination study?

7. In the Cp plots shown in Display 12.9 and Display 12.11, the model with all available explanatory variables falls on the line (with intercept 0 and slope 1), meaning that the value of Cp for this model is exactly p. Will this always be the case? Why?

8. What is the usual interpretation of the probability of an event E? How does a Bayesian interpretation differ?

9. How does the posterior function, discussed in Section 12.5, account for model-selection uncertainty?

Computational Exercises

10. A, B, and C are three explanatory variables in a multiple linear regression with $n = 28$ cases. Display 12.15 shows the residual sums of squares and degrees of freedom for all models.

DISPLAY 12.15 Data for Exercise 10

Model variables	Residual sum of squares	Degrees of freedom
None	8,100	27
A	6,240	26
B	5,980	26
C	6,760	26
AB	5,500	25
AC	5,250	25
BC	5,750	25
ABC	5,160	24

(a) Calculate the estimate of σ^2 for each model. (b) Calculate the adjusted R^2 for each model. (c) Calculate the Cp statistic for each model. (d) Calculate the BIC for each model. (e) Which model has (i) the smallest estimate of σ^2? (ii) the largest adjusted R^2? (iii) the smallest Cp statistic? (iv) the smallest BIC?

11. Using the residual sums of squares from Exercise 10, find the model indicated by forward selection. (Start with the model "None," and identify the single-variable model that has the smallest residual sum of squares. Then perform an extra-sum-of-squares F-test to see whether that variable is significant. If it is, find the two-variable model that includes the first term and has the smallest residual sum of squares. Then perform an extra-sum-of-squares F-test to see whether the additional variable is significant. Continue until no F-statistics greater than 4 remain for inclusion of another variable.)

12. Again referring to Exercise 10, calculate $\exp\{-\text{BIC} + \text{BIC}_{\text{min}}\}$ for each model. Add these and divide each by the sum. What is the resulting posterior distribution on the models?

13. Use the computer to simulate 100 data points from a normal distribution with mean 0 and variance 1. Store the results in a column called Y. Repeat this process 10 more times, storing results in $X_1, X_2, \ldots, X_{10}$. Notice that the Y should be totally unrelated to the explanatory variables. (a) Fit the regression of Y on all 10 explanatory variables. What is R^2? (b) What model is suggested by forward selection? (c) Which model has the smallest Cp statistic? (d) Which model has the smallest BIC? (e) What danger (if any) is there in using a variable selection technique when the number of explanatory variables is a substantial proportion of the sample size?

14. Blood–Brain Barrier. Using the data in Display 11.3 (file case1102), perform the following variable-selection techniques to find a subset of the covariates—days after inoculation, tumor weight, weight loss, initial weight, and sex—for explaining log of the ratio of brain tumor antibody count to liver antibody count. (a) Cp plot (b) forward selection (c) backward elimination (d) stepwise regression.

15. Blood–Brain Barrier. Repeat Exercise 14, but include sacrifice time (treated as a factor with three levels), treatment, and the interaction of sex and treatment with the other explanatory variables.

16. Sex Discrimination. The analysis in this chapter focused on beginning salaries. Another issue is whether annual salary increases tended to be higher for males than for females. If an annual raise of $100r\%$ is received in each of N successive years of employment, the salary in 1977 is: $\text{Sal77} = \text{SalBeg} \times (1 + r)^N$. Seniority measures the number of months of employment, so the number of years of employment is $N = \text{seniority}/12$; and

$$\log(1 + r) = (12/\text{seniority}) \times (\text{Sal77}/\text{SalBeg}) = z(\text{say})$$

for each individual. Calculate z from beginning salary, 1977 salary, and seniority; then calculate r as $\exp(z) - 1$. Now consider r as a response variable (the average annual raise). (a) Use a two-sample t-test to see whether the distribution of raises is different for males than for females. (Is a transformation necessary?) (b) What evidence is there of a sex effect after the effect of age on average raise has been accounted for? (c) What evidence is there of a sex effect after the effects of age and beginning salary have been accounted for?

17. Pollution and Mortality. Display 12.16 shows the complete set of variables for the problem introduced in Exercise 11.23. The 15 variables for each of 60 cities are (1) mean annual precipitation (in inches); (2) percent relative humidity (annual average at 1 P.M.); (3) mean January temperature (in degrees Fahrenheit); (4) mean July temperature (in degrees Fahrenheit); (5) percentage of the population aged 65 years or over; (6) population per household; (7) median number of school years completed by persons of age 25 years or more; (8) percentage of the housing that is sound with all facilities; (9) population density (in persons per square mile of urbanized area); (10) percentage of 1960 population that is nonwhite; (11) percentage of employment in white-collar occupations; (12) percentage of households with annual income under \$3,000 in 1960; (13) relative pollution potential of hydrocarbons (HC); (14) relative pollution potential of oxides of nitrogen (NO_X); and (15) relative pollution potential of sulphur dioxide (SO_2). (See Display 11.22 for the city names.) It is desired to determine whether the pollution variables (13, 14, and 15) are associated with mortality, after the other climate and socioeconomic variables are accounted for. (*Note:* These data have problems with influential observations and with lack of independence due to spatial correlation; these problems are ignored for purposes of this exercise.)

- **(a)** With mortality as the response, use a Cp plot and the BIC to select a good-fitting regression model involving weather and socioeconomic variables as explanatory. To the model with the lowest Cp, add the three pollution variables (transformed to their logarithms) and obtain the p-value from the extra-sum-of-squares F-test due to their addition.
- **(b)** Repeat part (a) but use a sequential variable selection technique (forward selection, backward elimination, or stepwise regression). How does the p-value compare?

18. Suppose that a problem involves four explanatory variables (like the weather variables in Exercise 17). How many variables would be included in the corresponding saturated second-order model (Section 12.7.3)?

DISPLAY 12.16 Pollution and mortality data for 60 cities; first 6 of 60 rows

		Weather				Demographic								Pollution potentials		
SMSA	Mortality (d/100.000)	Precipitation (in/yr)	Humidity (annual av%)	January temperature (F)	July temperature (F)	Over 65(%)	Population per household	Education (years)	Housing (% sound)	Density (popn/mi²)	Nonwhite(%)	White collar(%)	Poor(%<$3,000)	Hydrocarbons	Nitrogen oxides	Sulfur dioxide
SJCA	790.733	13	71	49	68	7.0	3.36	12.2	90.7	2702	3.0	51.9	9.7	105	32	3
WIKS	823.764	28	54	32	81	7.0	3.27	12.1	81.0	3665	7.5	51.6	13.2	4	2	1
SDCA	839.709	10	61	55	70	7.3	3.11	12.1	88.9	3033	5.9	51.0	14.0	144	66	20
LAPA	844.053	43	54	32	74	10.1	3.28	9.5	79.2	3214	2.9	43.7	12.0	11	7	32
MIMN	857.622	25	58	12	73	9.2	3.28	12.1	83.1	2095	2.0	51.9	9.8	20	11	26
DATX	860.101	35	54	46	85	7.1	3.22	11.8	79.9	1441	14.8	51.2	16.1	1	1	1

19. In the expression (Section 12.7.3) for the number of subset models with p parameters, the index j of summation refers to the number of original variables in a particular model. If there are a total of five original explanatory variables, the total number of subset models with seven parameters that have exactly three original variables is the jth term in the sum, $C_{5,3} \times C_{6,3} = 10 \times 20 = 200$. The reasoning here is that there are 10 ways to select three of the original five variables; and that, with three variables, there are three quadratic plus three product terms, or a total of six second-order terms; so, to make up a model with seven parameters, you have the constant (1) and the original variables (3), so you need $7 - 1 - 3 = 3$ second-order terms from the six available. Problem: Let the original variables be a, b, c, d, and e. Select any three of them. Then write down all 20 of the seven-parameter models involving only those three original variables.

Data Problems

20. Galapagos Islands. The data in Display 12.17 come from a 1973 study. (Data from M. P. Johnson and P. H. Raven, "Species Number and Endemism: The Galapagos Archipelago Revisited," *Science* 179 (1973): 893–5.) The number of species on an island is known to be related to the island's area. Of interest is what other variables are also related to the number of species, after island area is accounted for, and whether the answer differs for native species and nonnative species. (*Note:* Elevations for five of the islands were missing and have been replaced by estimates for purposes of this exercise.)

21. Predicting Desert Wildflower Blooms. Southwestern U.S. desert wildflower enthusiasts know that a triggering rainfall between late September and early December and regular rains through March often lead to a good wildflower show in the spring. Display 12.18 is a partial listing of a data set that might help with the prediction. It includes monthly rainfalls from September to March and the subjectively rated quality of the following spring wildflower display for each of a number of years at each of four desert locations in the southwestern United States (Upland Sonoran Desert near Tucson, the lower Colorado River Valley section of the Sonoran Desert, the Baja California region of the Sonoran Desert, and the Mojave Desert). The quality of the display was judged subjectively with ordered rating categories of poor, fair, good, great, and spectacular. The column labeled

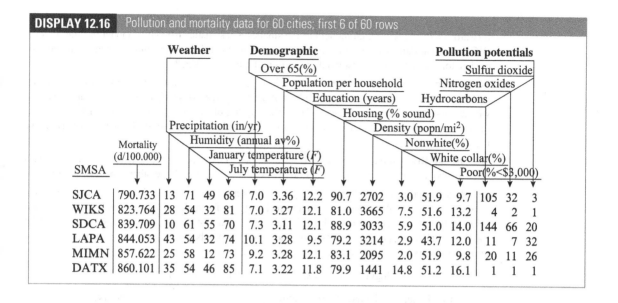

DISPLAY 12.17 Plant species and geography of the Galapagos Islands; first 5 rows of 30

Island	Observed species: Total	Observed species: Native	Area (km^2)	Elevation (m)	Distance(km): From nearest island	Distance(km): From Santa Cruz	Area of nearest island (km^2)
Baltra	58	23	25.09	332	0.6	0.6	1.84
Bartolome	31	21	1.24	109	0.6	26.3	572.33
Caldwell	3	3	0.21	114	2.8	58.7	0.78
Champion	25	9	0.10	46	1.9	47.4	0.18
Coamano	2	1	1.05	130	1.9	1.9	903.82

DISPLAY 12.18 Monthly rainfalls (inches) for seven months and rated quality of wildflower display in the spring, for multiple years in each of four desert regions; first 5 of 122 rows

Year	Region	Sep	Oct	Nov	Dec	Jan	Feb	Mar	Total	Rating	Score
1970	baja	0.27	0.00	0.00	0.00	0.00	0.04	0.12	0.43	Poor	0
1971	baja	1.35	0.00	0.00	0.12	0.00	0.04	0.00	1.51	Poor	0
1972	baja	0.00	0.01	0.01	0.04	0.00	0.00	0.00	0.06	Poor	0
1974	baja	0.00	0.00	0.00	0.00	0.04	0.00	0.00	0.04	Poor	0
1975	baja	0.00	0.04	0.00	0.00	0.00	0.00	0.00	0.04	Poor	0

Score is a numerical representation of the rating, with 0 for poor, 1 for fair, 2 for good, 3 for great, and 4 for spectacular. Although this is a made-up number and the suitability of regression for such a discrete response is questionable, an informal regression analysis might nevertheless be helpful for casual prediction. Analyze the data to find an equation for predicting the quality score from the monthly rainfalls. What is the predicted quality score for a season with these rainfall amounts in the Upland region: Sep: 0.45, Oct: 0.02, Nov: 0.80, Dec: 0.76, Jan: 0.17, Feb: 1.22, and Mar: 0.37? Also find a 95% prediction interval. Although the conditions are far from the ideal normality assumptions for the justification of the prediction interval, the rough prediction interval might be a useful way to clarify the precision with which the quality score can actually be predicted. (Data from Arizona-Sonora Desert Museum, "Wildflower Flourishes and Flops—a 50-Year History," www.desertmuseum.org/programs/flw_wildflwrbloom.html (July 25, 2011).)

22. **Bush–Gore Ballot Controversy.** Review the Palm Beach County ballot controversy description in Exercise 8.25. To estimate how much of Pat Buchanan's vote count might have been intended for Al Gore in Palm Beach County, Florida, that exercise required the fitting of a model for predicting Buchanan's count from Bush's count from all other counties in Florida (excluding Palm Beach), followed by the comparison of Buchanan's actual count in Palm Beach to a prediction interval. One might suspect that the prediction interval can be narrowed and the validity of the procedure strengthened by incorporating other relevant predictor variables. Display 12.19 shows the first few rows of a data set containing the vote counts by county in Florida for Buchanan and for four other presidential candidates in 2000, along with the total vote counts in 2000, the presidential vote counts for three presidential candidates in 1996, the vote count for Buchanan in his only other campaign in Florida—the 1996 Republican primary, the registration in Buchanan's Reform party, and the total

DISPLAY 12.19 Buchanan's 2000 presidential vote count, and predictor variables, in 67 Florida counties (first five rows)

County	Buchanan 2000	Gore 2000	Bush 2000	Nader 2000	Browne 2000	Total 2000	Clinton 1996	Dole 1996	Perot 1996	Buchanan 1996	Reform reg. 2000	Total reg. 2000
Alachua	262	47,300	34,062	3,215	658	85,235	40,144	25,303	8,072	2,151	91	120,867
Baker	73	2,392	5,610	53	17	8,072	2,273	3,684	667	73	4	12,352
Bay	248	18,850	38,637	828	171	58,486	17,020	28,290	5,922	1,816	55	92,749
Bradford	65	3,072	5,413	84	28	8,597	3,356	4,038	819	155	3	13,547
Brevard	570	97,318	115,185	4,470	643	217,616	80,416	87,980	25,249	7,927	148	283,680
...												

political party registration in the county. Analyze the data and write a statistical summary predicting the number of Buchanan votes in Palm Beach County that were not intended for him. It would be appropriate to describe any unverifiable assumptions used in applying the prediction equation for this purpose. (*Suggestion:* Find a model for predicting Buchanan's 2000 vote from other variables, excluding Palm Beach County, which is listed last in the data set. Consider a transformation of all counts.)

23. Intelligence and Class as Predictors of Future Income (Males only). In their 1994 book, *The Bell Curve: Intelligence and Class Structure in American Life*, psychologist Richard Hernstein and political scientist Charles Murray argued that a person's intelligence is a better predictor of success in life than is education and family's socioeconomic staus. The book was controversial mostly for its conclusions about intelligence and race, but also for the strength of its conclusions drawn from regression analyses on observational data using imperfect measures of intelligence and socioeconomic status. Hernstein and Murray used data from the National Longitudinal Survey of Youth (NLSY79). Display 12.20 lists the variables in the ex1223 data file on a subset of 2,584 individuals from the NLSY79 survey who were re-interviewed in 2006, who had paying jobs in 2005, and who had complete values for the listed variables, as previously described in Exercises 2.22 and 3.30. *For males only*, see whether intelligence (as measured by the ASVAB intelligence test score, *AFQT*, and its Components, *Word, Parag, Math*, and *Arith*) is a better predictor of 2005 income than education and socioeconomic status (as measured by the variables related to respondent's class and family education in 1979). Suggestion: First use exploratory techniques to decide on transformations of the response variable *Income2005*. Then, Stage 1: Use a variable selection procedure to find a subset of variables from this set: *Imagazine, Inewspaper, Ilibrary, MotherEd, FatherEd, FamilyIncome78*, and *Educ* to explain the distribution of *Income2005*. Note the percentage of variation explained by this model. Now use a variable selection procedure to find a subset starting with this same initial set of variables and the ASVAB variables. By how much is R^2 increased when the intelligence measures are included? Use an extra-sum-of-squares F-test to find a p-value for statistical significance of all the ASVAB variables in the final model (i.e., after accounting for *Educ* and the class variables). Stage 2 (reverse the order): Use a variable selection procedure to find a subset of variables from the set of all ASVAB variables. Note the percentage of variation explained by this model. Now use a variable selection procedure to find a subset from this same initial set of variables plus *Educ* and the family class, education, and 1979 income variables. By how much is R^2 increased when *Educ* and the class variables are included? Find a p-value for statistical significance of the class variables after accounting for education and ASVAB variables. Use the Stage 1 and Stage 2 results to draw conclusions about whether the class variables or the ASVAB variables are better predictors of 2005 income. (As noted in Exercise 3.30, the coded incomes greater than \$150,000 in NLSY79 were replaced in this data file by computer-simulated values to better match a true distribution of incomes.)

DISPLAY 12.20 Variables measured on 2,584 Americans who were between 14 and 22 in 1979, who took the Armed Services Vocational Aptitude Battery (AFVAB) of tests in 1981, who were available for re-interview in 2006, who had paying jobs in 2005, and who had complete records of the variables listed below

Variables Related to Family Class, Education, and Income in 1979

Imagazine	1 if any household member magazines regularly when respondent was about 14
Inewspaper	1 if any household member newspapers regularly when respondent was about 14
Ilibrary	1 if any household member had a library card when respondent was about age 14
MotherEd	Mother's years of education
FatherEd	Father's years of education
FamilyIncome78	Family's total net income in 1978

Personal Demographic Variables

Race	1 = Hispanic, 2 = Black, 3 = Non-Hispanic, Non-Black
Gender	Female or male
Educ	Years of education completed by 2006

Variables Related to ASVAB Test Scores in 1981

Science	Score on General Science component
Arith	Score on Arithmetic Reasoning component
Word	Score on Word Knowledge component
Parag	Score on Paragraph Comprehension component
Numer	Score on Numerical Operations component
Coding	Score on Coding Speed component
Auto	Score on Automotive and Shop Information component
Math	Score on Mathematics Knowledge component
Mechanic	Score on Mechanical Comprehension component
Elec	Score on Electronics Information component
AFQT	Armed Forces Qualifying Test Score (a combination of *Word, Parag, Math,* and *Arith*)

Variables Related to Life Success in 2006

Income2005	Total income from wages and salary in 2005
Esteem1	"I am a person of worth" 1: strongly agree, 2: agree, 3: disagree, 4: strongly disagree
Esteem2	"I have a number of good qualities" 1: strongly agree, 2: agree, 3: disagree, 4: strongly disagree
Esteem3	"I am inclined to feel like a failure" 1: strongly agree, 2: agree, 3: disagree, 4: strongly disagree
Esteem4	"I do things as well as others" 1: strongly agree, 2: agree, 3: disagree, 4: strongly disagree
Esteem5	"I do not have much to be proud of" 1: strongly agree, 2: agree, 3: disagree, 4: strongly disagree
Esteem6	"I take a positive attitude towards myself and others" 1: strongly agree, 2: agree, 3: disagree, 4: strongly disagree
Esteem7	"I am satisfied with myself" 1: strongly agree, 2: agree, 3: disagree, 4: strongly disagree
Esteem8	"I wish I could have more respect for myself" 1: strongly agree, 2: agree, 3: disagree, 4: strongly disagree
Esteem9	"I feel useless at times" 1: strongly agree, 2: agree, 3: disagree, 4: strongly disagree
Esteem10	"I think I am no good at all" 1: strongly agree, 2: agree, 3: disagree, 4: strongly disagree

24. Intelligence and Class as Predictors of Future Income (Females only). Repeat Exercise 23 for the subset of females.

25. Gender Differences in Wages. Display 12.21 is a partial listing of a data set with weekly earnings for 9,835 Americans surveyed in the March 2011 Current Population Survey (CPS). What evidence is there from these data that males tend to receive higher earnings than females with the same values of the other variables? By how many dollars or by what percent does the male distribution exceed the female distribution? Note that there might be an interaction between *Sex* and *Marital Status*. (Data from U.S. Bureau of Labor Statistics and U.S. Bureau of the Census: Current Population Survey, March 2011 http://www.bls.census.gov/cps_ftp.html#cpsbasic; accessed July 25, 2011.)

DISPLAY 12.21 Region of the United States (Northeast, Midwest, South, or West) where individual worked, Metropolitan Status (Metropolitan, Not Metropolitan, or Not Identified), Age (years), Sex (Male or Female), Marital Status (Married or Not Married), EdCode (corresponding roughly to increasing education categories), Education (16 categories), Job Class (Private, Federal Government, State Government, Local Government, or Private), and Weekly Earnings (in U.S. dollars) for 9,835 individuals surveyed in the March 2011 Current Population Survey; first 5 of 9,835 rows

Region	MetropolitanStatus	Age	Sex	MaritalStatus	Edcode	Education	JobClass	WeeklyEarnings
Northeast	Not Metropolitan	20	Male	Not Married	39	HighSchoolDiploma	Private	467.50
West	Metropolitan	59	Male	Married	43	BachelorsDegree	Private	1,269.00
West	Metropolitan	62	Male	Married	34	SeventhOrEighthGrade	Private	1,222.00
West	Metropolitan	39	Male	Married	39	HighSchoolDiploma	Private	276.92
South	Not Metropolitan	60	Female	Married	36	TenthGrade	Private	426.30

Answers to Conceptual Exercises

1. False. In the model with P and E, neither variable is significant (Display 12.6). But both are significant in the model *PER*.

2. False. Both income and rank are significant in the model *IR*. When T is included, however, neither is significant. (See Display 12.6.)

3. There are two reasons. (i) Percentage of takers and median rank explain so much of the variation that any additional effect of the other variables is hidden in the ordinary scatterplots. (ii) The questions of interest call for the examination of some of the variables after getting percentage of takers and median rank out of the way. The partial residual plots allow visual investigation into these questions.

4. (a) The estimated sex difference will probably be less after length of hair is accounted for. If there are differences between male and female hair lengths, then that variable is picking up the sex differences and the sex indicator variable will be less meaningful when it is included. (b) The males and females should be compared after adjustment for nondiscriminatory determinants of salary.

5. If X_2 explains some variation in Y in addition to what is explained by X_1, σ will be smaller, so the variance of the coefficient of X_1 will decrease.

6. (a) Use model selection tools to find a good-fitting model involving the weather and socioeconomic variables. Then include the pollution variables, using t- and F-statistics to judge importance.

As an alternative, employ the Bayesian strategy of averaging the pollution effects over a wide range of models. (b) A key weather variable (like humidity) may not be directly related to mortality, but it may interact with air pollution variables to affect mortality. A model selection method that chooses one "best" model may leave the key variable out. (See Exercise 17.)

7. Yes. Notice what happens to the formula for Cp when the model under investigation is also the "full" model: the second term is zero, so Cp $= p$.

8. The probability of E is taken to be the proportion of times when E occurs in a long run of trials. Bayesian statistics also interprets probability as a measure of belief. If M is a particular model among many, a Bayesian probability for M would be the proportion of one's total belief that is assignable to the belief that M is the correct model.

9. Estimates of the treatment effect can be made with all models involving confounding variables. The posterior function draws its conclusions based on all these estimates, with weights assigned according to the strength of the evidence supporting each model.

CHAPTER 18

Comparisons of Proportions or Odds

This chapter returns to the elementary setting of comparing two groups, but for the special case where the response measurement on each subject is binary (0 or 1). A binary variable is a way of coding a two-group categorical response—like dead or alive, or diseased or not diseased—into a number. Statistical analysis of binary responses leads to conclusions about two population proportions or probabilities. Alternatively, conclusions may be stated about odds—like the odds of death or the odds of disease.

The discussion here involves large-sample tests and confidence intervals for comparing two proportions or two odds. The next three chapters provide extensions that are analogous to the analysis of variance and regression extensions of the two-sample t-tools. In particular, logistic regression in Chapters 20 and 21 permits regression modeling with binary response variables.

18.1 CASE STUDIES

18.1.1 Obesity and Heart Disease—An Observational Study

There are two schools of thought about the relationship of obesity and heart disease. Proponents of a physiological connection cite several studies in North America and Europe that estimate higher risks of heart problems for obese persons. Opponents argue that the strain of social stigma brought on by obesity is to blame. One study sheds some light on this controversy, because it was conducted in Samoa, where obesity is not only common but socially desirable. As part of that study, the researcher categorized subjects as obese or not, according to whether their 1976 weight divided by the square of their height exceeded 30 kg/m^2. Shown in Display 18.1 are the numbers of women in each of these categories who died and did not die of cardiovascular disease (CVD) between 1976 and 1981. (Data from D. E. Crews, "Cardiovascular Mortality in American Samoa," *Human Biology* 60 (1988): 417–33.) Is CVD death in the population of Samoan women related to obesity?

DISPLAY 18.1 Cardiovascular deaths and obesity among women in American Samoa

	CVD death		
	Yes	No	Totals
Obese	16	2,045	2,061
Not obese	7	1,044	1,051
Totals	23	3,089	3,112

Statistical Conclusion

The proportion of CVD deaths among the obese women (0.00776 or 7.76 deaths per 1,000 women) was slightly higher than the corresponding proportion among nonobese women (0.00666). However, the estimated difference in population proportions, 0.00110, is small relative to its standard error (0.00325). The data are consistent with the hypothesis of equal proportions of CVD deaths in the populations of obese and nonobese Samoan women (one-sided p-value $=$ 0.37).

Scope of Inference

The sample was large, comprising approximately 60% of the adult population of Tutuila, the main island. The sample's age distribution was similar to the population's, so the 3,112 women may be assumed to constitute a representative sample from the population of all American Samoan women in 1976. Although the results cannot be extrapolated to populations of women outside of American Samoa, the small difference in CVD rates between obese and nonobese women in a population in which obesity is socially acceptable lends some credibility to the argument that the social stigma rather than the obesity itself may be responsible for the association of obesity and CVD in North America and Europe.

18.1.2 Vitamin C and the Common Cold—A Randomized Experiment

Linus Pauling, recipient of Nobel Prizes in Chemistry and in Peace, advocated the use of vitamin C for preventing the common cold. A Canadian experiment examined this claim, using 818 volunteers. At the beginning of the winter, subjects were randomly divided into two groups. The ***vitamin C*** group received a supply of vitamin C pills adequate to last through the entire cold season at 1,000 mg per day. The ***placebo*** group received an equivalent amount of inert pills. At the end of the cold season, each subject was interviewed by a physician who did not know the group to which the subject had been assigned. On the basis of the interview, the physician determined whether the subject had or had not suffered a cold during the period. The results are shown in Display 18.2. (Data from T. W. Anderson, D. B. W. Reid, and G. H. Beaton, "Vitamin C and the Common Cold," *Canadian Medical Association Journal* 107 (1972): 503–08.) Can the risk of a cold be reduced by using vitamin C?

DISPLAY 18.2 Vitamin C and the common cold

	Outcome		
	Cold	No Cold	Totals
Placebo	335	76	411
Vitamin C	302	105	407
Totals	637	181	818

Statistical Conclusion

Of the 411 subjects who took the placebo pill, 82% caught colds at some time during the winter. Among the 407 who took vitamin C pills, however, only 74% caught colds. The results provide strong evidence that the probability of catching cold was smaller for those who took vitamin C (one-sided p-value $= 0.0059$), lending strong support to Dr. Pauling's convictions. The odds of catching a cold while on the placebo regimen are estimated to be 1.10 times to 2.14 times as large as the odds for catching one while on the vitamin C regimen (approximate 95% confidence interval).

Scope of Inference

Because the treatments were randomly assigned to the available volunteers, the difference in cold rates in these groups can safely be attributed to differences in treatments, but it is important to understand exactly what the treatments entail.

One essential point is that the randomized experiment was *double-blind*, meaning that neither the subject nor the doctor evaluating the subject knew the treatment that the subject received. Without this precautionary step, knowledge of the treatment received would become another difference between the two groups and might be responsible for statistical differences. In other vitamin C experiments that were constructed to be double-blind, subjects correctly guessed—by taste—which treatment they were receiving. Surprisingly, the proportions with colds in that study depended significantly on which treatment the subjects *thought* they were receiving. Care was taken in the Canadian study to put citric acid and orange flavoring in the placebo pills, so the subjects were less likely to "break the blind."

18.1.3 Smoking and Lung Cancer—A Retrospective Observational Study

In an investigation of the association of smoking and lung cancer, 86 lung cancer patients (thought of as a random sample of all lung cancer patients at the hospitals studied) and 86 controls (thought of as a random sample of subjects without lung cancer from the same communities as the lung cancer patients) were interviewed about their smoking habits. Display 18.3 shows the numbers in each of these groups who did and did not smoke. (Data from H. F. Dorn, "The Relationship of Cancer of the Lung and the Use of Tobacco," *American Statistician* 8 (1954): 7–13.) Is there any evidence that the odds of lung cancer are greater for smokers than for nonsmokers? What is an estimate of the increased odds associated with smoking?

DISPLAY 18.3 Smoking and lung cancer

	Outcome		
	Cancer	Control	Totals
Smokers	83	72	155
Nonsmokers	3	14	17
Totals	86	86	172

Statistical Conclusion

The data provide convincing evidence that the odds of lung cancer are greater for smokers than for nonsmokers (approximate one-sided p-value $=$ 0.005). The odds of lung cancer for smokers are estimated to be 5.4 times the odds of lung cancer for nonsmokers (approximate 95% confidence interval: 1.5 times to 19.5 times).

Scope of Inference

This is an observational study, so whether smoking causes lung cancer cannot be addressed by the statistical analysis. Furthermore, the inference to a wider population is only appropriate to the extent that the samples of cancer patients and controls constitute random samples from populations with and without cancer.

These data are *retrospective*, meaning that samples were taken from each response group rather than from each explanatory group. The inferences allowed from this kind of sampling are discussed in Section 18.4.1.

18.2 INFERENCES FOR THE DIFFERENCE OF TWO PROPORTIONS

18.2.1 The Sampling Distribution of a Sample Proportion

Binary Response Variables

Suppose that a population of units is classified as either yes or no. For example, each member of the population of Samoan women described in Section 18.1.1 may be classified as yes if she died of cardiovascular disease in the six years of the study or no if she did not. Although this type of categorical response does not provide a numerical outcome, it is useful to manufacture one. For each member of the population a response variable, Y, is defined to be 1 for a yes and 0 for a no. Of course, yes and no may be replaced by any other label that distinguishes two levels of a categorical response.

Indicator variables with values of 0 or 1 have been used in previous chapters as *explanatory* variables in regression. They may also be used as *response* variables, in which case they are referred to as *binary response variables*. The average value of a binary response variable in a population is the proportion of members of the population that are classified as yes. The proportion of yes responses in the population is represented by the symbol π and is called the *population proportion*. If the population is hypothetical or if proportions are being compared in a randomized experiment, it is more common to refer to π as the *probability* of a yes response.

The Variance of a Binary Response Variable

Because a binary response variable can have only two numerical values, its variance is an exact function of its mean. Imagine what a listing of the responses for a population of $N = 1{,}000{,}000$ subjects might look like. If $\pi = 0.72$, then 720,000 responses would be 1's and the remaining 280,000 would be 0's. All the Y's would sum to 720,000, so the mean of the Y's would be $0.72 = \pi$. The population variance of a variable is the average of the squared difference of the Y's from their mean. In this population, the value of $(Y - \pi)^2$ is either $(1 - \pi)^2$ or $(0 - \pi)^2$, depending on whether Y is 1 or 0; so the set of all $(Y - \pi)^2$ values adds to $720{,}000(1 - \pi)^2 + 280{,}000(0 - \pi)^2$. Dividing by 1,000,000 produces the variance. Since 720,000 divided by 1,000,000 is π and 280,000 divided by 1,000,000 is $(1 - \pi)$, the operation yields

$$\begin{aligned}\text{Variance}(Y) &= \text{Mean}\{(Y - \pi)^2\} = \pi(1 - \pi)^2 + (1 - \pi)\pi^2 \\ &= \pi(1 - \pi)[(1 - \pi) + \pi] = \pi(1 - \pi).\end{aligned}$$

> *If Y is a binary response variable with population mean π, then*
>
> $$\textit{Variance}\,\{Y\} = \pi(1-\pi).$$

Two features make this unlike the models for response variables considered in earlier chapters: There is no additional parameter, like σ^2; and the variance is a function of the mean.

The Sample Total and the Binomial Distribution

Suppose that $\{Y_1, Y_2, \ldots, Y_n\}$ is a random sample from a population of binary response variables. The sum of these binary responses, $S = Y_1 + Y_2 + \cdots + Y_n$, is a count of the number of subjects for which $Y = 1$. The variable S has one of the oldest known and most intensively studied statistical distributions—the *binomial distribution*. S will have one of the values: $0, 1, \ldots, n$. The exact probability for S to be equal to the integer k is given by the formula

$$\Pr\{S = k\} = \frac{n!}{k!(n-k)!}\pi^k(1-\pi)^{n-k}.$$

The mean of S is $n\pi$; the variance of S is $n\pi(1-\pi)$; and the distribution of S is skewed except when $\pi = 1/2$. When the products $n\pi$ and $n(1-\pi)$ are both of moderate size (> 5), the normal distribution approximates binomial probabilities very closely—a fact that motivates many of the succeeding developments in this chapter.

The Sample Proportion

Let $\hat{\pi}$ represent the average of the binary responses in a random sample. Although it is an average, the symbol $\hat{\pi}$ is used rather than $\overline{Y}$ to emphasize that it is also the sample proportion. In the sample of 2,061 obese Samoan women, 16 died of CVD during the 6-year study period. The sample proportion $\hat{\pi}$ is therefore 0.00776.

The Sampling Distribution of the Sample Proportion

Since $\hat{\pi}$ is an average, standard statistical theory about the sampling distribution of an average applies:

> If $\hat{\pi}$ is a sample proportion based on a sample of size n from a population with population proportion π, then
>
> **1.** Mean$\{\hat{\pi}\} = \pi$.
> **2.** Var$\{\hat{\pi}\} = \pi(1-\pi)/n$.
> **3.** If n is large enough, the sampling distribution of $\hat{\pi}$ is approximately normal.

DISPLAY 18.4 Confidence interval for the proportion of cardiovascular disease deaths in the population of obese Samoan women

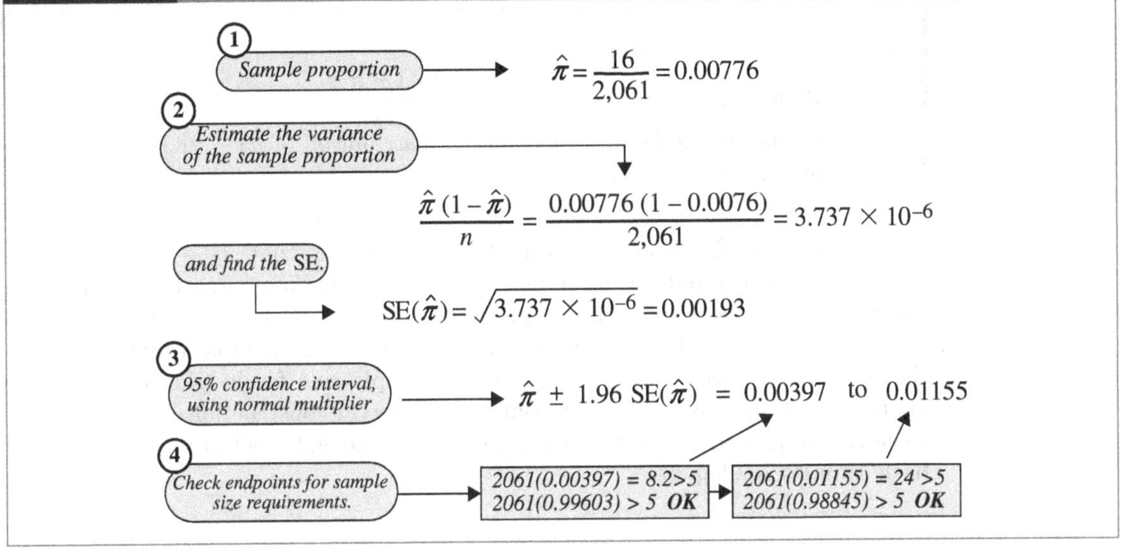

Procedures exist for drawing inferences based on the exact distribution of the sample proportion, but methods that rely on the normal approximation are adequate for many applications. *How large must the sample size be?* This is not an easy question to answer, because the answer depends on π. If π is near one-half, the sampling distribution is nearly normal for sample sizes as low as 5 to 10. If π is extremely close to one or zero, a sample size of 100 may not be adequate. For testing a hypothesis that π is some number π_0, it is generally safe to rely on the normal approximation if $n\pi_0 > 5$ and $n(1 - \pi_0) > 5$. A normal approximation confidence interval for π is generally safe if $n\pi_0 > 5$ and $n(1-\pi_0) > 5$ are satisfied by both the lower and upper endpoints of the confidence interval. A confidence interval, with these informal checks included, is demonstrated in Display 18.4. It is important to realize that the t-distribution is not involved here. The t-tools discussed earlier are based on normally distributed responses. In particular, the degrees of freedom attached to the t-tools are those associated with an estimate of σ. For binomial responses the variance is a known function of the mean, so there are no analogous degrees of freedom.

18.2.2 Sampling Distribution for the Difference Between Two Sample Proportions

The question of interest in the Samoan women study is whether the proportion of CVD deaths in the population of obese Samoan women differs from the proportion of CVD deaths in the population of nonobese Samoan women. What can be inferred about the difference in population proportions from the sample statistics? If $\hat{\pi}_1$ and

$\hat{\pi}_2$ are computed from independent random samples, the sampling distribution of their difference has these properties:

1. $\text{Mean}\{\hat{\pi}_2 - \hat{\pi}_1\} = \pi_2 - \pi_1$
2. $\text{Variance}\{\hat{\pi}_2 - \hat{\pi}_1\} = \dfrac{\pi_1(1-\pi_1)}{n_1} + \dfrac{\pi_2(1-\pi_2)}{n_2}$
3. If n_1 and n_2 are large, the sampling distribution of $\hat{\pi}_2 - \hat{\pi}_1$ is approximately normal.

These properties lead to a test of the hypothesis that $\pi_2 - \pi_1 = 0$ and to construction of a confidence interval for $\pi_2 - \pi_1$, based on a normal approximation. For the test, check whether $n_s\hat{\pi}_c$ and $n_s(1 - \hat{\pi}_c)$ are both greater than 5, where n_s is the smaller of the two sample sizes and π_c is the sample proportion when the two samples are combined into one. A simple rule for judging the adequacy of the normal approximation for confidence interval construction is not available. Usually, the procedure of checking whether $n\hat{\pi}$ and $n(1 - \hat{\pi})$ are larger than 5 in both samples is adequate.

Two Standard Errors for $\hat{\pi}_2 - \hat{\pi}_1$

The standard error of $\hat{\pi}_2 - \hat{\pi}_1$ is the square root of the variance of the sampling distribution, where the unknown parameters are replaced by their best estimates. Values of the best estimates, however, depend on what model is being fit. If the purpose is to get a confidence interval for $\pi_2 - \pi_1$, then the model being fit is the one with separate population proportions; π_1 is then estimated by $\hat{\pi}_1$, and π_2 is estimated by $\hat{\pi}_2$.

For confidence interval:

$$\text{SE}\,(\hat{\pi}_2 - \hat{\pi}_1) = \sqrt{\frac{\hat{\pi}_1(1-\hat{\pi}_1)}{n_1} + \frac{\hat{\pi}_2(1-\hat{\pi}_2)}{n_2}}$$

If, on the other hand, the purpose is to test the equality of the two population proportions, the null hypothesis model is that the two proportions are equal. The best estimate of both π_1 and π_2 in this model is $\hat{\pi}_c$, the sample proportion from the combined sample.

For testing equality:

$$\text{SE}_0\,(\hat{\pi}_2 - \hat{\pi}_1) = \sqrt{\frac{\hat{\pi}_c(1-\hat{\pi}_c)}{n_1} + \frac{\hat{\pi}_c(1-\hat{\pi}_c)}{n_2}}$$

The use of this different standard error for testing underscores the need to describe the sampling distribution of a test statistic supposing the null hypothesis is true.

18.2.3 Inferences About the Difference Between Two Population Proportions

Approximate Test for Equal Proportions

To the extent that the normal distribution adequately describes the sampling distribution of $\hat{\pi}_2 - \hat{\pi}_1$, the ratio

$$z\text{-ratio} = \frac{(\hat{\pi}_2 - \hat{\pi}_1) - (\pi_2 - \pi_1)}{\text{SE}(\hat{\pi}_2 - \hat{\pi}_1)}$$

has a standard normal distribution (a normal distribution with mean = 0 and standard deviation = 1). To test the null hypothesis $H\colon \pi_2 - \pi_1 = 0$, substitute the hypothesized value into the z-ratio, along with the sample information and the test version of the standard error. The hypothesis is rejected if the numerical result—the z-statistic—looks as though it did not come from a standard normal distribution:

$$z\text{-statistic} = \frac{(\hat{\pi}_2 - \hat{\pi}_1) - 0}{\text{SE}_0(\hat{\pi}_2 - \hat{\pi}_1)}.$$

The p-value is the proportion of a standard normal distribution that is more extreme than the observed z-statistic. The computations for the heart disease and obesity study are shown in Display 18.5.

Approximate Confidence Interval for the Difference Between Two Proportions

When the behavior of the z-ratio is adequately described by a standard normal distribution, the following procedure constructs a confidence interval with approximate coverage of $100(1-\alpha)\%$.

> *An approximate* $100(1-\alpha)\%$ *confidence interval for* $\pi_2 - \pi_1$ *is*
>
> $$(\hat{\pi}_2 - \hat{\pi}_1) \pm [z(1-\alpha/2) \times \text{SE}(\hat{\pi}_2 - \hat{\pi}_1)].$$

Here, $z(1-\alpha/2)$ is the $100(1-\alpha/2)$th percentile in the standard normal distribution. For a 95% confidence interval, for example, α is 0.05 and $z(0.975)$ is 1.96. A confidence interval for the vitamin C study is demonstrated in Display 18.6.

DISPLAY 18.5 Test for equality of two population proportions: obesity and CVD deaths

(1) *Sample proportions of CVD deaths*
- *group 2 (obese):* $\hat{\pi}_2 = \frac{16}{2{,}061} = 0.00776$
- *group 1 (not obese):* $\hat{\pi}_1 = \frac{7}{1{,}051} = 0.00666$

(2) *Difference* → $\hat{\pi}_2 - \hat{\pi}_1 = 0.00110$

(3) *Proportion from combined sample* → $\hat{\pi}_c = \frac{16 + 7}{2{,}061 + 1{,}051} = 0.00739$

(4) *Check adequacy of normal approximation.* → *1051(0.00739) = 7.8 > 5; 1051(0.99261) > 5* ***OK***

(5) *Standard error (test version)* → $\sqrt{\frac{(0.00739)(0.99261)}{2{,}061} + \frac{(0.00739)(0.99261)}{1{,}051}} = 0.00325$

(6) *z-statistic* → $z = \frac{0.00110}{0.00325} = 0.340$ (7) *p-value* → $\Pr(Z > 0.340) = \mathbf{0.37}$

Conclusion: There is no evidence that the two proportions differ.

DISPLAY 18.6 95% confidence interval for the difference in two population proportions: vitamin C and the common cold

(1) *Sample proportions who got colds*
- *group 2 (placebo):* $\hat{\pi}_2 = \frac{335}{411} = 0.815$
- *group 1 (vitamin C):* $\hat{\pi}_1 = \frac{302}{407} = 0.742$

(2) *Difference* → $\hat{\pi}_2 - \hat{\pi}_1 = 0.073$

(3) *Check adequacy of normal approximation.* → *411(0.815) > 5; 411(0.185) > 5; 407(0.742) > 5; 407(0.258) > 5* ***OK***

(4) *Standard error (C.I. version)* → $\sqrt{\frac{(0.815)(0.185)}{411} + \frac{(0.742)(0.258)}{407}} = 0.029$

(5) *Multiplier* → $z(0.975) = 1.96$ (6) *Interval half-width* → $(1.96)(0.029) = 0.057$

(7) *95% confidence interval for $\pi_1 - \pi_2$* → $0.073 \pm 0.057 =$ **0.016 to 0.130**

Conclusion: The probability of catching a cold after placebo exceeds the probability of catching a cold after vitamin C by an estimate of 0.073 (95% CI: 0.016 to 0.13).

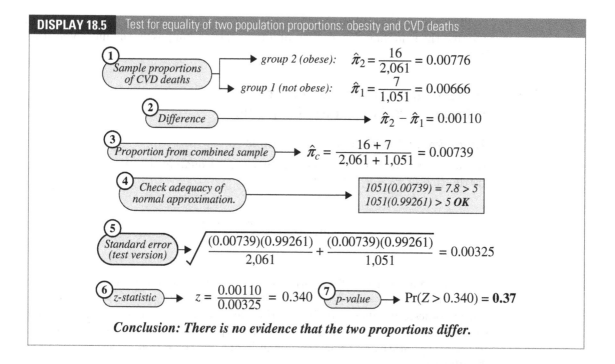

18.3 INFERENCE ABOUT THE RATIO OF TWO ODDS

18.3.1 A Problem with the Difference Between Proportions

There are difficulties with interpreting the difference between two population proportions or two probabilities. If π_1 is the probability of a disease in the absence of treatment, and if π_2 is the probability of a disease under treatment with a preventive drug, then a difference of 0.05 may be rather small if the true probabilities are 0.50 and 0.45, yet the same difference can have considerable practical significance if the true probabilities are 0.10 and 0.05. In both cases the disease rate can be reduced by 5% if the drug is used; but in the first case the treatment will prevent the disease in 1 out of 10 people who would have gotten the disease otherwise, whereas in the second case the treatment will prevent the disease in 1 out of 2 people who would have gotten it.

The point is that the *difference* between two population proportions may not be the best way to compare them. Differences tend to differ in meaning when the proportions are near 0.5 from when the proportions are near 0 or 1. An alternative to comparing proportions is to compare their corresponding odds. This makes sense over a wider range of possibilities and forms the basis for interpreting regression models for binary data.

18.3.2 Odds

If π is a population proportion (or probability) of yes outcomes, the corresponding *odds* of a yes outcome in the population are $\pi/(1-\pi)$, represented by the symbol ω (the Greek letter *omega*).

The *sample odds* are $\hat{\omega} = \hat{\pi}/(1-\hat{\pi})$. Instead of saying, for example, that the proportion of cold cases among the vitamin C group was 0.742, one can convey the same information by saying that the odds of getting a cold were 2.876 to 1. The number 2.876 is the ratio of the proportion of cold cases to the proportion of noncold cases, $2.876 = 0.742/0.258$. These odds make the 0.742 easier to visualize: There are 2.876 cold cases for every 1 noncold case (or about 23 cold cases for every 8 noncold cases). Odds are particularly useful in visualizing proportions that are close to 0 or 1.

It is customary to cite the larger number first when stating odds, so an event with chances of 0.95 has odds of 19 to 1 *in favor* of its occurrence while an event with chances of 0.05 has the same odds, 19 to 1, *against* it.

Here are some numerical facts about odds:

1. Whereas a proportion must be between 0 and 1, odds must be greater than or equal to 0 but have no upper limit.
2. A proportion of 1/2 corresponds to odds of 1. Some common nonscientific descriptions of this situation are "equal odds," "even odds," and "the odds are fifty-fifty."
3. If the odds of a yes outcome are ω, the odds of a no outcome are $1/\omega$.
4. If the odds of a yes outcome are ω, the probability of yes (or the population proportion) is $\pi = \omega/(1+\omega)$.

18.3.3 The Ratio of Two Odds

When two populations have proportions π_1 and π_2, with corresponding odds ω_1 and ω_2, a useful alternative to the difference in proportions is the *odds ratio*, $\phi = \omega_2/\omega_1$. (ϕ is the Greek letter *phi*.) If π_2 is larger than π_1, then ω_2 is larger than ω_1 and the odds ratio is greater than 1. If the odds ratio is some number—say, $\phi = 3$—then $\omega_2/\omega_1 = 3$ is equivalent to $\omega_2 = 3\omega_1$. In words, the odds of a yes outcome in the second group are three times the odds of a yes outcome in the first group. If, for example, $\omega_1 = 2.5$, then group 1 has 5 yes for every 2 no outcomes. With $\phi = 3$, group 2 would have 15 yes for every 2 no outcomes.

The sample odds ratio, or *estimated odds ratio* is $\hat{\phi} = \hat{\omega}_2/\hat{\omega}_1$. In the vitamin C and cold study, the odds of a cold in the placebo group are estimated as $\hat{\omega}_2 = 335/76 = 4.408$. (There were 4.408 people with colds for every 1 without a cold.) The odds of a cold in the vitamin C group are estimated as $\hat{\omega}_1 = 302/105 = 2.876$. The estimated odds ratio is therefore $4.408/2.876 = 1.53$. *The odds of getting a cold on the placebo regimen are estimated to be 1.53 times as large as the odds of getting a cold on the vitamin C regimen.* Equivalently, the odds of a cold on placebo are 53% greater than the odds of a cold on vitamin C.

When the counts are arranged in a 2 × 2 table, the odds ratio may quickly be estimated by dividing the product of the upper left and lower right counts by the product of the upper right and lower left counts, as shown in Display 18.7. One caution: The calculation in Display 18.7 provides the odds ratio estimate for getting a cold on the placebo regimen relative to the vitamin C regimen. To estimate the odds ratio for not getting a cold on the placebo regimen relative to the vitamin C regimen, one must invert the ratio in Display 18.7.

DISPLAY 18.7 Shortcut method for estimating the odds ratio in a 2 × 2 table

	Cold	No Cold	Odds on a Cold	
Placebo	335	76	*group 2: Placebo*	335 / 76 = 4.408
Vitamin C	302	105	*group 1: Vitamin C*	302 / 105 = 2.876

$Odds\ ratio = 4.408 / 2.876 = 1.53$

The numerator is the product of the diagonals.

Shortcut method: *Odds ratio* $= \dfrac{(335)(105)}{(76)(302)} = 1.53$

The denominator is the product of the off-diagonals.

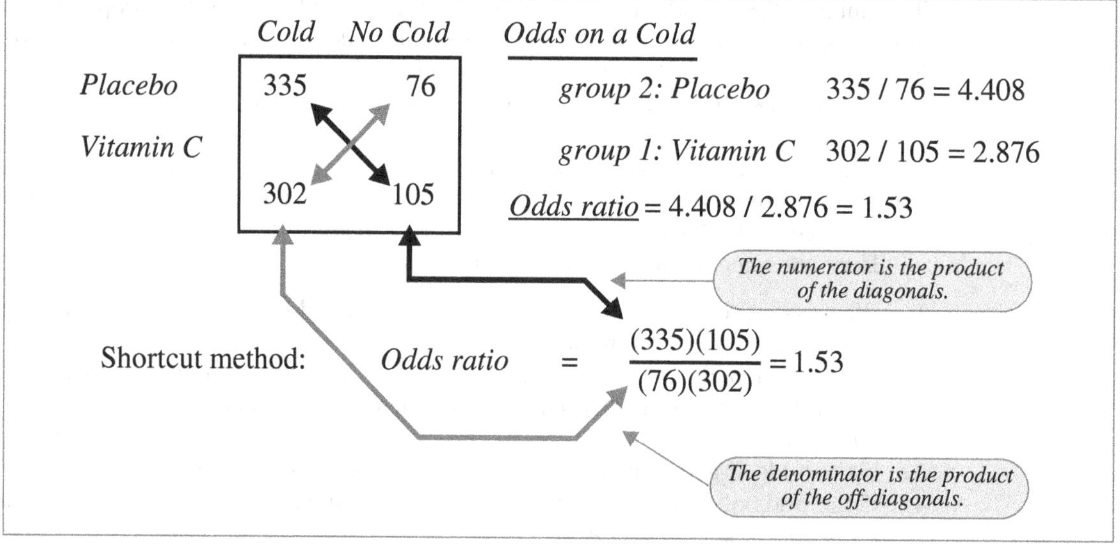

The odds ratio is a more desirable measure than the difference in population proportions, for three reasons:

1. In practice, the odds ratio tends to remain more nearly constant over levels of confounding variables.
2. The odds ratio is the *only* parameter that can be used to compare two groups of binary responses from a retrospective study (see Section 18.4).
3. The comparison of odds extends nicely to regression analysis (see Chapters 20 and 21).

18.3.4 Sampling Distribution of the Log of the Estimated Odds Ratio

If $\hat{\omega}_1$ and $\hat{\omega}_2$ are computed from independent samples, the sampling distribution of the *natural log of their ratio* has these properties:

1. $\text{Mean}\{\log(\hat{\omega}_2/\hat{\omega}_1)\} = \log(\omega_2/\omega_1)$ (approximately).
2. $\text{Var}\{\log(\hat{\omega}_2/\hat{\omega}_1)\} = [n_1\pi_1(1-\pi_1)]^{-1} + [n_2\pi_2(1-\pi_2)]^{-1}$ (approximately).
3. If n_1 and n_2 are large, the sampling distribution of $\log(\hat{\omega}_2/\hat{\omega}_1)$ is approximately normal.

Of course, inference is desired for the odds ratio, not for its logarithm (which has no useful interpretation). The sampling distribution of the log of the estimated odds ratio, however, is more closely approximated by the normal distribution than is the sampling distribution of the estimated odds ratio. Consequently, statistical inferences will be carried out for the log odds ratio and subsequently reexpressed in terms of odds ratios.

Statisticians do not have a good general guideline for checking whether the sample sizes are large enough in this problem. Most authorities agree that one can get away with smaller sample sizes here than for the difference of two proportions. Therefore, if the sample sizes pass the rough checks discussed in Section 18.2.2, they should be large enough to support inferences based on the approximate normality of the log of the estimated odds ratio, too.

Two Standard Errors for the Log of the Odds Ratio

The estimated variance is obtained by substituting sample quantities for unknowns in the variance formula in the preceding box. As before, the sample quantities used to replace the unknowns depend on the usage. For a confidence interval, π_1 and π_2 are replaced by their sample estimates.

Confidence interval:

$$\text{SE}[\log(\hat{\omega}_1/\hat{\omega}_2)] = \sqrt{\frac{1}{n_1\hat{\pi}_1(1-\hat{\pi}_1)} + \frac{1}{n_2\hat{\pi}_2(1-\hat{\pi}_2)}}$$

To test equality of odds in two populations, one must estimate the common proportion from the combined sample and compute the standard error based on it.

> *Testing equality:*
>
> $$\mathrm{SE}_0[\log(\hat{\omega}_1/\hat{\omega}_2)] = \sqrt{\frac{1}{n_1\hat{\pi}_c(1-\hat{\pi}_c)} + \frac{1}{n_2\hat{\pi}_c(1-\hat{\pi}_c)}}$$

If the odds are equal, then the odds ratio is 1 and the log of the odds ratio is 0. Therefore, a test of equal odds is carried out by testing whether the log of the odds ratio is 0. The test calculations for the heart disease and obesity data are shown in Display 18.8. The resulting p-value is nearly identical to that obtained with the z-test for equal proportions, as is the case when the sample sizes are large.

DISPLAY 18.8 Test for equality of two population odds: heart disease and obesity data

(1) *Estimate the odds on CVD death.*

→ *group 1 (obese):* $\hat{\omega}_1 = \frac{16}{2{,}045} = 0.00782$

→ *group 2 (not obese):* $\hat{\omega}_2 = \frac{7}{1{,}044} = 0.00670$

(2) *Odds ratio and its log* → $\hat{\phi} = \frac{0.00782}{0.00670} = 1.1669$ → $\log(\hat{\phi}) = 0.154$

(3) *Proportion from the combined sample* → $\hat{\pi}_c = \frac{16+7}{2{,}061 + 1{,}051} = 0.00739$

(4) *SE for the log odds ratio estimate (test version)*

$$\sqrt{\frac{1}{2{,}061(0.00739)(1-0.00739)} + \frac{1}{1{,}051(0.00739)(1-0.00739)}} = 0.443$$

(5) *z-statistic* → $z = \frac{0.154}{0.443} = 0.349$

(6) *one-side p-value* → $\Pr(Z > 0.349) = \mathbf{0.36}$

Conclusion: There is no evidence that the odds ratio differs from 1.

To get a confidence interval for the odds ratio, construct a confidence interval for the log of the odds ratio and take the antilogarithm of the endpoints. A shortcut formula for the standard error is the square root of the sum of the reciprocals of the four cell counts in the 2×2 table, as illustrated in step 2 of Display 18.9.

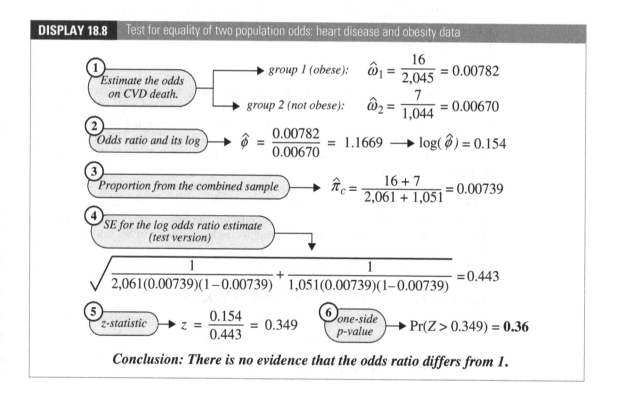

DISPLAY 18.9 Confidence interval for an odds ratio: vitamin C and cold data

	Cold	No Cold
Placebo	335	76
Vitamin C	302	105

(1) *Odds ratio and its log* → $\hat{\phi} = \frac{(335)(105)}{(76)(302)} = 1.533$ → $\log(\hat{\phi}) = 0.427$

(2) *Shortcut method for the SE of the log odds ratio* → $\sqrt{\frac{1}{335}+\frac{1}{76}+\frac{1}{302}+\frac{1}{105}} = 0.170$

(3) *95% interval for the log odds ratio* → $0.427 \pm (1.96)(0.170) = 0.094$ to 0.760

(4) *95% interval for the odds ratio* → exp(0.094) to exp(0.760); or **1.10 to 2.14**

Conclusion: The odds of a cold for the placebo group are estimated to be 1.53 times the odds of a cold for the vitamin C group (approximate 95% CI: 1.10 to 2.14).

18.4 INFERENCE FROM RETROSPECTIVE STUDIES

18.4.1 Retrospective Studies

Each of the case studies of Section 18.1 involves an explanatory factor with two levels and a response factor with two levels. In the obesity study, random samples were taken from both explanatory variable categories, obese and nonobese. In the vitamin C study, subjects were randomly assigned to an explanatory category, either vitamin C or placebo. Either plan is called *prospective sampling* because it sets the explanatory situation and then awaits the response.

In the smoking and lung cancer data of Section 18.1.3, the question of interest requires that the smoking behavior is explanatory and lung cancer is the response. However, the sampling was carried out for each level of the *response* factor—a sample of 86 lung cancer patients and a sample of 86 control subjects were obtained, and all subjects were asked whether they smoked. This is an example of *retrospective sampling*. Samples were taken for each level of the response variable and the levels of the explanatory factor were determined for subjects in these samples. A comparison of prospective and retrospective sampling is provided in Display 18.10.

Labeling of one of the factors a "response" and the other "explanatory" is a matter of choice. One could simply call lung cancer the explanatory variable and smoking the response, but comparing proportions of smokers in the populations of lung cancer patients and non–lung cancer patients is less pertinent than comparing

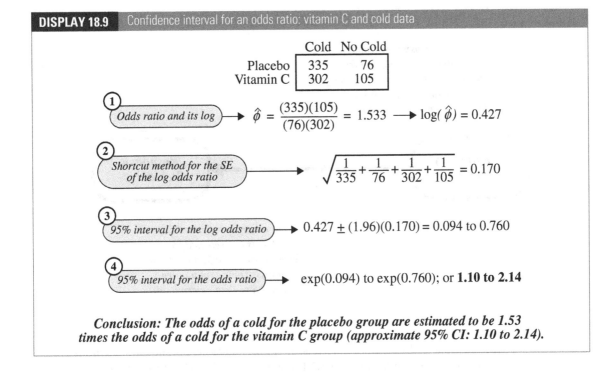

DISPLAY 18.10 Prospective and retrospective sampling

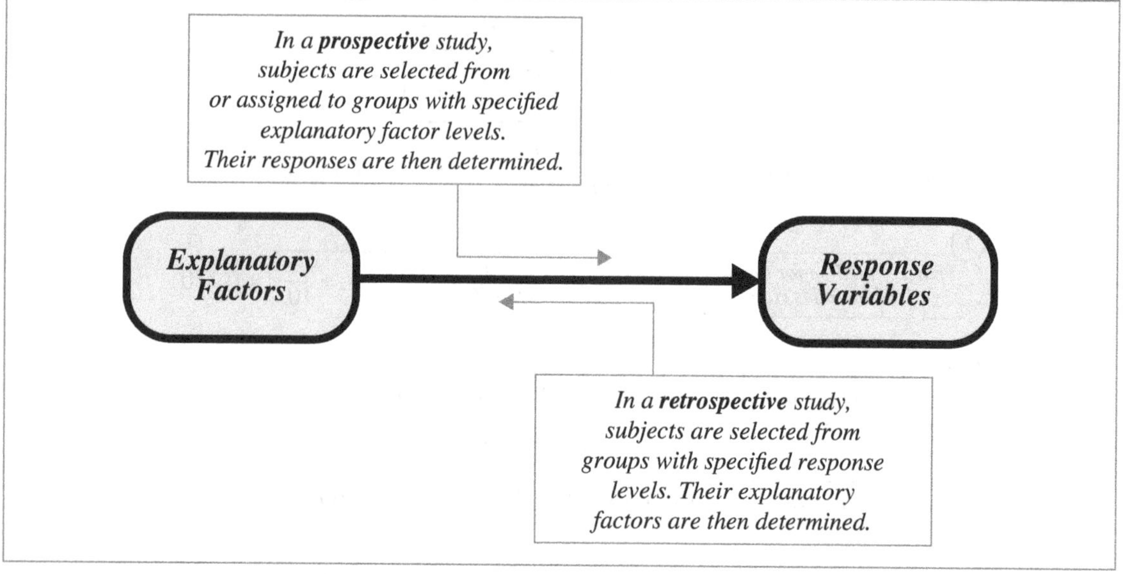
In a prospective study, subjects are selected from or assigned to groups with specified explanatory factor levels. Their responses are then determined.
Explanatory Factors
Response Variables
In a retrospective study, subjects are selected from groups with specified response levels. Their explanatory factors are then determined.

proportions of lung cancer victims in the populations of smokers and nonsmokers. The latter is of direct interest with regard to the health effects of smoking.

Reasons for Retrospective Sampling

In many areas of study, both prospective and retrospective studies play important roles. A prospective study that follows random samples of smokers and nonsmokers would be very useful, but a retrospective study can be accomplished without having to follow the subjects through their entire lifetimes. Furthermore, if the response proportions are small, then huge samples are needed in a prospective study to get enough yes outcomes to permit inferences to be drawn. This problem is bypassed in a retrospective study.

18.4.2 Why the Odds Ratio Is the Only Appropriate Parameter If the Sampling Is Retrospective

The odds ratio is the only parameter that describes binary response outcomes for the explanatory categories that can be estimated from retrospective data. The smoking and lung cancer data, for example, cannot be used to estimate the individual proportions of smokers and nonsmokers who get lung cancer or the difference between the proportions.

The odds ratio is the same regardless of which factor is considered the response. As a practical demonstration of this (without a formal proof), a hypothetical population of smokers and nonsmokers is shown in Display 18.11. Notice that the proportions of smokers among lung cancer patients and among controls tell nothing about the proportions of lung cancer patients among smokers and among

DISPLAY 18.11 Illustration that the odds ratio is the same regardless of which factor is thought of as the response, with a hypothetical population

	Lung cancer	No cancer
Smokers	1,000	2,000,000
Nonsmokers	4,000	16,000,000

$$\frac{\text{Odds of cancer among smokers}}{\text{Odds of cancer among nonsmokers}} = \frac{1{,}000/2{,}000{,}000}{4{,}000/16{,}000{,}000} = 2$$

$$\frac{\text{Odds that a cancer victim was a smoker}}{\text{Odds that a person without cancer was a smoker}} = \frac{1{,}000/4{,}000}{2{,}000{,}000/16{,}000{,}000} = 2$$

nonsmokers. Yet the odds ratio is the same regardless of which factor is thought of as the response: The odds of smoking among lung cancer patients is two times the odds of smoking among controls; and the odds of lung cancer among smokers is two times the odds of lung cancer among nonsmokers. Consequently, the odds ratio can be estimated either by observing cancer incidence in independent samples of smokers and nonsmokers or by observing proportions of smokers in independent samples of cancer victims and controls. The odds ratio is the only quantity pertaining to the prospective populations that can be estimated from retrospective studies.

Example—Smoking and Lung Cancer

From the data described in Example 18.1.3 the estimated odds ratio is 5.38, the natural log of the estimated odds ratio is 1.68, and the standard error of the estimated log odds ratio is 0.656 (confidence interval version). An approximate 95% confidence interval for the log odds ratio is 0.396 to 2.969. Therefore, the odds of lung cancer for smokers are estimated to be 5.38 times the odds of lung cancer for nonsmokers. An approximate 95% confidence interval for the odds ratio is 1.5 to 19.4. As usual, no causal interpretation can be drawn from the statistical analysis of observational data, although the data are consistent with the theory that smoking causes lung cancer.

18.5 SUMMARY

Populations of binary responses can be compared by estimating the difference in population proportions or the ratio of population odds. Tests and confidence intervals are based on the approximate normality of the difference in sample proportions or on the approximate normality of the logarithm of the estimated odds ratio. These techniques also apply to randomized experiments. The choice of whether to compare proportions or odds is subjective. The odds are often more appropriate if the proportions are close to 0 or 1. The proportions are more familiar to some audiences. If the data are sampled retrospectively, only the odds ratio can be estimated.

Obesity and Death from Cardiovascular Disease

Crews's study of obesity and heart disease deaths in American Samoa includes both women and men (see Exercise 9). For both women and men, the odds of CVD death are 15 to 20% higher among obese persons, but neither result is statistically significant. Confidence intervals are more explicit about whether any practically significant difference is likely. The similarity of results for men and women suggests asking whether a common odds ratio exists and is greater than 1.0. Tools for handling these questions are introduced in the next chapter.

Vitamin C and the Common Cold

The odds of getting a cold while using the placebo are estimated to be 4.4 to 1 in this Canadian study; use of vitamin C reduces those odds by approximately 35%. Increasingly, scientific journals and the popular press describe studies of risk factors in terms of multiplicative effects on odds ratios. This natural terminology comes from an increasing reliance on statistical analysis at the log-odds scale.

Smoking and Lung Cancer

A randomized experiment to determine a causal effect of smoking on lung cancer will never be performed. Nor does the time scale involved lend itself to prospective sampling. Retrospective sampling, however, is a powerful tool for studying associations in such situations. Here, the odds ratio of 5.4 to 1 offers strong evidence of some association between smoking and lung cancer.

18.6 EXERCISES

Conceptual Exercises

1. Obesity. For the heart disease and obesity data (Section 18.1.1), does the large p-value for the test of equal population proportions prove that 6-year cardiovascular disease death and obesity were unrelated over the 6-year study period in the population of American Samoan women? Why or why not? How does a confidence interval clarify the picture?

2. Vitamin C. The vitamin C and cold experiment (Section 18.1.2) was double-blind, meaning that neither the volunteers nor the doctors who interviewed them were told whether they were taking vitamin C or placebo. If many of the volunteers correctly guessed which group they were in, and if incidence of colds is highly associated with *perceived* treatment, do these considerations invalidate the assertion that cause-and-effect statements can be made from randomized experiments?

3. During investigation of the U.S. space shuttle *Challenger* disaster, it was learned that project managers had judged the probability of mission failure to be 0.00001, whereas engineers working on the project had estimated failure probability at 0.005. The difference between these two probabilities, 0.00499, was discounted as being too small to worry about. Is a different picture provided by considering odds? How is that interpreted?

4. Perry Preschool Project. In a 1962 social experiment, 123 3- and 4-year-old children from poverty-level families in Ypsilanti, Michigan, were randomly assigned either to a treatment group receiving 2 years of preschool instruction or to a control group receiving no preschool. The participants were followed into their adult years. The following table shows how many in each group were arrested for some crime by the time they were 19 years old. (Data reported in *Time*, July 29, 1991).

	Arrested for some crime?	
	Yes	No
Preschool	19	42
Control	32	30

(a) Is it possible to use these data to see whether preschool instruction can *cause* a lower rate of criminal arrest for some populations? (b) Is this a retrospective or a prospective study?

5. Violence Begets Violence. It is often argued that victims of violence exhibit more violent behavior toward others. To study this hypothesis a researcher searched court records to find 908 individuals who had been victims of abuse as children (11 years or younger). She then found 667 individuals, with similar demographic characteristics, who had not been abused as children. Based on a search through subsequent years of court records, she was able to determine how many in each of these groups became involved in violent crimes, as shown in the following table. (Data from C. S. Widom, "The Cycle of Violence," *Science* 244 (1989): 160.) (a) Is this a randomized experiment or an observational study? (b) Is this a prospective or a retrospective study? (c) Consider the populations of abused victims and controls from which these samples were selected. Let Y be a binary response variable taking the value 1 if an individual became involved in violent crime and 0 if not, in each of these populations. If π_1 and π_2 are the means of Y in the two populations, what is the variance of Y in each population? (d) What is the variance of the sampling distribution of $\hat{\pi}_1 - \hat{\pi}_2$? (e) Let Z be a binary response variable that takes on the value 1 if an individual did not become involved in violent crime and 0 if the individual did (so Z is $1 - Y$). What are the means of Z in the two populations, in terms of π_1 and π_2 (the means of Y)? (f) What are the variances of Z in the two populations, in terms of π_1 and π_2? (g) How does the analysis of the data differ if Z is used as a response variable instead of Y (in terms of estimating a difference in population proportions and estimating a ratio of odds)?

	Involved in a violent crime?	
	Yes	No
Abuse victim	102	806
Control	53	614

6. In Exercise 5, there are many ways of wording an odds ratio statement. For example:

(i) The odds of Yes among victims relative to controls.
(ii) The odds of Yes among controls relative to victims.
(iii) The odds of No among victims relative to controls.
(iv) The odds of No among controls relative to victims.
(v) The odds of victims among Yes relative to No.
(vi) The odds of victims among No relative to Yes.
(vii) The odds of controls among Yes relative to No.
(viii) The odds of controls among No relative to Yes.
(ix) The odds of victims among Yes relative to controls.

(a) Which, if any, of these are the same? (b) Do all of them make sense?

7. Alcohol and Breast Cancer. The following are partial results from a *case–control* study involving a sample of *cases* (women with breast cancer) and a sample of *controls* (demographically similar women without breast cancer). (Data from L. Rosenberg, J. R. Palmer, D. R. Miller, E. A. Clarke, and S. Shapiro, "A Case–Control Study of Alcoholic Beverage Consumption and Breast Cancer," *American Journal of Epidemiology* 131 (1990): 6–14.) The women were asked about their drinking habits, from which the following table was constructed. (a) For assessing the risk of drinking on breast cancer, which factor is the response? (b) Is this a retrospective or a prospective study? (c) Is this a randomized experiment or an observational study?

	Breast cancer	
	Cases	Controls
Fewer than 4 drinks per week	330	658
4 or more drinks per week	204	386

8. Salk Polio Vaccine. The Salk polio vaccine trials of 1954 included a double-blind experiment in which elementary school children of consenting parents were assigned at random to injection with the Salk vaccine or with a placebo. Both treatment and control groups were set at 200,000 because the target disease, infantile paralysis, was uncommon (but greatly feared). (Data from J. M. Tanur et al., *Statistics: A Guide to the Unknown*, San Francisco: Holden-Day, 1972.) (a) Is this a randomized experiment or an observational study? (b) Is this a retrospective or a prospective study? (c) Describe the population to which inferences can be made. (d) Would it make more sense to look at the difference in disease probabilities or the ratio of disease odds here?

	Infantile paralysis victim?	
	Yes	No
Placebo	142	199,858
Salk polio vaccine	56	199,944

Computational Exercises

9. Heart Disease and Obesity in American Samoan Men. The study described in Section 18.1.1 also included men. The data for men are shown in the accompanying table. (a) Compute (i) the sample proportions of CVD deaths for the obese and nonobese groups; (ii) the standard error for the difference in sample proportions; (iii) a 95% confidence interval for the difference in population proportions. (b) Find a one-sided p-value for the test of equal population proportions (using the standard error already computed). (c) Compute (i) the sample odds of CVD death for the obese and nonobese groups; (ii) the estimated odds ratio; (iii) the standard error of the estimated log odds ratio; (iv) a 95% confidence interval for the odds ratio. (d) Write a concluding sentence that incorporates the confidence interval from part (c).

	CVD death	
	Yes	No
Obese	22	1,179
Not obese	22	1,409

10. Salk Polio Vaccine. (a) For the data in Exercise 8 find a one-sided p-value for testing the hypothesis that the odds of infantile paralysis are the same for the placebo and for the Salk vaccine treatments. (b) An alternative method for testing equal proportions is to look only at the 198 children who contracted polio. Each of them had a 50% chance of being assigned to the Salk group or to the placebo group, unless group membership altered their chances. Under the null hypothesis of no difference in disease probabilities, the proportion of the 198 who are in either treatment group has a mean of 1/2. Is the observed proportion in the control group, 142/198, significantly larger than would be expected under the null hypothesis? Use the results about the sampling distribution of a single population proportion (in Section 18.2.1) to test this hypothesis. (This is a conditional test—conditional on the number getting the disease. It is useful here because the disease probabilities are so small.)

11. In the hypothetical example of Display 18.11, calculate each of the following quantities (i) in the population, (ii) in the expected results from prospective sampling, and (iii) in the expected results from retrospective sampling (with equal numbers of lung cancer patients and controls). (a) The proportion of lung cancer patients among smokers. (b) The proportion of lung cancer patients among nonsmokers. (c) The difference between parts (a) and (b). (d) Verify that retrospective sampling results cannot be used to estimate these population properties.

12. Suppose that the probability of a disease is 0.00369 in a population of unvaccinated subjects and that the probability of the disease is 0.001 in a population of vaccinated subjects.

- **(a)** What are the odds of disease without vaccine relative to the odds of disease with vaccine?
- **(b)** How many people out of 100,000 would get the disease if they were not treated?
- **(c)** How many people out of 100,000 would get the disease if they were vaccinated?
- **(d)** What proportion of people out of 100,000 who would have gotten the disease would be spared from it if all 100,000 were vaccinated? (This is called the *protection rate*.)
- **(e)** Follow the steps in parts (a)–(d) to derive the odds ratio and the protection rate if the unvaccinated probability of disease is 0.48052 and the vaccinated probability is 0.2. (The point is that the odds ratio is the same in the two situations, but the total benefit of vaccination also depends on the probabilities.)

13. An alternative to the odds ratio and the difference in probabilities is the relative risk, which is a common statistic for comparing disease rates between risk groups. If π_u and π_v are the probabilities of disease for unvaccinated and vaccinated subjects, the relative risk (due to not vaccinating) is $\rho = \pi_u/\pi_v$. (So if the relative risk is 2, the probability of disease is twice as large for unvaccinated as for vaccinated individuals.)

- **(a)** Show that the odds ratio is very close to the relative risk when very small proportions are involved. (You may do this by showing that the two quantities are quite similar for a few illustrative choices of small values of π_v and π_u.)
- **(b)** Calculate the relative risk of not vaccinating for the situation introduced at the beginning of Exercise 12.
- **(c)** Calculate the relative risk of not vaccinating for the situation described in Exercise 12(e).
- **(d)** Here is one situation with a relative risk of 50: The unvaccinated probability of disease is 0.0050 and the vaccinated probability of disease is 0.0001 (the probability of disease is 50 times greater for unvaccinated subjects than for vaccinated ones). Suppose in another situation the vaccinated probability of disease is 0.05. What unvaccinated probability of disease would imply a relative risk of 50? (Answer: None. This shows that the *range* of relative risk possibilities depends on the baseline disease rate. This is an undesirable feature of relative risk.)

14. The Pique Technique was developed because a target for solicitation is more likely to comply if mindless refusal is disrupted by a strange or unusual request. Researchers had young people ask 144 targets for money. The targets were asked either for a standard amount—a quarter—or an unusual amount—17 cents. Forty-three point one percent of those asked for 17 cents responded, compared to 30.6% of those asked for a quarter. Each group contained 72 targets. What do you make of that?

(Data from M. D. Santos, C. Leve, and A. R. Pratkanis, "Hey Buddy, Can You Spare Seventeen Cents?," *Journal of Applied Social Psychology* 24(9) (1994): 755–64.)

Data Problems

15. Perry Preschool Project. Analyze the data shown in Exercise 4 to determine whether a preschool program like the one used can lead to lower levels of criminal activity, and, if so, by how much.

16. Violence Begets Violence. Reconsider the data described in Exercise 5. The researcher concluded: "Early childhood victimization has demonstrable long-term consequences for ... violent criminal behavior." Conduct your own analysis of the data and comment on this conclusion. In particular, is there a statistically significant difference in population proportions, and is the strength of the causal implication of this statement justified by the data from this study?

17. Alcohol and Breast Cancer. In Exercise 7 the researcher is interested in the question of whether drinking can cause breast cancer. Analyze the data to see whether the data support the theory that it does, and quantify the association with an appropriate measure.

18. Hale-Bopp and Handedness. It is known that left-handed people tend to recall orientations of human heads or figures differently than right-handed people. To investigate whether there is a similar systematic difference in recollection of inanimate object orientation, researchers quizzed University of Oxford undergraduates on the orientation of the tail of the Hale-Bopp comet. The students were shown eight photographic pictures with the comet in different orientations (head of the comet facing left down, left level, left up, center up, right up, right level, right down, or center down) six months after the comet was visible in 1997. The students were asked to select the correct orientation. (The comet faced to the left and downward.) Shown below are the responses categorized as correct or not, shown separately for left- and right-handed students. Is there evidence that left- or right-handedness is associated with correct recollection of the orientation? If so, quantify the association. Write a brief statistical report of the findings. (Data from M. Martin, and G. V. Jones, "Hale-Bopp and Handedness: Individual Difference in Memory for Orientation," *American Psychological Science* 10 (1999): 267–70.)

Recollection of comet orientation

	Incorrect	Correct
Left-handed	149	48
Right-handed	129	68

19. Perry Preschool Project Results at Age 40. As discussed in Exercise 4, researchers in the mid-1960s identified 123 low-income, African-American children who were thought to be at high risk of school failure and randomly assigned 58 of them to participate in a high-quality preschool program at ages 3 and 4; the remaining 65 received no preschool education. One hundred nineteen of the study subjects were available for re-interview at age 40. The following two tables show results on two of the variables considered:

	Arrested 5 or more times by age 40		Made annual income of $20K or more by age 40	
	No	Yes	No	Yes
No Preschool	28	35	38	25
Preschool	36	20	22	34

Assess the evidence for an effect of preschool on these two variables and report the estimated odds ratios for 5 or more arrests and for annual income of $20K or more. (*Note*: Although the authors reported that random assignment was used, there was, in fact, some switching of groups after the initial assignment. For example, children placed in the preschool group were switched to the control group if their mothers could not transport them to preschool. Although the authors defend the results as just as good as if random assignment was made, the failure to follow through with exact randomization—even though it may have been unavoidable—causes a necessary extra debate about the causal conclusions from this study.)

Answers to Conceptual Exercises

1. No. There could be a difference, but the sample was too small to detect it. A confidence interval would reveal a set of possible values for the difference in population proportions that remain plausible in light of these data.

2. No, but in this case a confounding factor (whether the subject correctly guessed the treatment) arose after the randomization. The effect of treatment may be confounded with this factor, so causal conclusions must be made cautiously.

3. The odds of failure estimated by the project managers were 99,999 to 1 against. Engineers estimated them to be 199 to 1 against. The ratio is about 500. Thus, for every failure envisioned by the project managers, the engineers saw the possibility of 500 failures.

4. (a) Yes. (b) Prospective.

5. (a) Observational study. (b) Prospective. (c) $\pi_1(1-\pi_1)$ and $\pi_2(1-\pi_2)$. (d) $\pi_1(1-\pi_1)/908 + \pi_2(1-\pi_2)/667$. (e) $(1-\pi_1)$ and $(1-\pi_2)$. (f) $\pi_1(1-\pi_1)$ and $\pi_2(1-\pi_2)$ (the same as for Y). (g) The analysis will be the same, except that the sign of the difference in population proportions will be reversed and the odds ratio will be the reciprocal of the odds ratio based on Y.

6. (a) i = iv = v = viii; ii = iii = vi = vii (= the reciprocal of the value in the first group). (b) ix makes no sense.

7. (a) Breast cancer. (b) Retrospective. (c) Observational.

8. (a) Randomized experiment. (b) Prospective. (c) Elementary school children of consenting parents (in 1954). (d) There is nothing technically wrong with either measure in this prospective study, but the odds ratio offers a more suitable interpretation, since the probabilities are so small.

CHAPTER 19

More Tools for Tables of Counts

The data sets in the previous chapter can be displayed as 2 × 2 tables of counts, which list the numbers of subjects falling in each cross-classification of a row factor (such as obese or not obese) and a column factor (such as heart disease or no heart disease). This chapter elaborates on sampling schemes leading to tables of counts, the appropriate wording of hypotheses based on the sampling scheme and the question of interest, an exact test for 2 × 2 tables, and two extensions to larger tables.

Fisher's Exact Test is the gold standard of testing tools for 2 × 2 tables, appropriate under any of the sampling schemes that lead to 2 × 2 tables. It is exact in the sense that the p-value is based on a permutation distribution and requires no approximations. An approximate test, the *chi-squared test*, is also presented for 2 × 2 tables. It extends in a straightforward way to larger tables of counts. Sometimes several 2 × 2 tables of counts are used, with each separate table corresponding to a different level of a blocking or potentially confounding variable. If the odds ratio is the same in each 2 × 2 table—even though the individual odds may vary—the *Mantel–Haenszel Test* tests whether the common odds ratio is 1.

19.1 CASE STUDIES

19.1.1 Sex Role Stereotypes and Personnel Decisions—A Randomized Experiment

Psychologists performed an experiment on male bank supervisors attending a management institute, to investigate biases against women in personnel decisions. Among other tasks in their training course, the supervisors were asked to make a decision on whether to promote a hypothetical applicant based on a personnel file. For half of them, the application file described a female candidate; for the others it described a male. The files were identical in all other respects. Results on the promotion decisions for the two groups are shown in Display 19.1. (Data from B. Rosen and T. Jerdee, "Influence of Sex Role Stereotypes on Personnel Decisions," *Journal of Applied Psychology* 59 (1974): 9–14.) Was there a bias against female applicants, or can the observed difference in promotion rates be explained just as reasonably by the chance involved in random allocation of subjects to treatment groups?

DISPLAY 19.1 Decisions of 48 male supervisors to promote a fictitious candidate, listed separately for those receiving "male" and "female" personnel folders

	Promoted	
	Yes	No
Male	21	3
Female	14	10

Statistical Conclusion

These data provide strong evidence that the promotion rate is higher for male applicants than for female applicants. The one-sided p-value from Fisher's Exact Test is 0.0245.

Scope of Inference

Since randomization was used, the inference is that the sex listed on the applicant file was responsible for the difference observed with these 48 supervisors. The inference derives from a probability model based on the randomization in the following way. If there had been no sex effect, each supervisor would have reached the same decision regardless of the sex listed on the application. Fisher's Exact Test examines all possible assignments of applications to supervisors and finds that only 2.45% of them result in a difference as great as that observed. Since the 48 supervisors were not a random sample from any population, extension of the inference to any broader group is speculative.

19.1.2 Death Penalty and Race of Murder Victim—An Observational Study

In a 1988 law case, *McCleskey v. Zant*, lawyers for the defendant showed that, among black convicted murderers in Georgia, 35% of those who killed whites and 6% of those who killed blacks received the death penalty. Their claim, that racial discrimination was a factor in sentencing, was contested by lawyers for the state of Georgia. Murders with black victims, they argued, were more likely to be unaggravated "barroom brawls, liquor-induced arguments, or lovers' quarrels"—crimes that rarely result in the death sentence anywhere. Among white-victim murders, on the other hand, there was a higher proportion of killings committed in the course of an armed robbery or involving torture—which more often result in death sentences.

To examine the validity of this argument, researchers collected data on death penalty sentencing in Georgia, including the race of victim, categorized separately for six progressively serious types of murder. Category 1 comprises barroom brawls, liquor-induced arguments, lovers' quarrels, and similar crimes. Category 6 includes the most vicious, cruel, cold-blooded, unprovoked crimes. (Data from G. C. Woodworth, "Statistics and the Death Penalty," *Stats* 2 (1989): 9–12.) Display 19.2 shows, for each category, a 2 × 2 table categorizing convicted black murderers according to the race of their victim and whether they received the death penalty. Is there evidence that black murderers of white victims are more likely to receive the death penalty than black murderers of black victims, even after accounting for the level of aggravation?

DISPLAY 19.2 Death sentences for black convicted murderers of white and black victims in Georgia, by aggravation category

Aggravation level	Race of victim	Death penalty: Yes	Death penalty: No
1	White	2	60
	Black	1	181
2	White	2	15
	Black	1	21
3	White	6	7
	Black	2	9
4	White	9	3
	Black	2	4
5	White	9	0
	Black	4	3
6	White	17	0
	Black	4	0

Statistical Conclusion

The data provide convincing evidence that convicted black murderers were more likely to receive the death sentence if their victim was white than if their victim was black, even after accounting for the aggravation level of the crime (one-sided p-value $= 0.0004$ from a Mantel–Haenszel Test). The odds of a white-victim murderer receiving the death penalty are estimated to be 5.5 times the odds for a black-victim murderer in the same aggravation category.

Scope of Inference

Since the data are observational, one cannot discount the possibility that other confounding variables may be associated with both the victim of the crime and the sentence. However, it is safe to exclude the level of aggravation and chance variation as determinants of the disparity in sentencing.

19.2 POPULATION MODELS FOR 2 × 2 TABLES OF COUNTS

19.2.1 Hypotheses of Homogeneity and of Independence

The last chapter presented tests for the following identical hypotheses:

$$\text{H}_0\colon \pi_2 - \pi_1 = 0 \qquad \text{and} \qquad \text{H}_0\colon \omega_2/\omega_1 = 1,$$

where π_2 and π_1 are population proportions or probabilities and ω_2 and ω_1 are the corresponding odds. These are often called hypotheses of ***homogeneity*** because interest is focused on whether the distribution of the binary response is homogeneous—the same—across populations.

The hypothesis of ***independence***, appropriate for some sampling schemes, is used to investigate an association between row and column factors without specifying one of them as a *response*. Although the hypothesis can be expressed in terms of parameters, the actual parameters used differ for different sampling schemes, and it is more convenient to use words:

$$\text{H}_0\text{: The row categorization is independent of the column categorization.}$$

For example, individuals in a population may be categorized according to whether they favor legalized abortion or not and according to whether they favor the death penalty or not. It is natural to ask whether an association exists between the two factors—whether people who answer yes on one of the questions are more (or less) likely to answer yes on the other. If there is no association, the factors are independent. With the test of independence, the two factors are treated equally, and neither needs to be specified as the response.

One difference between hypotheses of homogeneity and of independence lies in whether one thinks of two populations, each with two categories of response, or of a single population with four categories of response (formed by two binary factors in a 2 × 2 table). The homogeneity hypothesis is appropriate for the first, and independence is appropriate for the second. Whether the two-population or single-population model (or both) is appropriate depends on the research question and on how the data were collected. In the next section, several sampling schemes are described, and the appropriateness of the two hypotheses is discussed for each.

19.2.2 Sampling Schemes Leading to 2 × 2 Tables

Poisson Sampling

In this scheme (named after the French mathematician Siméon Denis Poisson, 1781–1840), a fixed amount of time (or space, volume, money, etc.) is devoted to collecting a random sample from a single population, and each member of the population falls into one of the four cells of the 2 × 2 table. An example is the heart disease and obesity data of Section 18.1.1. In that case, the researchers spent a certain amount of time sampling health records in American Samoa and ended up with 3,112 women. The women were categorized as having been obese or not, and as having died of cardiovascular disease (in six years) or not. One easy way to recognize this sampling scheme is to observe that none of the marginal totals (see Display 19.3) is known in advance of data collection.

DISPLAY 19.3 Recognizing the sampling schemes and appropriate test hypotheses by noticing which marginal totals are fixed

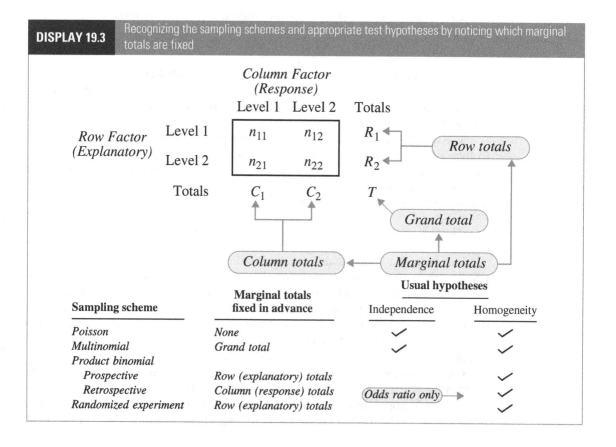

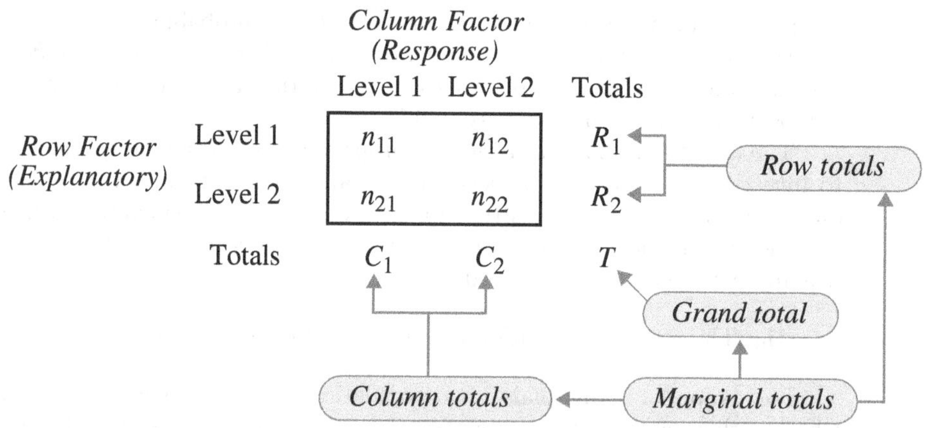

Sampling scheme	Marginal totals fixed in advance	Usual hypotheses: Independence	Usual hypotheses: Homogeneity
Poisson	*None*	✓	✓
Multinomial	*Grand total*	✓	✓
Product binomial			
Prospective	*Row (explanatory) totals*		✓
Retrospective	*Column (response) totals*	*Odds ratio only* →	✓
Randomized experiment	*Row (explanatory) totals*		✓

Multinomial Sampling

This scheme is very similar to Poisson sampling except that the total sample size is determined in advance. If a list of all women in American Samoa existed, the researcher could have selected a random sample of 5,000 from this list. The term *multinomial* means that each subject selected in the sample is categorized into one of several cells—in this case, one of four cells in the 2×2 table. The only marginal total known in advance of data collection is the grand total.

Prospective Product Binomial Sampling

This sampling scheme formed the basis for deriving the tools in the previous chapter. The two populations defined by the levels of the explanatory factor are identified, and random samples are selected from each. If separate lists of the populations of obese American Samoan women and nonobese American Samoan women were available, a random sample of 2,500 could be selected from each. The term *binomial* means that each woman (in each population) falls into one of *two* categories (see Section 18.2.1). *Product binomial* is a term whose statistical meaning reflects there being more than one binomial population from which independent samples have been taken. The sizes of the samples from each of the populations are chosen by the researcher.

Retrospective Product Binomial Sampling

Retrospective studies were discussed in Section 18.4.1. Random samples are selected from the subpopulations defined by each level of the response factor. The sampling scheme is technically the same as in the prospective product binomial, except that the roles of response and explanatory factors (according to the questions of interest) are reversed. The researcher specifies the sample totals to be obtained at each level of the response factor.

Randomized Binomial Experiment

In a randomized experiment, the subjects—regardless of how they were obtained—are randomly allocated to the two levels of the explanatory factor. Aside from the fact that randomization rather than random sampling is used, the setup here is essentially the same as in prospective product binomial sampling. The explanatory factor totals are fixed by the researcher.

The Hypergeometric Probability Distribution

Sometimes an experiment can be conducted in such a way that the row and column totals are *both* fixed by the researcher. In a classic example, a woman is asked to taste 14 cups of tea with milk and categorize them according to whether the tea or the milk was poured first. She is told that there were seven of each, so she tries to separate them into two groups of size seven. If a 2×2 table is formed, with the row factor being actual status for each of the 14 cups (milk first or tea first) and the column factor being guessed status (milk first or tea first), then each row and each column total will be seven. If the subject cannot correctly guess the order of milk and tea, then the seven cups she guesses to have had milk first are like a random sample from the 14 eligible cups. The hypergeometric probability distribution describes the probability of getting as many correct assignments as she did, given that she was just guessing.

Although this type of *hypergeometric experiment* is rare, the hypergeometric probability distribution is applied more generally. If interest is strictly focused on the odds ratio, statistical analysis may be conducted conditionally on the row and column totals that were observed. Therefore, no matter which sampling scheme produced the 2×2 table, one may use the hypergeometric distribution to calculate a p-value. This is known as ***Fisher's Exact Test***, which is described in more detail in Section 19.4.

19.2.3 Testable Hypotheses and Estimable Parameters

In prospective product binomial sampling and randomized experiments, it is very natural to draw inferences about the parameters π_1 and π_2 (or ω_1 and ω_2) describing the two binomial distributions. A test of homogeneity is used to determine whether the populations are identical. A confidence interval for $\pi_1 - \pi_2$ or for the odds ratio, ω_1/ω_2, is used to describe the difference. As discussed in Chapter 18, ω_1/ω_2 is the only comparison parameter that may be estimated if the sampling is retrospective.

With Poisson and multinomial sampling, information is available about the single population with four categories, and independence may be tested without specifying either row or column factor as a response. Some further options are also available. Either the row or the column may be designated as a response, thus dividing the population into two subpopulations for each level of the explanatory factor. Although the sample sizes may differ, the samples are indeed random samples from each subpopulation. Consequently, all inferential tools for prospective product binomial sampling may be used with Poisson and multinomial sampling as well. The choice of testing for independence or homogeneity should depend on whether treating one of the categories as a response is appropriate. A summary of the hypotheses that may be tested for each sampling scheme is shown in Display 19.3.

The heart disease and obesity study employed Poisson sampling. Since CVD death is the response factor, one would phrase inferences in terms of proportions of CVD deaths in the two subpopulations, thus testing homogeneity and providing a confidence interval for the difference in proportions or the ratio of odds.

Estimable Parameters

Although testing independence or estimating an odds ratio is often a primary goal, additional interest may be focused on other population features. The question of interest may involve the overall proportion of yes responses in the entire population or the proportions of units having different levels of the explanatory factor. For example, one might have interest in estimating the overall proportion of American Samoan women who were obese or the overall proportion of CVD deaths. These are marginal proportions, in the sense that estimates are constructed from ratios of

DISPLAY 19.4 Estimable parameters for sampling schemes leading to 2 × 2 tables of counts (the check mark indicates the feature is estimable)

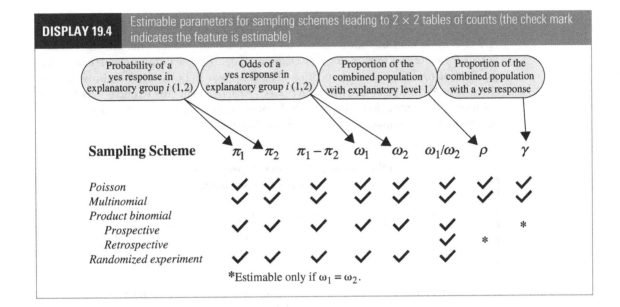

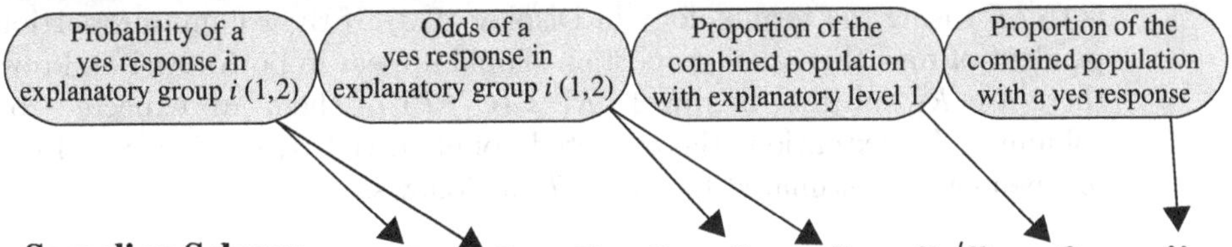

Sampling Scheme	π_1	π_2	$\pi_1 - \pi_2$	ω_1	ω_2	ω_1/ω_2	ρ	γ
Poisson	✓	✓	✓	✓	✓	✓	✓	✓
Multinomial	✓	✓	✓	✓	✓	✓	✓	✓
Product binomial								
Prospective	✓	✓	✓	✓	✓	✓		*
Retrospective						✓	*	
Randomized experiment	✓	✓	✓	✓	✓	✓		

*Estimable only if $\omega_1 = \omega_2$.

marginal totals in the table of counts. Display 19.4 indicates whether the proportion of the population in row 1 (call this population quantity ρ) and in column 1 (call this γ) can be estimated from the different sampling procedures.

19.3 THE CHI-SQUARED TEST

19.3.1 The Pearson Chi-Squared Test for Goodness of Fit

Karl Pearson (1857–1936) devised a general method for comparing fact with theory. His approach assumes that sampled units fall randomly into cells and that the chance of a unit falling into a particular cell can be estimated from the theory under test. The number of units that fall in a cell is the cell's *observed* count, and the number predicted by the theory to do so is the cell's *expected* count. Pearson combined these into the single statistic

$$X^2 = \sum \frac{(Observed - Expected)^2}{Expected}.$$

The sum is over all cells of the table. The name *Pearson chi-square statistic* attached to X^2 reflects its approximate sampling distribution, which is chi-squared if the theory (null hypothesis) is correct. The degrees of freedom are equal to (Number of cells) minus (Number of parameters estimated) minus 1.

19.3.2 Chi-Squared Test of Independence in a 2 × 2 Table

A 2 × 2 table contains four cells. The proportion of counts that fall in column 1 is C_1/T (using the terminology in Display 19.3). If the column proportion is independent of row, this same proportion should appear in both row 1 and row 2. Since there are R_1 subjects in row 1, $R_1 \times (C_1/T)$ of them are expected to fall into column 1. By extension, the expected count in cell (i, j), if row and column are independent, is estimated by R_iC_j/T, as follows.

Observed and Estimated Expected Cell Counts for Testing Independence

Cell (i, j)	$(1, 1)$	$(1, 2)$	$(2, 1)$	$(2, 2)$
Observed:	n_{11}	n_{12}	n_{21}	n_{22}
Expected (est.):	R_1C_1/T	R_1C_2/T	R_2C_1/T	R_2C_2/T

The pattern for determining an estimate for an expected cell count is to multiply the marginal totals in the same row and the same column together, then divide by the grand total.

Testing the hypothesis of independence involves the following steps:

1. Estimate the expected cell counts from the preceding formulas.
2. Compute the Pearson X^2 statistic from the formula in the previous box.
3. Find the p-value as the proportion of values from a chi-squared distribution on one degree of freedom that are greater than X^2, using a calculator or computer program with chi-squared distribution probabilities.

A small p-value provides evidence that the row and column categorizations are not independent. The example in Display 19.5 tests independence of the death penalty sentence and the race of the victim, using all six aggravation levels combined.

19.3.3 Equivalence of Several Tests for 2 × 2 Tables

The Chi-Squared Test for Homogeneity

Since the expected cell counts one obtains under the independence hypothesis are also the expected cell counts under the homogeneity hypothesis, the chi-squared test of independence in Section 19.3.2 is also a test of homogeneity. The chi-squared statistic is identical to the square of the Z-statistic for testing equal population proportions (see Section 18.2.3). Furthermore, the X^2 distribution on one degree of freedom is identical to the distribution of the square of a standard normal variable. Consequently, the p-values from the chi-squared test of independence and the chi-squared test of homogeneity are identical to the two-sided p-value from the Z-test of equal population proportions.

DISPLAY 19.5 Chi-squared test of independence between death penalty and race of victim (ignoring aggravation level of the crime)

Observed Counts

		Death Penalty Yes	Death Penalty No	Totals
Race of Victim	White	45	85	130
	Black	14	218	232
	Totals	59	303	362

Expected (Independence)

	Death Penalty Yes	Death Penalty No
White	21.19	108.81
Black	37.81	194.19

(1) *Compute expected cell counts from the marginal totals. For example, 21.19 = (130)(59)/(362).*

(2) *Compute the X^2 statistic.*

$$X^2 = \frac{(45-21.19)^2}{21.19} + \frac{(85-108.81)^2}{108.81} + \frac{(14-37.81)^2}{37.81} + \frac{(218-194.19)^2}{194.19}$$

$$= 49.89$$

(3) *Compute the p-value from a chi-squared on 1 d.f. (χ^2_1) probability.* → $\Pr(\chi^2_1 > 49.89) < 0.0001$

Conclusion: There is convincing evidence that death penalty sentence and race of victim are not independent.

Calculation and Continuity Correction

The most convenient formula for the chi-squared statistic from a 2×2 table is the following:

$$X^2 = \frac{T(n_{11}n_{22} - n_{12}n_{21})^2}{R_1 R_2 C_1 C_2}.$$

The statistic has approximately a chi-squared distribution with one degree of freedom, under the independence or homogeneity hypothesis. The approximation is premised on there being large cell counts; all expected counts larger than 5 is the customary check for validity.

An adjustment—the *continuity correction*—improves the approximation when sample sizes are not large:

$$X^2 = \frac{T(|n_{11}n_{22} - n_{12}n_{21}| - T/2)^2}{R_1 R_2 C_1 C_2}.$$

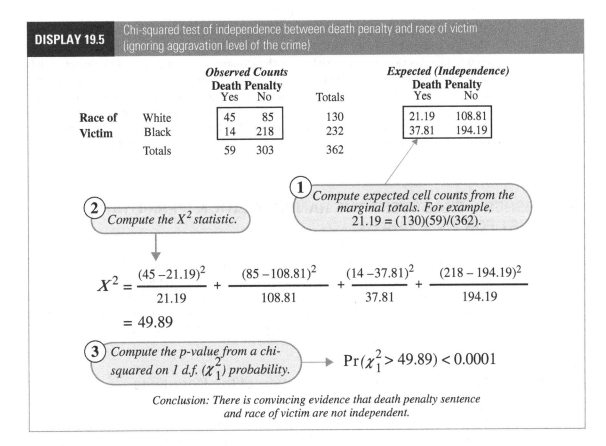

Limitations

The chi-squared tests for independence and homogeneity extend to $r \times c$ tables of counts (see Section 19.6.1). Although the chi-squared test is one of the most widely used of statistical tools, it is also one of the least informative. First, the only product is a p-value; there is no associated parameter to describe the degree of dependence. Second, the alternative hypothesis—that row and column are not independent—is very general. When more than two rows and columns are involved, there may be a more specific form of dependence to explore (see Chapters 21 and 22).

19.4 FISHER'S EXACT TEST: THE RANDOMIZATION (PERMUTATION) TEST FOR 2 $\times$ 2 TABLES

Although there are several statistical motivations for using Fisher's Exact Test, it is explained here as a randomization test for the difference in sample proportions. It is identical to the randomization test presented in Chapter 4, except that the responses are binary and the statistic of interest is the difference in sample proportions. Fisher's Exact Test is appropriate for any sample size, for tests of homogeneity or of independence.

19.4.1 The Randomization Distribution of the Difference in Sample Proportions

This section illustrates the randomization distribution of the difference of sample proportions on a small, hypothetical example. While it is more convenient to use the formulas in the next section or a computer program to obtain the same p-value, the approach here is helpful for visualizing the randomization distribution on which the calculations are based. The p-value has a concise meaning related to the probabilities involved in the random allocation of subjects to treatment groups.

Consider a small, hypothetical version of the sex role stereotype study of Section 19.1.1. Suppose each of seven men is randomly assigned to one of two groups such that four are in group 1 and three are in group 2. Each man decides whether to promote the applicant, based on the personnel file. Suppose that the men in group 1 are A, B, C, and D, while E, F, and G are in group 2. Suppose further that A, B, C, and E decide to promote while the others decide not to promote. Then the proportions of promotions in groups 1 and 2 are, respectively, 3/4 and 1/3. This evidence suggests that group 1 has a larger promotion rate than group 2; but the observed difference may be the result of the chance allocation of subjects to groups. The p-value measures how likely that explanation is.

The randomization distribution of the difference in sample proportions is a listing of the differences in proportions, given the observed outcomes, for every possible allocation of the men into groups of size four and three. This randomization distribution is shown in Display 19.6. The groupings in Display 19.6 are ordered according to the difference between promotion proportions.

DISPLAY 19.6 The randomization distribution of the difference in two sample proportions

① *List the subjects with their observed responses.*

Subject code	Actual group	Response
A	1	1
B	1	1
C	1	1
D	1	0
E	0	1
F	0	0
G	0	0

② *List all possible ways that the subjects can be arranged into two groups.*

③ *For each arrangement, calculate the difference between the sample proportions; e.g., (2/4 – 2/3) = –0.167.*

④ *The p-value is the proportion of groupings with differences in proportions ≥ 0.417.*

One-sided p-value = 13/35 = 0.37

Hypothetical grouping: Group 1	Hypothetical grouping: Group 2	Number of 1's: in 1	Number of 1's: in 2	Difference in sample proportions
ADFG	BCE	1	3	−0.750
BDFG	ACE	1	3	−0.750
CDFG	ABE	1	3	−0.750
DEFG	ABC	1	3	−0.750
ABDF	CEG	2	2	−0.167
ABDG	CEF	2	2	−0.167
ABFG	CDE	2	2	−0.167
ACDF	BEG	2	2	−0.167
ACDG	BEF	2	2	−0.167
ACFG	BDE	2	2	−0.167
ADEF	BCG	2	2	−0.167
ADEG	BCF	2	2	−0.167
AEFG	BCD	2	2	−0.167
BCDF	AEG	2	2	−0.167
BCDG	AEF	2	2	−0.167
BCFG	ADE	2	2	−0.167
BDEF	ACG	2	2	−0.167
BDEG	ACF	2	2	−0.167
BEFG	ACD	2	2	−0.167
CDEF	ABG	2	2	−0.167
CDEG	ABF	2	2	−0.167
CEFG	ABD	2	2	−0.167
ABCD	**EFG**	**3**	**1**	**0.417**
ABCF	DEG	3	1	0.417
ABCG	DEF	3	1	0.417
ABDE	CFG	3	1	0.417
ABEF	CDG	3	1	0.417
ABEG	CDF	3	1	0.417
ACDE	BFG	3	1	0.417
ACEF	BDG	3	1	0.417
ACEG	BDF	3	1	0.417
BCDE	AFG	3	1	0.417
BCEF	ADG	3	1	0.417
BCEG	ADF	3	1	0.417
ABCE	DFG	4	0	1.000

The promotion decision for each man remains the same, but his group status is allowed to vary. Therefore, the differences in observed proportions in the final column are due entirely to the different group assignments. The one-sided p-value from the randomization test is the proportion of groupings that lead to values of $\hat{\pi}_1 - \hat{\pi}_2$ greater than or equal to the one actually observed ($\hat{\pi}_1 - \hat{\pi}_2 = 0.417$). Since 13 groupings out of the 35 lead to differences as great as or greater than the one observed, the p-value is 0.37. Interestingly, if the alternative hypothesis had been two-sided in this example, the p-value might have been defined as the proportion of groupings that have $\hat{\pi}_1 - \hat{\pi}_2$ greater than or equal to 0.417 in magnitude. The p-value in this case is 0.49, which is quite different from twice the one-sided p-value.

This motivation suggests an appealing interpretation of p-values from randomized, binomial experiments—as the probability that randomization alone is responsible for the observed disparity between proportions. If an applicant's sex had no effect in the sex role stereotype study, the promotion decisions would be unaffected by whether a male or female folder was received. However, the probability that chance allocations led to a disparity in proportions as great as or greater than the observed difference—assuming decisions are unaffected by sex of the applicant—is only 0.0245.

19.4.2 The Hypergeometric Formula for One-Sided p-Values

Although the randomization distribution can always be listed as in Display 19.6, the number of possible outcomes can be quite large even for small sample sizes. There is, however, a short-cut formula for the p-value, based on the *hypergeometric probability function*:

1. Focus on the upper left-hand cell of the table in Display 19.1. (It makes no difference which cell is selected, but the customary choice is the upper left-hand cell, and the instructions that follow are based on that choice.)
2. Determine the *expected* number of counts in the upper left-hand cell of the table, as in the computation of the chi-squared statistic: $Expected = R_1C_1/T$.
3. Identify values (k) for the upper left-hand cell of the table that are as extreme as or more extreme than the observed value n_{11} in evidence against the null hypothesis. If n_{11} exceeds *Expected*, the values are $k = n_{11}, n_{11} + 1, \ldots, \min(R_1, C_1)$ (i.e., the smaller of the totals in the first row and first column). If n_{11} is less than *Expected*, the values are $k = 0, 1, \ldots, n_{11}$.
4. For each value of k, determine the proportion of possible randomizations leading to that value, according to the hypergeometric probability formula:

$$\Pr\{k\} = \frac{R_1!R_2!C_1!C_2!}{T!k!(R_1 - k)!(C_1 - k)!(R_2 - C_1 + k)!}.$$

   (Recall that $5! = 5 \times 4 \times 3 \times 2 \times 1$ and that $0!$ is taken to be 1.)
5. Calculate the p-value by adding up these probabilities. Display 19.7 illustrates the calculations for the sex role stereotype data.

19.4.3 Fisher's Exact Test for Observational Studies

Fisher's Exact Test has been presented here as a randomization test based on the statistic $\hat{\pi}_1 - \hat{\pi}_2$. If the data are observational, however, it is thought of as a permutation test, which is a useful interpretation when the entire population has been sampled or when the sample is not random. The permutation test here is also

DISPLAY 19.7 Fisher's Exact Test: calculations for testing equal promotion probabilities for bank supervisors given "male" and "female" application folders

(1) *Select a cell from observed table.*

	Promoted? Yes	No	Totals
Male	21	3	24
Female	14	10	24
Totals	35	13	48

(2) *Determine expected cell count.* → $Expected = \frac{(24)(35)}{48} = 17.5$

(3) *Identify the set of possible cell counts (k) that are as extreme or more extreme than those observed (given the marginal totals).* → *Observed* (21) > *Expected* (17.5)

$min(35,24) = 24$

$k = 21, 22, 23, 24$

(4) *Calculate hypergeometric probabilities.* →

$$\Pr\{21\} = \frac{24!\,24!\,35!\,13!}{48!\,21!\,3!\,14!\,10!} = 0.02058$$

$$\Pr\{22\} = \frac{24!\,24!\,35!\,13!}{48!\,22!\,2!\,13!\,11!} = 0.00357$$

$$\Pr\{23\} = \frac{24!\,24!\,35!\,13!}{48!\,23!\,1!\,12!\,12!} = 0.00034$$

$$\Pr\{24\} = \frac{24!\,24!\,35!\,13!}{48!\,24!\,0!\,11!\,13!} = 0.00001$$

(5) *Add the probabilities.* → One-sided p-value = **0.02450**

Conclusion: There is substantial evidence that the observed division of favorable promotion decisions was not the result of random allocation.

equivalent to a sampling distribution, thus allowing inference about population parameters with random samples from the Poisson, multinomial, or product binomial sampling schemes. As demonstrated, the calculations can be used to obtain exact p-values for tests of equal population proportions, of equal population odds, or for independence.

19.4.4 Fisher's Exact Test Versus Other Tests

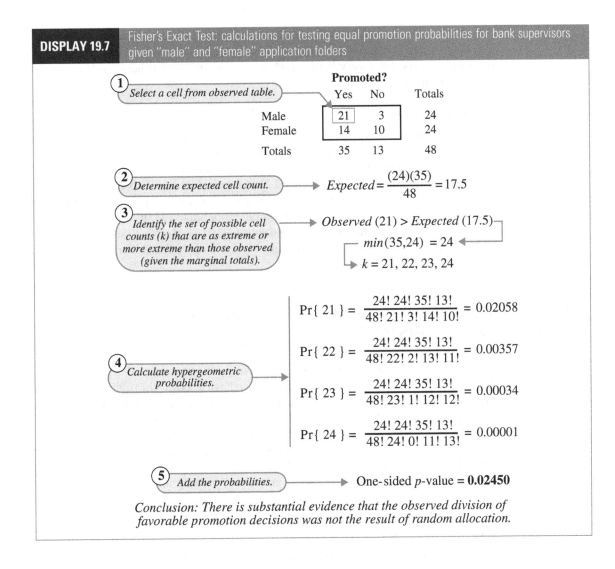

If Fisher's test gives exact p-values for all hypotheses, why should one ever consider using the Z-test for equal population proportions, the Z-test for equal population odds, or the chi-squared test for independence? If a test is all that is desired and if a computer program is available, there is indeed no reason to use a test other than Fisher's. But confidence intervals are commonly desired. Furthermore, if the sample sizes are large, not only are the permutation calculations infeasible but also the various tests differ very little, so choosing one over the others tends not to be a matter of major concern.

Report p-Values

It is important to provide a p-value rather than simply to state whether the p-value exceeds or fails to exceed some prespecified significance level (such as 0.05). It is especially important in categorical data analysis. For example, consider the test of whether a flipped coin has probability 1/2 of turning up heads. Based on 10 tosses, the probability of getting eight or more heads is 0.055 and the probability of getting nine or more heads is 0.011 (from binomial probability calculations—see Section 18.2.1). Thus, the null hypothesis would be rejected at the 5% level of significance if nine or more heads were obtained, but only 1.1% of all possible outcomes lead to such a rejection.

19.5 COMBINING RESULTS FROM SEVERAL TABLES WITH EQUAL ODDS RATIOS

Cause-and-effect conclusions cannot be drawn from observational studies, because the effect of confounding variables associated with both the indicator variable (defining the populations) and the binary response cannot be ruled out. As with regression, however, tools are available for taking the next step: conducting comparisons that account for the effect of a confounding variable. The Mantel–Haenszel procedure may be used for this purpose when several 2×2 tables are involved, one for each level of a confounding variable, if the odds ratio is the same for all levels of the confounding variable. It is also appropriate for randomized block experiments in which a separate 2×2 table exists for each block.

19.5.1 The Mantel–Haenszel Excess

The *excess* is a name given to the observed count minus the expected count (according to the hypothesis) in one of the cells of the 2×2 table. It is only necessary to consider the excess in one of the cells, and the convention used here will be to focus on the upper left cell. If

$$Observed = n_{11}$$

then

$$Expected = \frac{R_1 C_1}{T}$$

and

$$Excess = Observed - Expected.$$

The relevant properties of the excess are displayed in the following box.

> When the null hypothesis is correct,
>
> $$\text{Mean}\{\textit{Excess}\} = 0 \qquad \text{and} \qquad \text{Variance}\{\textit{Excess}\} = \frac{R_1 R_2 C_1 C_2}{TT(T-1)}.$$
>
> And when, in addition, the sample sizes are large, *the sampling distribution of the excess is approximately normal.*

An approximate p-value is obtained in the usual way, based on a Z-test. This is illustrated in Display 19.8 in connection with the sex role stereotype data. The one-sided p-value is 0.012, which is somewhat smaller than the exact one-sided p-value of 0.024.

DISPLAY 19.8 Excess test as an approximation to Fisher's Exact Test: calculations for testing equal promotion probabilities for bank supervisors given "male" and "female" application folders

	Promoted? Yes	No	Totals
Male	21	3	24
Female	14	10	24
Totals	35	13	48

(1) *Observed* $= 21$

(2) *Expected* $= \dfrac{(24)(35)}{48} = 17.5$

Difference

(3) *Excess* $= 3.5$

(4) *Variance* $= \dfrac{(24)(24)(35)(13)}{(48)(48)(47)} = 2.42$

(5) *Z-statistic* $= \dfrac{3.5}{\sqrt{2.42}} = 2.25$

(6) One-sided p-value $=$ **0.012**

The essential features of this statistic are as follows:

1. For 2×2 tables of counts, the excess is an approximation to Fisher's Exact Test.
2. The p-value is close but not identical to the one from the Z-test for equal population proportions and the chi-squared test for independence.

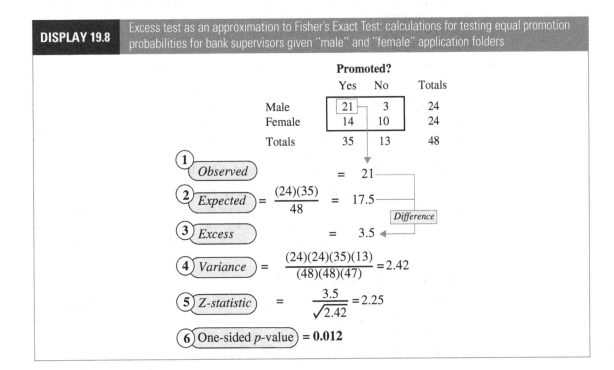

3. The excess statistics from several 2×2 tables can be added up to get combined evidence in 2×2 tables for each of several levels of a confounding or blocking factor.

The Mantel–Haenszel excess test provides nothing new for analyzing 2×2 tables, but it is the only test that can be combined over several tables.

19.5.2 The Mantel–Haenszel Test for Equal Odds in Several 2×2 Tables

The Mantel–Haenszel testing procedure applies when several 2×2 tables are involved—one for each level of a confounding or blocking factor. Such is the case in the death penalty and race study of Section 19.1.2, where the confounding variable is the aggravation level of the crime. The proportion of death penalty sentences is expected to be larger for higher aggravation categories, but the relative odds—the odds of death penalty for white-victim murderers relative to the odds of death penalty for black-victim murderers—may nonetheless remain the same for all aggravation levels. If so, the Mantel–Haenszel test is appropriate for evaluating the hypothesis that the common odds ratio is 1.

The test statistic, a single *Excess* statistic, is the sum of the excesses from the individual tables. The procedure for running the test is as follows:

1. Select one cell in the 2×2 table, where the excess will be computed.
2. Compute *Excess* and Var{*Excess*} for that cell in each table separately.
3. The combined *Excess* is the sum of the individual Excesses.
4. The variance of the combined *Excess* is the sum of the individual variances. The standard deviation is the square root of this.
5. A Z-statistic is the combined *Excess* divided by its standard deviation.

This procedure is demonstrated in Display 19.9. Even after the aggravation level of the crime is accounted for, convincing evidence remains that the odds of a black murderer's getting the death penalty if the victim was white are greater than the odds of death penalty if the victim was black (one-sided p-value $= 0.0004$).

Notes About the Mantel–Haenszel Test

1. One assumption is that the odds ratio is the same in each 2×2 table. For one thing, the hypothesis that the odds ratio is 1 does not make sense if the odds ratio differs in the different 2×2 tables. Fortunately, even though the odds and the proportions may differ considerably in the different tables, the relative odds often remain roughly constant. A way to check the assumption is given in Chapter 21.
2. The p-value is approximate, based on large sample size. Fortunately, the total sample size of all tables is the important measure. A safe guideline is that the sum of the expected cell counts (added over all tables) should be at least 5, for each of the four cell positions.
3. The Mantel–Haenszel test treats the different levels of the confounding variable as nominal, ignoring any quantitative structure. It is possible to use *logistic*

regression (see Chapters 20 and 21) to include aggravation level, for example, as a numerical explanatory variable.

4. The Mantel–Haenszel test is appropriate for prospective and retrospective observational data, as well as for randomized experiments.
5. Often a chi-squared version of the Mantel–Haenszel test is used. It is equal to the square of the Z-statistic.

19.5.3 Estimate of the Common Odds Ratio

The Mantel–Haenszel theory also provides an estimate of the common odds ratio from all tables combined. The estimated odds ratio in a single 2×2 table is $(n_{11}n_{22})/(n_{12}n_{21})$. In aggravation level 1 of Display 19.9, for example, this value is $[(2)(181)]/[(60)(1)] = 6.03$. To get an estimated odds ratio from all tables together,

DISPLAY 19.9 Mantel–Haenszel calculations for death penalty and race data from Section 19.1.2

(1) *Select a cell.* → White-victim murderers given the death penalty

(2) *Find the excess and its variance for each table.*

Aggravation level	Race of victim	Death penalty: Yes	Death penalty: No	Expected	Excess	Variance
1	White	2	60	0.762	1.238	0.564
	Black	1	181			
2	White	2	15	1.308	0.692	0.699
	Black	1	21			
3	White	6	7	4.333	1.667	1.382
	Black	2	9			
4	White	9	3	7.333	1.667	1.007
	Black	2	4			
5	White	9	0	7.313	1.688	0.640
	Black	4	3			
6	White	17	0	17.000	0.000	0.000
	Black	4	0			

(3) *Add the excesses and their variances.* → 6.952 4.292

(4) *Calculate the Z-statistic.* → $Z\text{-statistic} = \dfrac{6.951}{\sqrt{4.292}} = 3.355$

(5) *One-sided p-value from the normal distribution* → $\Pr\{Z > 3.356\} = \mathbf{0.0004}$

Conclusion: Nearly 7 more white-victim murderers received the death penalty than would have been expected if the odds of receiving the death penalty were the same for white-victim and black-victim murders. This cannot reasonably be attributed to chance.

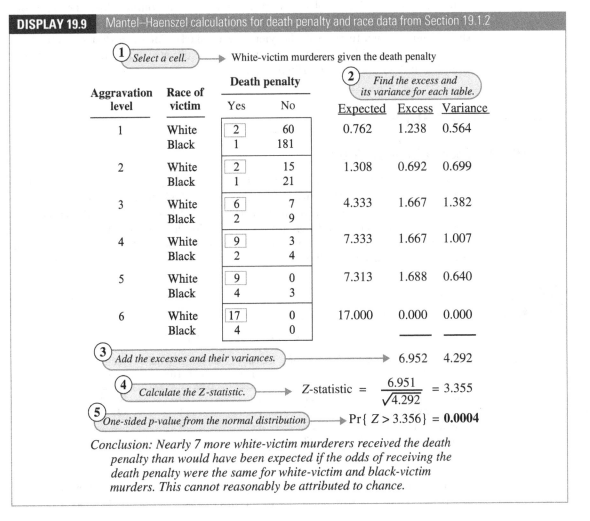

one sums the numerators in these expressions, weighted by the reciprocal of their table totals. Then one sums the denominators, similarly weighted. The estimate of the common odds ratio from several tables is

$$\hat{\omega} = \frac{\text{Sum of } \{n_{11}n_{22}/T\} \text{ over all tables}}{\text{Sum of } \{n_{12}n_{21}/T\} \text{ over all tables}}.$$

Use of this estimate assumes that the odds ratio is the same in all tables.

Example—Death Penalty and Race

The Mantel–Haenszel estimated odds ratio from the death penalty data is

$$\hat{\omega} = \frac{\frac{(2)(181)}{244} + \frac{(2)(21)}{39} + \frac{(6)(9)}{24} + \frac{(9)(4)}{18} + \frac{(9)(3)}{16} + \frac{(17)(0)}{21}}{\frac{(60)(1)}{244} + \frac{(15)(1)}{39} + \frac{(7)(2)}{24} + \frac{(3)(2)}{18} + \frac{(0)(4)}{16} + \frac{(0)(4)}{21}} = 5.49.$$

After the aggravation level of the crime has been accounted for, the odds of a black killer's receiving the death penalty are estimated to be 5.49 times greater if the victim was white than if the victim was black. Software that computes this estimate also provides a standard error or a confidence interval. The method is complex and will not be described here.

19.6 RELATED ISSUES

19.6.1 $r \times c$ Tables of Counts

More general tables arise from counts of subjects falling into cross-classifications of several factors, each with many levels. Display 19.10 shows a table listing the number of homicides of children by their parents, over 10 years in Canada, categorized according to the parent–offspring sex combination and according to age category. (Data from M. Daly and M. Wilson, "Evolutionary Social Psychology and Family Homocide," *Science* 242 (1988): 519–24.)

DISPLAY 19.10 Parent–offspring homicides in Canada, 1974 to 1983, by sexes of the parent killer and the child killed, and by age class of the child

	Age classification of child				
Sex of parent/child	Infantile (0–1)	Oedipal (2–5)	Latency (6–10)	Circumpubertal (11–16)	Adult (≥17)
Male/Male	24	21	21	29	104
Male/Female	17	27	10	14	47
Female/Male	53	21	19	9	8
Female/Female	50	27	5	4	15

The sampling scheme for this example is Poisson, so one should either test for independence of the row and column factors or declare one of these as a response factor and test for homogeniety of the (multinomial) responses across levels of the explanatory factor. For this problem it is unnecessary to specify either factor as response.

The chi-squared test of independence is conducted just as for 2×2 tables; that is, the *observed* cell count in row i and column j is n_{ij}, and the *expected* cell count (supposing that the null hypothesis of independence or homogeneity is true) is the row total times the column total, divided by the grand total: $R_i C_j / T$. The chi-squared statistic is the sum, for all cells in the table, of

$$(Observed - Expected)^2 / Expected.$$

The p-value is found by comparing this test statistic to a chi-squared distribution on $(r-1) \times (c-1)$ degrees of freedom, where r is the number of rows and c is the number of columns in the table.

19.6.2 Higher-Dimensional Tables of Counts

Some additional structure is present in the table because the four levels of the row factor correspond to the cross-classifications of parent (male or female) and child (male or female). The table should therefore more properly be considered a $2 \times 2 \times 5$ table of counts. The chi-squared analysis has been historically used for this type of table. An initial screening test assesses independence between all three factors. If the independence hypothesis is rejected, the chi-square statistic is decomposed into a set of component chi-squared statistics with smaller degrees of freedom, each testing different aspects of the dependency structure. This is accomplished by means of a series of slices and lumpings of categories. Carried out fully, the original chi-squared statistic is decomposed into chi-squared statistics (with one degree of freedom), each of which tests a hypothesis of interest.

Because the analysis sequence resembles an analysis of variance so closely, it suggests that there may be a comparable regression approach. There is. Chapter 22 introduces Poisson log-linear regression, which offers a more flexible approach to analyzing tables of counts, using regression techniques. For example, age of the offspring may be modeled either as a factor or as a numerical explanatory variable.

19.7 SUMMARY

Fisher's Exact Test can be used for testing in 2×2 tables, no matter how the data arose, no matter which of the two hypotheses is being tested, and no matter what the sample sizes are. The interpretation of the p-value, however, should be customized to the particular hypothesis, which is chosen according to the question of interest and the sampling scheme. For large sample sizes, the easier-to-compute approximations are adequate.

Pearson's chi-squared statistic is a convenient approximation to the exact test. A principal advantage of the Pearson method is that it permits testing independence in larger tables. The principal disadvantage is that it is a test procedure only.

The Mantel–Haenszel Test compares the odds in two groups after accounting for the effect of a blocking or confounding variable. If the blocking variable is numerical, however, there is some merit in attempting a more specific model, using logistic regression as in Chapter 21. Logistic regression can also be used to test for the condition that the odds ratio is the same in all the 2×2 tables: if it isn't, the Mantel–Haenszel test is inappropriate.

Sex Role Stereotypes in Promotions

Fisher's Exact Test is appropriate for testing whether the probability of promotion is the same regardless of the sex of the applicant. The p-value in this case, 0.0245, is exact and provides an inferential measure tied directly to the probabilities of assignment in the randomized experiment.

Death Penalty Sentencing Study

The Mantel–Haenszel procedure applied to these data computes an overall excess of death penalty sentences given to murderers of white victims. *Excess* means the additional number of murderers receiving the death penalty over what would be expected if death penalty and race of victim are independent. The excess is a direct measure of the practical significance of the disparity in sentencing and provides the basis for the measure of statistical significance.

19.8 EXERCISES

Conceptual Exercises

1. Sex Role Stereotypes. (a) From this study, what can be inferred about the attitudes of bank managers toward promotion of females? (b) How would you respond to someone who asserted that the higher proportion of the "male" folders promoted occurred simply by chance, because the tougher managers happened to be placed in the "female" group?

2. Death Penalty and Race. (a) Lawyers for the defense showed that only 6% of black-victim murderers received the death penalty, while 35% of white-victim murderers received the death penalty. They claimed that the pattern of sentencing was racially biased. How should the (statistically knowledgeable) lawyers for the state respond? (b) What can the lawyers for the defense say to rebut the state lawyers' response? (c) How should the lawyers for the state respond to this further analysis?

3. (a) Why is the hypothesis $H_0: \pi_1 - \pi_2 = 0$ identical to the hypothesis $H_0: \omega_1/\omega_2 = 1$. (b) Does this mean that retrospective data can be used to estimate $\pi_1 - \pi_2$?

4. Coffee Drinking and Sexual Habits. From a sample of 13,912 households in Washtenaw County, Michigan, 2,993 persons were identified as being 60 years old or older. Of these, 1,956 agreed to participate in a study. The following table is based on data from the married women in this sample who answered the questions about whether they were sexually active and whether they drank coffee. (Data from A. C. Diokno et al., "Sexual Function in the Elderly," *Archives of Internal Medicine* 150 (1990): 197–200.) (a) What type of sampling scheme was this? To what population can inferences be made? (b) Are these data more useful for investigating whether coffee drinking causes decreased sexual activity or whether sexual activity causes decreased coffee drinking? (c) If it is desired to treat sexual activity as a categorical response, does the inequality of the row totals create a problem?

	Sexually active	
	Yes	No
Coffee drinker	15	25
Not coffee drinker	115	70

5. Crab Parasites. Crabs were collected from Yaquina Bay, Oregon, during June 1982 and categorized by their species (Red or Dungeness) and by whether a particular type of parasite was found. (a) What type of sampling would you call this? (b) What tool would you use to determine whether the parasite is more likely to be present on a Red Crab than on a Dungeness Crab?

	Parasite present	
	Yes	No
Red Crab	5	312
Dungeness Crab	0	503

6. Hunting Strategies. A popular way to synthesize results in a research area is to assemble a bibliography of all similar studies and to use a random sample of them for statistical analysis. The following table categorizes a random sample of 50 (out of 1,000) published references relating to the hunting success of predators. Each article was categorized according to whether the animal under discussion hunted alone or in groups and whether it hunted single or multiple prey. (Data from R. E. Morris, Oregon State University Department of Fisheries and Wildlife.) (a) What type of sampling would you call this? (b) What is the population to which inferences can be made?

	Prey sought	
	Single	Multiple
Hunt alone	17	12
Hunt in groups	14	7

7. (a) In what way is the relationship between the normal and chi-squared distributions similar to the relationship between the t- and F-distributions? (b) How could one obtain a p-value for the chi-squared test of independence in a 2×2 table, using the standard normal (and not the chi-squared) distribution?

8. (a) In investigating the odds ratio from a randomized experiment, why must one account for the effects of an additional variable (with the Mantel–Haenszel procedure, for example), if it is important? (b) What further reason is there if the data are observational?

9. The mathematics and physics departments of Crane and Eagle colleges (fictitious institutions) had a friendly competition on a standard science exam. Of the two math departments, the Crane students had a higher percentage of passes; of the two physics departments, the Crane students again had a higher percentage of passes. But for the two departments combined, the students at Eagle had a higher percentage of passes. (No joke: this is an example of *Simpson's paradox*.) The results are shown in the following table. Which college "won"? Why? Can you explain the apparent paradox in nonstatistical language?

	Mathematics		Physics	
	Pass	Fail	Pass	Fail
Crane College	12	6	8	14
Eagle College	20	16	1	3

Computational Exercises

10. Coffee Drinking and Sexual Habits. Find the p-value for the chi-squared test of independence in the coffee drinking and sexual habits data (Exercise 4).

11. Crab Parasites. Find the one-sided p-value for testing whether the odds of a parasite's being present are the same for Dungeness Crab as for Red Crab (Exercise 5), using Fisher's Exact Test. (Notice that only one hypergeometric probability needs to be computed.)

12. Hunting Strategy. For the hunting strategy data of Exercise 6, find the p-value for testing independence by (a) the chi-squared test of independence, and (b) Fisher's Exact Test.

13. Smoking and Lung Cancer. For the data in Section 18.1.3, test whether the odds of lung cancer for smokers are equal to the odds of lung cancer for nonsmokers, using (a) the excess statistic and (b) Fisher's Exact Test.

14. Mantel–Haenszel Test for Censored Survival Times: Lymphoma and Radiation Data. Central nervous system (CNS) lymphoma is a rare but deadly brain cancer; its victims usually die within one year. Patients at Oregon Health Sciences University all received chemotherapy administered after a new treatment designed to allow the drugs to penetrate the brain's barrier to large molecules carried in the blood. Of 30 patients as of December 1990, 17 had received only this chemotherapy; the other 13 had received radiation therapy prior to their arrival at OHSU. The table in Display 19.11 shows survival times, in months, for all patients. Asterisks refer to censored times; in these cases, the survival time is known to be at least as long as reported, but how much longer is not known, because the patient was still living when the study was reported. (Data from E. A. Neuwelt et al., "Primary CNS Lymphoma Treated with Osmotic Blood-brain Barrier Disruption: Prolonged Survival and Preservation of Cognitive Function," *Journal of Clinical Oncology* 9(9) (1991): 1580–90.)

The Mantel–Haenszel procedure can be used to construct a test for equality of survival patterns in the two groups. It involves constructing separate 2×2 tables for each month in which a death occurred, as shown in Display 19.12. The idea is to compare the numbers surviving and the numbers dying in the two groups, accounting for the length of time until death. The data are arranged in several 2×2 tables, as in Display 19.12. The censored observations are included in all tables for which the patients are known to be alive and are excluded from the others. Compute the Mantel–Haenszel test statistic (equivalently called the *log-rank statistic*), and find the one-sided p-value. What can be inferred from the result?

15. Bats. Bat Conservation International (BCI) is an organization that supports research on bats and encourages people to provide bat roost boxes (much like ordinary birdhouses) in backyards and gardens. BCI phoned people across Canada and the United States who were known to have put out boxes, to determine the sizes, shapes, and locations of the boxes and to identify whether the boxes were being used by bats. Among BCI's findings was the conclusion, "Bats were significantly more

DISPLAY 19.11 Survival times (months) for two groups of lymphoma patients

Status (dead or alive) and survival times as of December 1990 for CNS lymphoma patients with and without prior radiation

Radiation		No radiation	
Months	Status	Months	Status
3	D	3	D
3	D	7	D
10	D	9	D
11	D	9	D
12	D	16	D
18	D	*17	A
21	D	*18	A
22	D	*20	A
23	D	24	D
33	D	*25	A
36	D	*30	A
48	D	32	D
*62	A	*38	A
		41	D
		*54	A
		*71	A
		*99	A

*Censored survival time: patient was alive at end of study; actual survival time is unknown, but is at least as large as what is listed.

likely to move into houses during the first season if they were made of old wood. Of comparable houses first made available in spring and eventually occupied, eight houses made of old wood were all used in the first season, compared with only 32 of 65 houses made of new wood." (Data from "The Bat House Study," *Bats* 11(1) (Spring 1993): 4–11.) The researchers required that, to be significant, a chi-square test meet the 5% level of significance. (a) Is the chi-squared test appropriate for these data? (b) Evaluate the data's significance, using Fisher's Exact Test.

16. Vitamin C and Colds. Pursuing an experiment similar to the one in Section 18.1.2, skeptics interviewed the same 800 subjects to determine who knew and who did not know to which group they had been assigned. Vitamin C, they argued, has a unique bitter taste, and those familiar with it could easily recognize whether their pills contained it. If so, the blind was broken, and the possibility arose that subjects' responses may have been influenced by the knowledge of their group status. The following table presents a fictitious view of how the resulting study might have turned out. Use the Mantel–Haenszel procedure (a) to test whether the odds of a cold are the same for the placebo and vitamin C users, after accounting for group awareness, and (b) to estimate the common odds ratio.

Vitamin C and colds data

	Knew their group		Did not know	
	Cold	No cold	Cold	No cold
Placebo	266	46	62	26
Vitamin C	139	29	157	75

DISPLAY 19.12 Several 2 × 2 tables, formed from the censored and uncensored survival times

	Radiation group		No radiation group	
Month after diagnosis	Known to survive beyond this month	Known to die after this many months	Known to survive beyond this month	Known to die after this many months
3	11	2	16	1
7	11	0	15	1
9	11	0	13	2
10	10	1	13	0
11	9	1	13	0
12	8	1	13	0
16	8	0	12	1
18	7	1	10	0
21	6	1	9	0
22	5	1	9	0
23	4	1	9	0
24	4	0	8	1
32	4	0	5	1
33	3	1	5	0
36	2	1	5	0
41	2	0	3	1
48	1	1	3	0

Data Problems

17. Alcohol Consumption and Breast Cancer—A Retrospective Study. A 1982–1986 study of women in Toronto, Ontario, assessed the added risk of breast cancer due to alcohol consumption. (Data from Rosenberg et al., "A Case–Control Study of Alcoholic Beverage Consumption and Breast Cancer," *American Journal of Epidemiology* 131 (1990): 6–14.) A sample of confirmed breast cancer patients at Princess Margaret Hospital who consented to participate in the study was obtained, as was a sample of cancer-free women from the same neighborhoods who were close in age to the cases. The following tables show data only for women in two categories at the ends of the alcohol consumption scale: those who drank less than one alcoholic beverage per month (3 years before the interview), and those that drank at least one alcoholic beverage per day. The women are further categorized by their body mass, a possible confounding factor. What evidence do these data offer that the risk of breast cancer is greater for heavy drinkers ($\geq$ 1 drink/day) than for light drinkers ($<$ 1 drink/month)?

Alcohol consumption and breast cancer study

	<21 kg/m^2		21–25 kg/m^2		>25 kg/m^2	
	Cases	Controls	Cases	Controls	Cases	Controls
At least one drink per day	38	52	65	147	30	42
Less than one drink per month	26	61	94	153	56	102

18. The Donner Party. In 1846 the Donner party became stranded while crossing the Sierra Nevada mountains near Lake Tahoe. (Data from D. J. Grayson, "Donner Party Deaths: A Demographic Assessment," *Journal of Anthropological Research* 46 (1990): 223–42. See also Section 20.1.1.) Shown in the following tables are counts for male and female survivors for six age groups. Is there evidence that the odds of survival are greater for females than for males, after the possible effects of age have been accounted for?

	Age category											
	15–19		20–29		30–39		40–49		50–59		60–69	
	Lived	Died	Lived	Died	Lived	Died	Lived	Died	Lived	Died	Lived	Died
Male	1	1	5	10	2	4	2	1	0	1	0	3
Female	1	0	6	1	2	0	1	3	0	1	0	0

19. Tire-Related Fatal Accidents and Ford Sports Utility Vehicles. The table in Display 19.13 shows the numbers of compact sports utility vehicles involved in fatal accidents in the United States between 1995 and 1999, categorized according to travel speed, make of car (Ford or other), and cause of accident (tire-related or other). From this table, test whether the odds of a tire-related fatal accident depend on whether the sports utility vehicle is a Ford, after accounting for travel speed. For this subset of fatal accidents, estimate the excess number of Ford tire-related accidents. (This is a subset of data described more fully in Exercise 20.18.).

DISPLAY 19.13 Numbers of sports utility vehicles (1995 and newer) involved in fatal traffic accidents between 1995 and 1999, excluding those struck by other vehicles, those for which travel speed was unavailable, and those involved in alcohol-related accidents

		Cause of accident	
Speed (m.p.h.)	Make	Other	Tire
0–40	Ford	171	0
	Other	465	2
41–55	Ford	243	0
	Other	753	3
56–65	Ford	98	3
	Other	325	3
>66	Ford	108	21
	Other	273	8

20. Diet Wars I. In the study of three different diets for losing weight (Exercise 6.23), 50 subjects who began the study dropped out before the two-year period ended. Only one of the 45 women dropped out, while 49 of the original 277 men dropped out. What evidence does this provide that the women were more likely to complete the study?

21. Diet Wars II. In the study of different diets for losing weight (Exercises 6.23 and 20 above), there appear to have been many more experimental subjects that dropped out from the low-carbohydrate diet group than from the other two diet groups. The numbers are shown in the following table:

	Dropped Out	Completed	Total
Low-Fat Diet	10	94	104
Mediterranean Diet	16	83	99
Low-Carbohydrate Diet	24	85	109
Total	50	262	312

What evidence is there of different drop-out rates in the three diets?

22. Diet Wars III. In Exercise 20 above, it appears that women were more likely to complete the study than men. In Exercise 21 above, it appears that there was a higher tendency to drop out from the low-carbohydrate diet than from the other two diets. Use the table below to assess the evidence that women were more likely to complete the study, after accounting for possible differences in dropout rates in the different diets. (Use the Mantel–Haenszel procedure to perform the test, but comment on the degree to which the p-value can be trusted.)

		Dropped Out	Completed	Total
Low-Fat	**Women**	0	15	15
	Men	10	79	89
Mediterranean	**Women**	0	15	15
	Men	16	78	94
Low-Carbohydrate	**Women**	1	14	15
	Men	23	71	94
		50	272	322

23. Who Looks After the Kids? Exercises 10.31 and 17.13 considered data on clutch volume and body mass to investigate the similarities of dinosaurs with several groups of currently living animals. One issue concerning the validity of the study is the selection of the bird species in the set of currently living animals. Were species (i) selected at random? (ii) stratified by group and selected within group at random? (iii) selected based on what information was available? or (iv) selected to support the study conclusions? Ignore the possibility of (iv). If the answer is (i) or (ii) then that would not cause any problem with the analyses performed, but (iii) could potentially result in biased sampling and therefore incorrect conclusions. Furthermore, (iii) seems like the most probable form of sampling of the modern groups and almost certainly the form of sampling among the dinosaur groups. One way to assess the likelihood (iii) for birds is to compare the numbers of species from each of the 29 orders in the study with the known total number of species in each of the orders, as shown in Display 19.14. If the selection of birds had been at random, the expected proportion of species in the study from one particular order, n, is the proportion of all species in that order ($N/9{,}866$) times the total number of species in the sample (414). That is, the expected number in each sample, if random sampling were used, is $(N/9{,}866) \times 414$. Calculate the expected numbers and compare the observed numbers with them using Pearson's chi-square statistic. It will have $29 - 1 = 28$ d.f. What do you conclude?

Answers to Conceptual Exercises

1. (a) Any inference to a population other than the 48 males in the training course is speculative because these subjects were not randomly selected from any population. (b) They are possibly correct. But the chance of such an extreme randomization is about 2 out of 100 (the p-value).

2. (a) They argue that no inference that the victims' races cause different death penalty rates can be drawn from this observational data. Potentially confounding variables, such as aggravation category, might be associated with both the race of the victim and the likelihood of a death sentence. (b) The

DISPLAY 19.14 Total number of bird species in each of 29 orders and number in the sample from each order

Order of Species	Total Number in this Order, N	Sample Number in this Order, n
Anseriformes	150	5
Apodiformes	424	3
Caprimulgiformes	118	8
Casuariiformes	4	2
Charadriiformes	343	62
Ciconiiformes	114	14
Coliiformes	6	1
Columbiformes	309	2
Coraciiformes	213	8
Cuculiformes	159	3
Falconiformes	308	31
Galbuliformes	54	2
Galliformes	281	24
Gruiformes	204	21
Musophagiformes	23	1
Opisthodomiformes	1	0
Passeriformes	5,920	179
Pelecaniformes	65	6
Phoenicopteriformes	5	1
Piciformes	344	2
Podicipediformes	22	1
Procellariiformes	108	5
Psittaciformes	353	1
Pterocliformes	16	3
Sphenisciformes	17	1
Strigiformes	209	2
Struthioniformes	10	5
Tinamiformes	47	19
Trogoniformes	39	2
Total	**9,866**	**414**

defense lawyers compare the odds of white-victim and black-victim death penalty sentences, after accounting for a conspicuous confounding variable—aggravation level (using the Mantel–Haenszel procedure). (c) The state lawyers respond that further confounding variables may exist, whether they can be identified or not, that might explain the difference.

3. (a) If $\pi_1 = \pi_2$, then $\pi_1/(1-\pi_1) = \pi_2/(1-\pi_2)$ (so their ratio is 1). This algebra shows that if two probabilities are equal then their corresponding odds are also equal. Similar algebra working in the reverse direction shows that if two odds are equal then their corresponding probabilities are also equal. (b) No. Unless $\omega_1/\omega_2 = 1$, the value of ω_1/ω_2 does not identify the value of $\pi_1 - \pi_2$.

4. (a) It is not clear whether the sampling scheme was Poisson or multinomial. The 225 women are a sample from a population of married women of age 60 or older in the particular county, who agreed to participate in the study and who answered the questions about sexual activity and coffee drinking. Extrapolating to any other population introduces the potential for bias. (b) Although the observational data can be examined for consistency with the causal theories, the statistical analysis alone cannot be used to imply causation, one way or another. (c) No.

5. (a) Poisson. (b) Fisher's Exact Test, since there is a problem with small cell counts.

6. (a) Multinomial. (b) The population of 1,000 published articles about hunting success.

7. (a) The F-distribution on 1 and k degrees of freedom describes the distribution of the square of a variable with the t-distribution on k degrees of freedom. Similarly, the chi-square distribution on 1 degree of freedom describes the square of a variable with the standard normal distribution. (b) Take the square root of X^2, and double the area under a standard normal curve to the right of this.

8. (a) Accounting for the confounding variable in the analysis will make the inference about the association under study more precise. (b) The study's lack of control over the confounding variable at the design stage provides an added incentive.

9. Among math students, 67% passed at Crane, compared to 56% at Eagle. Among physics students, 36% passed at Crane, compared to 25% at Eagle. Among all students (ignoring department), 50% passed at Crane compared to 53% at Eagle. Crane won. The paradox arises because the math students had an easier time with their exam than the physics students had with theirs, coupled with the fact that Eagle had a very high proportion of participants from its math department (90%, compared to 45% at Crane). Crane won because it won both match-ups involving students with supposedly comparable training.

CHAPTER 20

Logistic Regression for Binary Response Variables

Of considerable importance in many fields of application are statistical regression models in which the response variable is binary—meaning it takes on one of two values, 0 or 1. The zero-or-one response is usually an artificial device for modeling a categorical variable with two categories. If, for example, Y is a response taking on the value 1 for study subjects who developed lung cancer and zero for those who did not, then the mean of Y is a probability of developing lung cancer. Regression models for binary responses, therefore, are used to describe probabilities as functions of explanatory variables.

Linear regression models are usually inadequate for binary responses because they can permit probabilities less than 0 and greater than 1. While there are several alternatives, the most important one is logistic regression. This has a form in which a usual regression-like function of explanatory variables and regression coefficients describes a function of the mean (a function of the probability in this case) rather than the mean itself. The particular function is the logarithm of the odds.

Interpretations of logistic regression coefficients are made in terms of statements about odds and odds ratios. Although the model, estimation, and inference procedure differ from those of ordinary regression, the practical use of logistic regression analysis with a statistical computing program parallels the usual regression analysis closely.

20.1 CASE STUDIES

20.1.1 Survival in the Donner Party—An Observational Study

In 1846 the Donner and Reed families left Springfield, Illinois, for California by covered wagon. In July, the Donner Party, as it became known, reached Fort Bridger, Wyoming. There its leaders decided to attempt a new and untested route to the Sacramento Valley. Having reached its full size of 87 people and 20 wagons, the party was delayed by a difficult crossing of the Wasatch Range and again in the crossing of the desert west of the Great Salt Lake. The group became stranded in the eastern Sierra Nevada mountains when the region was hit by heavy snows in late October. By the time the last survivor was rescued on April 21, 1847, 40 of the 87 members had died from famine and exposure to extreme cold.

Display 20.1 shows the ages and sexes of the adult (over 15 years) survivors and nonsurvivors of the party. These data were used by an anthropologist to study the theory that females are better able to withstand harsh conditions than are males. (Data from D. K. Grayson, "Donner Party Deaths: A Demographic Assessment," *Journal of Anthropological Research* 46 (1990): 223–42.) For any given age, were the odds of survival greater for women than for men?

Statistical Conclusion

While it generally appears that older adults were less likely than younger adults to survive, the data provide highly suggestive evidence that the age effect is more pronounced for females than for males (the two-sided p-value for interaction of gender and age is 0.05). Ignoring the interaction leads to an informal conclusion about the overall difference between men and women: The data provide highly suggestive evidence that the survival probability was higher for women (in the Donner Party) than for the men (two-sided p-value $= 0.02$ from a drop-in-deviance test). The odds of survival for women are estimated to be 5 times the odds of survival for similarly aged men (95% confidence interval: 1.2 times to 25.2 times, from a drop-in-deviance-based confidence interval).

Scope of Inference

Since the data are observational, the result cannot be used as proof that women were more apt to survive than men; the possibility of confounding variables cannot be excluded. Furthermore, since the 45 individuals were not drawn at random from any population, inference to a broader population is not justified.

20.1.2 Birdkeeping and Lung Cancer—A Retrospective Observational Study

A 1972–1981 health survey in The Hague, Netherlands, discovered an association between keeping pet birds and increased risk of lung cancer. To investigate birdkeeping as a risk factor, researchers conducted a *case–control* study of patients in 1985 at four hospitals in The Hague (population 450,000). They identified 49 cases

DISPLAY 20.1 Sex, age, and survival status of Donner Party members, 15 years or over

Name	Sex	Age	Survival
Antonio	male	23	no
Breen, Mary	female	40	yes
Breen, Patrick	male	40	yes
Burger, Charles	male	30	no
Denton, John	male	28	no
Dolan, Patrick	male	40	no
Donner, Elizabeth	female	45	no
Donner, George	male	62	no
Donner, Jacob	male	65	no
Donner, Tamsen	female	45	no
Eddy, Eleanor	female	25	no
Eddy, William	male	28	yes
Elliot, Milton	male	28	no
Fosdick, Jay	male	23	no
Fosdick, Sarah	female	22	yes
Foster, Sarah	female	23	yes
Foster, William	male	28	yes
Graves, Eleanor	female	15	yes
Graves, Elizabeth	female	47	no
Graves, Franklin	male	57	no
Graves, Mary	female	20	yes
Graves, William	male	18	yes
Halloran, Luke	male	25	no
Hardkoop, Mr.	male	60	no
Herron, William	male	25	yes
Noah, James	male	20	yes
Keseberg, Lewis	male	32	yes
Keseberg, Phillipine	female	32	yes
McCutcheon, Amanda	female	24	yes
McCutcheon, William	male	30	yes
Murphy, John	male	15	no
Murphy, Lavina	female	50	no
Pike, Harriet	female	21	yes
Pike, William	male	25	no
Reed, James	male	46	yes
Reed, Margaret	female	32	yes
Reinhardt, Joseph	male	30	no
Shoemaker, Samuel	male	25	no
Smith, James	male	25	no
Snyder, John	male	25	no
Spitzer, Augustus	male	30	no
Stanton, Charles	male	35	no
Trubode, J.B.	male	23	yes
Williams, Baylis	male	24	no
Williams, Eliza	female	25	yes

of lung cancer among patients who were registered with a general practice, who were age 65 or younger, and who had resided in the city since 1965. They also selected 98 controls from a population of residents having the same general age structure. (Data based on P. A. Holst, D. Kromhout, and R. Brand, "For Debate: Pet Birds as an Independent Risk Factor for Lung Cancer," *British Medical Journal* 297 (1988): 13–21.) Display 20.2 shows the data gathered on the following variables:

FM = Sex (1 = F, 0 = M)
AG = Age, in years
SS = Socioeconomic status (1 = High, 0 = Low), determined by occupation of the household's principal wage earner
YR = Years of smoking prior to diagnosis or examination
CD = Average rate of smoking, in cigarettes per day
BK = Indicator of birdkeeping (caged birds in the home for more than 6 consecutive months from 5 to 14 years before diagnosis (cases) or examination (controls)

Age and smoking history are both known to be associated with lung cancer incidence. After age, socioeconomic status, and smoking have been controlled for, is an additional risk associated with birdkeeping?

Statistical Conclusion

The odds of lung cancer among birdkeepers are estimated to be 3.8 times as large as the odds of lung cancer among nonbirdkeepers, after accounting for the effects of smoking, sex, age, and socioeconomic status. An approximate 95% confidence interval for this odds ratio is 1.7 to 8.5. The data provide convincing evidence that increased odds of lung cancer were associated with keeping pet birds, even after accounting for the effect of smoking (one-sided p-value = 0.0008).

Scope of Inference

Inference extends to the population of lung cancer patients and unaffected individuals in The Hague in 1985. Statistical analysis of these observational data cannot be used as evidence that birdkeeping causes the excess cancer incidence among birdkeepers, although the data are consistent with that theory. (As further evidence in support of the causal theory, the researchers investigated other potential confounding variables [beta-carotene intake, vitamin C intake, and alcohol consumption]. They also cited medical rationale supporting the statistical associations: People who keep birds inhale and expectorate excess allergens and dust particles, increasing the likelihood of dysfunction of lung macrophages, which in turn can lead to diminished immune system response.)

20.2 THE LOGISTIC REGRESSION MODEL

The response variables in both case studies—*survival* in the Donner Party study and *lung cancer* in the birdkeeping study—are *binary*, meaning that they take on

DISPLAY 20.2 Birdkeeping and lung cancer data: subject number (*SN*), lung cancer (*LC* = 1 for lung cancer patients, 0 for controls), sex (*FM* = 1 for females), socioeconomic status (*SS* = 1 for high), birdkeeping (*BK* = 1 if yes), age (*AG*), years smoked (*YR*), and cigarettes per day (*CD*)

SN	LC	FM	SS	BK	AG	YR	CD
1	1	0	0	1	37	19	12
2	1	0	0	1	41	22	15
3	1	0	1	0	43	19	15
4	1	0	0	1	46	24	15
5	1	0	0	1	49	31	20
6	1	0	1	0	51	24	15
7	1	0	1	1	52	31	20
8	1	0	0	0	53	33	20
9	1	0	0	1	56	33	10
10	1	0	1	0	56	26	25
11	1	0	0	0	56	35	40
12	1	0	0	0	56	36	25
13	1	0	0	1	56	36	20
14	1	0	0	1	57	39	25
15	1	0	0	0	58	38	20
16	1	0	0	0	58	35	25
17	1	0	1	0	58	42	30
18	1	0	0	1	59	39	20
19	1	0	1	1	59	40	15
20	1	0	0	1	60	38	15
21	1	0	0	0	61	28	15
22	1	0	1	1	62	39	20
23	1	0	0	0	62	43	20
24	1	0	0	1	62	40	15
25	1	0	1	1	63	41	40
26	1	0	0	0	63	45	20
27	1	0	0	1	63	41	10
28	1	0	0	1	64	42	20
29	1	0	0	1	64	44	15
30	1	0	1	1	64	47	16
31	1	0	1	1	64	13	30
32	1	0	0	1	64	42	20
33	1	0	0	0	64	32	3
34	1	0	1	0	65	45	10
35	1	0	1	1	65	43	30
36	1	0	0	1	66	50	25
37	1	0	0	1	66	47	10
38	1	1	0	1	44	22	15
39	1	1	0	1	46	24	15
40	1	1	0	1	47	25	25
41	1	1	0	1	49	27	20
42	1	1	0	1	49	23	20
43	1	1	0	1	50	28	20
44	1	1	0	1	54	33	6
45	1	1	0	1	58	37	20
46	1	1	0	1	60	38	15
47	1	1	0	1	61	0	0
48	1	1	0	0	63	29	20
49	1	1	0	0	64	40	25

SN	LC	FM	SS	BK	AG	YR	CD
50	0	0	0	0	37	16	2
51	0	0	0	1	38	20	20
52	0	0	1	0	40	13	25
53	0	0	1	1	42	21	8
54	0	0	1	0	42	17	15
55	0	0	0	0	43	25	25
56	0	0	1	0	46	24	20
57	0	0	0	1	47	28	40
58	0	0	0	1	49	28	10
59	0	0	0	1	49	15	10
60	0	0	1	0	51	5	4
61	0	0	1	0	51	17	10
62	0	0	1	1	52	30	37
63	0	0	0	1	52	28	25
64	0	0	1	0	53	29	10
65	0	0	0	0	53	19	15
66	0	0	0	1	55	39	15
67	0	0	0	0	55	41	30
68	0	0	0	0	55	18	10
69	0	0	0	1	56	36	20
70	0	0	0	1	56	22	25
71	0	0	0	0	56	39	25
72	0	0	0	0	56	32	30
73	0	0	1	0	56	19	5
74	0	0	0	1	57	24	8
75	0	0	0	0	57	35	15
76	0	0	0	0	57	24	15
77	0	0	0	0	58	38	20
78	0	0	0	0	58	39	20
79	0	0	0	0	58	22	10
80	0	0	0	0	58	15	40
81	0	0	0	1	59	36	15
82	0	0	0	1	59	35	20
83	0	0	1	0	59	41	12
84	0	0	0	0	59	37	15
85	0	0	1	0	59	7	1
86	0	0	0	0	59	34	20
87	0	0	1	1	60	25	15
88	0	0	1	1	60	39	12
89	0	0	1	0	60	34	1
90	0	0	1	0	60	0	0
91	0	0	0	0	61	42	12
92	0	0	0	0	61	43	20
93	0	0	0	1	62	38	20
94	0	0	1	0	62	0	0
95	0	0	1	0	62	14	30
96	0	0	0	0	62	44	30
97	0	0	0	0	62	28	18
98	0	0	1	1	63	0	0

SN	LC	FM	SS	BK	AG	YR	CD
99	0	0	0	1	63	0	0
100	0	0	0	0	63	22	15
101	0	0	0	0	63	22	20
102	0	0	1	0	63	41	20
103	0	0	0	0	63	20	15
104	0	0	1	0	63	43	20
105	0	0	1	0	63	42	10
106	0	0	0	1	64	40	20
107	0	0	0	1	64	40	10
108	0	0	0	1	64	46	20
109	0	0	0	0	64	41	6
110	0	0	1	0	64	39	25
111	0	0	1	0	64	39	20
112	0	0	0	0	64	45	20
113	0	0	0	0	64	36	15
114	0	0	0	0	64	44	20
115	0	0	0	0	65	44	6
116	0	0	1	0	65	30	20
117	0	0	1	0	65	47	45
118	0	0	0	0	65	46	20
119	0	0	0	0	65	34	10
120	0	0	0	1	66	38	25
121	0	0	1	0	66	42	18
122	0	0	0	0	66	0	0
123	0	0	1	0	67	0	0
124	0	1	0	0	43	21	20
125	0	1	0	1	45	0	0
126	0	1	0	1	46	24	4
127	0	1	0	1	46	0	0
128	0	1	1	1	46	16	5
129	0	1	1	0	47	0	0
130	0	1	0	1	49	25	15
131	0	1	0	1	49	0	0
132	0	1	0	0	49	25	15
133	0	1	0	0	49	27	20
134	0	1	0	1	50	0	0
135	0	1	0	0	51	23	12
136	0	1	0	1	53	9	10
137	0	1	1	0	55	24	15
138	0	1	0	1	58	34	15
139	0	1	0	1	60	36	25
140	0	1	0	0	60	0	0
141	0	1	0	0	61	0	0
142	0	1	1	1	62	0	0
143	0	1	0	1	62	20	10
144	0	1	1	0	63	0	0
145	0	1	0	0	64	39	20
146	0	1	1	0	64	0	0
147	0	1	0	0	65	7	2

values of either 0 or 1. Both studies involve several explanatory variables, some categorical and some numerical. Logistic regression is an appropriate tool in such situations.

20.2.1 Logistic Regression as a Generalized Linear Model

A generalized linear model (GLM) is a probability model in which the mean of a response variable is related to explanatory variables through a regression equation. Let $\mu - \mu\{Y|X_1, \ldots, X_p\}$ represent the mean response. The regression structure is linear in unknown parameters $(\beta_0, \beta_1, \ldots, \beta_p)$:

$$\beta_0 + \beta_1 X_1 + \cdots + \beta_p X_p;$$

and some specified function of μ—called the *link function*—is equal to the regression structure

$$g(\mu) = \beta_0 + \beta_1 X_1 + \cdots + \beta_p X_p.$$

The strength of a generalized linear model and analysis comes from the regression equation. Virtually all of the regression tools developed in Chapters 7 through 11 carry over with minor modification, allowing users to relate explanatory information to a much wider class of response variables. The special variables—indicator variables for categorical factors, polynomial terms for curvature, products for interactions—provide flexible and sensible ways to incorporate and evaluate explanatory information.

Link Functions

The appropriate link function depends on the distribution of the response variable. The natural choice of link for a normally distributed response, for example, is the identity link, $g(\mu) = \mu$, leading to ordinary multiple linear regression. Other response variables such as binary responses or counts have different natural links to the regression structure.

The Logit Link for Binary Responses

The natural link for a binary response variable is the *logit*, or log-odds, function. The symbol π (rather than μ) is used for the population mean, to emphasize that it is a proportion or probability. The logit link is $g(\pi) = \text{logit}(\pi)$, which is defined as $\log[\pi/(1-\pi)]$, and logistic regression is

$$\text{logit}(\pi) = \beta_0 + \beta_1 X_1 + \cdots + \beta_p X_p.$$

The inverse of the logit function is called the *logistic function*. If $\text{logit}(\pi) = \eta$, then

$$\pi = \exp(\eta)/[1 + \exp(\eta)].$$

Display 20.3 illustrates how the logit and logistic functions connect the regression structure (represented by η) with the population proportion.

DISPLAY 20.3 The standard logistic function

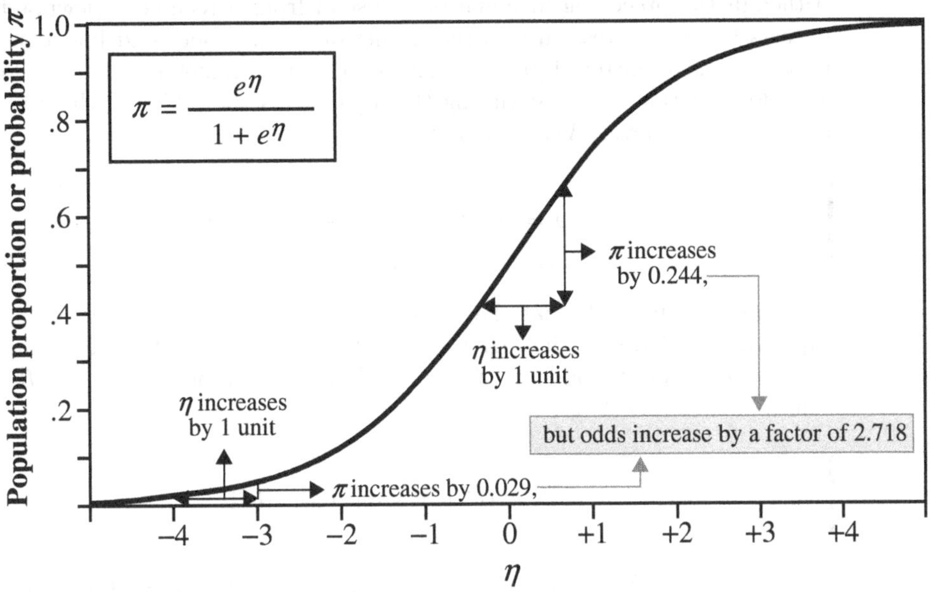

Nonconstant Variance

In normal regression, $\text{Var}\{Y|X_1,\ldots,X_p\} = \sigma^2$, where σ^2 is a constant that does not depend on the explanatory variable values. As was shown in Section 18.2.1, however, the variance of a population of binary response variables with mean π is $\pi(1-\pi)$. Therefore, the mean and variance specifications of the logistic regression model are as given in the following box.

Logistic Regression Model

$$\mu\{Y|X_1,\ldots,X_p\} = \pi; \qquad \text{Var}\{Y|X_1,\ldots,X_p\} = \pi(1-\pi);$$
$$\text{logit}(\pi) = \beta_0 + \beta_1 X_1 + \cdots + \beta_p X_p.$$

Logistic regression is a kind of nonlinear regression, since the equation for $\mu\{Y|X_1,\ldots,X_p\}$ is not linear in β's. Its nonlinearity, however, is solely contained in the link function—hence the term *generalized linear*. The implication of this is that the regression structure (the part with the β's) can be used in much the same way as ordinary linear regression. The model also differs from ordinary regression in variance structure: $\text{Var}\{Y|X_1,\ldots,X_p\}$ is a function of the mean and contains no additional parameter (like σ^2).

20.2.2 Interpretation of Coefficients

Although the preceding formulation is useful from a technical viewpoint because it casts logistic regression into the framework of a generalized linear model, it is important as a matter of practical interpretation to remember that the logit is the log odds function. Exponentiating the logit yields the odds. So the odds of a *yes* response at the levels $X_1, \ldots, X_p$ are

$$\omega = \exp(\beta_0 + \beta_1 X_1 + \cdots + \beta_p X_p).$$

For example, the odds that the response is 1 at $X_1 = 0, X_2 = 0, \ldots, X_p = 0$ are $\exp(\beta_0)$, provided that the regression model is appropriate for these values of X's.

The ratio of the odds at $X_1 = A$ relative to the odds at $X_1 = B$, for fixed values of the other X's, is

$$\omega_A/\omega_B = \exp[\beta_1(A - B)].$$

In particular, if X_1 increases by 1 unit, the odds that $Y = 1$ will change by a multiplicative factor of $\exp(\beta_1)$, other variables being the same.

Example—Donner Party

The fit of a logistic regression model to the Donner Party data, where π represents survival probability, gives

$$\text{logit}(\hat{\pi}) = 1.63 - 0.078\ age + 1.60\ fem.$$

For comparing women 50 years old (A) with women 20 years old (B), the odds ratio is estimated to be $\exp[-0.078(50 - 20)] = 0.096$, or about 1/10. So the odds of a 20-year-old woman surviving were about 10 times the odds of a 50-year-old woman surviving. Comparing a woman ($fem = 1 = A$) with a man ($fem = 0 = B$) of the same age, the estimated odds ratio is $\exp(1.60[1 - 0]) = 4.95$—a woman's odds were about 5 times the odds of survival of a man of the same age.

Retrospective Studies

In logistic regression problems, probabilities for binary responses are modeled prospectively as functions of explanatory variables, as if responses were to be measured at set levels of the explanatory factors. In some studies—particularly those in which the probabilities of yes responses are very small—independent samples are drawn retrospectively from the populations with different response outcomes. As discussed in connection with the case of a single explanatory factor with two levels (Section 18.4), prospective probabilities cannot be estimated from these retrospective samples. Yet retrospective odds ratios are the same as prospective odds ratios, so they can be estimated from retrospective samples. In a logistic regression model for a retrospective study, the estimated intercept is not an estimate of the prospective intercept. However, estimates of coefficients of explanatory variables from a retrospective sample estimate the corresponding coefficients in the prospective model and may be interpreted in the same way. For example, it is possible with retrospective samples of lung cancer patients and of patients with no lung cancer to make a statement such as: The odds of lung cancer are estimated to increase by a factor of 1.1 for each year that a person has smoked. This result has tremendous bearing on medical case–control studies and the field of epidemiology.

20.3 ESTIMATION OF LOGISTIC REGRESSION COEFFICIENTS

In generalized linear models, the method of least squares is replaced by the *method of maximum likelihood*, which is outlined here for the logistic regression problem.

20.3.1 Maximum Likelihood Parameter Estimation

When values of the parameters are given, the logistic regression model specifies how to calculate the probability that any outcome (such as $y_1 = 1$, $y_2 = 0$, $y_3 = 1, \ldots$) will occur. A convenient way to express the model begins with the convention that raising any number to the power of 0 results in the number 1 ($x^0 = 1$, for any x.) Then the probability model for a single binary response Y can be written as

$$\Pr\{Y = y\} = \pi^y (1 - \pi)^{1-y},$$

where the substitution of $y = 1$ results in π and the substitution of $y = 0$ results in $(1 - \pi)$. Now suppose that there are n such responses, with π_i being the parameter for $Y_i (i = 1, \ldots, n)$. If the responses are independent, their probabilities multiply to give the probability

$$\Pr\{Y_1 = y_1, \ldots, Y_n = y_n\} = \prod_{i=1}^{n} \pi_i^{y_i} (1 - \pi_i)^{1-y_i}$$

for the outcomes $y_1, \ldots, y_n$.

Examples—Donner Party

Suppose that this model is plausible for the Donner Party data. It is as though a coin flip of fate had occurred for each party member. If the ith flip came up heads—which it would do with probability π_i—the ith member survived. Suppose, too, that an additive logistic model is correct, with the following hypothetical parameter values:

$$\text{logit}(\pi_i) = 1.50 - 0.080\, age_i + 1.25\, fem_i.$$

Then, referring to Display 20.1, Antonio (23-year-old male: $age_1 = 23$, $fem_1 = 0$) would have

$$\text{logit}(\pi_1) = 1.50 - (0.080 \times 23) + (1.25 \times 0) = -0.340,$$

so

$$\pi_1 = \exp(-0.340)/[1 + \exp(-0.340)] = 0.416.$$

Antonio had a 0.416 chance of surviving. Mary Breen (40-year-old female) would have a survival probability of 0.389; Patrick Breen (40-year-old male) would have a survival probability of 0.154; and so on down the list to Eliza Williams, with her survival probability of 0.679.

Now the probability for any combination of independent outcomes can be calculated by multiplying individual probabilities. For example, the probability that all persons survive (all the y_i are $= 1$) is $(0.416)(0.389)(0.154)\cdots(0.679) = \exp(-53.3631)$. Conversely, the probability that all persons die (all $y_i = 0$) is $(1-0.416)(1-0.389)(1-0.154)\cdots(1-0.679) = \exp(-26.1331)$. The probability that all the women survive but none of the men do is $(1-0.416)(0.389)(1-0.154)\cdots(0.679) = \exp(-21.1631)$; while the probability that the individuals survive who actually did survive is $(1-0.416)(0.389)(0.154)\cdots(0.679) = \exp(-26.1531)$. The best way to make these calculations is to add the logarithms of the individual probabilities.

The outcome in which everyone survives may be rosier, but it holds little interest because it did not actually occur. What is of particular interest is the single outcome that did occur. When attention is focused on that one outcome, the probability formula assumes a different role.

Likelihood

The same formula can be used to calculate the probability of the observed outcome under various possible choices for the β's. The probability of the known outcome, calculated earlier as $\exp(-26.1531)$ when $\beta_0 = 1.50$, $\beta_1 = -0.080$, and $\beta_2 = 1.25$, is called the *likelihood* of the values (1.50, –0.080, 1.25) for the three parameters.

> *The **likelihood** of any specific values for the parameters is the probablity of the actual outcome, calculated with those parameter values.*

According to this terminology, one set of parameter values is more likely than another set if it gives the observed outcome a higher probability of occurrence.

Maximum Likelihood Estimation

Examples of likelihood values (on the logarithmic scale) for several possible values of β_0, β_1, and β_2 in the logistic regression model for the Donner Party data are shown in Display 20.4. Notice that the largest likelihood listed is for the values $\beta_0 = 1.63$, $\beta_1 = -0.078$, and $\beta_2 = 1.60$. These values are in fact the most likely of all possible; they are called the *maximum likelihood estimates* of the parameters. As

DISPLAY 20.4 Some possible parameter values and their log-likelihoods, for the Donner Party data (logit of probability of survival = $\beta_0 + \beta_1 age + \beta_2 fem$)

β_0	β_1	β_2	*log*(likelihood)
1.50	−0.050	1.25	−27.7083
		1.80	−28.7931
	−0.08	1.25	−26.1531
		1.80	−25.7272
1.70	−0.050	1.25	−29.0467
		1.80	−30.3692
	−0.08	1.25	−25.7972
		1.80	−25.6904
1.63	−0.078	1.60	−25.6282

a guide for how to select parameter estimates, the *maximum likelihood method* says to choose as estimates of parameters the values that assign the highest probability to the observed outcome.

Notes on Calculation

Although a computer routine could be used to search through many possible parameter values to find the combination that produces the largest likelihood (or equivalently, the largest log-likelihood), this is unnecessary. As with the method of least squares, calculus provides a way to find the maximizing parameter values. Unlike least squares for linear regression, however, calculus does not provide closed-form expressions for the answers. Some iterative computational procedures are used. Statistical computer packages include these procedures; they will not be discussed further here.

Properties of Maximum Likelihood Estimators

Statistical theory confirms that the maximum likelihood method leads to estimators that generally have good properties.

Properties of Maximum Likelihood Estimates

If a model is correct and the sample size is large enough, then:

1. The maximum likelihood estimators are essentially unbiased.
2. Formulas exist for estimating the standard deviations of the sampling distributions of the estimators.
3. The estimators are about as precise as any nearly unbiased estimators that can be obtained.
4. The shapes of the sampling distributions are approximately normal.

The upshot of properties (1) and (3) is that the maximum likelihood principle leads to pretty good estimates, at least when the sample size is large. Properties (1), (2), and (4) provide a way to carry out tests and construct confidence intervals. Both of these aspects are dampened somewhat by the requirement that the sample size be large enough. For small samples, it is best to label confidence intervals and test results as approximate.

20.3.2 Tests and Confidence Intervals for Single Coefficients

The properties imply that each estimated coefficient β_j in logistic regression has a normal sampling distribution, approximately, and therefore that

$$Z\text{-ratio} = (\hat{\beta}_j - \beta_j)/[\text{SE}(\hat{\beta}_j)]$$

has a standard normal distribution (approximately).

The standard error is the estimated standard deviation of the sampling distribution of the estimator. Its computation is carried out by the statistical computer package. The known likely values of the standard normal distribution may be used to construct confidence intervals or to obtain p-values for tests about the individual coefficients.

A confidence interval is the estimate plus and minus the half-width, which is a z-multiplier times the standard error of the estimate; and the z-multiplier is the appropriate percentile of the standard normal distribution. For a 95% confidence interval, it is the 97.5th percentile. Similarly, a test statistic is the ratio of the estimate minus the hypothesized value to the standard error, and the p-value is obtained by comparing this to a standard normal distribution. A test based on this approximate normality of maximum likelihood estimates is referred to as *Wald's test*. Since the t-theory only applies to normally distributed response variables, t-distributions are not involved here.

Example of Wald's Test for a Single Coefficient (Donner Party Data)

Display 20.5 demonstrates how Wald's test can be used to determine whether the log odds of survival are associated with age differently for men than for women. The display shows typical output from a logistic regression fit to a model that includes age, sex, and the interaction between sex and age. Notice that the summaries of the estimated coefficients, their standard errors, and the z-statistics (for the tests that each coefficient is zero) parallel the standard summary for multiple regression analysis. The quantity at the bottom labeled *deviance* is new, but it is analogous to the sum of squared residuals in regression (and is discussed in Section 20.4).

DISPLAY 20.5 Wald's test for the hypothesis that the coefficient of the interaction term is zero in the logistic regression of survival (1 or 0) on *age*, *fem* (= 1 for females), and *age* × *fem*: Donner Party data

Variable	Coefficient	Standard error	*z*-statistic
Constant	0.318	1.131	0.28
age	−0.032	0.035	−0.92
fem	6.927	3.354	2.06
age×*fem*	−0.162	0.093	−1.73
Deviance = 47.346		**Degrees of freedom** = 41	

From the normal distribution → Two-sided *p*-value = 2 × Pr(Z > 1.73) = **0.085**

Conclusion: There is suggestive but inconclusive evidence of an interaction.

Example of a Confidence Interval for a Coefficient (Donner Party Data)

Display 20.6 shows the results of fitting the additive logistic regression of survival on age and the indicator variable for sex. Although some suggestive evidence was found in support of an interactive effect of sex and age, the coefficient of *fem* in this parallel lines model is a convenient approximation to an average difference between the sexes. The antilogarithm of the coefficient of the indicator variable should be used in a summarizing statement. A confidence interval should be obtained for the coefficient of *fem*, and the antilogarithm of the endpoints should be taken as a confidence interval for the odds ratio.

Example of Confidence Interval for a Multiple of a Coefficient (Donner Party Data)

From the output in Display 20.6, the estimated coefficient of age is −0.078. A 95% confidence interval is −0.078 ± (1.96 × 0.037), which is −0.1505 to −0.0055. By taking antilogarithms of the estimate and the interval endpoints, one obtains the following summary statement results: It is estimated that the odds of survival change by a factor of 0.92 for each one year increase in age (95% confidence interval for multiplicative odds factor is 0.86 to 0.99).

Another question that may be asked is "What are the odds of survival for a 55-year-old relative to the odds of survival for a 30-year-old?" In the model fit in Display 20.6, the log odds change by −0.078 for each extra year of age. Therefore, for 25 extra years of age, the log odds of survival change by 25 × −0.078 or −1.95. Similarly, a 95% confidence interval for the change in log odds of survival for an additional 25 years of age (based on the results of the preceding paragraph) is 25 × −0.1505 to 25 × −0.0055, or −3.76 to −0.137. By back-transforming this estimate and confidence interval, the summary statement is this: It is estimated that the odds of survival for a 55-year-old are 0.14 times the odds of survival for a 30-year-old (95% confidence interval for odds ratio is 0.02 to 0.87).

DISPLAY 20.6 Confidence intervals for the odds of survival for females divided by the odds of survival for males, accounting for age, from the model without interaction (Donner Party data)

Variable	Coefficient	Standard error	*z*-statistic
Constant	1.633	1.105	1.48
age	−0.078	0.037	−2.11
fem	1.597	0.753	2.10
Deviance = 51.256		Degrees of freedom = 42	

95% Confidence interval for the coefficient of fem → $1.597 \pm (1.96 \times 0.753) = 0.121$ to 3.073 ($z(0.975)$)

Take antilogarithms of endpoints to get interval for the odds ratio. → $exp(0.121)$ to $exp(3.073) = 1.13$ to 21.60

Conclusion: The odds of survival for females are estimated to have been 4.9 times the odds of survival for males of similar age (95% CI: 1.1 times to 21.6 times).

20.4 THE DROP-IN-DEVIANCE TEST

Comparing a Full to a Reduced Model

As in ordinary regression, one must often judge the adequacy of a *reduced* model relative to a *full* model. The reduced model is the special case of the full model obtained by supposing that the hypothesis of interest is true. Typically, the hypothesis is that several of the coefficients in the full model equal zero. For example, if three indicator explanatory variables are used to model the effects of four treatments on the probability of survival of heart disease patients, the hypothesis that no difference exists between treatments is equivalent to the hypothesis that the coefficients of the three indicator variables all equal zero. The reduced model is the logistic regression model excluding those three terms.

The Likelihood Ratio Test

The extra-sum-of-squares F-test was used for testing the adequacy of a reduced model relative to a full one in linear regression. A similar test associated with likelihood analysis is the *likelihood ratio test* (*LRT*). The likelihood of the parameters in a full model is evaluated at its maximum, giving $\text{LMAX}_{\text{full}}$, say. The maximum of the likelihood within the reduced model, $\text{LMAX}_{\text{reduced}}$, is calculated in a similar manner. The general form of the *likelihood ratio test statistic* is

$$LRT = 2 \times \log(\text{LMAX}_{\text{full}}) - 2 \times \log(\text{LMAX}_{\text{reduced}}).$$

When the reduced model (and therefore the null hypothesis) is correct, the *LRT* has approximately a chi-square distribution with degrees of freedom equal to the difference between the numbers of parameters in the full and reduced models. Inadequacies of the reduced model tend to inflate the *LRT* statistic. The approximate p-value, therefore, is the proportion of the chi-square distribution that is greater than *LRT*.

The *LRT* depends on the mathematical form of the likelihood function, which we will not present here. As with the extra-sum-of-squares F-test, the easiest calculation of the *LRT* is often achieved by fitting the full and reduced models and calculating *LRT* manually from the maximized values of the likelihood, which are supplied as part of the computer output.

For reasons that will be more clear in the next chapter, some statistical packages provide a quantity called the *deviance* rather than the maximized likelihood, where

$$\textit{deviance} = \text{constant } - 2 \times \log(\text{LMAX})$$

and the constant, which will also be described in the next chapter, is the same for both the full and reduced models. It follows then that

$$LRT = \textit{deviance}_{\text{reduced}} - \textit{deviance}_{\text{full}}.$$

The likelihood ratio test for generalized linear models is accomplished, therefore, by separately fitting the full and reduced models, extracting the deviances of these from the computer output, calculating *LRT* from this last equation, and finding the p-value as the proportion of a chi-squared distribution that is greater than *LRT*. For this reason and since $\textit{deviance}_{\text{reduced}}$ is never smaller than $\textit{deviance}_{\text{full}}$, the likelihood ratio test applied to generalized linear models is often referred to as the *drop-in-deviance test*. Furthermore, as will be evident in the next chapter, the deviance may be thought of as a sum of squared residuals, so the test statistic is the amount by which the sum of squared residuals for the reduced model exceeds that for the full model.

Example—Birdkeeping

The goal of the birdkeeping and lung cancer analysis is to see whether increased probability of lung cancer is associated with birdkeeping, even after other factors, such as smoking, are accounted for. In the logistic model shown in connection with step 1 of Display 20.7, the log of the odds of getting lung cancer are explained by regression terms with sex of the individual (*FM*), age (*AG*), socioeconomic status (*SS*), years the individual has smoked (*YR*), and the indicator variable for birdkeeping (*BK*, which is 1 for birdkeepers). The question of interest here is a hypothesis that the coefficient of *BK* is zero. The reduced model, therefore, is the same model, but without *BK* in the list of explanatory variables. The steps in the drop-in-deviance test are detailed in Display 20.7.

DISPLAY 20.7 A drop-in-deviance test for the association of the odds of lung cancer with birdkeeping, after accounting for other factors

(1) *Fit the **full model**:* $\text{logit}(\pi) = \beta_0 + \beta_1 FM + \beta_2 AG + \beta_3 SS + \beta_4 YR + \beta_5 BK.$

Variable	Coefficient	Standard error	z-statistic	Two-sided p-value
Constant	−1.4063	1.7173	−0.8189	0.4128
FM	0.5213	0.5296	0.9844	0.3249
AG	−0.0463	0.0349	−1.3263	0.1847
SS	0.1321	0.4642	0.2846	0.7760
YR	0.0829	0.0249	3.3335	0.0009
BK	1.3349	0.4091	3.2627	0.0011

Deviance = 155.24 **Degrees of freedom** = 141

(2) *Fit the **reduced model**:* $\text{logit}(\pi) = \beta_0 + \beta_1 FM + \beta_2 AG + \beta_3 SS + \beta_4 YR.$

Variable	Coefficient	Standard error	z-statistic	Two-sided p-value
Constant	0.1031	1.5762	0.0653	0.9479
FM	0.7214	0.5027	1.4349	0.1513
AG	−0.0632	0.0337	−1.8733	0.0610
SS	−0.0564	0.4367	−0.1293	0.8971
YR	0.0873	0.0248	3.5174	0.0004

Deviance = 166.53 **Degrees of freedom** = 142

(3) *Calculate the **drop in deviance** and the **drop in degrees of freedom.***

166.53 − 155.24 → **Drop in Deviance** = 11.29

142 − 141 → **Drop in d.f.** = 1

(4) *Determine the p-value.* → $\Pr(\chi_1^2 > 11.29) = \mathbf{0.0008}$

Conclusion: There is strong evidence of an association between birdkeeping and lung cancer, after accounting for sex, age, status, and smoking years.

Notes on the Drop-in-Deviance Test

1. As with the extra-sum-of-squares F-test, a few special cases of the drop-in-deviance test deserve mention. A test of whether all coefficients (except the constant term) are zero is obtained by taking the model with no explanatory

variables as the reduced model. Similarly, the significance of a single term may be tested by taking the reduced model to be the full model minus the single term (as demonstrated in Display 20.7). This is *not* the same as Wald's test for a single coefficient described in Section 20.3. If the two give different results, the drop-in-deviance test has the more reliable p-value.

2. Confidence intervals can be constructed for a single coefficient from the theory of the drop-in-deviance test. Computation of the drop-in-deviance confidence interval, which is also called the likelihood ratio confidence interval or the profile likelihood confidence interval, requires a special program. Fortunately, since this confidence interval is generally more reliable than the Wald confidence interval (estimate plus-and-minus 1.96 standard errors for a 95% confidence interval) when they differ, many generalized linear model routines in statistical computer packages now include such a program.

20.5 STRATEGIES FOR DATA ANALYSIS USING LOGISTIC REGRESSION

The strategy for data analysis with binary responses and explanatory variables is similar to the general strategy for data analysis with regression, as shown in the chart in Display 9.9. Identifying the questions of interest is the important starting point. The data analysis should then revolve around finding models that fit the data and allow the questions to be answered through inference about parameters.

A main difference from ordinary regression is that scatterplots and residual plots are usually of little value, since only two distinct values of the response variables are possible. Fortunately, there is no burden to check for nonconstant variance or outliers, so some functions of usual residual analysis are unnecessary. Model terms and the adequacy of the logistic model, on the other hand, must be checked. Informal testing of extra terms, such as squared terms and interaction terms, plays an important role in this regard.

20.5.1 Exploratory Analysis

Graphical Methods

A plot of the binary response variable versus an explanatory variable is not worthwhile, since there are only two distinct values for the response variable. Although no graphical approach can be prescribed for all problems, it is occasionally useful to examine a scatterplot of one of the explanatory variables versus another, with codes to indicate whether the response is 0 or 1. An example is shown in Display 20.10.

Grouping

Some data sets, particularly large ones, may contain *near replicates*—several observations with very similar values of the explanatory variables. As part of an informal analysis to see which explanatory variables are important and to obtain some graphical display, it may be useful to group these together. For example, ages may be grouped into five-year intervals. A sample proportion is obtained as the proportion of observations with binary response of 1 in the group, and the logit transformation can be applied to this sample proportion so that scatterplots may be drawn. The sample logit can be plotted against the midpoint of the grouped version of the explanatory variable, for instance. An example of this approach is shown in Display 20.11.

Informal Testing of Extra Model Terms

One can also construct preliminary logistic regression models that include polynomial terms, interaction terms, and the like to judge whether the desired model needs expansion to compensate for very specific problems. This is probably the most useful of the three approaches for exploratory data analysis on binary responses.

Model Selection

As in normal regression, the search for a suitable model may encompass a wide range of possibilities. The Bayesian Information Criterion (BIC) and Akaike Information Criterion (AIC) extend to generalized linear models as follows:

$$\begin{aligned} \text{BIC} &= \text{Deviance} + \log(n) \times p \\ \text{AIC} &= \text{Deviance} + 2 \times p \end{aligned}$$

where n is the sample size and p is the number of regression coefficients in the model. As in Chapter 12 for ordinary regression, models that produce smaller values of these fitting criteria should be preferred. AIC and BIC differ in their degrees of penalization for number of regression coefficients, with BIC usually favoring models with fewer terms. Using BIC, the uncertainty involved in model selection can be described by Bayes' posterior probabilities, exactly as in Section 12.4.2.

20.6 ANALYSES OF CASE STUDIES

20.6.1 Analysis of Donner Party Data

Whether women and men had different survival chances can best be answered in the inferential model,

$$\text{logit}(\pi) = \beta_0 + \beta_1 fem + \beta_2 age,$$

which provides control for the effects of age while specifying a simple difference between the logits for women and for men. Is this model defensible? There are two possible problems: first, the logit of survival probability may not be linear in age;

DISPLAY 20.8 Quadratic logistic regression model to assess fit for Donner Party survival

Variable	Coefficient	Standard error	z-statistic	p-value
Constant	−3.318	3.940	−0.84	0.40
fem	0.265	10.430	0.03	0.98
age	0.183	0.227	0.81	0.42
$fem \times age$	0.300	0.694	0.43	0.66
age^2	−0.0028	0.0030	−0.94	0.35
$fem \times age^2$	−0.0074	0.0107	−0.69	0.49

Deviance = 45.361 **Degrees of freedom** = 39

and second, even if the logit is linear, the slope may not be the same for males and females.

Since one cannot directly assess the proposed model from sample logits, the best alternative is to embed the preceding model in a larger model that allows for departures from these critical assumptions. Quadratic terms, for example, provide evidence of one kind of departure from a straight line.

Display 20.8 shows the output from the fit of the logistic regression of survival on *age*, *fem*, and three extra terms that assess different aspects of model inadequacy. This model allows for distinct quadratic curves for male and female log odds of survival. One way to pare this down is to use a directed backward elimination technique on the extra terms. *Directed* here refers to examining the least desirable terms first (those that would most complicate the statement of results, for example). It is not necessary to have an exact order for examination, but a suitable starting point is the $fem \times age^2$ term. Since its coefficient has a large p-value, it may be dropped; and the resulting model is refit. (Remember that the p-values change when a term is dropped.) The p-value for the coefficient of age^2 in the model without $fem \times age^2$ is 0.26, so it is dropped.

The resulting model is the one examined in Display 20.5, which provided suggestive evidence of an interactive effect. Since the interactive effect is weak and since it would complicate the statement of the results, one might choose to ignore it (realizing that more data would be needed to assess it carefully) and to report the resulting linear additive model—cautiously—as a general indication of the difference between the males and females. The result is

$$\text{logit}(\hat{\pi}) = \underset{(1.105)}{1.633} - \underset{(0.037)}{0.078}age - \underset{(0.753)}{1.597}fem.$$

A plot of the fitted model is shown in Display 20.9. A statement of the summary is provided in Section 20.1.1.

20.6.2 Analysis of Birdkeeping and Lung Cancer Data

The goal is to examine the odds of lung cancer for birdkeepers relative to persons with similar demographic and smoking characteristics who do not keep pet birds.

DISPLAY 20.9 Fit of logistic regression model for Donner Party Survival, with *age* and *fem* (1 = Female) as explanatory variables

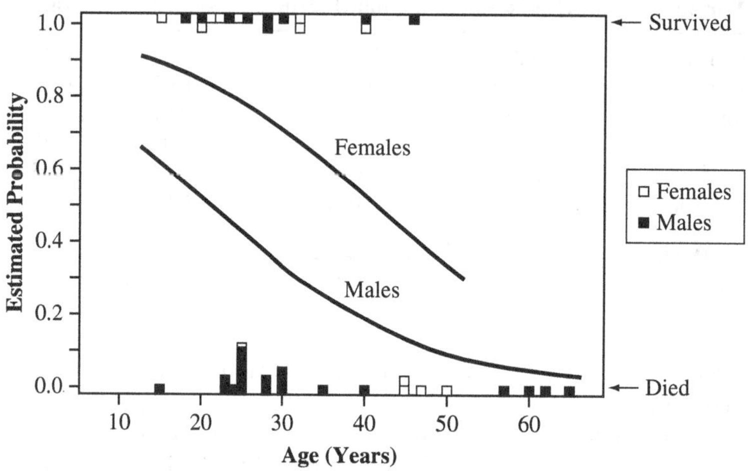

Effect of Smoking

The first stage of the analysis is to find an adequate model to explain the effect of smoking. Display 20.10 is a coded scatterplot of the number of years that a subject smoked, versus the individual's age. The plotting symbols show whether the individual is a birdkeeper (triangle) or not (circle) and whether the subject is one of the lung cancer cases (filled symbol) or not (unfilled symbol).

One striking feature of the plot is that, in any vertical column of points in the plot, more dark symbols appear at the top of the column than at the bottom. This indicates that, for similarly aged subjects, the persons with lung cancer are more likely to have been relatively long-time smokers. To investigate the effect of birdkeeping on this plot, one must look at certain regions of the plot—particularly horizontal slices—to compare subjects with similar smoking history and to ask whether a higher proportion of triangles tend to be filled symbols than unfilled symbols. For example, the only lung cancer patient among subjects who did not smoke (for which years of smoking is zero) happened to be a birdkeeper. Similarly, among individuals who smoked for about 10 years, the only lung cancer patient was a birdkeeper. For the upper regions of the scatterplot, it is a bit harder to make a comparison. Slight visual evidence suggests that the proportion of birdkeepers is higher among the lung cancer patients than among the controls.

Display 20.11 is a plot of sample logits, $\log[\overline{\pi}/(1-\overline{\pi})]$, versus years smoking, where $\overline{\pi}$ is the proportion of lung cancer patients among subjects grouped into similar years of smoking (0, 1–20, 21–30, 31–40, and 41–50). The midpoint of the interval was used to represent the years of smoking for the members of the group.

Evidently, the logit increases with increasing years of smoking. The trend is predominantly linear, but some departure from linearity may be present. This matter can be resolved through informal testing of a model with a quadratic term.

DISPLAY 20.10 Coded scatterplot of years of smoking versus age of subject: triangles represent birdkeepers, circles are nonbirdkeepers, and filled symbols are subjects with lung cancer

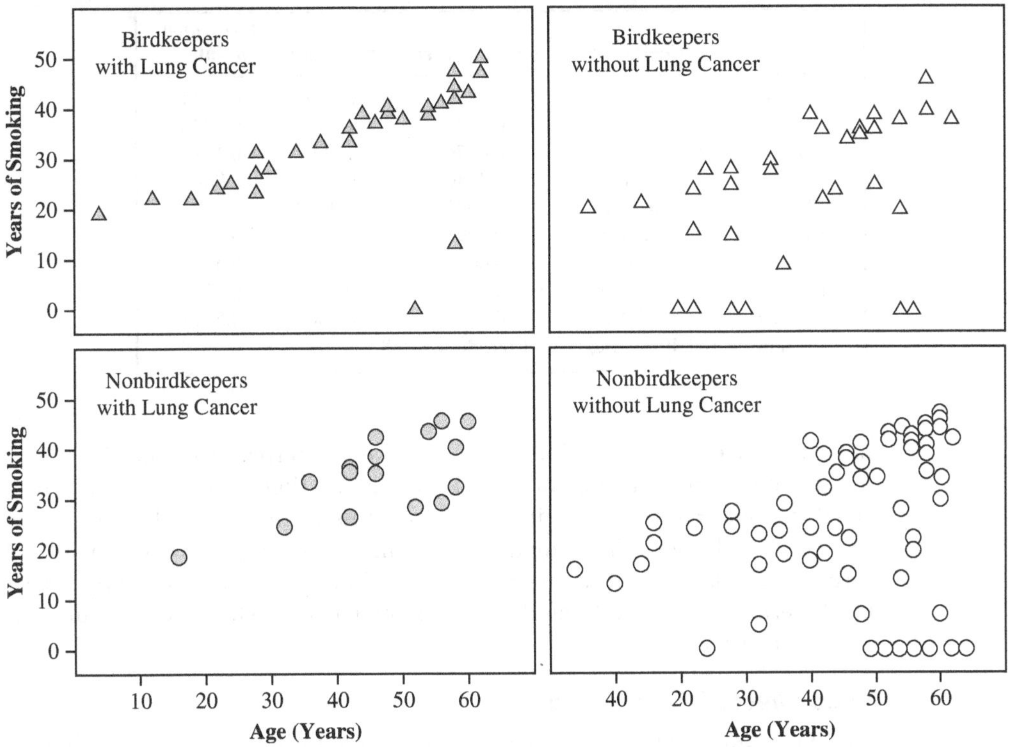

DISPLAY 20.11 Scatterplot of sample logits versus years smoking for data grouped by intervals of years of smoking

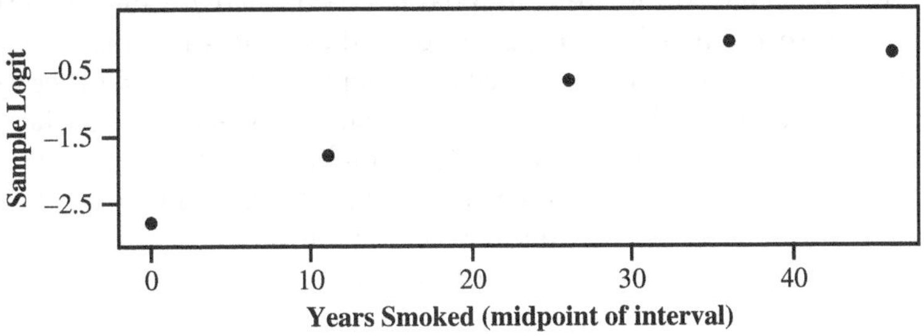

Logistic regression models for the proportion of lung cancer cases as related to the smoking variables, age, and sex may be structured in several ways. Display 20.12 shows the results of one fit, which includes years of smoking, cigarettes per day, and squares and interaction of these. One approach to paring down the model is to

DISPLAY 20.12 Logistic regression of lung cancer incidence with rich model; from the birdkeeping and lung cancer case–control study

Variable	Coefficient	Standard error	z-statistic	Two-sided p-value
Constant	−0.5922	2.2835	−0.2593	0.7954
FM	0.7947	0.5175	1.5359	0.1246
AG	−0.0568	0.0376	−1.5097	0.1311
SS	−0.0173	0.4440	−0.0390	0.9689
YR	0.0278	0.0921	0.3023	0.7624
CD	0.1306	0.0994	1.3143	0.1887
YR^2	0.0011	0.0019	0.6083	0.5430
CD^2	−0.0016	0.0020	−0.8006	0.4234
$YR \times CD$	−0.0013	0.0023	−0.5806	0.5615

Deviance = 164.49 **Degrees of freedom** = 138

use a selective backward elimination of the extra terms involving the squares and interaction of the smoking variables. The interaction term can be dropped (two-sided p-value = 0.5615). Refitting without this term reveals that the squared *YR* term can be dropped. The squared *CD* term can be dropped next. Finally, the *CD* term itself is not needed when *YR* is included in the model. Thus it appears that the single term *YR* adequately models the smoking effect, when the matching variables are also considered.

Extra Effect of Birdkeeping

The final step adds the indicator variable for birdkeeping into the model with the final list of other explanatory variables already present. The estimated model is summarized in the first table of Display 20.7. The drop in deviance—11.29 with 1 degree of freedom—provides convincing evidence that birdkeeping has a strong association with lung cancer, even after the effects of smoking, sex, age, and socioeconomic status are accounted for (two-sided p-value = 0.0008).

The estimate of the coefficient of birdkeeping is 1.3349, so the odds of lung cancer for birdkeepers are estimated to be exp(1.3349)—or, in other words, 3.80—times as large as the odds of lung cancer for nonbirdkeepers, after the other explanatory variables are accounted for. A 95% confidence interval for the coefficient of the birdkeeping indicator variable is 0.5330 to 2.1368. Consequently, a 95% confidence interval for the odds of lung cancer for birdkeepers relative to the odds of lung cancer for nonbirdkeepers is 1.70 to 8.47.

20.7 RELATED ISSUES

20.7.1 Matched Case–Control Studies

Matched *case–control* studies are a special type of retrospective sampling in which controls are matched with cases on the basis of equality of certain variables. Only one or two variables are generally used to match subjects, because overmatching can bias important comparisons. The number of controls matched to each case is usually one to three; little is gained from including more. Variables used in the matching process and their possible interactions with other variables should be entered as covariates in logistic regression analyses, much as block effects are included in the analysis of a randomized block experiment. The purpose is to exercise control in the analysis, as well as in the design. The coefficients of these variables cannot be interpreted as odds ratios, and the intercept measures no meaningful feature. Analysis of matched case–control studies is accomplished by using conditional likelihood logistic regression, which differs somewhat from the methods of this chapter.

20.7.2 Probit Regression

There are several alternatives to logistic regression for binary responses. Any function $F(\pi)$ that has characteristics similar to the logit function—steadily increasing from $-\infty$ to ∞ as π goes from 0 to 1—could easily be chosen as a link. One popular alternative is to choose $F(\pi)$ to be the inverse of the cumulative standard normal probability distribution function; that is, $F(\pi)$ equals the 100πth percentile in the standard normal distribution. This choice leads to ***probit regression***. The results from probit regression are similar to those from logistic regression, at least for values of π between 0.2 and 0.8.

20.7.3 Discriminant Analysis Using Logistic Regression

Logistic regression may be used to predict future binary responses. On the basis of a training set, a logistic model can be estimated. When applied to a particular set of explanatory variable values for which the binary response is unknown, the estimated probability π may be used as an estimate of the probability that the response is 1.

Logistic regression prediction problems are similar in structure to problems in discriminant analysis, which were discussed in Section 17.5.1. In fact, logistic regression is preferable to standard DFA solutions when the explanatory variables have nonnormal distributions (when they are categorical, for example). Logistic regression is nearly as efficient as the standard tools when the explanatory variables do have normal distributions. Therefore, when discrimination between two groups is the goal, logistic regression should always be considered as an appropriate tool.

When estimated from a retrospective sample, logistic regression cannot be used to estimate the probability of group status. If the odds of being a case in the population are known (from some other source) to be ω_p, however, and if the odds based on a subject's explanatory variable values and on the logistic regression (estimated from the retrospective sample) are estimated to be ω_r, then the odds that the subject is a case are $(n_0/n_1)\omega_p\omega_r$, where n_0 and n_1 are the number of controls and the number of cases, respectively, in the sample.

20.7.4 Monte Carlo Methods

The term *Monte Carlo methods* refers to a broad class of computer algorithms that attempt to learn about the behavior of some random process by repeated

DISPLAY 20.13 100 Monte Carlo simulations of 20-year performance trajectories of an initial $10,000 investment when the annual rate of return is normally distributed with mean 5.4% and standard deviation 8.9 percentage points, with the average, and 10th and 90th percentiles of the simulated 20-year returns indicated

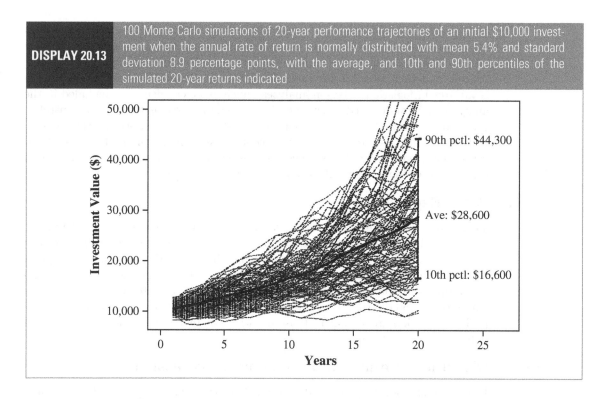

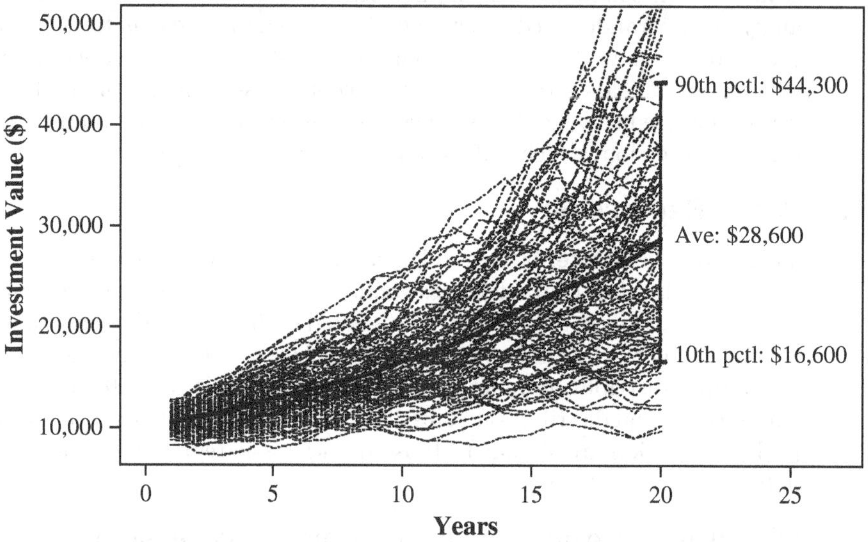

probabilistic simulation of that process. The name comes from the Monte Carlo Casino in recognition of the common importance of probability in the simulation methods and in gambling.

A simple application is a Monte Carlo simulation of possible outcomes of investments in an uncertain economic future. If a person invests $10,000 in an investment that returns a 5.4% annual yield, then the value of the investment after 20 years, assuming reinvestment of all earnings, is $10,000 \times 1.054^{20}$, which is $28,629. In real life, though, investment yields can be highly variable from year to year. What possible values can the investor expect if the annual return is 5.4% on average but has a standard deviation of 8.9 percentage points? Display 20.13 shows the results of 100 Monte Carlo simulations of 20-year performance trajectories of a $10,000 investment. In each year of each 20-year trajectory, the annual yield for the year was determined by a computer-simulated draw from a normal probability distribution with mean 5.4% and standard deviation 8.9 percentage points. With this display, the investor gets an idea of the range of possible outcomes of the investment based on the historical variability of the annual rate of return.

In statistics, Monte Carlo methods are useful for studying the properties of tests and confidence intervals when the ideal assumptions upon which they are mathematically derived are not met. Display 3.4, for example, shows the results of a Monte Carlo investigation into the behavior of 95% confidence intervals for the difference in two normal population means when, in fact, the population distributions are not normal. The computer was instructed to repeatedly draw samples from two nonnormal populations (with population shapes indicated in the histograms above the table), compute 95% confidence intervals for the difference in means from each of these pairs of samples, and keep track of the success rate of these intervals. In some of the cases the actual success rate was very close to the desired coverage of 95%, which indicated that the method was doing what it was supposed to be doing. The Monte Carlo investigation helped clarify that the confidence interval and the two-sample t-test are robust against certain departures from the normality assumption. Display 3.5 shows the results of a similar Monte Carlo investigation into the importance of the equal-variance assumption.

Data analysts sometimes use Monte Carlo methods to see whether a particular statistical tool is appropriate for a particular data problem. In a typical investigation of this kind, the researcher will instruct a computer to repeatedly simulate data sets that resemble the actual one, use the tool of interest (like a confidence interval for a particular parameter) on each of the simulated data sets, and keep track of some property of the tool (like the actual success rate of the confidence interval). If the Monte Carlo property matches the theoretical property, then the researcher can be reasonably confident that the tool is appropriate for the given conditions.

This can be useful when using a "large sample" test or confidence interval associated with maximum likelihood estimation. The large sample theory does not reveal how large of n is large enough. Are the p-value and confidence interval for the sex effect in the Donner Party problem, for example, justified given that the sample size is only 45? To answer this, the computer was instructed to repeatedly simulate 0's and 1's, for death and survival, based on probabilities derived from the regression coefficients estimated from the data. That is, the estimated coefficients played the role of the true coefficients in the Monte Carlo investigation. The values of all the explanatory variables were kept the same. For 946 of the 1,000 samples, the drop-in-deviance-based 95% confidence interval successfully included the value of the sex coefficient upon which the simulations were based. The Monte Carlo success rate of 94.6% provides some reassurance that the methodology is sound for this sample size.

20.8 SUMMARY

The logistic regression model describes a population proportion or probability as a function of explanatory variables. It is a nonlinear regression model, but of the special type called ***generalized linear model***, where the logit of the population proportion is a linear function of regression coefficients. If β_3 is the coefficient of an explanatory variable in a logistic regression model, each unit increase of X_3 is associated with an $\exp(\beta_3)$-fold increase in the odds. This interpretation is particularly useful if X_3 is an indicator variable used to distinguish between two groups. Then $\exp(\beta_3)$ is the odds ratio for the two groups, such as the odds of cancer for birdkeepers relative to the odds of cancer for nonbirdkeepers.

Logistic regression analysis for binary responses proceeds in much the same way as regular regression analysis. Since it is difficult to learn much from plots or from residual analysis, model checking is based largely on fitting models that include extra terms (such as squared terms or interaction terms) whose significance would indicate shortcomings of the "target" model. Tests and confidence intervals for single coefficients may be carried out in a familiar way by comparing the coefficient estimate to its standard error. Tests of several coefficients are carried out by the drop-in-deviance chi-squared test. The drop in deviance is the sum of squared deviance residuals from the reduced model minus the sum of squared deviance residuals from the full model.

Donner Party

Given the likely inadequacy of the independence assumption and the inappropriateness of generalizing inferences from these data to any broader population, the analysis is necessarily informal. Nevertheless, it is useful to fit a logistic regression model for survival as a function of sex and age. Initial model fitting requires some examination into the linearity of the effect of age on the logit and of the interaction between sex and age. The coefficient of the sex indicator variable in the additive logistic regression model may be used to make a statement about the relative odds of survival of similarly aged males and females.

Birdkeeping and Lung Cancer

Initial tentative model fitting and model building are used to determine an adequate representation of the smoking variables for explaining lung cancer. Once years of smoking were included in the model, no further smoking variable was found significant when added. Then the variables associated with socioeconomic status and age were included, and interactions were explored. Finally, the effect of birdkeeping was inferred by adding the birdkeeping indicator variable into the model. A drop-in-deviance test established that birdkeeping had a significant effect, even after smoking was accounted for.

20.9 EXERCISES

Conceptual Exercises

1. Donner Party. One assumption underlying the correct use of logistic regression is that observations are independent of one another. (a) Is there some basis for thinking this assumption might be violated in the Donner Party data? (b) Write a logistic regression model for studying whether survival of a party member is associated with whether another adult member of the party had the same surname as that subject, after the effect of age has been accounted for. (c) Prepare a table of grouped responses that may be used for initial exploration of the question in part (b).

2. Donner Party. (a) Why should one be reluctant to draw conclusions about the ratio of male and female odds of survival for Donner Party members over 50? (b) What confounding variables might be present that could explain the significance of the sex indicator variable?

3. Donner Party. From the Donner Party data, the log odds of survival were estimated to be $1.6 - (0.078 \times age) + (1.60 \times fem)$, based on a binary response that takes the value 1 if an individual survived and with *fem* being an indicator variable that takes the value 1 for females. (a) What would be the estimated equation for the log-odds of survival if the indicator variable for sex were 1 for males and 0 for females? (b) What would be the estimated equation for the log-odds of perishing if the binary response were 1 for a person who perished and 0 for a person who survived?

4. Birdkeeping. (a) Describe the retrospective sampling of the birdkeeping data. (b) What are the limitations of logistic regression from this type of sampling?

5. Since the term *regression* refers to the mean of a response variable as a function of one or more explanatory variables, why is it appropriate to use the word "regression" in the term *logistic regression* to describe a proportion or probability as a function of explanatory variables?

6. Give two reasons why the simple linear regression model is usually inappropriate for describing the regression of a binary response variable on a single explanatory variable.

7. How can logistic regression be used to test the hypothesis of equal odds in a 2×2 table of counts?

8. How can one obtain from the computer an estimate of the log-odds of $y = 1$ at a chosen configuration of explanatory variables, and its standard error?

Computational Exercises

9. Donner Party. It was estimated that the log odds of survival were $3.2 - (0.078 \times age)$ for females and $1.6 - (0.078 \times age)$ for males in the Donner Party. (a) What are the estimated probabilities of survival for men and women of ages 25 and 50? (b) What is the age at which the estimated probability of survival is 50% (i) for women and (ii) for men?

10. It was stated in Section 20.2 that if ω_A and ω_B are odds at A and B, respectively, and if the logistic regression model holds, then $\omega_A/\omega_B = \exp[\beta_1(A - B)]$. Show that this is true, using algebra.

11. Space Shuttle. The data in Display 20.14 are the launch temperatures (degrees Fahrenheit) and an indicator of O-ring failures for 24 space shuttle launches prior to the space shuttle *Challenger* disaster of January 28, 1986. (See the description in Section 4.1.1 for more background information.) (a) Fit the logistic regression of *Failure* (1 for failure) on *Temperature*. Report the estimated coefficients and their standard errors. (b) Test whether the coefficient of *Temperature* is 0, using Wald's test. Report a one-sided p-value (the alternative hypothesis is that the coefficient is negative; odds of failure decrease with increasing temperature). (c) Test whether the coefficient of *Temperature* is 0, using the drop-in-deviance test. (d) Give a 95% confidence interval for the coefficient of

DISPLAY 20.14 Launch temperature (degrees Fahrenheit) and incidence of O-ring failure for 24 space shuttle launches

Temperature	Failure	Temperature	Failure	Temperature	Failure
53	Yes	68	No	75	No
56	Yes	69	No	75	Yes
57	Yes	70	No	76	No
63	No	70	Yes	76	No
66	No	70	Yes	78	No
67	No	70	Yes	79	No
67	No	72	No	80	No
67	No	73	No	81	No

Temperature. (e) What is the estimated logit of failure probability at 31°F (the launch temperature on January 28, 1986)? What is the estimated probability of failure? (f) Why must the answer to part (e) be treated cautiously? (*Answer*: It is a prediction outside the range of the available explanatory variable values.)

12. Muscular Dystrophy. Duchenne Muscular Dystrophy (DMD) is a genetically transmitted disease, passed from a mother to her children. Boys with the disease usually die at a young age; but affected girls usually do not suffer symptoms, may unknowingly carry the disease, and may pass it to their offspring. It is believed that about 1 in 3,300 women are DMD carriers. A woman might suspect she is a carrier when a related male child develops the disease. Doctors must rely on some kind of test to detect the presence of the disease. The data in Display 20.15 are levels of two enzymes in the blood, creatine kinase (CK) and hemopexin (H), for 38 known DMD carriers and 82 women who are not carriers. (Data from D. F. Andrews and A. M. Herzberg, *Data*, New York: Springer-Verlag, 1985.) It is desired to use these data to obtain an equation for indicating whether a woman is a likely carrier.

DISPLAY 20.15 Values of creatine kinase (CK) and hemopexin (H) for 5 of the 82 controls (C = 0) and 5 of the 38 muscular dystrophy carriers (C = 1) in the muscular dystrophy data file

Controls (C = 0)		Muscular dystrophy carriers (C = 1)	
CK	H	CK	H
52	83.5	167	89
20	77	104	81
28	86.5	30	108
30	104	65	87
40	83	440	107

(a) Make a scatterplot of H versus log(CK); use one plotting symbol to represent the controls on the plot and another to represent the carriers. Does it appear from the plot that these enzymes might be useful predictors of whether a woman is a carrier? (b) Fit the logistic regression of carrier on CK and CK-squared. Does the CK-squared term significantly differ from 0? Next fit the logistic regression of carrier on log(CK) and $[\log(\text{CK})]^2$. Does the squared term significantly differ from 0? Which scale (untransformed or log-transformed) seems more appropriate for CK? (c) Fit the logistic regression of carrier on log(CK) and H. Report the coefficients and standard errors. (d) Carry out a drop-in-deviance test for the hypothesis that neither log(CK) nor H are useful predictors of whether a woman is a carrier. (e) Typical values of CK and H are 80 and 85. Suppose that a suspected carrier has values of 300 and 100. What are the odds that she is a carrier relative to the odds that a woman with typical values (80 and 85) is a carrier?

13. Donner Party. Consider the Donner Party females (only) and the logistic regression model $\beta_0 + \beta_1 age$ for the logit of survival probability. If A^* represents the age at which the probability of survival is 0.5, then $\beta_0 + \beta_1 A^* = 0$ (since the logit is 0 when the probability is one-half). This implies that $\beta_0 = -\beta_1 A^*$. The hypothesis that $A^* = 30$ years may be tested by the drop-in-deviance test with the following reduced and full models for the logit:

$$\text{Reduced:} \quad -\beta_1 30 + \beta_1 age = 0 + \beta_1(age - 30)$$
$$\text{Full:} \quad \beta_0 + \beta_1 age$$

To fit the reduced model, one must subtract 30 from the ages and drop the intercept term. The drop-in-deviance test statistic is computed in the usual way. Carry out the test that $A^* = 30$ for the Donner Party females. Report the two-sided p-value.

14. Donner Party. The estimate in Exercise 13 of A^* is $-b_0/b_1$. A 95% confidence interval for A^* can be obtained by finding numbers A_L^* and A_U^* below and above this estimate such that two-sided p-values for the tests $A^* = A_L^*$ and $A^* = A_U^*$ are both 0.05. This can be accomplished by trial and error. Choose several possible values for A_L^*, and follow the testing procedure described in the preceding exercise until a value is found for which the two-sided p-value is 0.05. Then repeat the process to find the upper bound A_U^*. Use this procedure to find a 95% confidence interval for the age at which the Donner Party female survival probability is 0.5.

Data Problems

15. Spotted Owl Habitat. A study examined the association between nesting locations of the Northern Spotted Owl and availability of mature forests. Wildlife biologists identified 30 nest sites. (Data from W. J. Ripple et al., "Old-growth and Mature Forests Near Spotted Owl Nests in Western Oregon," *Journal of Wildlife Management* 55: 316–18.) The researchers selected 30 other sites at random coordinates in the same forest. On the basis of aerial photographs, the percentage of mature forest (older than 80 years) was measured in various rings around each of the 60 sites, as shown in Display 20.16. (a) Apply two-sample t-tools to these data to see whether the percentage of mature forest is larger at nest sites than at random sites. (b) Construct a binary response variable that indicates whether a site is a nest site. Use logistic regression to investigate how large an area about the site has importance in distinguishing nest sites from random sites on the basis of mature forest. (Notice that this was a case–control study.) You may wish to transform the ring percentages to circle percentages.

DISPLAY 20.16 A subset of the Spotted Owl data: percentages of mature forest (>80 years) in successive rings around sites in western Oregon forests (5 of 30 random sites and 5 of 30 nest sites)

Randomly chosen sites, Outer radius of ring (km)							Spotted owl nest sites, Outer radius of ring (km)						
0.91	*1.18*	*1.40*	*1.60*	*1.77*	*2.41*	*3.38*	*0.91*	*1.18*	*1.40*	*1.60*	*1.77*	*2.41*	*3.38*
26.0	33.3	25.6	19.1	31.4	24.8	17.9	81.0	83.4	88.9	92.9	80.6	72.0	44.4
100.0	92.7	90.1	72.8	51.9	50.6	41.5	80.0	87.3	93.3	81.6	85.0	82.8	63.6
32.0	22.2	38.3	39.9	22.1	20.2	38.2	96.0	74.0	76.7	66.2	69.1	84.5	52.5
43.0	79.7	61.4	81.2	47.7	69.6	54.8	82.0	79.6	91.3	70.7	75.6	73.5	66.8
74.0	61.8	48.6	67.4	74.9	66.0	55.8	83.0	80.6	88.9	79.6	55.8	62.9	52.7

16. Bumpus Natural Selection Data. Hermon Bumpus analyzed various characteristics of some house sparrows that were found on the ground after a severe winter storm in 1898. Some of the sparrows survived and some perished. The data on male sparrows in Display 20.17 are survival status (1 = survived, 2 = perished), age (1 = adult, 2 = juvenile), the length from tip of beak to tip of tail (in mm), the alar extent (length from tip to tip of the extended wings, in mm), the weight in grams, the length of the head in mm, the length of the humerus (arm bone, in inches), the length of the femur (thigh bones, in inches), the length of the tibio-tarsus (leg bone, in inches), the breadth of

DISPLAY 20.17 A subset of data on 51 male sparrows that survived (SV = 1) and 36 that perished (SV = 2) during a severe winter storm: Age (*AG*) is 1 for adults, 2 for juveniles; *TL* is total length; *AE* is alar extent; *WT* is weight; *BH* is length of beak and head; *HL* is length of humerus; *FL* is length of femur; *TT* is length of tibio-tarsus; *SK* is width of skull; and *KL* is length of keel of sternum

SV	*AG*	*TL*	*AE*	*WT*	*BH*	*HL*	*FL*	*TT*	*SK*	*KL*
1	1	154	241	24.5	31.2	0.687	0.668	1.022	0.587	0.830
1	1	160	252	26.9	30.8	0.736	0.709	1.180	0.602	0.841
1	1	155	243	26.9	30.6	0.733	0.704	1.151	0.602	0.846
1	1	154	245	24.3	31.7	0.741	0.688	1.146	0.584	0.839
1	1	156	247	24.1	31.5	0.715	0.706	1.129	0.575	0.821
2	1	162	247	27.6	31.8	0.731	0.719	1.113	0.597	0.869
2	1	163	246	25.8	31.4	0.689	0.662	1.073	0.604	0.836
2	1	161	246	24.9	30.5	0.739	0.726	1.138	0.580	0.803
2	1	160	242	26.0	31.0	0.745	0.713	1.105	0.600	0.803
2	1	162	246	26.5	31.5	0.720	0.696	1.092	0.606	0.809

the skull in inches, and the length of the sternum in inches. (A subset of this data was discussed in Exercise 2.21.)

Analyze the data to see whether the probability of survival is associated with physical characteristics of the birds. This would be consistent, according to Bumpus, with the theory of natural selection: those that survived did so because of some superior physical traits. Realize that (i) the sampling is from a population of grounded sparrows, and (ii) physical measurements and survival are both random. (Thus, either could be treated as the response variable.)

17. Catholic Stance. The Catholic church has explicitly opposed authoritarian rule in some (but not all) Latin American countries. Although such action could be explained as a desire to counter repression or to increase the quality of life of its parishioners, A. J. Gill supplies evidence that the underlying reason may be competition from evangelical Protestant denominations. He compiled measures of: (1) *Repression* = Average civil rights score for the period of authoritarian rule until 1979; (2) *PQLI* (Physical Quality of Life Index in the mid-1970s) = Average of life expectancy at age 1, infant mortality, and literacy at age 15+; and (3) *Competition* = Percentage increase of competitive religious groups during the period 1900–1970. His goal was to determine which of these factors could be responsible for the church being pro-authoritarian in six countries and anti-authoritarian in six others. (Data from A. J. Gill, "Rendering unto Caesar? Religious Competition and Catholic Political Strategy in Latin America, 1962–1979," *American Journal of Political Science* 38(2) (1994): 403–25.)

To what extent do these measurements (see Display 20.18) distinguish between countries where the church took pro- and anti-authoritarian positions? Is it possible to determine the most influential variable(s) for distinguishing the groups?

18. Fatal Car Accidents Involving Tire Failure on Ford Explores. The Ford Explorer is a popular sports utility vehicle made in the United States and sold throughout the world. Early in its production, concern arose over a potential accident risk associated with tires of the prescribed size when the vehicle was carrying heavy loads, but the risk was thought to be acceptable if a low tire pressure was recommended. The problem was apparently exacerbated by a particular type of Firestone tire that was overly prone to tread separation, especially in warm temperatures. This type of tire was a common one used on Explorers in model years 1995 and later. By the end of 1999 more than

DISPLAY 20.18 Religious competition, quality of life, and repression in 12 predominantly Catholic Latin American countries

Catholic church stance	Country	PQLI	Repression	Competition (% increase)
Pro-authoritarian	Argentina	85	5.3	2.7
	Bolivia	39	4.3	4.1
	Guatemala	54	3.5	6.3
	Honduras	53	3.0	3.1
	Paraguay	75	5.2	2.1
	Uruguay	86	4.7	1.2
Anti-authoritarian	Brazil	66	4.8	12.0
	Chile	79	5.0	15.5
	Ecuador	69	3.7	2.9
	El Salvador	64	4.4	5.5
	Nicaragua	55	4.3	5.6
	Panama	79	5.7	4.4

DISPLAY 20.19 All 1995 and later model compact sports utility vehicles involved in fatal accidents in the United States between 1995 and 1999, excluding vehicles that were struck by other vehicles and those involved in alcohol-related accidents; first 5 of 2,321 rows

Type of sports utility vehicle (1 if Ford, 0 if not)	Vehicle age (years)	Number of passengers	Cause of fatal accident (1 if tires, 0 if not)
0	1	0	0
0	1	0	0
0	1	0	0
0	1	0	0
0	1	1	0

30 lawsuits had been filed over accidents that were thought to be associated with this problem. U.S. federal data on fatal car accidents were analyzed at that time, showing that the odds of a fatal accident being associated with tire failure were three times as great for Explorers as for other sports utility vehicles. Additional data from 1999 and additional variables may be used to further explore the odds ratio. Display 20.19 lists data on 1995 and later model compact sports utility vehicles involved in fatal accidents in the United States between 1995 and 1999, excluding those that were struck by another car and excluding accidents that, according to police reports, involved alcohol. It is of interest to see whether the odds that a fatal accident is tire-related depend on whether the vehicle is a Ford, after accounting for age of the car and number of passengers. Since the Ford tire problem may be due to the load carried, there is some interest in seeing whether the odds associated with a Ford depend on the number of passengers. (Suggestions: (i) Presumably, older tires are more likely to fail than newer ones. Although tire age is not available, vehicle age is an approximate substitute for it. Since many car owners replace their tires after the car is 3 to 5 years old, however, we may expect the odds of tire failure to increase with age up to some number of years, and then to perhaps decrease after that.

(ii) If there is an interactive effect of Ford and the number of passengers, it may be worthwhile to present an odds ratio separately for 0, 1, 2, 3, and 4 passengers.) The data are from the National Highway Traffic Safely Administration, Fatality Analysis Reporting System (http://www-fars.nhtsa.dot.gov/).

19. Missile Defenses. Following a successful test of an anti-ballistic missile (ABM) in July 1999, many prominent U.S. politicians called for the early deployment of a full ABM system. The scientific community was less enthusiastic about the efficacy of such a system. G. N. Lewis, T. A. Postol, and J. Pike ("Why National Missile Defense Won't Work," *Scientific American, 281*(2): 36–41, August 1999) traced the history of ABM testing, reporting the results shown in Display 20.20. Do these data suggest any improvement in ABM test success probability over time?

DISPLAY 20.20 Date of ABM test, months after the first test, and result of the test; partial listing

Date	Months	Result
Mar-83	0	Failure
Jun-83	3	Failure
Dec-83	9	Failure
Jun-84	15	Success
. . .		
Jul-99	196	Success

20. Factors Associated with Self-Esteem. Reconsider the NLSY79 data in Exercise 12.23. The variable *Esteem1* is the respondent's degree of agreement in 2006 with the statement "I feel that I'm a person of worth, at least on equal basis with others," with values of 1, 2, 3, and 4 signifying strong agreement, agreement, disagreement, or strong disagreement, respectively. Construct a new binary response variable from this, which takes on the value 1 for strong agreement and 0 for agreement, disagreement, or strong disagreement. Explore the dependence of the probability that a person has positive self-esteem (as indicated by a response of strong agreement on *Esteem1*) on these explanatory variables: log annual income in 2005 (*Income2005*), intelligence as measured by the AFQT score in 1981 (*AFQT*), years of education (*Educ*), and gender (*Gender*).

Answers to Conceptual Exercises

1. (a) Yes. Members within the same family may have been likely to share the same fate; and if so, the binary responses for members within the same family would not be independent of one another. (It is difficult to investigate this possibility or to correct it for this small data set. Further discussion of this type of violation is provided in the next chapter.) (b) $\text{logit}(\pi) = \beta_0 + \beta_1 age + \beta_2 sur$, where $sur = 1$ if the individual had a surname that was shared by at least one other adult party member, and 0 if not. (c) Here is one possibility:

Age group	Number in group with shared surname	Proportion surviving	Number in age group without shared surname	Proportion surviving
15–25	13	0.625	8	0.375
26–45	10	0.800	7	0.000
>45	6	0.160	1	0.000

2. (a) There were no females over 50. Any comparisons for older people must be based on an assumption that the model extends to that region. Such an assumption cannot be verified with these data. (b) If males and females had different tasks and if survival was associated with task, the tasks would be a confounding variable.

3. (a) $3.2 - 0.078\ age - 1.6\ male$. (b) Same equation but with all coefficients having opposite signs.

4. A sample of 49 individuals from all those with lung cancer was taken from a population. These were the "cases." Another sample of 98 individuals was taken from a similar population of individuals who did not have lung cancer. (b) Logistic regression may be used, with the 0–1 response representing lung cancer; but the intercept cannot be interpreted.

5. The mean of a binary response is a proportion (or a probability, if the populations are hypothetical).

6. (i) Proportions must fall between 0 and 1, and lines cross these boundaries. (ii) The variance is necessarily nonconstant.

7. Use a binary response to distinguish levels of the response category and an indicator variable to distinguish the levels of the explanatory category. The coefficient of the indicator variable is the log odds of one group minus the log odds of the other group. Carry out inference about that coefficient.

8. Subtract the specified value of each variable from all the variable values to create a new set of variables centered (zeroed) at the desired configuration. Run a logistic regression analysis on the new set, and record the "constant" coefficient and its standard error (see Section 10.2).

CHAPTER 21

Logistic Regression for Binomial Counts

This chapter extends logistic regression models to the case where the responses are proportions from counts. These may be grouped binary responses (such as the proportion of subjects of a given age who develop lung cancer) or a proportion based on a count for each subject (such as the proportion of 100 cells from each subject that have chromosome aberrations). Although the response variable is more general, the logistic regression model is the same as in Chapter 20. The population proportion or probability is modeled, through the logit link, by a linear function of regression coefficients.

One new aspect of the analysis for this type of response is that some care must be taken in checking distributional assumptions. (This was not a problem for ungrouped binary responses.) Some additional tools for data analysis aid in this checking: scatterplots of the sample logits, residual analysis, and a goodness-of-fit test.

21.1 CASE STUDIES

21.1.1 Island Size and Bird Extinctions—An Observational Study

Scientists agree that preserving certain habitats in their natural states is necessary to slow the accelerating rate of species extinctions. But they are divided on how to construct such reserves. Given a finite amount of available land, is it better to have many small reserves or a few large ones? Central to the debate on this question are observational studies of what has happened in island archipelagos, where nearly the same fauna tries to survive on islands of different sizes.

In a study of the Krunnit Islands archipelago, researchers presented results of extensive bird surveys taken over four decades. They visited each island several times, cataloging species. If a species was found on a specific island in 1949, it was considered to be at risk of extinction for the next survey of the island in 1959. If it was not found in 1959, it was counted as an "extinction," even though it might reappear later. The data on island size, number of species at risk to become extinct, and number of extinctions are shown in Display 21.1. (Data from R. A. Väisänen and O. Järvinen, "Dynamics of Protected Bird Communities in a Finnish Archipelago," *Journal of Animal Ecology* 46 (1977): 891–908.)

The data are plotted in Display 21.2. Each island is indicated by a single dot. The horizontal measurement shows the island's area on a logarithmic scale. The

DISPLAY 21.1 Island area, number of bird species present in 1949, and number of these not present in 1959: the Krunnit Islands study

Island	Area (km^2)	Species at risk	Extinctions
Ulkokrunni	185.80	75	5
Maakrunni	105.80	67	3
Ristikari	30.70	66	10
Isonkivenletto	8.50	51	6
Hietakraasukka	4.80	28	3
Kraasukka	4.50	20	4
Länsiletto	4.30	43	8
Pihlajakari	3.60	31	3
Tyni	2.60	28	5
Tasasenletto	1.70	32	6
Raiska	1.20	30	8
Pohjanletto	0.70	20	2
Törö	0.70	31	9
Luusiletto	0.60	16	5
Vatunginletto	0.40	15	7
Vatunginnokka	0.30	33	8
Tiirakari	0.20	40	13
Ristikarenletto	0.07	6	3

DISPLAY 21.2 Extinctions of bird species versus island size in the Krunnit Island archipelago

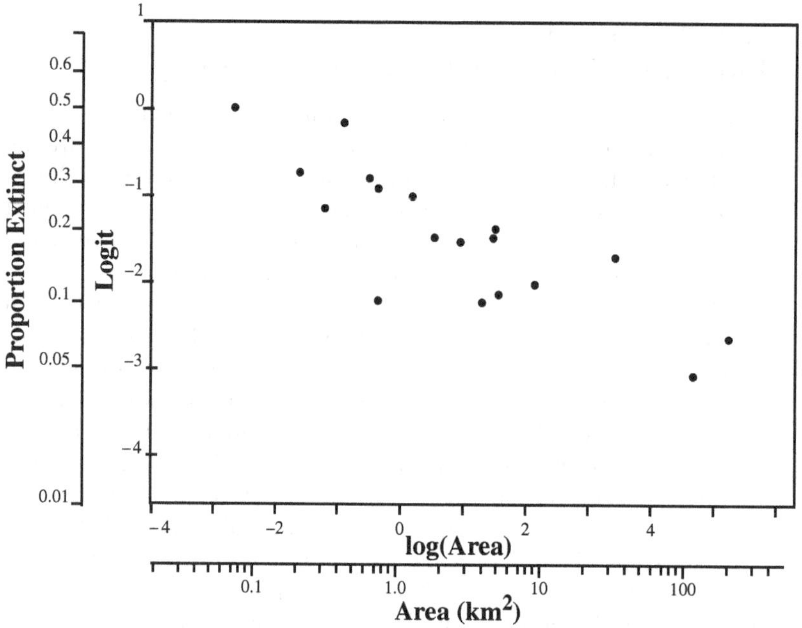

vertical measurement is the logit of the proportion of the island's at-risk occasions that resulted in an extinction. The scatterplot shows dramatically that larger islands had smaller extinction rates. Another striking feature of the scatterplot is how well a straight line would approximate the apparent relationship.

Statistical Conclusion

The strong linear component in the relationship is undeniable. Each 50% reduction in island area is associated with a 23% increase in the odds of extinction (95% confidence interval: 13.6% to 31.5%).

Two difficulties with the statistical analysis are evident: It assumes that each species found on an island has the same chance of becoming extinct, and it assumes that extinctions occur independently. Although these assumptions are highly questionable, the linearity in the scatterplot and the conclusiveness of the p-values make any inaccuracy in p-values and standard errors fairly unimportant.

Scope of Inference

Since the data are observational, confounding variables associated with both island size and the event of a species extinction may be responsible for the observed association in this data. In addition, these islands are not a sample of any larger population of islands. Therefore, although the data are consistent with the theory that the odds of extinction decrease with increasing island size, the statistical analysis alone cannot be used to infer that the same relationship would exist in other islands or habitat areas or that island size is a cause of extinction in these islands.

21.1.2 Moth Coloration and Natural Selection—A Randomized Experiment

Population geneticists consider clines particularly favorable situations for investigating evolutionary phenomena. A cline is a region where two color morphs of one species arrange themselves at opposite ends of an environmental gradient, with increasing mixtures occurring between. Such a cline exists near Liverpool, England, where a dark morph of a local moth has flourished in response to the blackening of tree trunks by smoke pollution from the mills. The moths are nocturnal, resting during the day on tree trunks, where their coloration acts as camouflage against predatory birds. In Liverpool, where tree trunks are blackened by smoke, a high percentage of the moths are of the dark morph. One encounters a higher percentage of the typical (pepper-and-salt) morph as one travels from the city into the Welsh countryside, where tree trunks are lighter. J. A. Bishop used this cline to study the intensity of natural selection. Bishop selected seven locations progressively farther from Liverpool. At each location, Bishop chose eight trees at random. Equal numbers of dead (frozen) light (Typicals) and dark (*Carbonaria*) moths were glued to the trunks in lifelike positions. After 24 hours, a count was taken of the numbers of each morph that had been removed—presumably by predators. The data are shown in Display 21.3. (Data from J. A. Bishop, "An Experimental Study of the Cline of

DISPLAY 21.3 Numbers of light (Typicals) morph and dark (*Carbonaria*) morph moths placed by researchers and numbers of these removed by predators, at each of seven locations of varying distances from Liverpool

Location	Distance from Liverpool (km)	Morph	Number of moths placed	Number removed
Sefton Park	0.0	light	56	17
		dark	56	14
Eastham Ferry	7.2	light	80	28
		dark	80	20
Hawarden	24.1	light	52	18
		dark	52	22
Loggerheads	30.2	light	60	9
		dark	60	16
Llanbedr	36.4	light	60	16
		dark	60	23
Pwyllglas	41.5	light	84	20
		dark	84	40
Clergy Mawr	51.2	light	92	24
		dark	92	39

Industrial Melanism in *Biston betularia* [Lepidoptera] Between Urban Liverpool and Rural North Wales," *Journal of Animal Ecology* 41 (1972): 209–43.)

The question of interest is whether the proportion removed differs between the dark morph moths and the light morph moths and, more importantly, whether this difference depends on the distance from Liverpool. If the relative proportion of dark morph removals increases with increasing distance from Liverpool, that would be evidence in support of survival of the fittest, via appropriate camouflage.

Statistical Conclusion

The logit of the proportion of moths removed, of each morph at each location, is plotted in Display 21.4 against the distance of the location from Liverpool. From this it appears that the odds that a moth will be removed by a predator increase with increasing distance from Liverpool for the dark morph and decrease with increasing distance from Liverpool for the light morph. A logistic regression analysis confirms that the odds of removal for the dark morph, relative to the odds of removal for the light morph, increase with increasing distance from Liverpool (one-sided p-value $= 0.0006$ for interaction of morph and distance). The odds of removal increase by an estimated 2% for each kilometer in distance from Liverpool for the

DISPLAY 21.4 Proportions of moths of two color morphs taken by predators at seven locations near Liverpool, England

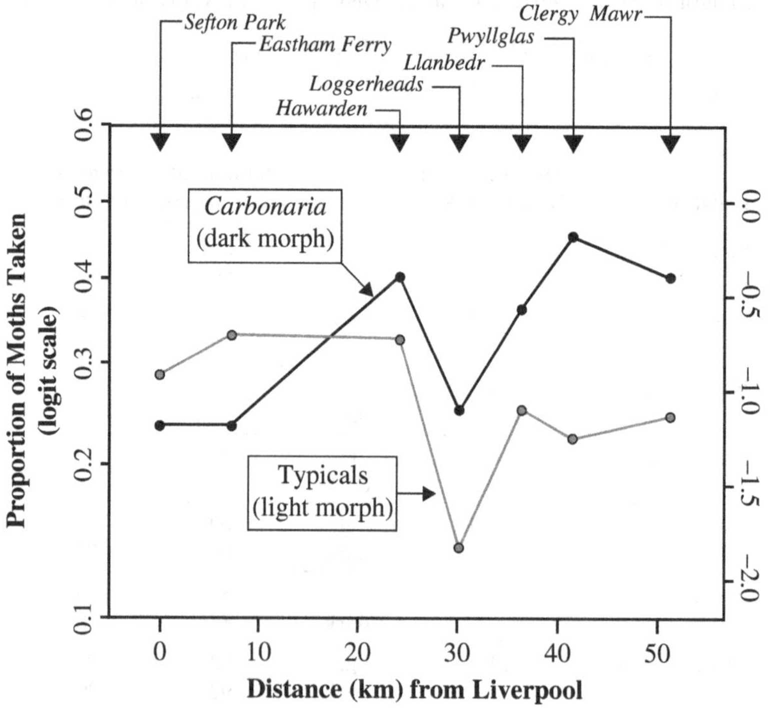

dark morph; and they decrease by an estimated 1% per kilometer for the light morph.

Scope of Inference

This is not a fully randomized experiment, since the moths were not randomly assigned to trees and to positions on trees. It can be argued, however, that the haphazard assignment of moths to trees is just as good as randomization. The possibility of a confounding variable seems slight.

21.2 LOGISTIC REGRESSION FOR BINOMIAL RESPONSES

21.2.1 Binomial Responses

As noted in Section 18.2.1, a binomial count is the sum of independent binary responses with identical probabilities. If m binary responses constitute a random sample from a population of binary responses with population proportion π, their sum (which must be an integer between 0 and m) has the *binomial*(m, π) distribution. The observed proportion of 1's among the m binary responses is Y/m and is called a *binomial proportion*. For this reason, m is called the *binomial denominator*. (Although it may seem natural to think of m as a sample size, the term *sample size* is reserved for the number of binomial proportions in the study.)

In regression settings, the ith response variable Y_i is binomial(m_i, π_i). The binomial denominators m_i need not be the same, but they must be known. In the first example, Y_i is the number of extinctions on island i, out of m_i bird species at risk for extinction. In the second example, Y_i is the number of moths taken, out of the m_i moths of a particular morph, that were placed at one of the locations.

The response variables here are the binomial counts (or the corresponding proportions), not the individual binary indicators that contribute to those counts. As in the previous chapter, however, a logistic regression model is used to describe π_i as a function of explanatory variables and unknown regression coefficients.

A Note About Counted and Continuous Proportions

A binomial response variable is a *count* that can acquire the integer values between 0 and m. The data may be listed either as the counts or (equivalently) as proportions Y/m, but in both cases m must be known. The term *counted proportion* refers to the proportion of 1's in m binary responses. This must be distinguished from a *continuous proportion* of amounts, such as the proportion of fat (by weight) that is saturated fat, or the proportion of body mass composed of water. In continuous proportions, the numerator and denominator are not integers (except possibly due to rounding), and no m is involved. When they function as response variables, continuous proportions must be handled with ordinary regression methods, possibly after a transformation. The methods of this chapter should never be used with continuous proportions as responses, or with arbitrary and fictitious values of m such as 100.

21.2.2 The Logistic Regression Model for Binomial Responses

As described in Chapter 20, logistic regression is a special case of generalized linear models. The elements of such a model are a distribution for the response variable and a function that links the mean of the distribution to the explanatory variables. A binomial logistic regression with two explanatory variables is specified as follows:

Binomial Logistic Regression Model

Y is binomial(m, π).

$$\text{logit}(\pi) = \beta_0 + \beta_1 X_1 + \beta_2 X_2.$$

The interpretation of the logistic model here is exactly the same as in Chapter 20. The model shows how the population proportion or probability π depends on the explanatory variables through a nonlinear link function. But the nonlinear part is completely captured by the logit of π. Logistic regression on binary responses is a special case of the binomial model in which all m_i's are 1.

Krunnit Islands Example

Let Y_i represent the number of extinctions on island i, out of m_i species at risk, for islands $i = 1, 2, \ldots, 18$. For example, for $i = 1$, the island Ulkokrunni had $Y_1 = 5$ extinctions out of $m_1 = 75$ species at risk. If the logit of the probability of extinctions is assumed to be linear in X = log of island area (Ulkokrunni has log area $X_1 = \log(185.50\text{ km}^2) = 5.2231$), the logistic regression model would specify that $\text{logit}(\pi_1) = \beta_0 + \beta_1 5.2231$ for Ulkokrunni, and that a similar expression would describe the proportions of extinctions for other islands. If β_1 in the model is zero, the odds of extinction (and hence the probability of extinction) are not associated with island area.

21.3 MODEL ASSESSMENT

21.3.1 Scatterplot of Empirical Logits Versus an Explanatory Variable

Associated with observation i is a response proportion, $\overline{\pi}_i = Y_i/m_i$. The observed logits, or *empirical logits*, are $\log[\overline{\pi}_i/(1 - \overline{\pi}_i)]$, which may also be written as $\log[Y_i/(m_i - Y_i)]$. Plots of the observed logits versus one or more of the explanatory variables are useful for visual examination, paralleling the role of ordinary scatterplots in linear regression. The scatterplot in Display 21.2, for example, suggests that the logit of extinction probability is linear in log of island area.

If the m_i's are small for some observations, there may be only a few possible values for the observed proportion, making the display less informative. If some of the observed proportions are 0 or 1, a modification is necessary because the logit function is undefined at these values. For purposes of gaining a useful visual display, add a small amount to the numerator and denominator, such as 0.5, so that all the logits, $\log[(Y_i + 0.5)/(m_i - Y_i + 0.5)]$, are defined. (No such addition is required for fitting the model, however.)

21.3.2 Examination of Residuals

Two types of residuals are widely used for logistic regression for binomial counts: deviance residuals and Pearson residuals. The *deviance residual* is defined so that the sum of all n squared deviance residuals is the deviance statistic, which was introduced in Section 20.4 and which will be clarified in the next section. This type of residual measures a kind of discrepancy in the likelihood function due to the fit of the model at each observation. The formula is shown in the following box for a binomial response based on a binomial denominator of size m_i. The part of the definition multiplying the square root, $\text{sign}(Y_i - m_i\hat{\pi}_i)$, is either +1 (if Y_i is above its estimated mean) or −1 (if below).

Deviance Residual

$$Dres_i = \text{sign}(Y_i - m_i\hat{\pi}_i)\sqrt{2\left\{Y_i \log\left(\frac{Y_i}{m_i\hat{\pi}_i}\right) + (m_i - Y_i)\log\left(\frac{m_i - Y_i}{m_i - m_i\hat{\pi}_i}\right)\right\}}$$

The *Pearson residual* is more transparent. It is defined as an observed binomial response variable minus its estimated mean, divided by its estimated standard deviation. Thus, it is constructed to have, at least roughly, mean 0 and variance 1.

Pearson Residual

$$Pres_i = \frac{Y_i - m_i\hat{\pi}_i}{\sqrt{m_i\hat{\pi}_i(1 - \hat{\pi}_i)}}$$

Although the two often exhibit little difference, the Pearson residuals are easier to understand and therefore are more useful for casual examination. The deviance residuals, on the other hand, have the advantage of being directly connected to drop-in-deviance tests. Either set of residuals can be examined for patterns in plots against suspected explanatory variables.

For large values of the binomial denominator, both types of residuals tend to behave as if they came from a standard normal distribution—if the model is correct. Paying close attention to those that exceed 2 in magnitude identifies possible outliers. If the m_i's are small—less than 5, say—the relative sizes of the residuals can be compared, but decisions based on comparison to the standard normal distribution should be avoided.

Example—Krunnit Islands Extinctions

The Pearson and deviance residuals from the fit to the Krunnit Islands extinction data are shown in Display 21.5. The logistic regression model for estimating the proportions of extinctions is: $\text{logit}(\hat{\pi}) = -1.196 - 0.297 \times \log(\text{Area})$. For this example, the two types of residuals differ little, and no evidence of outlying observations is present.

21.3.3 The Deviance Goodness-of-Fit Test

When a linear regression problem such as the insulating fluid example of Section 8.1.2 has replicate observations at each configuration of the explanatory variable(s), one can compare estimates of group means derived from some model to group sample averages in a test of goodness-of-fit for the model. The problems of this chapter are analogous because one compares estimates of proportions derived from a model to the sample proportions. The comparison is like an extra-sum-of-squares test for comparing the model of interest to a general model with a separate population proportion for each observation. Informally, the null hypothesis is that the model of interest is adequate, and the alternative is that more structure is needed to fit the data adequately. More precisely,

Model of interest: $\text{logit}(\pi_i) = \beta_0 + \beta_1 x_{i1} + \cdots$ (p parameters);
Saturated model: $\text{logit}(\pi_i) = \alpha_i$ ($i = 1, 2, \ldots, n$, with n different parameters).

DISPLAY 21.5 Residuals from the logistic regression of extinction probability on log of island area: Krunnit Islands data (Raw Residual is $Y/m - \hat{\pi}$)

Island	Observed proportion	Model estimate	Raw residual	Pearson residual	Deviance residual
Ulkokrunni	0.067	0.061	0.006	0.24	0.23
Maakrunni	0.045	0.070	−0.026	−0.82	−0.87
Ristikari	0.152	0.099	0.053	1.44	1.35
Isonkivenletto	0.118	0.138	−0.020	−0.42	−0.43
Hietakraasukka	0.107	0.159	−0.052	−0.76	−0.80
Kraasukka	0.200	0.162	0.038	0.46	0.45
Länsiletto	0.186	0.164	0.022	0.39	0.39
Pihlajakari	0.097	0.171	−0.075	−1.10	−1.18
Tyni	0.179	0.185	−0.007	−0.09	−0.09
Tasasenletto	0.188	0.205	−0.018	−0.25	−0.25
Raiska	0.267	0.223	0.044	0.58	0.57
Pohjanletto	0.100	0.252	−0.152	−1.56	−1.72
Törö	0.290	0.252	0.039	0.50	0.49
Luusiletto	0.313	0.260	0.052	0.48	0.47
Vatunginletto	0.467	0.284	0.182	1.57	1.50
Vatunginnokka	0.242	0.302	−0.059	−0.74	−0.76
Tiirakari	0.325	0.328	−0.003	−0.04	−0.04
Ristikarenletto	0.500	0.400	0.100	0.50	0.50

The drop in deviance—due to the $n-p$ additional parameters in the full model that are not included in the reduced model—is the sum of squared deviance residuals from the model of interest (the *deviance statistic*).

If the trial model fits *and the denominators are all large*, the deviance statistic has approximately a chi-squared distribution on $n - p$ degrees of freedom.

$$\textit{Approximate goodness of fit test p-value} = Pr(\chi^2_{n-p} > \textit{Deviance Statistic})$$

Attempts to assess the model with this approach when a substantial proportion of the m_i's are less than 5 can lead to misleading conclusions.

The deviance statistic and its p-value are included in standard output from logistic regression programs. Large p-values indicate that either the model is appropriate or the test is not powerful enough to pick up the inadequacies. Small p-values can mean one of three things: The model for the mean is incorrect (for example, the set of explanatory variables is insufficient); the binomial is an inadequate model for the response distribution; or there are a few severely outlying observations.

In view of the various possibilities just described, it is often difficult to make strong conclusions solely on the basis of the deviance goodness-of-fit test. Further discussion of the case where the binomial model is inadequate for the response distribution is given in Section 21.5 on extra-binomial variation. An example of the deviance goodness-of-fit test is provided in Display 21.6. Notice that the deviance statistic in Display 21.6 is the sum of squared deviance residuals from Display 21.5.

DISPLAY 21.6 Deviance goodness-of-fit test for Krunnit Islands data

Generalized linear model

Specifies logistic regression →

Response:	*Extinctions*
Distribution:	binomial
Binomial m:	*atrisk*
Link:	logit
Explanatory:	log(Area)

with model → $\text{logit}(\pi) = \beta_0 + \beta_1 \times \log(\text{Area})$

Computer output:

Variable	Coefficient	Standard error	z-statistic
Constant	−1.196	0.118	−10.10
log(Area)	−0.297	0.054	−5.42

Deviance = 12.06   Degrees of freedom = 16

Goodness-of-Fit-Test: p-value = $\Pr\{\chi^2_{16} > 12.06\}$ = **0.74**
Conclusion: There is no evidence that the model is inadequate.

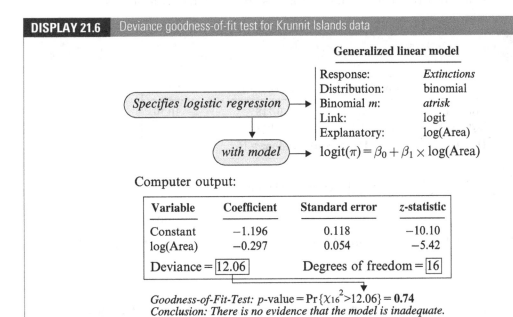

21.4 INFERENCES ABOUT LOGISTIC REGRESSION COEFFICIENTS

21.4.1 Wald's Tests and Confidence Intervals for Single Coefficients

Maximum likelihood estimates of the logistic regression coefficients and their standard errors are also included in standard output for logistic regression. Their use, interpretation, and properties are the same as in Chapter 20. The estimated coefficients are approximately normally distributed, so confidence intervals and Wald's tests for single coefficients are accomplished in the same way. The normal approximation is adequate if either the overall sample size n is large or the binomial denominators are all large. It is difficult to state any precise rules about how large a sample size or denominator is large enough, so it is best to label the standard errors and the inferential statements as *approximate*.

Example—Krunnit Islands Extinctions

The output in Display 21.7 is from the logistic regression of the proportion of extinctions on the log of island area. This is the fit to the model $\text{logit}(\pi) = \beta_0 + \beta_1 \times \log(\text{Area})$, where π is the extinction probability.

Is extinction probability associated with island size? If β_1 is zero, the probability of extinction does not depend on area. The question, therefore, is addressed through a test of the hypothesis that $\beta_1 = 0$. The approximate two-sided p-value, as shown in Display 21.7, is extremely small, confirming the visual evidence in the scatterplot that extinction probabilities are associated with island area.

DISPLAY 21.7 Wald's test and confidence interval for a single coefficient, from the logistic regression of extinction probability on log of island area: Krunnit Islands data

Variable	Coefficient	Standard error	z-statistic
Constant	−1.196	0.118	−10.10
log(Area)	−0.297	0.054	−5.42

Deviance = 12.06 **Degrees of freedom** = 16

(1) *Test whether the coefficient β_1 is zero.* Approximate two-sided p-value $= 2 \times \Pr(Z > 5.42) = 6 \times 10^{-8}$

(2) *Approximate 95% confidence interval for β_1.* $-0.297 \pm 1.96 \times 0.054 = -0.403$ to -0.191 (1.96 = $z(0.975)$)

Interpretation of the Coefficient of a Logged Explanatory Variable

Since the logit is the log of the odds, a unit increase in an explanatory variable is associated with a multiplicative change in the odds by a factor of $\exp(\beta)$, where

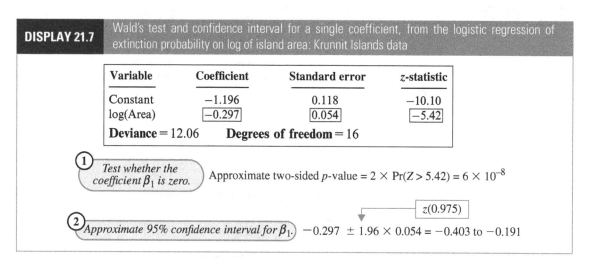

β is its coefficient in the logistic regression. If the explanatory variable is $\log(X)$, then (as in ordinary regression) the interpretation is most conveniently worded in terms of a two-fold, ten-fold, or some other k-fold change in X. For each doubling of X, for example, $\log(X)$ increases by $\log(2)$ units, so that the log odds change by $\beta \times \log(2)$ units. The odds, therefore, change by a factor of $\exp[\beta \times \log(2)] = 2^{\beta}$. The result is stated generally as follows:

> If $logit(\pi) = \beta_0 + \beta \times \log(X)$, then each k-fold increase in X is associated with a change in the odd by a multiplicative factor of k^{β}.

A confidence interval for k^{β} extends from k raised to the lower endpoint to k raised to the upper endpoint of a confidence interval for β. As shown in the accompanying Krunnit Islands example, some customizing of this result may improve the wording of the conclusion.

Example—Krunnit Islands Extinctions

The estimated coefficient of log island area is -0.297. Based on the preceding results with $k = 2$, each doubling of island area is associated with a change in the odds of extinction by an estimated factor of $2^{-0.297}$, or 0.813. It is estimated that the odds of extinction for an at-risk species on an island of area $2A$ is only 81.3% of the odds of extinction for such a species on an island of area A.

Since the effect of reduced habitat area on the odds of extinction is of interest, a researcher might prefer to express the results of the study as the change in odds associated with a *reduction* in island area. With $k = 0.5$, for example, each halving of island area is associated with a multiplicative change in the odds of extinction of $0.5^{-0.297}$, or 1.23. The odds of extinction of an island with area $A/2$ are 123% of the odds of extinction of an island of area A. Since an approximate 95% confidence interval for β_1 is -0.403 to -0.191, an approximate 95% confidence interval for the factor associated with a halving of island area is $0.5^{-0.403}$ to $0.5^{-0.191}$, which is 1.14 to 1.32.

Finally, a 123% change can also be expressed as a 23% increase. Therefore, each halving of area is associated with an increase in the odds of extinction by an estimated 23%. An approximate 95% confidence interval for the percentage increase in odds is 14% to 32%.

21.4.2 The Drop-in-Deviance Test

Drop-in-deviance tests are extra-sum-of-squares tests based on the deviance residuals, as discussed in Section 20.4.2. The issues involved here are the same as those described in Chapter 20 for binary response models. For binomial responses the tests are appropriate if either the overall sample size n is large or the binomial denominators are all large. Studies suggest that the drop-in-deviance test performs well even when the sample size or the denominators are fairly small; again, however, it is appropriate to label the inferences *approximate*.

Example—Krunnit Islands

In the previous section, Wald's test was used to test whether β_1 is zero, resulting in a two-sided p-value of 6×10^{-8}. Alternatively, the drop-in-deviance test can be used to assess the same hypothesis. The reduced model has β_1 set to 0. Its deviance is obtained by fitting a logistic regression model without any explanatory variables (but including a constant term). From computer output (not shown), this deviance is found to be 45.34, with 17 degrees of freedom. In Display 21.7, the deviance statistic from a fit of the model with log(Area) included is 12.06, with 16 degrees of freedom. The deviance dropped by 33.28, with a drop of one degree of freedom. The associated p-value from a chi-square distribution is 8.0×10^{-9}. This is a two-sided p-value, and it provides essentially the same result as that given by Wald's test.

Wald's Test or Drop-in-Deviance Test?

If the hypothesis involves several coefficients, the drop-in-deviance test is the only practical alternative. For a hypothesis that a single coefficient is zero, either test can be used. The approximate p-value from the drop-in-deviance test is generally more accurate than the approximate p-value from the Wald's test. The latter, however, is easier to obtain and is usually satisfactory for conducting tests of secondary importance (such as for informally testing extra terms in an exploratory model-building stage) and when the p-value is clearly large or clearly very small.

21.5 EXTRA-BINOMIAL VARIATION

A binomial count is the sum of independent binary responses with identical π's. If it is the sum of m_i such binary responses, its distribution is known to have mean $m_i \pi_i$ and variance $m_i \pi_i (1 - \pi_i)$. If the binary trials are not independent, or the π's for the binary responses are not the same, or important explanatory variables are not included in the model for π, then the response counts will not have binomial distributions. Typically, the variance of the Y_i's will be greater than expected from a binomial distribution. The terms *extra-binomial variation* and *overdispersion* describe the inadequacy of the binomial model in these situations.

When the response variance is larger than the binomial variance, the regression parameter estimates based on the binomial model will not be seriously biased, but their standard errors will tend to be smaller than they should be; in addition, p-values will tend to be too small, and confidence intervals will tend to be too narrow. Thus, if extra-binomial variation exists, alternatives to the usual inference tools must be found.

When in doubt, it is safer to suppose that extra-binomial variation is present than to ignore it. The consequences of assuming the presence of extra-binomial variation when there is actually none are minor. Inferences may be less precise, but they will not be misleading. More severe consequences result from assuming that the responses are binomial when they are not.

21.5.1 Extra-Binomial Variation and the Logistic Regression Model

Several procedures allow for extra-binomial variation; but one, the *quasi-likelihood approach*, is particularly easy to use and works in most situations. Unlike the maximum likelihood approach, this method does not require complete specification of the distribution of the response variables. It is based only on the mean and the variance, and these are taken to match the binomial model, except for the presence of a *dispersion parameter*. As shown in the following box, the dispersion parameter, represented by ψ, accounts for extra-binomial variation. If ψ is greater than 1, the variance is greater than the binomial variance. However, ψ does not represent a variance in this model; it represents a multiplicative departure from binomial variation.

Logistic Regression with Extra-Binomial Variation

for quasi-likelihood approach (with two explanatory variables)

$$\mu\{Y_i|X_{1i}, X_{2i}\} = m_i\pi_i; \qquad \text{logit}(\pi_i) = \beta_0 + \beta_1 X_{1i} + \beta_2 X_{2i};$$
$$\text{Var}\{Y_i|X_{1i}, X_{2i}\} = \psi m_i \pi_i (1 - \pi_i)$$

21.5.2 Checking for Extra-Binomial Variation

There are three tasks in checking for extra-binomial variation: considering whether extra-binomial variation is likely in the particular responses; examining the deviance goodness-of-fit test after fitting a rich model; and examining residuals to see whether a large deviance statistic may be due to one or two outliers rather than to extra-binomial variation.

Considering Whether Extra-Binomial Variation Is Present

In thinking about whether extra-binomial variation is likely, ask the following questions. Are the binary responses included in each count unlikely to be independent? Are observations with identical values of the explanatory variables likely to have different π's? Is the model for π somewhat naive? A yes answer to any of these questions should make a researcher cautious about using the binomial model. Even so, extra-binomial variation is rarely a serious problem when the binomial denominators are small (less than five, say). It is more likely to be a problem for larger denominators.

Extra-binomial variation may be present in the moth coloration responses for a number of reasons:

1. There may be a lack of independence of binary responses (moth removals) at each site. For example, if predators succeed in finding one moth, their success in finding moths of the same coloration may increase.

2. Binary responses at each site may have different probabilities. If a site has variation in tree color, moths placed on different trees at the site may have different odds of being taken.
3. The model may be incorrect. Two sites equally distant from Liverpool do not necessarily have the same level of tree coloration, so the model in which the logit of π is a linear function of distance from Liverpool may be naive.

Examining the Goodness-of-Fit Test

A large deviance goodness-of-fit test statistic (Section 21.3.3) indicates that the binomial model with specified explanatory variables is not supported by the data. If it can be established that the set of explanatory variables is suitable, however, the large deviance statistic probably indicates extra-binomial variation. This should be examined at an early stage in the analysis by fitting a rich model that includes all the terms that might be important (possibly including polynomial and interaction terms).

Examining Residuals

Since the deviance statistic is the sum of squared deviance residuals, it is useful to examine the deviance residuals themselves to see whether a few of these are responsible for a large deviance statistic. If so, the analysis should focus on the outlier problem rather than on the extra-binomial variation problem.

21.5.3 Quasi-Likelihood Inference when Extra-Binomial Variation Is Present

The *quasi-likelihood* approach was developed in such a way that the *maximum quasi-likelihood estimates* of parameters are identical to the maximum likelihood estimates, but standard errors are larger and inferences are adjusted to account for the extra-binomial variation.

One possible estimate of the dispersion parameter is the deviance statistic divided by its degrees of freedom:

$$\hat{\psi} = \frac{\text{Deviance}}{\text{Degrees of freedom}}.$$

The degrees of freedom are $n - p$, the sample size minus the number of parameters in the model for the mean. Apart from the degrees of freedom adjustment, this amounts to a sample variance of the deviance residuals. It should be approximately equal to 1 if the data are binomial, and larger than 1 if the data contain more variation than would be expected from the binomial distribution.

The standard errors of the maximum quasi-likelihood estimates are $\sqrt{\hat{\psi}}$ times the standard errors for the maximum likelihood estimates. The greater the variation in the responses, the greater the uncertainty in the estimates of the parameters.

Inference Based on Asymptotic Normality

Tests and confidence intervals for single coefficients may be accomplished in the usual way with the estimated coefficients and their adjusted standard errors. The inferences are based on the approximate normality of the estimators in situations involving either large sample sizes or large denominators. A test statistic for a hypothesis that β is 7, say, is $t = (\hat{\beta} - 7)/\mathrm{SE}(\hat{\beta})$. The theory suggests that this be compared to a standard normal distribution. Some statisticians, however, prefer to compare it to a t-distribution on $n - p$ degrees of freedom. Similarly, a confidence interval for β may be taken as $\hat{\beta} \pm [t_{\mathrm{df}}(1 - \alpha/2) \times \mathrm{SE}(\hat{\beta})]$, where $t_{\mathrm{df}}(1 - a/2)$ is the appropriate multiplier from the t-distribution on $n - p$ degrees of freedom. Although no theory justifies using t-distributions, in this setting the practice parallels that of ordinary regression models, since the standard error involves an estimate of the unknown ψ on $n - p$ degrees of freedom. And although the normal and t-distributions usually yield very similar results, when there is a difference, the approach using the t is more conservative.

Drop-in-Deviance F-Tests

It is also reasonable to consider the drop in the deviance as a way to test whether a single coefficient is zero or whether several coefficients are all zero—as well as to test other hypotheses that can be formulated in terms of a reduced and a full model. Since there is an additional parameter in the response variance, however, the drop in deviance must be standardized by the estimate of ψ. The resulting statistic is compared to an F-distribution, rather than to a chi-square. This F-test is equivalent to the extra-sum-of-squares F-test for ordinary regression, except that deviance residuals are used rather than ordinary residuals:

Drop-in-Deviance F-Test when Extra-Binomial Variation Is Present

$$F\text{-stat} = \frac{\text{Drop in deviance}/d}{\hat{\psi}},$$

where d represents the number of parameters in the full model minus the number of parameters in the reduced model, and $\hat{\psi}$ is the estimate of the dispersion parameter from the full model. The p-value is $\Pr(F_{d,n-p} > F\text{-stat})$, where p is the number of parameters in the full model. No solid theory justifies using this test, but it is a sensible approach, motivated by the analogies to ordinary linear regression.

Notes

1. Many statisticians recommend using the sum of squared Pearson residuals divided by $(n - p)$ rather than the sum of squared deviance residuals divided by $(n - p)$ to estimate ψ. Although this issue has not been completely resolved, it is always wiser to use one of these approaches than to assume that $\psi = 1$ when extra-binomial variation is present.

2. Deviance and Pearson residuals from the quasi-likelihood fit are identical to those from the maximum likelihood fit. They may be examined for peculiarities, but they should no longer be compared to a standard normal distribution for outlier detection.
3. Some statistical computer programs allow the user to request quasi-likelihood analysis for binomial-like data. It is evident from the preceding discussion, however, that the quasi-likelihood inferences can be computed with standard output from a binomial logistic regression routine.

21.6 ANALYSIS OF MOTH PREDATION DATA

In the moth study, two color morphs were placed on trees at various distances from Liverpool. The question is whether the relative rate of removal depends on tree color. Distance is used as a surrogate for tree color. A logistic regression model can isolate a difference that changes steadily with distance by including the product variable *dark* × *dist*, where *dark* is an indicator of dark morph and *dist* is the actual distance. The model

$$\text{logit}(\pi) = \beta_0 + \beta_1 dark + \beta_2 dist + \beta_3 (dark \times dist)$$

enables the analyst to answer the question directly. The constant term β_0 is the log odds for removal of the light morph at Sefton Park (for which $dist = 0$); $\beta_0 + \beta_1$ is the log odds for removal of the dark morph at Sefton Park. So $\beta_1 = (\beta_0+\beta_1)-\beta_0$ is the log of the odds *ratio* for removal of the dark morph relative to the light morph, at Sefton Park. The log odds for removal of the light morph at a site that is D kilometers distant from Sefton Park is $\beta_0 + \beta_2 D$; and for the dark morph, it is $(\beta_0 + \beta_1) + (\beta_2 + \beta_3)D$. Their difference, $\beta_1 + \beta_3 D$, is the log of the odds ratio for removal of the dark morph relative to the light morph at distance D from Sefton Park. The rate of change in the log odds ratio with distance is

$$[(\beta_1 + \beta_3 D) - \beta_1]/D = \beta_3.$$

Drop-in-Deviance Test for Effect of Distance on Relative Removal Rates

If $\beta_3 = 0$, the relative odds that the light morph will be taken are the same at all locations as the odds that the dark morph will be taken. But if $\beta_3 \neq 0$, the relative odds change with distance. So the question can be addressed by testing the hypothesis that $\beta_3 = 0$, in the context of the preceding model. This may be accomplished by using either Wald's test in the fit of the model or the drop in deviance associated with including the product term in a model that includes only *dark* and *dist*. The latter approach is illustrated in Display 21.8, where the answer to the question seems convincing.

DISPLAY 21.8 The drop-in-deviance chi-squared test for the interaction term in the logistic regression of *removed* (number of moths removed out of the number *placed*) on *dark*, *dist* (distance from Liverpool), and *dark* × *dist*: moth coloration data

$$\text{logit}(\pi) = \beta_0 + \beta_1 dark + \beta_2 dist + \beta_3 dark \times dist$$

(1) *Fit full model.*

Variable	Coefficient	Standard error	z-statistic
Constant	−0.718	0.190	−3.77
dark	−0.411	0.275	−1.50
dist	−0.00929	0.00579	−1.60
dark × *dist*	0.0278	0.0081	3.44

Deviance = 13.230 **Degrees of freedom** = 10

(2) *Fit reduced model.* $\text{logit}(\pi) = \beta_0 + \beta_1 dark + \beta_2 dist$

Variable	Coefficient	Standard error	z-statistic
Constant	−1.137	0.157	−7.25
dark	0.404	0.139	2.90
dist	0.0053	0.0040	1.33

Deviance = 25.161 **Degrees of freedom** = 11

(3) *Calculate the drop in deviance.* → 25.161 − 13.230 = 11.931
d.f.= 11 − 10 = 1

(4) *Determine the p-value.* → $p\text{-value} = \Pr(\chi_1^2 > 11.931) = \mathbf{0.0006}$

Conclusion: Substantial evidence indicates that the relative odds of removal of the two color morphs change steadily with distance from Liverpool.

Checking for Extra-Binomial Variation

Extra-binomial variation may be suspected in this example, based on the study design. Therefore, a close look at the goodness-of-fit deviance statistic from the full model is in order. Overall, the magnitude of the deviance statistic, 13.230, is not large for its degrees of freedom, 10. The p-value for goodness-of-fit, 0.23, shows no substantial problem.

On the other hand, there is no reason to anticipate a linear relationship between logit of removal probability and distance from Liverpool; and the scatterplot of sample logits versus distance (in Display 21.4) suggests departures from a straight line. The issue is clarified by examining the individual deviance residuals in Display 21.9.

DISPLAY 21.9 Deviance residuals from fit of the full model in the moth study

Location	Distance from Liverpool (km)	Morph	Binomial proportion	Model estimate	Deviance residual
Sefton Park	0.0	light	0.304	0.328	−0.39
		dark	0.250	0.244	0.10
Eastham Ferry	7.2	light	0.350	0.313	0.70
		dark	0.250	0.270	−0.40
Hawarden	24.1	light	0.346	0.281	1.03
		dark	0.423	0.336	1.31
Loggerheads	30.2	light	0.150	0.269	−2.21
		dark	0.267	0.361	−1.56
Llanbedr	36.4	light	0.267	0.258	0.15
		dark	0.383	0.388	−0.08
Pwyllglas	41.5	light	0.238	0.249	−0.24
		dark	0.476	0.411	1.21
Clergy Mawr	51.2	light	0.261	0.233	0.63
		dark	0.424	0.455	−0.59

One residual—for the light morph at Loggerheads—is somewhat extreme. It might be ignored except that the residual for the dark morph at Loggerheads is also extreme in the same direction. Meanwhile, both residuals at Hawarden are mildly extreme in the opposite direction. These features suggest that each location has unique characteristics—that the differences in removal probabilities at different locations may be due to factors besides the distance from Liverpool.

Adjustment for Extra-Binomial Variation

The quasi-likelihood procedure for adjusting the inference based on the Wald's test appears in Display 21.10. The test statistic for the coefficient of the interaction term (which was 3.44 for the binomial model) is 2.99 from the quasi-likelihood fit.

The drop-in-deviance test is calculated from the same deviances that were obtained from the maximum likelihood fit, but the F-version of the statistic is used to allow for extra-binomial variation:

Drop in deviance/drop in d.f.

$$F\text{-statistic} = \frac{(11.931)/(1)}{(13.230)/10} = 9.02.$$

Deviance/d.f.

The p-value is $\Pr(F_{1,10} > 9.02) = 0.013$.

DISPLAY 21.10 Quasi-likelihood (QL) inference about distance-by-morph interaction, based on maximum likelihood (ML) output: moth coloration data

① *Obtain standard ML analysis.*

ML and Q estimates of regression parameters are the same.

Variable	Coefficient	Standard error	Quasi-likehood standard error	*t*-statistic
Constant	−0.718	0.190	0.219	−3.28
dark	−0.411	0.275	0.316	−1.30
dist	−0.00929	0.00579	0.00666	−1.39
dark × *dist*	0.0278	0.0081	0.0093	2.99

Deviance = 13.230 **Degrees of freedom** = 10

② *Estimate the dispersion parameter.* → $\hat{\psi} = \frac{13.230}{10} = 1.323;\ \sqrt{\hat{\psi}} = \sqrt{1.323} = 1.150$

③ *Multiply ML standard errors by* $\sqrt{\hat{\psi}}$ *to get QL standard errors.*

④ *Divide coefficients by QL standard errors to get adjusted Wald's statistics.*

⑤ *p-value for* H: $\beta_3 = 0$ → Two-sided *p*-value = $2 \times \Pr(t_{10} > 2.99)$ = **0.014**

Conclusion: The data provide strong evidence that the interaction is nonzero.

Location Treated as a Factor with Seven Levels

An alternative, less restrictive model lets the logits for the light morphs be arbitrary at each location, but the difference between the dark morph logit and the light morph logit changes linearly with distance:

$$\text{logit}(\pi) = \mathit{LOCATION} + \mathit{dark} + (\mathit{dark} \times \mathit{dist}).$$

LOCATION contains six indicator variables to represent the seven locations, but the interaction among them is still modeled with the numerical explanatory variable *dist* multiplied by the indicator variable for the dark species.

This model may be a bit more realistic, since it allows the overall log odds of removal to differ for different locations (that is, it does not restrict them to falling on a straight line function of distance). The interaction of interest, however, continues to be represented by a single term: the difference in log odds of removal between the light and dark morphs is modeled to fall on a straight line function of distance from Liverpool. The output is reported in Display 21.11. Here, cause for concern about extra-binomial variation has greatly diminished, since the *p*-value from the goodness-of-fit test is 0.72. Furthermore, this approach sharpens the inference about

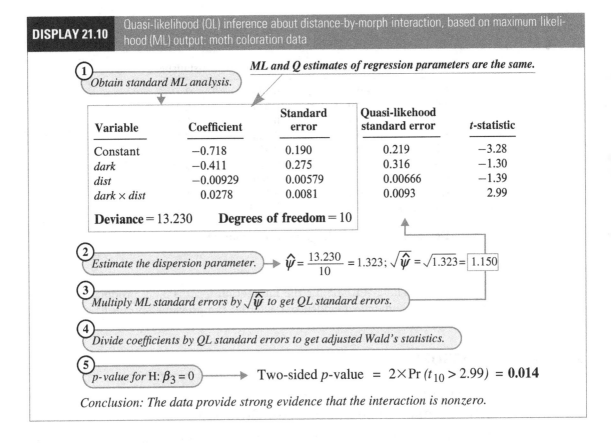

DISPLAY 21.11 Logistic regression of proportion of moths removed on location (treated as a factor with seven levels), morph, and a distance × morph interaction

Variable	Coefficient	Standard error	z-statistic
Constant	−0.767	0.246	−3.11
EF	0.020	0.274	0.74
HW	0.163	0.305	0.54
LH	−0.798	0.330	−2.42
LB	−0.282	0.322	−0.88
PG	−0.241	0.314	−0.68
CM	−0.434	0.340	−1.28
dark	−0.406	0.275	−1.47
dark × *dist*	0.0277	0.0081	3.43

Deviance = 2.876 **Degrees of freedom** = 5

the key issue: The z-statistic for the interaction term is 3.43, producing a smaller p-value—more convincing evidence—than did the previous approach.

21.7 RELATED ISSUES

21.7.1 Why the Deviance Changes when Binary Observations Are Grouped

If a number of distinct binary responses exist for each combination of explanatory variables, the data analyst may adopt either of two approaches: Fit logistic regression models with the individual binary variables as responses (for example, the 1 or 0 response for each of the 968 moths in the coloration study); or group the observations according to distinct explanatory variable combinations and use as responses the binomial counts—the sums of the binary responses at each distinct explanatory variable combination (for example, the number of moths out of m_i placed, for each of the 14 combinations of type and location). The inferential results are identical; but because the latter approach reduces the data set to a smaller size, it is often more convenient.

As a check, one could use both approaches to fit the same logistic regression model; this would confirm that tests and confidence intervals for coefficients are identical. One feature of the output differs, however: the deviance statistic. The deviance statistic is the *likelihood ratio* test for comparing the fitted model to a saturated model—a model that has a distinct parameter for each observation. The test statistic is different in the two approaches because there are different saturated models. With binary responses, the saturated model has a separate probability parameter for each subject (968 parameters in the moth example). With grouped responses, an *observation* is defined as the binomial count at a distinct explanatory variable combination, so the saturated model has only 14 parameters.

It should not be surprising or alarming, therefore, that the deviance statistic differs in these two approaches, even though all other aspects of estimation, testing, and confidence intervals remain the same. This circumstance is mentioned because the different deviances sometimes cause confusion. The deviance goodness-of-fit test remains inappropriate for binary responses.

21.7.2 Logistic Models for Multilevel Categorical Responses

Multinomial Logits

Suppose that each subject falls into one of three categories of a categorical response variable—for instance, HIV negative, HIV positive but without AIDS symptoms, and HIV positive with AIDS symptoms. Let π_1, π_2, and π_3 be the probabilities of falling into the three categories. It might be desired to see how these probabilities depend on an explanatory variable, such as number of sexual contacts with an infected partner. Since the probabilities sum to 1, any two of them determine the third. Consequently, the analyst need only indicate how any two of them depend on the explanatory variables. Multinomial logits may be defined as follows:

$$\theta_2 = \log(\pi_2/\pi_1)$$
$$\theta_3 = \log(\pi_3/\pi_1).$$

In the example, θ_2 is the log of the odds of being HIV positive relative to being HIV negative, and θ_3 is the log odds of being HIV positive with AIDS symptoms, relative to being HIV negative. Each θ gives the log odds relative to the first category.

If there is no ordering in the response categories, the multinomial logistic regression model has a separate logistic regression for each of the two logits—for example,

$$\log(\pi_{2i}/\pi_{1i}) = \beta_0 + \beta_1 contacts_i$$

and

$$\log(\pi_{3i}/\pi_{1i}) = \alpha_0 + \alpha_1 contacts_i.$$

The different Greek letters for regression coefficients are needed to distinguish the regression coefficients for the two logits.

Some statistical computer packages produce the estimated coefficients and standard errors needed for this model. The packages may differ in their definition of multinomial logits—for example, the choice of the reference level may differ—so some care must be exercised to ascertain what model is being fit by the package. Drop-in-deviance tests are available as before.

Ordered Categorical Responses

The levels of a categorical response commonly have some internal order. Thus the categories just discussed might be considered as ranging essentially from no disease to some disease to severe disease. Because the multinomial logistic regression model has no features that can capture this ordering, it wastes the information from this valuable structure.

Some models do account for such ordering. Their implementation depends on whether they are featured in available statistical computer packages. See, for example, Exercise 21.21. The models listed here represent the case involving three response categories; but they extend in a straightforward manner to cases involving more than three categories, and in fact they are most helpful for handling cases involving substantially more categories.

The *common slopes* model stipulates that the coefficients of the explanatory variable (but not the intercepts) in the multinomial model are the same for each multinomial logit:

$$\log(\pi_{2i}/\pi_{1i}) = \beta_0 + \beta_1 contacts_i$$

and

$$\log(\pi_{3i}/\pi_{1i}) = \alpha_0 + \beta_1 contacts_i.$$

This implies that the odds of being HIV positive relative to being HIV negative and the odds of having AIDS symptoms relative to being HIV negative increase at the same rate with increasing numbers of sexual contacts. This model is appropriate if the odds of having AIDS relative to the odds of being HIV positive do not depend on the number of sexual contacts.

The *linear trends over response categories* model is

$$\log(\pi_{ji}/\pi_{1i}) = \beta_0 + \beta_1(j-1)contacts_i,$$

for j equal to 2 up to the total number of categories (3, in the example). In this model, the odds of having AIDS symptoms relative to being HIV negative increase with the number of sexual contacts at twice the rate of the odds of being HIV positive relative to being HIV negative. This model has the same number of parameters as the common slopes model; but it emphasizes, for example, the greater chance of being in the next highest category with increasing levels of an explanatory variable.

The *proportional odds model* defines logistic regression models for *cumulative probabilities*. If γ_j represents the probability that a subject will respond in category j or less, then $\gamma_1 = \pi_1, \gamma_2 = \pi_1 + \pi_2$, and $\gamma_3 = 1$. Models can then be specified for each cumulative probability except the last:

$$\log[\gamma_1/(1-\gamma_1)] = \beta_0 + \beta_1 contacts_i$$
$$\log[\gamma_2/(1-\gamma_2)] = \alpha_0 + \beta_1 contacts_i.$$

The first model is for the odds of being HIV negative. The second is for the odds of being AIDS-free. (Here, unlike in the preceding models, the coefficient of *contacts* would be negative.)

21.7.3 The Maximum Likelihood Principle

The *maximum likelihood (ML) principle*, attributed to the British statistician Sir Ronald Fisher, is the approach statisticians predominantly use to select estimation procedures. When they say that a weighted average is preferable to an unweighted one in a specific situation, their justification is often premised on the ML principle. An estimation procedure that has figured prominently in this book—the *least squares principle*—arises directly from the maximum likelihood principle when the response distribution is assumed to be normal.

The ML principle states that the parameters chosen for a model should be those that yield the highest probability of observing precisely what was observed. Application of the ML principle to many statistical problems entails differential calculus and often some formidable algebra. Fortunately, computers handle such problems routinely, which eliminates the need to fathom the mathematics. Such is the case with logistic regression analysis.

This section describes the ML solution to one specific problem (not a logistic regression problem), which was chosen for its simplicity so that the principle itself could be explained.

The Bucket Problem

A bucket contains 12 marbles. Each marble is either red or white, but the numbers of red and white marbles are not known. To gain some information on the bucket's composition, you are allowed to select and view five of them. What inference about the bucket's composition can be drawn from the sample information?

Notation The following symbols represent the features of this game:

N = Number of marbles in the bucket (= 12)
θ = Number of red marbles in the bucket (unknown)
n = Sample size = Number of marbles selected for viewing (= 5)
x = Number of red marbles in the sample.

A rule is desired for estimating θ once x is known.

Probability Model Imagine that the 12 marbles are numbered consecutively, 1–12, and that the first θ of them are red. A sample of five marbles thus consists of five distinct marble numbers—for example, {2, 5, 6, 9, 12}. If the bucket contained $\theta = 7$ reds, this sample would have $x = 3$ red marbles (numbers 2, 5, and 6).

Assume that the sample selection process ensures that all 792 possible samples of size 5 from the 12 have an equal chance of selection (1/792). Then for any given number of red marbles in the bucket, one can determine what the chances are of drawing the numbers of red marbles actually obtained in the sample. These chances form the rows of Display 21.12.

For example, if no red marbles were in the bucket to begin with (row 1), the chances are certain (100%) that the sample will have no red marbles. If instead the bucket contained two red marbles (row 3), there is slightly better than a 50–50 chance that the sample will include exactly one red marble; the chances of obtaining no red marbles in this case are double the chances of selecting both red marbles.

The numbers in Display 21.12 come from the hypergeometric probability distribution (previously encountered in the discussion of Fisher's Exact Test). The hypergeometric formula for the probabilities that x will take on each of its possible values when the bucket has θ red marbles is not given here. Each row shows

DISPLAY 21.12 Probability distributions and likelihood functions for the bucket with 12 marbles and five draws

Number (x) of Red Marbles Found in the Sample (columns); **Number (θ) of Red Marbles Originally in the Bucket** (rows)

θ \ x	0	1	2	3	4	5
0	1	0	0	0	0	0
1	0.5833	0.4167	0	0	0	0
2	0.3182	0.5303	0.1515	0	0	0
3	0.1591	0.4773	0.3182	0.0454	0	0
4	0.0707	0.3535	0.4243	0.1414	0.0101	0
5	0.0265	0.2210	0.4419	0.2652	0.0442	0.0012
6	0.0076	0.1136	0.3788	0.3788	0.1136	0.0076
7	0.0012	0.0442	0.2652	0.4419	0.2210	0.0265
8	0	0.0101	0.1414	0.4243	0.3535	0.0707
9	0	0	0.0454	0.3182	0.4773	0.1591
10	0	0	0	0.1515	0.5303	0.3182
11	0	0	0	0	0.4167	0.5833
12	0	0	0	0	0	1

When you know how many red marbles are in the bucket, look across that row to see how many might be observed in the sample. (Rows are ***probability functions.****)*

When you know how many red marbles were in the sample, look down that column to infer how many might have been in the bucket. (Columns are ***likelihood functions.****)*

the chances of different outcomes to the sampling experiment when the bucket composition is known.

For the inferential problem—making a statement about the likely values of θ based on the observed sample x—the *columns* of Display 21.12 are important. After the sampling is done and three reds (for example) are observed, one can look at the entries in the column headed by $x = 3$ to see how to use the sample information. Clearly, $\theta = 0, 1$, and 2 are impossible. There must have been at least three reds in the bucket, because three showed up in the sample. Similarly, since two whites appeared in the sample, there must be at least that many whites in the bucket. Thus, the column also rules out $\theta = 11$ and 12.

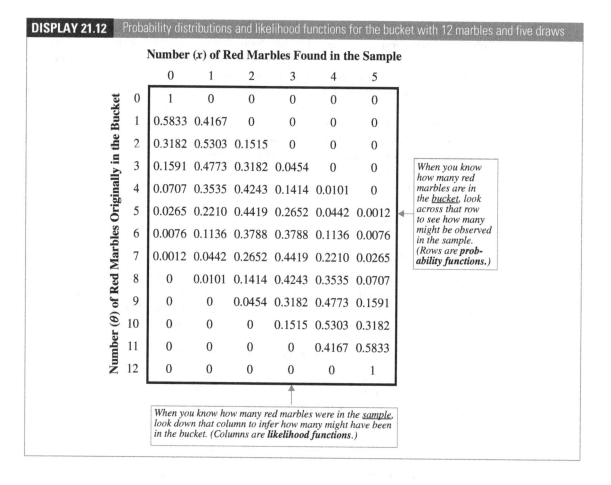

Can more be said? It is possible that the bucket began with three reds and all were selected; and it is also possible that the bucket began with seven reds and three were selected. But although both are possible, the likelihoods of these two cases are not equal. If only three reds were in the bucket, finding three in the sample would be a relatively *rare* event—an outcome expected to occur in fewer than one in 20 trials. But if seven reds were in the bucket, finding three in the sample would be a *common* occurrence, to be expected in nearly half of all such trials. Since common events are expected and rare events are not, it is natural to look at the possibilities and—having observed three reds in the sample—conclude that it is more likely that the bucket began with seven reds.

The probabilities in Display 21.12 are called *likelihoods* when viewed down a column. The column as a whole is a *likelihood function*, meaning that it is a function of the unknown parameter θ. Each observed number of reds in the sample gives rise to a different likelihood function. Adding probabilities across a row always yields a sum of 1.0; however, the same is *not* true for adding down a column. Therefore, likelihoods must not be confused with probabilities. It is not appropriate to say that, if $x = 3$, the probability that $\theta = 7$ is 0.4419. On the other hand, likelihoods can be expressed in a relative sense. For $x = 3$, the statement "$\theta = 7$ is 10 times more likely than $\theta = 3$" is a legitimate interpretation.

The Maximum Likelihood Estimator for the Bucket Problem

To make a best guess at θ, according to the maximum likelihood principle—as the name implies—one should select the θ for which the likelihood is largest. This amounts to guessing that θ has *the value that gives the highest probability of observing what was, in fact, observed.*

Suppose that a particular sample yielded $x = 3$. In the likelihood column for "$x = 3$," the largest likelihood, 0.4419, corresponds to $\theta = 7$. Then the best (ML) guess is that the number of reds originally in the bucket was seven.

Examining all columns in Display 21.12 leads to a rule describing which θ to guess for each outcome x. Display 21.13 shows this *maximum likelihood estimator* for the bucket problem.

DISPLAY 21.13 The maximum likelihood estimator for the unknown number of red marbles in a bucket of 12, based on a sample of five

Observed number of reds in sample (x):	0	1	2	3	4	5
$\hat{\theta}$ = ML estimate of number of reds in bucket:	0	2	5	7	10	12

Notes About Maximum Likelihood Estimation

1. The probability distribution of the response variables must be specified entirely, except for the values of the unknown parameters. This contrasts to the method of least squares and the method of quasi-likelihood, which only require knowledge of the mean and variance of the response distribution.
2. Display 21.13 also provides the sampling distribution of the ML estimator. The rows of the table show the probabilities for the sample outcomes, and hence for the values of the estimator. In more complex situations, considerable effort is required to determine the sampling distribution of an ML estimator.

3. The ML principle itself does not guarantee that the ML estimator has suitable properties, but theoretical statisticians have shown that the resulting procedure does indeed lead to desirable estimators.

21.7.4 Bayesian Inference

This textbook generally follows the "frequentist" school, in which statements about parameters and uncertainties arise directly from sampling distributions: "confidence level," "estimates," "standard errors," and "p-values." Members of the Bayesian school—introduced in Chapter 12—present statements about parameters as probability distributions and their descriptors: means, standard deviations, etc. For example, a frequentist inference about the slope of a regression line might read:

> The slope is estimated to be 0.37 g/sec, with a 95% confidence interval from 0.15 to 0.59 g/sec. There is highly suggestive evidence against the hypothesis that the slope is zero or negative (one-sided p-value = 0.03).

A Bayesian inference about the same parameter might read as follows:

> The mean slope estimate is 0.35 g/sec (standard deviation = 0.09 g/sec). There is a 95% chance that the slope is between 0.17 and 0.53 g/sec, and the chance that the slope is zero or negative is 0.028.

A picture of the probability distribution may also be included.

Bayesian inferences are constructed as "inverse probabilities," according to a theorem derived by Rev. Thomas Bayes (1702–1761). The theorem demonstrates how probabilities for parameter values judged prior to examining the data (D) are converted logically to probabilities for parameter values judged after seeing the data. In its simplest form, suppose there is a single unknown parameter, θ. What is known *a priori* about θ is summarized by a *prior* probability function, $f(\theta)$. For each θ, one determines the probability of obtaining D; this is represented by $\Pr\{D|\theta\}$, the *likelihood* function discussed in the previous section. Bayes' Theorem provides the *posterior* probability function of θ from the data as

BAYES' THEOREM

The posterior *probablity θ, given the data, is proportional to the* ***product*** *of the* prior *probablity of θ and the* likelihood *of θ given the data.*

The Bucket Problem

Reconsider the bucket with its 12 red and white marbles, Section 21.7.3. Suppose you know that the filler of the bucket rolled a pair of dice and put in as many red marbles as appeared on the two faces showing. If so, there is no chance that the bucket would contain 0 or 1 red marble. The chance it contained 2 is the chance that both dice showed 1, which is $1/36 = 0.0278$. It contained 3 red marbles if the dice show $(1, 2)$ or $(2, 1)$; the chance is $2/36 = 0.0556$. Continuing, you find that the chance the bucket contained θ red marbles is $f(\theta) = (6 - |\theta - 7|)/36$ for $\theta = 2, 3, \ldots, 12$.

Display 21.12 showed the probabilities, $Pr\{D|q\}$, for any outcome of the draw and for all possible numbers of red marbles. So, suppose the bucket is mixed, five marbles are drawn, and only one of the five is red. Then Display 21.14 shows the steps required to apply Bayes' Theorem. The resulting posterior probability function appears in Display 21.15. Notice there is no chance that the bucket began with fewer than 2 or more than 8 red marbles. The most probable value is 4 reds, but 3 and 5 are quite probable as well. The mean of the distribution is 4.2 red marbles, and its standard deviation is 2.1 red marbles.

DISPLAY 21.14 Bayes' Theorem applied to the Bucket Problem when one red marble appears in five draws. The bucket is composed by the toss of a pair of dice.

1. Write down the possible values for the unknown parameter.

2. Specify probabilities for each value prior to viewing the data.

3. Gather the data, D: One red marble in five draws.

4. Determine the probability of the data for each parameter value.

5. Calculate the products of the prior and likelihood values.

	Prior	**Likelihood**		**Posterior**
θ	$f(\theta)$	$Pr\{1\ red \mid \theta\}$	*Product*	$Pr\{\theta \mid 1\ red\}$
0	0	0	0	0
1	0	0.4167	0	0
2	0.0278	0.5303	0.0147	0.1230
3	0.0556	0.4773	0.0265	0.2213
4	0.0833	0.3535	0.0295	0.2459
5	0.1111	0.2110	0.0246	0.2049
6	0.1389	0.1136	0.0158	0.1317
7	0.1667	0.0442	0.0074	0.0615
8	0.1389	0.0101	0.0014	0.0117
9	0.1111	0	0	0
10	0.0833	0	0	0
11	0.0556	0	0	0
12	0.0278	0	0	0

Sum = 0.1198

6. Add the products.

7. Divide each product by the sum, so the posterior probabilities add to 1.

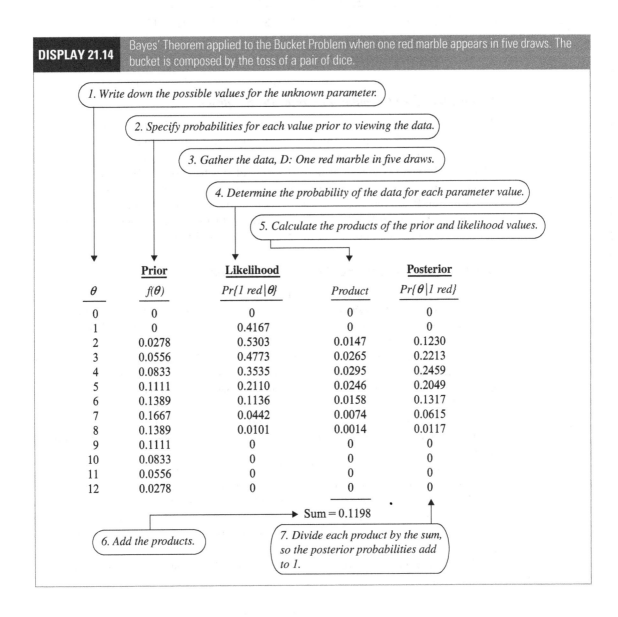

DISPLAY 21.15 Bayes' posterior distribution of the number of red marbles, from a die-throwing prior and one red marble in five draws

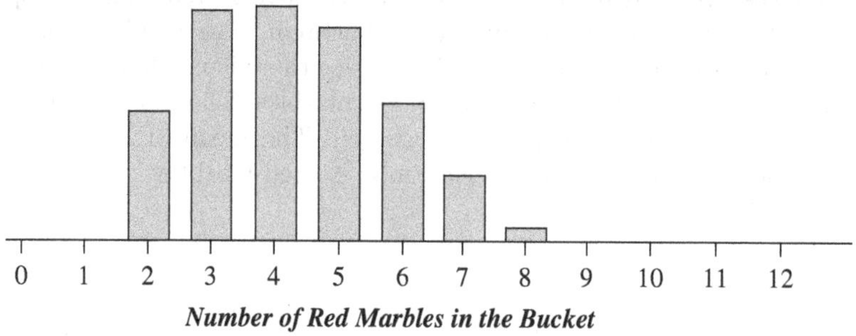

Interpretation of Bayesian Posterior Probabilities

The bucket example was unusual because the method of filling the bucket—rolling a pair of dice—was known. Without such knowledge, it may seem impossible to assign prior probabilities. This has prevented Bayesian inference from achieving more widespread usage. One way around the difficulty is to allow $f(\theta)$ to represent one's belief about θ, expressed as a probability distribution. There are good aspects to this solution: Researchers can incorporate legitimate prior information they possess. But the posterior probabilities have no interpretation as frequencies of occurrence and must also be interpreted as measures of personal belief.

Sex Role Stereotypes and Personnel Decisions

As a second example, reconsider the experiment where male bank supervisors decided promotions based on files with male and female names, Section 19.1.1. Of 24 files with male names, 21 were promoted, whereas only 14 of 24 with female names were promoted. In Chapter 19, this example was used to illustrate Fisher's exact test. No confidence interval for the odds ratio (the odds of promotion of a man relative to those of a woman) was offered because the normal distribution method fails to give a reasonable approximation to the sampling distribution.

A Bayesian solution to inference about the odds ratio requires a prior distribution and the likelihood. The mechanics that follow are offered without derivation; the reader may wish to skip to Display 21.16, which shows the posterior distribution for the odds ratio.

With no real prior information, the common choice is to use an "ignorance" prior for $\log(\omega)$ which assigns the same amount of probability to all intervals of equal length. That is, $f(\omega) = 1/\omega$, for $\omega > 0$. This is not a proper prior, in that it adds to infinity rather than 1. Nevertheless, one may think of this as a proper prior that is nearly flat throughout the range where there is any likelihood of the parameter.

The likelihood function is an example of the "noncentral hypergeometric" distribution:

$$\Pr\{21 \text{ promotions to males, given there were 35 promotions in all } |\omega\}$$
$$= C_{24,21}C_{24,11}\omega^{21} / \sum_{k=11}^{24} C_{24,k}C_{24,35-k}\omega^{k}$$

and Display 21.16 shows the product of prior and likelihood.

DISPLAY 21.16 Bayes' posterior distribution for the odds of promotion of a male applicant relative to the odds of promotion of a female applicant

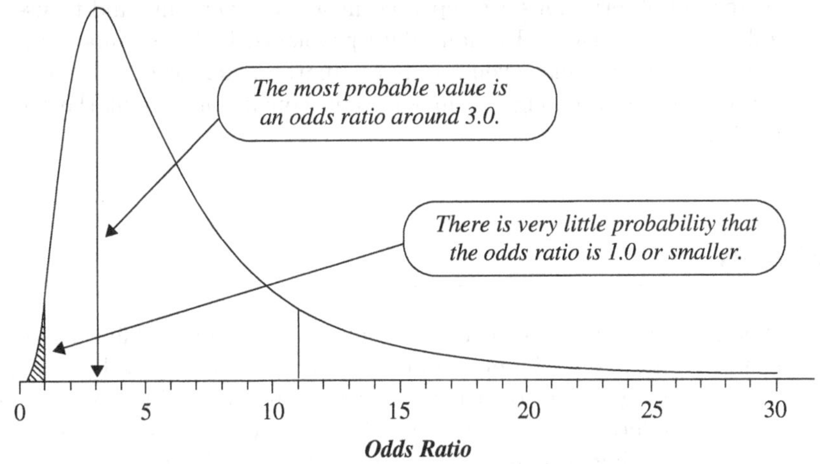
The most probable value is an odds ratio around 3.0.
There is very little probability that the odds ratio is 1.0 or smaller.
0
5
10
15
20
25
30
Odds Ratio

The posterior distribution conveys all the information about the odds ratio. The posterior probability that the odds ratio is 1.0 or smaller is the shaded area, as a fraction of the total area, approximately 0.015. The most probable odds ratio is around 3.0, but values as high as 5–10 are not at all improbable.

21.7.5 Generalized Estimating Equations

Generalized estimating equations (GEEs) offer a useful approach for regression analysis of repeated measures or any response variables that may be dependent within identifiable clusters. The approach can be used for inference about ordinary linear models and for generalized linear models, and it captures the within-subject or within-group dependencies by imposing a correlation structure.

For binomial repeated measures, a researcher using a GEE module in a statistical computer program specifies the logistic regression model in the usual way, specifies the name of the variable that indicates *subject* (or *group* identification variable, so that observations with the same value of the group indicator variable are thought to be correlated with each other), and selects a model for the within-subject (or within-group) correlation structure. Two useful choices for correlation structure are "exchangeable" and "AR(1)" models. In the exchangeable model, the observations for a given subject are taken to be correlated with each other and the correlation of any two of them (after accounting for the effects of the explanatory variables) is the same. In the AR(1) (autoregressive of lag 1) model, the correlation of observations within a subject is thought to be induced by an AR(1) time series (see Section 15.3). In this model, observations taken closer together in time are thought to be more highly correlated with each other than observations taken farther apart in time. In both of these models there is a single additional parameter—a correlation coefficient or a first serial correlation coefficient—which is assumed to be the same for all subjects or clusters.

A good starting point for modeling correlation is to use the AR(1) model if the correlated observations are repeated measures over time and to use the exchangeable model otherwise. In many data problems, it doesn't matter too much which model is used because even an approximate model for correlation structure solves the problem well enough to make useful conclusions about the logistic regression model.

21.8 SUMMARY

Binomial logistic regression is a special type of generalized linear model in which the response variable is binomial and the logit of the associated probability is a linear function of regression coefficients. The case in which all the binomial denominators are 1 was referred to as binary logistic regression and was discussed in Chapter 20.

In analyzing proportions, one should begin, if possible, with scatterplots of the sample logits versus explanatory variables. The model-building stage consists of checking for extra-binomial variation and then finding an appropriate set of explanatory variables. The main inferential statements are based on Wald's z-test or the associated confidence interval for a single coefficient, or on the drop-in-deviance chi-squared test for a hypothesis about one or several coefficients. If extra-binomial variation is suspected, these tests are replaced by t-tests and F-tests, respectively.

Krunnit Islands Data

The analysis comes into focus with a plot of empirical logits versus the log of island area. It is evident from Display 21.2 that the log odds of extinction are linear in the log of island area. This can be confirmed by fitting the model and observing that the deviance statistic is not large (the p-value from the deviance goodness-of-fit test is 0.74), and also by testing for the significance of a squared term in log island area. The remainder of the analysis only requires that the inferential statement about the association between island area and log odds of extinction be worded on the original scale.

Moth Coloration Data

A plot of the empirical logits versus distance from Liverpool, with different codes for the light and dark moths, revealed initial evidence of an interactive effect—particularly that the odds of light morph removal decrease with increasing distance from Liverpool while the odds of dark morph removal increase with increasing distance from Liverpool. Based on the scatterplot (Display 21.4), the linearity of the logistic regression on distance is questionable, particularly since the logits drop rather dramatically at Loggerheads. There seems to be good reason to believe that distance from Liverpool is not entirely adequate as a surrogate for darkness of trees. Two avenues for dealing with this problem were suggested. One involved using quasi-likelihood analysis, which incorporated the model inadequacy into a dispersion parameter. The other, more appealing model, however, included location as a factor with seven levels (rather than relying on their distance from Liverpool as a single numerical explanatory variable) but retained the product of the morph indicator variable and distance so that a single parameter represented the interactive effect of interest. This model fit well and maximum likelihood analysis was used to draw inferences about the effect.

21.9 EXERCISES

Conceptual Exercises

1. **Krunnit Islands Extinctions.** Logistic regression analysis assumes that, after the effects of explanatory variables have been accounted for, the responses are independent of each other. Why might this assumption be violated in these data?

2. **Moth Coloration.** (a) Why might one suspect extra-binomial variation in this problem? (b) Why does the question of interest pertain to an interaction effect? (c) How (if at all) can the response variable be redefined so that the question of interest is about a "main effect" rather than an interaction?

3. Which of the following are not *counted proportions*? (a) The proportion of cells with chromosome aberrations, out of 100 examined. (b) The proportion of body fat lost after a special diet. (c) The proportion of 10 grams of offered food eaten by a quail. (d) The proportion of a mouse's four limbs that have malformations.

4. Suppose that Y has a binomial(m, π) distribution. (a) Is the variance of the binomial count Y an increasing or decreasing function of m? (b) Is the variance of the binomial proportion Y/m an increasing or decreasing function of m?

5. To confirm the appropriateness of the logistic regression model $\text{logit}(\pi) = \beta_0 + \beta_1 x$, it is sometimes useful to fit $\text{logit}(\pi) = \beta_0 + \beta_1 x + \beta_2 x^2$ and test whether β_2 is zero. (a) Does the reliability of this test depend on whether the denominators are large? (b) Is the test more relevant when the denominators are small?

6. What is the quasi-likelihood approach for extending the logistic regression model to account for a case in which more variation appears than was predicted by the binomial distribution?

7. **Malformed Forelimbs.** In a randomized experiment, mouse fetuses were injected with 500, 750, or 1,000 mg/kg of acetazolamide on day 9 of gestation. (Data from L. B. Holmes, H. Kawanishi, and A. Munoz, "Acetazolamide: Maternal Toxicity, Pattern of Malformations, and Litter Effect," *Teratology* 37 (1988): 335–42.) On day 18 of gestation, the fetus was removed and examined for malformations in the front legs. Display 21.17 shows the results. (a) Suppose that the response

variable for each mouse fetus is the number of forelimbs with malformations. Explain how logistic regression can be used to study the relationship between malformation probability and dose. (b) What is the sample size? (c) Is it necessary to worry about extra-binomial variation? (d) Is it appropriate to use the deviance goodness-of-fit test for model checking? (e) Is it appropriate to obtain a confidence interval for the effect of dose? (f) How would the resulting confidence interval be used in a sentence summarizing the statistical analysis?

DISPLAY 21.17 Dose of acetazolamide and number of malformed forelimbs in mouse fetuses

Dose (mg/kg)	Number of malformed forelimbs			Sample size
	0	1	2	
500	137	40	15	192
750	81	68	42	191
1,000	87	49	69	205

8. If your computer software always produces binary response deviances, how can you determine the lack-of-fit chi-square for the moth data? (See Section 21.7.1.)

Computational Exercises

9. Moth Coloration. For the moth coloration data in Section 21.1.1, consider the response count to be the number of light moths removed and the binomial denominator to be the total number of moths removed (light and dark) at each location. (a) Plot the logit of the proportion of light moths removed versus distance from Liverpool. (b) Fit the logistic regression model with distance as the explanatory variable; then report the estimates and standard errors. (c) Compute the deviance goodness-of-fit test statistic, and obtain a p-value for testing the adequacy of the model.

10. Death Penalty and Race of Victim. Reconsider the data of Display 19.2 involving death penalty and race of victim. Reanalyze these data using logistic regression. The response variable is the number of convicted murderers in each category who receive the death sentence, out of the m convicted murderers in that category. (a) Plot the logits of the observed proportions versus the level of aggravation. The logit, however, is undefined for the rows where the proportion is 0 or 1, so compute the empirical logit $= \log[(y + 0.5)/(m - y + 0.5)]$ and plot this versus aggravation level, using different plotting symbols to distinguish proportions based on white and black victims. (b) Fit the logistic regression of death sentence proportions on aggravation level and an indicator variable for race of victim. (c) Report the p-value from the deviance goodness-of-fit test for this fit. (d) Test whether the coefficient of the indicator variable for race is equal to 0, using the Wald's test. (e) Construct a confidence interval for the same coefficient, and interpret it in a sentence about the odds of death sentence for white-victim murderers relative to black-victim murderers, accounting for aggravation level of the crime.

11. Death Penalty and Race of Victim. Fit a logistic regression model as in Exercise 10, but treat aggravation level as a factor. Include the products of the race variable with the aggravation level indicators to model interaction, and determine the drop in deviance for including them. This constitutes a test of the equal odds ratio assumption, which could be used to check the assumption for the Mantel–Haenszel test. What can you conclude about the assumption?

12. Vitamin C and Colds. Reconsider the data in Display 18.2 from the randomized experiment on 818 volunteers. Let Y_i represent the number of individuals with colds in group i, out of m_i. Let X_i be an indicator variable that takes on the value 1 for the placebo group and 0 for the vitamin C group. Then the observed values for the two groups are:

Group (i)	Individuals with Colds (Y)	Group size (m)	Placebo indicator (X)
1	335	411	1
2	302	407	0

Consider the logistic regression model in which Y_i is binomial(π_i, m_i) and logit$(\pi_i) = \beta_0 + \beta_1 X_i$. (a) Using a computer package, obtain an estimate of and a confidence interval for β_1. (b) Interpret the results in terms of the odds of cold for the placebo group relative to the odds of a cold for the vitamin C group. (c) How does the answer to part (b) compare to the answer in Display 18.9?

13. Vitamin C. Between December 1972 and February 1973, a large number of volunteers participated in a randomized experiment to assess the effect of large doses of vitamin C on the incidence of colds. (Data from T. W. Anderson, G. Suranyi, and G. H. Beaton, "The Effect on Winter Illness of Large Doses of Vitamin C," *Canadian Medical Association Journal* 111 (1974): 31–36.) The subjects were given tablets to take daily, but neither the subjects nor the doctors who evaluated them were aware of the dose of vitamin C contained in the tablets. Shown in Display 21.18 are the proportion of subjects in each of the four dose categories who did not report any illnesses during the study period.

DISPLAY 21.18 Vitamin C and colds

Daily dose of vitamin C (g)	Number of subjects	Number with no illnesses	Proportion with no illnesses
0	1,158	267	0.231
0.25	331	74	0.224
1	552	130	0.236
2	308	65	0.211

(a) For each of the four dose groups, calculate the logit of the estimated proportion. Plot the logit versus the dose of vitamin C. (b) Fit the logistic regression model, logit$(\pi) = \beta_0 + \beta_1$ *dose*. Report the estimated coefficients and their standard errors. Report the p-value from the deviance goodness-of-fit test. Report the p-value for a Wald's test that β_1 is 0. Report the p-value for a drop-in-deviance test for the hypothesis that β_1 is 0. (c) What can be concluded about the adequacy of the binomial logistic regression model? What evidence is there that the odds of a cold are associated with the dose of vitamin C?

14. Spock Conspiracy Trial. Reconsider the proportions of women on venires in the Boston U.S. district courts (case study 5.2). Analyze the data by treating the number of women out of 30 people on a venire as a binomial response. (a) Do the odds of a female on a venire differ for the different judges? Answer this with a drop-in-deviance chi-square test, comparing the full model with judge as a factor to the reduced model with only an intercept. (b) Do judges A–F differ in their probabilities of selecting females on the venire? Answer this with a drop-in-deviance chi-square test by comparing the full model with judge as a factor to the reduced model that has an intercept and an indicator variable for Spock's judge. (c) How different are the odds of a woman on Spock's judge's venires from the odds on the other judges? Answer this by interpreting the coefficients in the binomial logistic regression model with an intercept and an indicator variable for Spock's judge.

Data Problems

15. Belief Accessibility. Increasingly, politicians look to public opinion surveys to shape their public stances. Does this represent the ultimate in democracy? Or are seemingly scientific polls being rigged by the manner of questioning? Psychologists believe that opinions—expressed as answers to questions—are usually generated at the time the question is asked. Answers are based on a quick sampling of relevant beliefs held by the subject, rather than a systematic canvas of all such beliefs. Furthermore, this sampling of beliefs tends to overrepresent whatever beliefs happen to be most accessible at the time the question is asked. This aspect of delivering opinions can be abused by the pollster. Here, for example, is one sequence of questions: (1) "Do you believe the Bill of Rights protects personal freedom?" (2) "Are you in favor of a ban on handguns?" Here is another: (1) "Do you think something should be done to reduce violent crime?" (2) "Are you in favor of a ban on handguns?" The proportion of yes answers to question 2 may be quite different depending on which question 1 is asked first.

To study the effect of *context questions* prior to a *target question*, researchers conducted a poll involving 1,054 subjects selected randomly from the Chicago phone directory. To include possibly unlisted phones, selected numbers were randomly altered in the last position. (Data from R. Tourangeau, K. A. Rasinski, N. Bradburn, and R. D'Andrade, "Belief Accessibility and Context Effects in Attitude Measurement," *Journal of Experimental Social Psychology* 25 (1989): 401–21.) The data in Display 21.19 show the responses to one of the questions asked concerning continuing U.S. aid to the Nicaraguan Contra rebels. Eight different versions of the interview were given, representing all possible combinations of three factors at each of two levels. The experimental factors were CONTEXT, MODE, and LEVEL. CONTEXT refers to the type of context questions preceding the question about Nicaraguan aid. Some subjects received a context question about Vietnam, designed to elicit reticence about having the United States become involved in another foreign war in a third-world country. The other context question was about Cuba, designed to elicit anti-communist sentiments. MODE refers to whether the target question immediately followed the context question or whether there were other questions scattered in between. LEVEL refers to two versions of the context question. In the HIGH LEVEL the question was worded to elicit a higher level of agreement than in the LOW LEVEL wording.

Analyze these data to answer the following questions of interest: (a) Does the proportion of favorable responses to the target question depend on the wording (LEVEL) of the question? If so, by how much? (b) Does the proportion depend on the context question? If so, by how much? (c) Does the proportion depend on the context question to different extents according to whether the target and context questions are scattered? (*Hint:* Use indicator explanatory variables for the factors.)

16. Aflatoxicol and Liver Tumors in Trout. An experiment at the Marine/Freshwater Biomedical Sciences Center at Oregon State University investigated the carcinogenic effects of aflatoxicol, a metabolite of Aflatoxin B1, which is a toxic by-product produced by a mold that infects cottonseed meal, peanuts, and grains. Twenty tanks of rainbow trout embryos were exposed to one of five doses

DISPLAY 21.19 Opinion polls and belief accessibility: CONTEXT refers to the context of the question preceding the target question about U.S. aid to the Nicaraguan Contra rebels; MODE is "scattered" if the target question was not asked directly after the context question; and LEVEL refers to the wording of the question ("high" designed to elicit a higher favorable response)

GROUP:	1	2	3	4	5	6	7	8
CONTEXT:	Vietnam	Cuba	Vietnam	Cuba	Vietnam	Cuba	Vietnam	Cuba
MODE:	scattered	scattered	scattered	scattered	not	not	not	not
LEVEL:	high	high	low	low	high	high	low	low
NUMBER (m):	132	132	132	131	132	131	132	132
PERCENTAGE IN FAVOR OF CONTRA AID:	20.5	31.1	28.0	38.9	34.1	48.9	23.5	45.5

of Aflatoxicol for one hour. The data in Display 21.20 (from George Bailey and Jerry Hendricks) represent the numbers of fish in each tank and the numbers of these that had liver tumors after one year. Describe the relationship between dose of Aflatoxicol and odds of liver tumor. It is also of interest to determine the dose at which 50% of the fish will get liver tumors. (*Note: Tank effects* are to be expected, meaning that tanks given the same dose may have slightly different π's. Thus, one should suspect extra-binomial variation.)

DISPLAY 21.20 Aflatoxicol and liver tumors in trout (four tanks at each dose)

Dose (ppm)	Number of trout with liver tumors (Y)/Number in tank (m)			
0.010	9/87	5/86	2/89	9/85
0.025	30/86	41/86	27/86	34/88
0.050	54/89	53/86	64/90	55/88
0.100	71/88	73/89	65/88	72/90
0.250	66/86	75/82	72/81	73/89

17. Effect of Stress During Conception on Odds of a Male Birth. The probability of a male birth in humans is about 0.51. It has previously been noticed that lower proportions of male births are observed when offspring are conceived at times of exposure to smog, floods, or earthquakes. Danish researchers hypothesized that sources of stress associated with severe life events may also have some bearing on the sex ratio. To investigate this theory they obtained the sexes of all 3,072 children who were born in Denmark between January 1, 1980 and December 31, 1992 to women who experienced the following kinds of severe life events in the year of the birth or the year prior to the birth: death or admission to hospital for cancer or heart attack of their partner or of their other children. They also obtained sexes on a sample of 20,337 births for mothers who did not experience these life stress episodes. Shown in Display 21.21 are the percentages of boys among the births, grouped according to when the severe life event took place. Notice that for one group the exposure is listed as taking place during the first trimester of pregnancy. The rationale for this is that the stress associated with the cancer or heart attack of a family member may well have started before the recorded time of death or hospital admission. Analyze the data to investigate the researchers' hypothesis. Write a summary of statistical findings. (Data from D. Hansen et al., "Severe Periconceptional Life Events and the Sex Ratio in Offspring: Follow Up Study Based on Five National Registers," *British Medical Journal,* 319 (1999): 548–49.)

18. HIV and Circumcision. Researchers in Kenya identified a cohort of more than 1,000 prostitutes who were known to be a major reservoir of sexually transmitted diseases in 1985. It was determined that more than 85% of them were infected with human immunodeficiency virus (HIV)

DISPLAY 21.21 Percentages of boys born to mothers in a control group and to mothers exposed to severe life events, at various times in relation to the birth

Group	Time of stress event	Number of births	% boys
Control	(none)	20,337	51.2
Exposed	13–16 mo. prior	71	52.1
Exposed	7–12 mo. prior	789	49.6
Exposed	0–6 mo. prior	1,922	48.9
Exposed	1st trimester	290	46.0

in February 1986. The researchers then identified men who acquired a sexually transmitted disease from this group of women after the men sought treatment at a free clinic. Display 21.22 shows the subset of those men who did not test positive for the HIV on their first visit and who agreed to participate in the study. The men are categorized according to whether they later tested positive for HIV during the study period, whether they had one or multiple sexual contacts with the prostitutes, and whether they were circumcised. Describe how the odds of testing positive are associated with number of contacts and with whether the male was circumcised. (Data from D. W. Cameron et al., "Female to Male Transmission of Human Immunodeficiency Virus Type 1: Risk Factors for Seroconversion in Men," *The Lancet* (1989): 403–07.)

DISPLAY 21.22 Number of Kenyan men who tested positive for HIV, categorized according to two possible risk factors

	Single contact with prostitutes		**Multiple contact with prostitutes**	
	Circumcised	Uncircumcised	Circumcised	Uncircumcised
Tested positive for HIV	1	5	5	13
Number of men	46	27	168	52

19. Meta-Analysis of Breast Cancer and Lactation Studies. Meta-analysis refers to the analysis of analyses. When the main results of studies can be cast into 2 × 2 tables of counts, it is natural to combine individual odds ratios with a logistic regression model that includes a factor to account for different odds from the different studies. In addition, the odds ratio itself might differ slightly among studies because of different effects on different populations or different research techniques. One approach for dealing with this is to suppose an underlying common odds ratio and to model between-study variability as extra-binomial variation. Display 21.23 shows the results of 10 separate case–control studies on the association of breast cancer and whether a woman had breast fed children. How much greater are the odds of breast cancer for those who did not breast feed than for those who did breast feed? (Data gathered from various sources by Karolyn Kolassa as part of a Master's project, Oregon State University.)

20. Clever Hans Effect. Exercise 14.19 described a randomized block experiment conducted on psychology students to examine the effects of expectations—induced in the experiment by falsely telling some of the students that their rats were bred to be good maze learners—on the students' success in training the rats. The number of successful maze runs, out of 50, was recorded for each student on each of five consecutive days. Data file ex2120 repeats the data but in a revised format, with the responses on the five days listed on separate rows in the variable *Success*. There are 60 rows corresponding to five responses for each of the 12 students in the study. Also listed are the students reported expectation of success (*PriorExp*) prior to the randomization and commencement

DISPLAY 21.23 Counts from 10 case–control studies investigating the association between cancer and whether women had breast fed children (lactated); first 4 of 20 rows

Study		**Breast cancer**	**No breast cancer**
1	Did not lactate	107	226
	Lactated	352	865
2	Did not lactate	244	489
	Lactated	574	1,059

of the study, a student ID (*Student*), *Treatment* (*bright* if students were told their rats were bright and *dull* if not), and *Day* of the experiment. Analyze the data to judge the effect of the treatment on the students' success. (It isn't possible to simultaneously estimate a *Treatment* Effect and a *Student* Effect. Try including *PriorExp* instead of *Student* as a way of modeling the possible student effect. To more specifically deal with dependence of observations on the same student, use GEE, as discussed in Section 21.7.5. Be alert for possible overdispersion.)

21. Predicting Quality of Spring Desert Wildflower Display (Ordered Categorical Response). Review the description of the desert wildflower data problem in Exercise 12.21. The response for each desert–year combination is the subjectively rated quality of the spring wildflower display, with ordered response categories *poor, fair, good, great*, and *spectacular*. Use a computer routine for regression on ordered categorical responses (such as proportional odds logistic regression with polr in R or proc logistic in SAS) to fit a regression model with wildflower display quality as the ordered categorical response and total rainfall in the months September to March as an explanatory variable.

22. Environmental Voting. The data file ex0126 includes the number of pro-environment and anti-environment votes cast by each member of the U.S. House of Representatives in 2005, 2006, and 2007 (according to the League of Conservation Voters). Describe the disparity between Republican and Democratic representatives on the probability of casting a pro-environment vote after accounting for *Year*. Ignore representatives from parties other than *D* and *R*.

Answers to Conceptual Exercises

1. Neighboring islands are likely to be more similar in the types of species present and in the likelihood of extinction for each species, than islands farther apart. Another possibility is that extinctions of (migratory) shorebirds can occur in another part of the world and simultaneously affect all or several of the Krunnit Islands populations.

2. (a) Sites the same distance from Liverpool may have different probabilities of removal for dark and light morphs because the blackening of the trees may depend on more than simple distance from Liverpool. (b) The question is not whether the probability of removal depends on the covariate distance from Liverpool or on the factor morph, but whether the probability of removal depends on distance from Liverpool to a different extent for light morphs than for dark morphs. (c) Let m be the number of moths removed at a location. Let Y be the number of these that are dark morphs. How does the proportion of removed moths that are dark morphs depend on distance from Liverpool?

3. Choices (b) and (c).

4. (a) Increasing. (b) Decreasing.

5. (a) Partly. The Wald's test and the drop-in-deviance test both can be used if either the sample size is large or the binomial denominators are large. (b) The test is useful as an informal device for assessing the model in each case. It may be more relevant when the denominators are small, however, since few alternatives are available for model checking in this case.

6. The assumed variance is $\psi m_i \pi_i (1 - \pi_i)$. The dispersion parameter ψ is introduced to allow for variance greater than (or possibly less than) that expected from a binomial response.

7. (a) Suppose that Y_i is binomial$(2, \pi_i)$ with logit$(\pi_i) = \beta_0 + \beta_1 dose_i$ for $i = 1, \ldots, 588$. Explore the adequacy of this model. If it fits, make inferences about β_1. (b) 588. (c) Probably not, since the denominators are all so small. (d) No, the denominators are too small. (e) Yes. (f) "The odds of a malformation are estimated to increase by (lower limit) to (upper limit) for each 100 mg/kg of acetazolamide (95% confidence interval)." (*Note:* If L and U are the lower and upper endpoints of a confidence interval for β_1, then the interval in this sentence is exp$(100L)$ and exp$(100U)$.)

8. Compare the deviance from the model of interest to the deviance from the full model, with a separate mean for each morph at each location.

CHAPTER 22

Log-Linear Regression for Poisson Counts

The number of traffic accidents at an intersection over a year, the number of shrimp caught in a 1-kilometer tow of a net, the number of outbreaks of an infectious disease in a county during a year, and the number of flaws on a computer chip are examples of responses that consist of integer counts. Such counts differ from binomial counts of successes in a fixed number of trials, where each trial contributes either +1 or 0 to the total. They are counts of the occurrences of some event over time or space, without a definite upper bound. The *Poisson probability distribution* is often useful for describing the population distribution of these types of counts. A related regression model—Poisson log-linear regression—specifies that the logarithms of the means of Poisson responses are linear in regression coefficients.

Poisson log-linear regression, like binomial logistic regression, is a generalized linear model. The tools for analysis are similar to those described in Chapters 20 and 21. Furthermore, the tools extend to account for responses that are Poisson-like but possess excess variation, through addition of a dispersion parameter and use of quasi-likelihood analysis.

22.1 CASE STUDIES

22.1.1 Age and Mating Success of Male Elephants—An Observational Study

Although male elephants are capable of reproducing by 14 to 17 years of age, young adult males are usually unsuccessful in competing with their larger elders for the attention of receptive females. Since male elephants continue to grow throughout their lifetimes, and since larger males tend to be more successful at mating, the males most likely to pass their genes to future generations are those whose characteristics enable them to live long lives. Joyce Poole studied a population of African elephants in Amboseli National Park, Kenya, for 8 years. Display 22.1 shows the number of successful matings and ages (at the study's beginning) of 41 male elephants. (Data from J. H. Poole, "Mate Guarding, Reproductive Success and Female Choice in African Elephants," *Animal Behavior* 37 (1989): 842–49.) What is the relationship between mating success and age? Do males have diminished success after reaching some optimal age?

DISPLAY 22.1 Age at beginning of study and number of successful matings for 41 African elephants

Age	Matings	Age	Matings	Age	Matings
27	0	33	3	39	1
28	1	33	3	41	3
28	1	33	3	42	4
28	1	33	2	43	0
28	3	34	1	43	2
29	0	34	1	43	3
29	0	34	2	43	4
29	0	34	3	43	9
29	2	36	5	44	3
29	2	36	6	45	5
29	2	37	1	47	7
30	1	37	1	48	2
32	2	37	6	52	9
33	4	38	2		

Statistical Conclusion

The average number of successful matings increased with age, for male elephants between 25 and 50 years old (Display 22.2). The data provide no evidence that the mean reached some maximum value at any age less than 50 (the one-sided p-value for significance of a quadratic term in the log-linear regression is 0.33). The estimated mean number of successful matings (in 8 years) for 27-year-old males was 1.31, and the mean increased by a factor of 2.00 for each 10-year increase in age up to about 50 years (95% confidence interval for the multiplicative factor: 1.52 to 2.60).

DISPLAY 22.2 Counts of successful matings versus age for 41 male African elephants (jittered to avoid overlapping points): the curve is the estimated log-linear regression model for the mean count

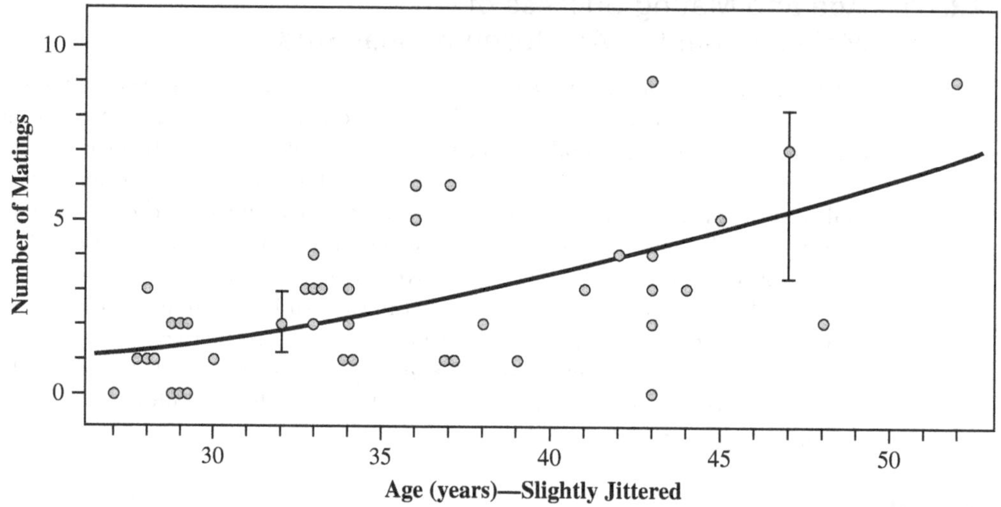

Scope of Inference

There may be some bias in these results, due to measurement error in estimating the age of each elephant and lack of independence of the counts for different males. Bias would also result if successful matings were not correctly attributed. These problems have only minor effects on the relatively informal conclusions obtained. Attempting to apply inferences from these data to a wider elephant population would require careful attention to the method of sampling used.

22.1.2 Characteristics Associated with Salamander Habitat

The Del Norte Salamander (*plethodon elongates*) is a small (5–7 cm) salamander found among rock rubble, rock outcrops, and moss-covered talus in a narrow range of northwest California. To study the habitat characteristics of the species and particularly the tendency of these salamanders to reside in dwindling old-growth forests, researchers selected 47 sites from plausible salamander habitat in national forest and parkland. Randomly chosen grid points were searched for the presence of a site with suitable rocky habitat. At each suitable site, a 7 meter by 7 meter search area was examined for the number of salamanders it contained. Shown in Display 22.3 are the counts of salamanders at the sites, along with the percentage of forest canopy cover and the age of the forest in years. How is the size of the salamander count related to canopy cover and forest age? (Data from H. H. Welsh and A. J. Lind, *Journal of Herpetology* 29(2) 1995: 198–210.)

DISPLAY 22.3 Salamanders found in a 49 m^2 area, percentage of canopy cover, and forest age in 47 northern California sites

Site	Salamanders	% cover	Forest age	Site	Salamanders	% cover	Forest age
1	13	85	316	25	1	46	30
2	11	86	88	26	1	80	215
3	11	90	548	27	1	86	586
4	9	88	64	28	1	88	105
5	8	89	43	29	1	92	210
6	7	83	368	30	0	0	0
7	6	83	200	31	0	1	4
8	6	91	71	32	0	3	3
9	5	88	42	33	0	5	2
10	5	90	551	34	0	8	10
11	4	87	675	35	0	9	8
12	3	83	217	36	0	11	6
13	3	87	212	37	0	14	49
14	3	89	398	38	0	17	29
15	3	92	357	39	0	24	57
16	3	93	478	40	0	44	59
17	2	2	5	41	0	52	78
18	2	87	30	42	0	77	50
19	2	93	551	43	0	78	320
20	1	7	3	44	0	80	411
21	1	16	15	45	0	86	133
22	1	19	31	46	0	89	60
23	1	29	10	47	0	91	187
24	1	34	49				

Statistical Conclusion

After accounting for canopy cover, there is no evidence that the mean number of salamanders depends on forest age (p-value $= 0.78$, from $F = 0.53$, with 6 and 35 d.f., from a quasi-likelihood fit to a log-linear model). Two distinct canopy cover conditions were evident: closed canopy with percentage >70% and open canopy with percentage <70%. The mean number of salamanders followed different curved relationships under the two conditions (p-value $= 0.0185$ from a t-test). A graph of the resulting fit appears in Display 22.4.

Scope of Inference

Inference to the totality of suitable sites should be possible, because of the manner of site selection. Causal inference is not possible from this observational study. Furthermore, with this kind of study involving animal counts, one must be concerned about issues of *detectability*. The detectability of an animal is the chance it will be found and counted given that it is present at the site. If detectability is related to variables used to model the mean animal counts and if its influence is not taken

DISPLAY 22.4 Salamander numbers vs. percentage canopy cover, with Poisson regression fit

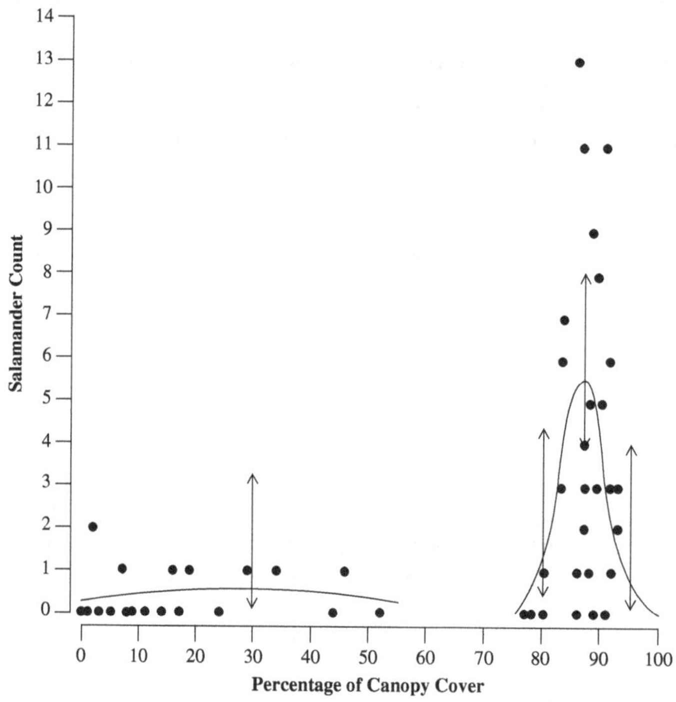

into account, the results can be quite misleading. Here, for example, more open canopy conditions could force the salamanders to seek relief from sunlight by going deeper in the rock piles where they are more difficult to find.

22.2 LOG-LINEAR REGRESSION FOR POISSON RESPONSES

22.2.1 Poisson Responses

The response variables in the two case studies are counts—the count of successful elephant matings in 8 years of observation, and the count of salamanders in a 7 m square plot. Counts of events or objects occurring in specified time intervals or in specified geographical areas differ from binomial counts.

The Poisson Probability Distribution

Siméon Denis Poisson's treatise on probability theory, "Recherches sur la probabilité des jugements en matière criminelle et en matière civile" (Research on the probability of judgements in criminal and in civil matters) (1837), examines the form that the binomial distribution takes for large numbers of trials with small probabilities of success on any given trial. The limiting form of the distribution has no upper limit to the count of successes; the probability of obtaining Y successes is given by the formula

$$\text{Probability}\{Y\} = \exp(-\mu) \times \mu^Y / Y!, \quad \text{for any } Y = 0, 1, 2, \ldots,$$

where μ is the mean. The distributions described by this mathematical rule have continued to carry Poisson's name.

Since Poisson's time, the distribution has been used to model a wide variety of situations. It is most appropriate for counts of rare events that occur at completely random points in space or time. Its general features make it a reasonable approximation to a wider class of problems involving count data where spread increases with the mean.

The Poisson distribution has several characteristic features:

1. The variance is equal to the mean.
2. The distribution tends to be skewed to the right, and this skewness is most pronounced when the mean is small.
3. Poisson distributions with larger means tend to be well-approximated by a normal distribution.

22.2.2 The Poisson Log-Linear Model

With a count response and explanatory variables, the Poisson log-linear regression model specifies that the distribution of the response is Poisson and that the log of the mean is linear in regression coefficients:

A Poisson Log-Linear Regression Model

Y is Poisson, with $\mu\{Y|X_1, X_2\} = \mu$

$$\log(\mu) = \beta_0 + \beta_1 X_1 + \beta_2 X_2.$$

This is a generalized linear model with a Poisson response distribution. The link function, which relates the response mean to the explanatory variables, is the logarithm.

A hypothetical example is illustrated in Display 22.5. The distribution of the response, as a function of a single explanatory variable X, is Poisson; and with μ representing $\mu\{Y|X\}$, the log-linear model is $\log(\mu) = -1.7 + 0.20X$. Poisson probability histograms are displayed at $X = 5, 14$, and 20. Notice that the regression is nonlinear in X, as expected, since $\mu\{Y|X\} = \exp(-1.7 + 0.20X)$. In addition, the variance is greater for distributions with larger means. The skewness of the Poisson distribution is substantial when the mean is small, but it diminishes with increasing mean.

Poisson Regression or Transformation of the Response?

Taking the logarithm of the response helps to straighten out the relationship. For the situation in Display 22.5, $\mu\{\log(Y)|X\}$ is approximately a straight line function

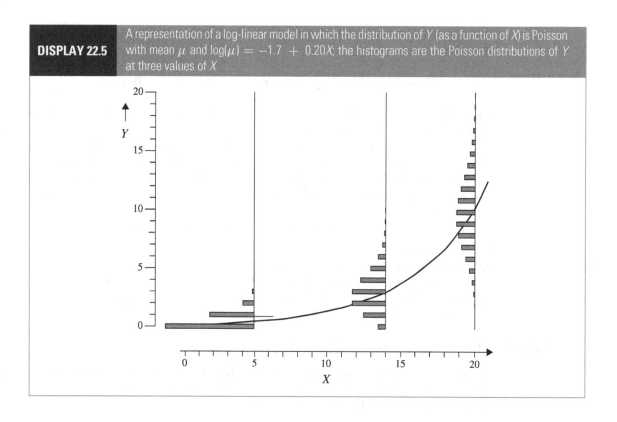

DISPLAY 22.5 A representation of a log-linear model in which the distribution of Y (as a function of X) is Poisson with mean μ and $\log(\mu) = -1.7 + 0.20X$; the histograms are the Poisson distributions of Y at three values of X

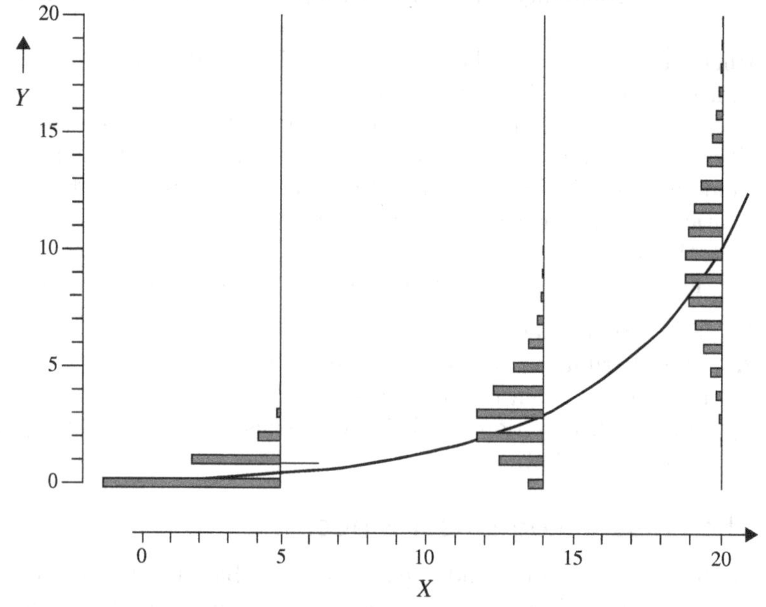

of X. On the other hand, the variance is still nonconstant after this transformation. The transformation that stabilizes the variance is the square root: $\text{Var}\{Y^{1/2}|X\}$ is approximately constant. Both transformations, therefore, are somewhat unsatisfactory. Poisson log-linear regression offers a more suitable approach and does not involve a transformation of the response.

Example—Elephant Matings

The response variable is the number of successful matings, and the explanatory variable is the elephant's age. If the Poisson log-linear model is used, the numbers of successful matings are taken to be Poisson distributed, and the mean number of counts as a function of age is specified by the log-linear model:

$$\log(\mu) = \beta_0 + \beta_1 age \qquad \text{or equivalently,} \qquad \mu = \exp(\beta_0 + \beta_1 age).$$

Thus an increase of one year in age is associated with a K-fold change in the mean number of matings, where $K = \exp(\beta_1)$. The estimated log-linear model is drawn on the scatterplot in Display 22.2. The estimate of β_1 is 0.069, so an increase in 1 year of age is estimated to be associated with a 1.07-fold change in the mean number of successful matings (a 7% increase in the mean for each additional year of age).

22.2.3 Estimation by Maximum Likelihood

Maximum likelihood fitting by a Poisson log-linear regression computer package yields a familiar display of coefficients and standard errors. A fit of a quadratic log-linear model to the elephant mating data appears in Display 22.6. The tools involved in the maximum likelihood analysis, including residuals, the deviance goodness-of-fit test, Wald's tests and confidence intervals, and the drop-in-deviance test are used in the same way as for binomial logistic regression. These are discussed in the next few sections and demonstrated in connection with the case studies.

DISPLAY 22.6 Poisson log-linear regression of number of successful matings on age and age-squared, from observations on 41 male elephants: $\log(\mu) = \beta_0 + \beta_1 age + \beta_2 age^2$

Variable	Coefficient	Standard error	z-statistic	Two-sided p-value
Constant	−2.857	3.036	0.941	0.3467
age	0.136	0.158	0.861	0.3894
age^2	−0.00086	0.00201	0.427	0.6692

Deviance = 50.83 **d.f.** = 38

22.3 MODEL ASSESSMENT

22.3.1 Scatterplot of Logged Counts Versus an Explanatory Variable

Initial exploration often involves plotting the logarithms of responses versus one or more explanatory variables, using graphical techniques such as coding and jittering. Display 22.2 shows the number of matings (not the log of the number of matings) versus elephant age. This is useful in the summary, but it does not indicate the appropriateness of the model. A plot of the log of the number of counts versus age would better illustrate the linearity on the log scale. As with empirical logits, however, some small amount like 1/2 needs to be added to the counts if some of them are zero, so that the logarithm can be computed.

22.3.2 Residuals

The issues involved in using residuals are similar to those for binomial logistic regression. The formal definition of a deviance residual—as the maximum possible value of the logarithm of the likelihood at observation i minus the value of the logarithm of the likelihood estimated from the current model—remains unchanged. However, the definition leads to the following somewhat different formula for Poisson regression:

Deviance Residual for Poisson Regression

$$Dres_i = \text{sign}(Y_i - \hat{\mu}_i)\sqrt{2\left[Y_i \times \log\left(\frac{Y_i}{\hat{\mu}_i}\right) - Y_i + \hat{\mu}_i\right]}.$$

As before, practitioners with access to computer software can disregard the unpleasant appearance of this formula. Subsequent uses of these residuals are straightforward and resemble uses of standard residuals in ordinary regression.

A Pearson residual is an observed response minus its estimated mean, divided by its estimated standard deviation. Since the mean and the variance are both μ for Poisson response variables, the formula for a Pearson residual is

Pearson Residual for Poisson Regression

$$Pres_i = \frac{Y_i - \hat{\mu}_i}{\sqrt{\hat{\mu}_i}}.$$

Typically, the analyst examines one set of residuals, not both, depending on what is available in the computer package being used. The deviance residuals are somewhat more reliable for detecting outliers, but the Pearson residuals are easier to interpret. Usually, the two sets of residuals are similar. More important differences occur when they are used for extra-sum-of-squares tests to compare models.

If the Poisson means are large, the distributions of both deviance and Pearson residuals are approximately standard normal. Thus, problems are indicated if more than 5% of residuals exceed 2 in magnitude or if one or two residuals greatly exceed 2. If the Poisson means are small (<5)—as they are in both case studies—neither set of residuals follows a normal distribution very well, so comparison to a standard normal distribution provides a poor look at lack of fit. Display 22.7 shows a plot of the deviance residuals for the salamander study. The overall pattern of large-magnitude residuals is evident.

22.3.3 The Deviance Goodness-of-Fit Test

A deviance goodness-of-fit test provides an informal assessment of the adequacy of a fitted model. The term *informal* is used here to emphasize that the p-value is sometimes unreliable. The test is not particularly good at detecting model inadequacies, and it needs to be used in conjunction with plots and tests of model terms for assessing the adequacy of a particular fit. A *large* p-value indicates either that the model is adequate or that insufficient data are available to detect inadequacies. A *small* p-value may indicate any of three things: the model for the mean is incorrect (for example, because the set of explanatory variables is insufficient); that the

DISPLAY 22.7 Deviance residuals versus fitted mean salamander counts from the fit of a log-linear regression on forest age, age squared, canopy cover, cover squared, age times cover, an indicator of canopy closure >70%, and the products of the indicator with all previous variables

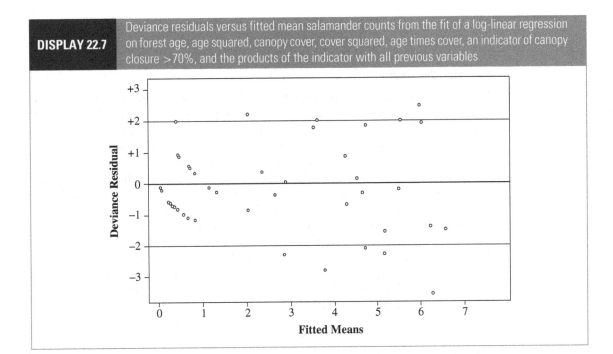

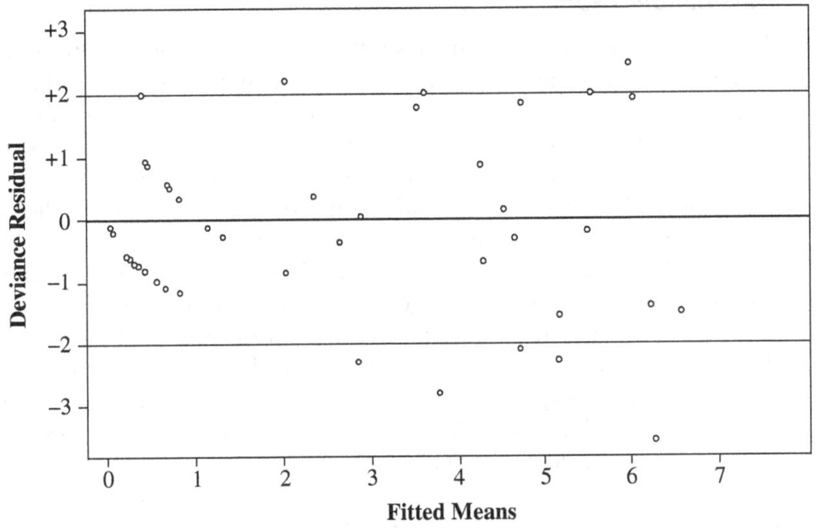

Poisson is an inadequate model for the response distribution; or that a few severely outlying observations are contaminating the data.

The deviance goodness-of-fit test is similar to the lack-of-fit F-test in normal regression. It compares the fit of a model to that of a saturated model in which each observation's mean is estimated by the observation itself. The p-value is obtained by comparing the deviance statistic to a chi-squared distribution on $n - p$ degrees of freedom, where p is the number of parameters in the model:

Deviance Statistic = Sum of Squared Deviance Residuals

$$\text{Approximate } p\text{-value} = \Pr(\chi^2_{n-p} > \text{Deviance Statistic}).$$

The chi-square approximation to the null distribution of the deviance statistic is valid when the Poisson means are large. If a substantial proportion of the cases have estimated means of less than 5, it is potentially misleading to rely on the deviance test as a tool for model assessment. Other tools, particularly testing of extra terms that represent model inadequacy, should be used instead.

Elephant Matings Example

The deviance statistic, 50.83, is provided in the output of Display 22.6 for the log-linear regression of number of matings on age and age-squared. As Display 22.2 makes clear, however, most of the estimated means are less than 5 and many are less than 2. Thus, it is not prudent to compare 50.83 to a chi-square distribution as a goodness-of-fit test for this model with these data.

Salamander Example

The objective of the salamander study is to describe the relationship of salamander numbers to forest age and canopy cover. A matrix of scatterplots, Display 22.9, suggests either a very simple situation or perhaps a very complex one. There is a substantial break in canopy cover percentage, separating closed canopy (percentage >70) from open canopy (percentage <60) conditions. The simple explanation would be that the mean salamander count depends only on this dichotomy, with no additional structure. Complex models could involve curvatures and interactions.

Initially, a rich, complex model is fit. The model for the log-mean is a saturated second-order model in forest age and canopy cover (including squares and their product), with possibly differing coefficients in open and closed canopy regimes. The residuals from this model—shown in Display 22.7—combine to give a deviance of 89.18 with 35 degrees of freedom. As a rough test of goodness of fit, one may compare this deviance to a chi-square distribution with 35 degrees of freedom. Since $\Pr(\chi^2_{35} > 89.13)$ is less than 0.0001, this would constitute substantial evidence that the Poisson log-linear model does not fit well. (This is "rough" because the adequacy of the approximation based on large Poisson means is questionable in this problem.) Furthermore, as is evident in Display 22.7, the large deviance statistic is not due to one or two outlying observations; many (10 of 47) residuals fall outside the normal −2 to +2 limits. This leads to the conclusion that variability exceeds what is predicted by the Poisson distribution.

The quasi-likelihood approach (Section 22.5) allows for such extra-Poisson variation. It is an approximate procedure, and here the approximation results from the difficulty imposed by small counts. Quasi-likelihood is, however, the most appropriate tool available, so it is used to draw inferences about individual terms in the model. An "approximate" warning is included in all inferential statements.

22.3.4 The Pearson Chi-Squared Goodness-of-Fit Test

Another goodness-of-fit test statistic with the same purpose as the deviance goodness-of-fit test statistic is the sum of squared Pearson residuals. Its p-value, too, is obtained by comparing the observed statistic to a chi-squared distribution on $n - p$ degrees of freedom. Although the deviance goodness-of-fit test is generally preferred, the Pearson test is important for historical reasons and for its current widespread use to analyze tables of counts.

22.4 INFERENCES ABOUT LOG-LINEAR REGRESSION COEFFICIENTS

With the Poisson log-linear regression model, tests and confidence intervals for parameters are based on maximum likelihood analysis. The Wald procedure applies to tests and confidence intervals for single coefficients. The drop-in-deviance test compares a full to a reduced model, particularly for hypotheses that several regression coefficients are simultaneously zero.

22.4.1 Wald's Test and Confidence Interval for Single Coefficients

The estimated regression coefficients and their standard errors are calculated by the maximum likelihood method. The sampling distributions of the estimated coefficients are approximately normal if either the sample size is large or the Poisson means are large. Tests and confidence intervals are constructed with the estimates, their standard errors, and tabled values of the standard normal distribution.

Example—Elephant Matings

A main question asked is whether the mean number of successful matings tends to peak at some age and then decrease, or whether the mean continues to increase with age. The inclusion of an age-squared term in the model addresses this question. If a peak occurs at some age, the coefficient of the age-squared term is not zero. Therefore, the test for a peak is accomplished through a test that the coefficient of age-squared is zero. Display 22.6 indicates that the two-sided p-value is 0.67. There is no evidence of any need for a quadratic term in the log-linear model.

Display 22.8 shows the fit of the data without the quadratic term. A confidence interval for the coefficient of age is constructed in the usual way. The estimate of the coefficient, 0.0687, is interpreted as follows: a 1-year increase in age is associated with a multiplicative change in the mean number of successful matings by a factor of $\exp(0.0687) = 1.071$. In words: Each additional year of age is associated with a 7.1% increase in the mean number of successful matings. A 95% confidence interval for the coefficient of age is 0.0417 to 0.0956. Antilogarithms produce a 95% confidence interval for the multiplicative factor by which the mean changes with one year of age—1.043 to 1.100.

22.4.2 The Drop-in-Deviance Test

The drop-in-deviance test compares a full model to a reduced version. For the test of a single coefficient, the analyst may use either the drop-in-deviance test or

DISPLAY 22.8 Poisson log-linear regression of number of matings on age

Coefficients	Estimate	Standard error	z-statistic	p-value
Intercept	−1.582	0.545	−2.905	0.0037
age	0.0687	0.0138	4.996	<0.0001

Deviance = 51.01 **d.f.** = 39

95% Confidence interval for coefficient of age:

$0.0687 \pm (1.96)(0.0138) = 0.0417$ *to* 0.0956

Conclusion: The mean number successful matings increases by 7.1% for each additional year of age; 95% confidence interval is 4.3% to 10.0%.

Wald's test. Wald's test is more convenient, but the drop-in-deviance test is more reliable. Its approximate p-value is more accurate, especially if the sample size and the Poisson means are not large. A common practice is to adopt Wald's test for a casual determination of significance or when the test's p-value is very small or very large, and to use the drop-in-deviance test if the test is an especially important aspect of the analysis or when the Wald's test p-value does not provide a conclusive result.

Like other tests based on extra sums of squared residuals, this test is appropriate when a full model is deemed to fit well and when the hypothesis of interest places restrictions on some of the coefficients. The drop-in-deviance test statistic is the deviance statistic from a fit to the reduced model minus the deviance statistic from the full model. This extra sum of squared residuals is the variability explained by the terms that are included in the full model but absent from the reduced model. If the reduced model is adequate, the sampling distribution of the drop in deviance has an approximate chi-squared distribution with degrees of freedom equal to the number of unknown coefficients in the full model minus the number of unknown coefficients in the reduced model.

Example—Elephant Matings

The Wald's test for the significance of the age-squared term was included in the output of the fit to the model with age and age-squared (Display 22.6). To obtain the more accurate p-value from the drop-in-deviance test, fit both models and compute the drop in deviance. The deviance from the fit without age-squared is 51.01. The deviance from the fit with age-squared is 50.83. The drop in deviance due to including age-squared in the model that already has age, therefore, is 0.18. The p-value (two-sided) is the probability that a chi-squared distribution on 1 degree of freedom is greater than 0.18. The answer, 0.67, is essentially the same as that provided by the two-sided p-value from the Wald's test.

22.5 EXTRA-POISSON VARIATION AND THE LOG-LINEAR MODEL

22.5.1 Extra-Poisson Variation

The Poisson distribution is one of many distributions that assign probabilities to outcomes for counted responses. It is a good choice, especially for processes that generate events uniformly over time or space. Sometimes, however, unmeasured effects, clustering of events, or other contaminating influences combine to produce more variation in the responses than is predicted by the Poisson model. Using Poisson log-linear regression when such extra-Poisson variation is present has three consequences: parameter estimates are still roughly unbiased; but standard errors are smaller than they should be; and tests give smaller p-values than are truly warranted by the data. The last two consequences are problems that must be addressed.

The quasi-likelihood approach, as shown in Chapter 21, is a method for extending the model to allow for this possibility. It is based on a slight modification of the model for the mean and variance of the response. In particular, the model specifies only the mean and the variance of the responses (not the entire distribution), and it allows an additional parameter to account for more variability than is predicted by the Poisson distribution. For the case of two explanatory variables, the model for extra-Poisson variation is

A Log-Linear Model That Permits Extra-Poisson Variation

$$\mu\{Y_i|X_{1i}, X_{2i}\} = \mu_i$$

$$\text{Var}\{Y_i|X_{1i}, X_{2i}\} = \psi\mu_i$$

$$\log(\mu_i) = \beta_0 + \beta_1 X_{1i} + \beta_2 X_{2i}.$$

The extra parameter ψ, which again is referred to as the *dispersion parameter*, captures the extra variation. If the variance of the responses is like the Poisson variance, then $\psi = 1$. If the variance is larger or smaller than the mean, then ψ will be larger or smaller than 1.

22.5.2 Checking for Extra-Poisson Variation

There are four ways to check for extra-Poisson variation:

1. Think about whether extra-Poisson variation is likely.
2. Compare the sample variances to the sample averages computed for groups of responses with identical explanatory variable values.
3. Examine the deviance goodness-of-fit test after fitting a rich model.
4. Examine residuals to see if a large deviance statistic may be due to outliers.

A combination of these usually decides the issue. When the situation is dubious, it is safest to assume that extra-Poisson variation is present.

Consider the Situation

Extra-Poisson variation should be expected when important explanatory variables are not available, when individuals with the same level of explanatory variables may for some reason behave differently, and when the events making up the count are clustered or spaced systematically through time or space, rather than being randomly spaced. However, if the means are all fairly small—less than 5, say—there is seldom reason to worry about extra-Poisson variation.

Plot Variances Against Averages

If responses are grouped by identical values of the explanatory variables, a sample variance and a sample average can be computed for each group. If the responses are Poisson, the sample variances should be approximately equal to the averages. Plotting variances against averages can reveal gross exceptions, and it can also suggest whether the quasi-likelihood model of proportionality is preferable. Plots based on grouping subjects with similar, but not identical, explanatory variable values can be misleading, because the grouping itself induces extra-Poisson variability.

Fit a Rich Model

If the Poisson means are large, the deviance goodness-of-fit test can be examined after a fit to a rich model (one with terms that reflect possible departures from the proposed model). The idea is that including unnecessary terms will not have much effect on the residuals, and a large deviance statistic will indicate inadequacies with the Poisson assumption rather than inadequacies with insufficient explanatory variable terms.

Look for Outliers

The deviance statistic might be large because of one or two outliers. Therefore, it is appropriate to examine the individual deviance residuals in addition to the sum of their squares. An individual deviance residual should not, however, be judged an outlier simply because its size exceeds 2.0. Outliers should not look like they belong to the tail end of an otherwise natural distribution.

In the salamander study, for example, Display 22.7 contains several individual deviances that do exceed the normal limit. But these appear natural in the context of the distribution of the other deviance residuals. In such cases, the quasi-likelihood method provides a more conservative assessment than the maximum likelihood method. It leads to less powerful inferences than the maximum likelihood method when the Poisson model is correct, but Display 22.7 strongly suggest that the Poisson model does not adequately describe these data. The Poisson model follows from an assumption that individual salamanders are distributed *independently* across the region of interest. If the study plots contained family units, the assumption would be violated and lack-of-fit to the Poisson model would be expected.

22.5.3 Inferences when Extra-Poisson Variation Is Present

Tests and confidence intervals are carried out in the same way as they were for extra-binomial variation in Section 21.5.3:

1. An estimate of ψ is the sum of squared deviance residuals divided by the residual degrees of freedom—the sample size minus the number of parameters in the mean.
2. The quasi-likelihood standard errors are the maximum likelihood standard errors multiplied by the square root of the estimate of ψ. Ideally, the quasi-likelihood approach can be requested from the statistical computing program, so standard errors will be properly adjusted. If not, the user must manually compute the estimate of ψ and multiply the given standard errors by the square root of this estimate.
3. All t-tests and confidence intervals are based on the coefficient estimates and the adjusted standard errors. It is prudent to use the t-distribution rather than the standard normal distribution as a reference, even though there is no established theory for this practice. The degrees of freedom are $(n - p)$.

4. The drop in deviance per drop in degrees of freedom is divided by the estimate of ψ to form an F-statistic. This is analogous to the extra-sum-of-squares F-test in ordinary regression, except that deviance residuals (rather than ordinary residuals) are used.

Example—Salamander Habitat Study

Having established that there is extra-Poisson variation, the next step in the analysis is to use quasi-likelihood methods for building a model. A quadratic model for investigation of the influence of forest age and canopy cover is

$$\log(\mu) = \beta_0 + \beta_1 age + \beta_2 cover + \beta_3 age^2 + \beta_4 cover^2 + \beta_5 age \times cover.$$

However, there are two distinct canopy cover conditions (Display 22.9) that may give rise to different quadratic models. If *closed* is an indicator variable for closed

DISPLAY 22.9 Scatterplots of Del Norte salamander counts with percentage of canopy cover and forest age

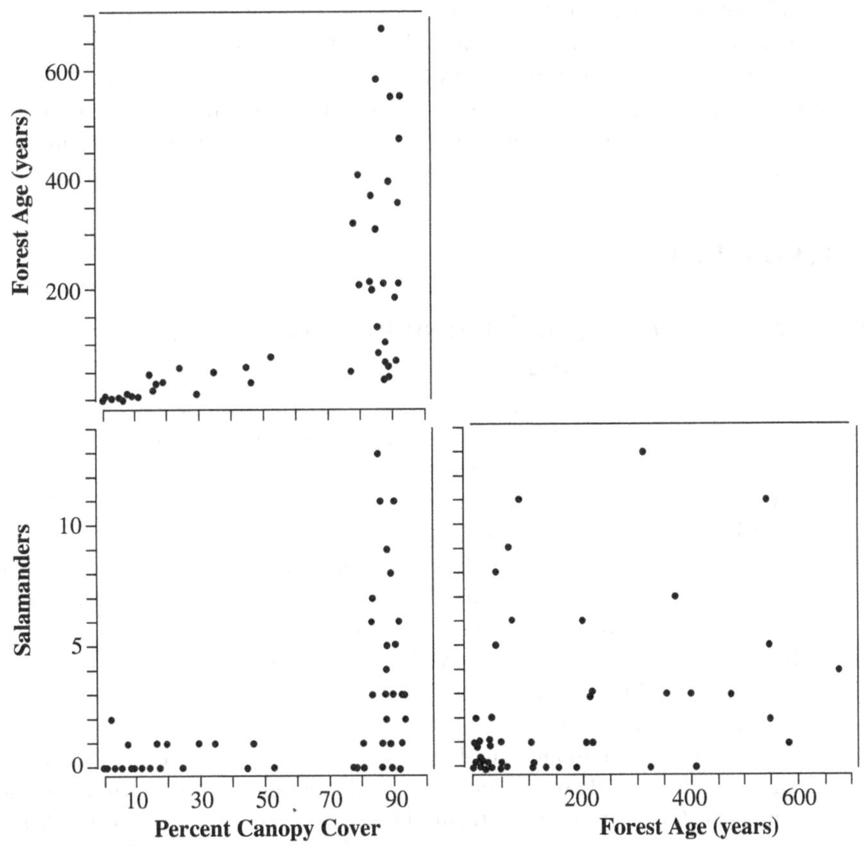

canopy (*cover* > 70%), the separate quadratic model comes from the addition of terms

$$\beta_6 closed + \beta_7 age \times closed + \beta_8 cover \times closed + \beta_9 age^2 \times closed$$
$$+ \beta_{10} cover^2 \times closed + \beta_{11} age \times cover \times closed$$

to the expression above. Now the coefficients in the first expression describe the association of salamander numbers with forest age and canopy cover *in open canopy conditions*. The coefficients in the additional expression describe how the individual coefficients differ between closed and open canopy conditions.

The quasi-likelihood t-statistics (Display 22.10) for testing the significance of all terms involving forest age are all nonsignificant, which suggests that forest age may not be important once canopy cover is accounted for. To test that hypothesis, all terms involving forest age are dropped and the model refit. Display 22.10 illustrates the process, and the models are compared with a quasi-likelihood drop in deviance F-test. The approximate p-value of 0.7843 confirms that forest age may safely be discarded.

The significance of the coefficient of $cover^2 \times closed$ indicates that there is different curvature in the two conditions. Fitting separate quadratics requires the inclusion of the remaining terms. The full fit of this model is plotted in Display 22.4. Approximate 95% confidence limits are obtained by fitting the same model after centering canopy cover at chosen values, then reading off the intercept coefficient and its standard error, constructing an interval, and exponentiating its endpoints.

22.6 RELATED ISSUES

22.6.1 Log-Linear Models for Testing Independence in Tables of Counts

The Poisson distribution also arises in the discussion of tables of counts (Chapter 19). There, Poisson sampling means that a fixed amount of effort is used to sample subjects, who are then categorized into one of the cells in the table. In contrast to the other sampling schemes, neither the marginal totals nor the grand total is known before the data are collected.

It is possible to analyze Poisson counts in tables of counts by treating the cell counts as Poisson response variables and employing a log-linear model with factors representing the different dimensions in the table. The Samoan obesity and heart disease data, for example, involved four cells in a table consisting of two rows and two columns. The means of the counts may be described by a log-linear model, as shown in Display 22.11.

The log-linear model involves obesity as its row factor and CVD death as its column factor. If *obese* and *death* act independently to determine the counts, there is no need for an interaction term. Therefore, a test for the $obese \times death$ interaction in the log-linear model provides a test of independence in the table of counts.

DISPLAY 22.10 Informal testing of extra terms, with a drop-in-deviance F-test, to find a good fitting model; Del Norte salamander data

① *Fit a rich model, including interactions and quadratic terms.*

Response: *salamanders*
Explanatory variables: *age, cover, age*^{2}*, cover*^{2}*, age × cover, closed, age × closed, cover × closed, age*^{2} *× closed, cover*^{2} *× closed, age × cover × closed*
Distribution: (Quasi-likelihood, with variance = $\psi\mu$)
Link: *logarithm*

Variable	Coefficient	Standard error	t-statistic	p-value
Constant	−1.4512	1.4131	−1.0270	0.3115
age	0.0468	0.1591	0.2939	0.7706
cover	0.0554	0.2129	0.2601	0.0116
age^2	−0.0026	0.0046	−0.4515	0.5780
cover^2	−0.0018	0.0065	−0.2837	0.7783
age × cover	0.0028	0.0091	0.3097	0.7586
closed	−274.266	102.930	−2.6646	0.0116
age × closed	−0.0793	0.1628	−0.4871	0.6292
cover × closed	6.4465	2.4102	2.6746	0.1130
age^2 *× closed*	0.0026	0.0046	0.5608	0.5785
cover^2 *× closed*	−0.0363	0.0155	−2.3437	0.0249
age × cover × closed	−0.0024	0.0091	−0.2654	0.7923

Deviance = 89.1778 **d.f.** = 35

② *Fit the inferential model, with extra terms deleted.*

Response: *salamanders*
Explanatory variable: *cover, cover*^{2}*, closed, cover × closed, cover*^{2} *× closed*
Distribution: (Quasi-likelihood, with variance = $\psi\mu$)
Link: *logarithm*

Variable	Coefficient	Standard error	t-statistic	p-value
Constant	−1.2575	1.2609	−0.9973	0.3245
cover	0.0471	0.1299	0.3625	0.7188
cover^2	−0.0008	0.0025	−0.3294	0.7435
closed	−218.641	85.260	−2.5644	0.0141
cover × closed	5.0538	1.9717	2.5632	0.0141
cover^2 *× closed*	−0.0285	0.0116	−2.4532	0.0185

Deviance = 97.2252 **d.f.** = 41

③ *Test adequacy of the inferential model.*

$$F\text{-statistic} = \frac{(97.2252 - 89.1778)/6}{(89.1778)/35} = 0.5264; \qquad p\text{-value} = 0.7483$$

Conclusion: There is no evidence of a need for the extra terms included in model 1.

DISPLAY 22.11 Log-linear analysis of a 2 × 2 table of counts—Samoan obesity and heart disease (see Section 18.1.1)

CVD Death

	Yes	No
Obese	16	2045
Not obese	7	1044

Data for Regression

count	*obese*	*death*
16	1	1
2045	1	0
7	0	1
1044	0	0

Indicator for row (obese) *Indicator for column* (death)

MODEL: Y_i is Poisson with mean μ_i, ($i = 1, 2, 3, 4$)

$$\log(\mu_i) = \beta_0 + \beta_1 obese_i + \beta_2 death_i + \beta_3 obese_i \times \text{death}_i$$

A test independence of row and column factors is a test that the interaction term in the log-linear model is zero.

For a table with r rows and c columns, a total of $(r - 1)(c - 1)$ terms are needed to describe all row × column interactions. These terms may be tested with an extra-sum-of-squared-residuals test. Although some controversy surrounds the question of which residuals are best, common practice is to use Pearson residuals for the analysis of tables of counts. Since the full model—with the interaction terms—is the saturated model (where all residuals are zero), the test statistic for independence equals the sum of the squared Pearson residuals from the log-linear fit to the model without interaction. The test statistic is compared to a chi-square distribution on $(r - 1)(c - 1)$ degrees of freedom.

Notes About Log-Linear Analysis of Tables of Counts

1. The Poisson log-linear model describes the mean count in terms of the row and column factors, and these two factors are treated equally. This is unlike ordinary regression, where the distinction between response and explanatory variables is important.
2. For 2 × 2 tables of counts, the Poisson analysis with the Pearson statistic is identical to the chi-squared test presented in Section 19.3.2. Both forms may also be applied to more complex tables of counts.
3. The log-linear model approach parallels ordinary regression and analysis of variance, with independence between two factors being tested through the significance of their interaction terms.

22.6.2 Poisson Counts from Varying Effort

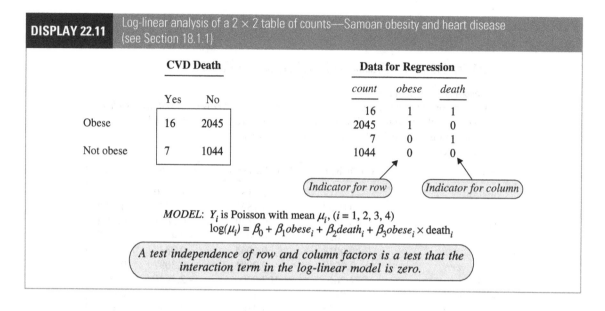

In the previous examples, the response counts were obtained by observing some process for a fixed amount of effort. All salamander counts came from searching 7 m square plots; all male elephants were observed for 8 years; and the CVD death counts for the Samoan women were based on a 7-year observation period. That is, each count arises from the same amount of sampling effort. In many situations that arise, however, counts are based on different amounts of effort. It is natural that higher sampling effort yields higher mean counts, and the analysis should concentrate on some measure of average counts per unit of effort, or count rates.

The easiest way to include effort in a Poisson regression equation is to include log(*effort*) as an additional explanatory variable in the log-linear model:

When the Mean Count Is an Occurence Rate Times Effort

$$\mu(Y_i) = \mu_i = effort_i \times rate_i$$

A log-linear model for rate is:

$$\log(rate_i) = \beta_0 + \beta_1 X_{1i} + \beta_2 X_{2i} + \cdots.$$

So, a model for the mean count is:

$$\log(\mu_i) = \log(effort_i) + \beta_0 + \beta_1 X_{1i} + \beta_2 X_{2i} + \cdots.$$

Effort is typically the time of observation during which each count is made, but it may also be an area. If the effort is the same for all observations in the data set, the term involving effort is a constant and can be absorbed into β_0.

The coefficient of log(*effort*) is known to be 1. In generalized linear models, a term with a known coefficient can be inserted into the regression equation. This term is called an *offset*. The mechanism for inserting it depends on the statistical package used. If a Poisson log-linear regression program is available that does not have the capability of including an offset, include log(*effort*) as an explanatory variable with the others having an unknown coefficient. This generally has little effect on the other estimates.

If scatterplots are drawn, it is necessary to plot *rate* (*count*/*effort*) or the log of rate—instead of raw counts—versus explanatory variables. In the analysis, a check for extra-Poisson variation remains necessary. Its presence can be handled in the usual way, using the quasi-likelihood method.

22.6.3 Negative Binomial Regression

The Poisson is often too simple of a probability distribution for nonnegative integer responses, especially when the counts are large. The quasi-likelihood version, with extra dispersion parameter, helps in extending its usefulness.. Another extension is the negative binomial distribution. Whereas a Poisson distribution has a single unknown parameter, μ, the negative binomial has an additional parameter ϕ, which allows greater flexibility in modeling count variation. If a response variable has mean μ, the following are the variances implied by the Poisson model, the quasi-Poisson model with extra dispersion parameter Ψ, and the negative binomial distribution:

Model	**Variance**
Poisson	μ
Poisson (quasi)	$\Psi\mu$
Negative binomial	$\mu(1 + \phi\mu)$

Because the negative binomial distribution is thought to be a natural model for extra-Poisson variation and because maximum likelihood can be used for regression inference based on it, negative binomial log-linear regression has always been an attractive alternative to Poisson log-linear regression. Until recently, however, easy-to-use computer programs for its use have not been available. Fortunately, statistical computer programs have now incorporated modules for negative binomial regression and these can be used in the same way as the GLM routines for Poisson log-linear regression. The strategies for data analysis remain the same, except there is no need to investigate extra-Poisson variation because it is automatically included in the model. Tests and confidence intervals are based on Wald and drop-in-deviance (also known as likelihood ratio) theory, with the same issues as for Poisson. There is no guarantee that the negative binomial model is appropriate for all nonnegative integer response variables, but it offers greater flexibility than Poisson and has proven itself useful in a wide variety of applications.

22.7 SUMMARY

Poisson regression should be considered whenever the response variable is a count of events that occur over some observation period or spatial area. The main feature of the Poisson distribution is that the variance is equal to the mean. Consequently, if ordinary regression is used, the residual plot will exhibit a funnel shape. In the Poisson log-linear regression model, the distribution of the response variables for each combination of the explanatory variables is thought to be Poisson, and the logarithms of the means of these distributions are modeled by formulas that are linear functions of regression coefficients. The method of maximum likelihood is used to estimate the unknown regression coefficients.

Elephant Mating Data

A scatterplot of the number of successful matings versus age for the 41 male elephants indicates the inadequacy of a straight-line regression and of a constant variance assumption for these data. Linear regression after a log transformation of number of matings is possible, but the zeros in the data set are a nuisance in this approach. Poisson log-linear regression provides an attractive alternative. The counts are too small to justify the deviance goodness-of-fit test, but plots and informal testing of extra terms imply that the model with the log of the mean count as a straight-line function of age is adequate. One question of interest is whether the mean count of successful matings peaks at some age. This can be investigated by testing the significance of an age-squared term. The drop-in-deviance test does not provide any evidence of a quadratic effect.

Salamander Data

Residuals from the fit of a rich model were larger than expected, strongly suggesting extra-Poisson variability. Therefore, quasi-likelihood methods were used in model selection and in reporting results. There were two distinct canopy cover regimes. A quasi-likelihood F-test indicated that forest age was not associated with mean counts, after taking into account canopy cover. The fit of the reduced model included a highly significant difference between quadratic effects of canopy, so the descriptive model was selected as having separate quadratics in low cover and high cover situations.

22.8 EXERCISES

Conceptual Exercises

1. Elephant Mating. Both the binomial and the Poisson distributions provide probability models for random counts. Is the binomial distribution appropriate for the number of successful matings of the male African elephants?

2. What is the difference between a log-linear model and a linear model after the log transformation of the response?

3. If a confidence interval is obtained for the coefficient of an indicator variable in a log-linear regression model, do the antilogarithms of the endpoints describe a confidence interval for a ratio of means or for a ratio of medians?

4. Elephant Mating. In Display 22.2 the spread of the responses is larger for larger values of the mean response. Is this something to be concerned about if Poisson log-linear regression is used?

5. Elephant Mating. From the estimated log-linear regression of elephants' successful matings on age (Display 22.8), what are the mean and the variance of the distribution of counts of successful matings (in 8 years) for elephants who are aged 25 years at the beginning of the observation period? What are the mean and the variance for elephants who are aged 45 years?

6. (a) Why are ordinary residuals—$(Y_i - \hat{\mu}_i)$—not particularly useful for Poisson regression? (b) How are the Pearson residuals designed to deal with this problem?

7. Consider the deviance goodness-of-fit test. (a) Under what conditions is it valid for Poisson regression? (b) When it is valid, what possibilities are suggested by a small p-value? (c) When it is valid, what possibilities are suggested by a large p-value?

8. (a) Why is it more difficult to check the adequacy of a Poisson log-linear regression model when the counts are small than when they are large? (b) What tools are available in this situation?

9. (a) How does the drop-in-deviance test for Poisson log-linear regression resemble the extra-sum-of-squares test in ordinary regression? (b) How does it differ?

10. (a) How does the drop-in-deviance test for the quasi-likelihood version of the log-linear regression model resemble the extra-sum-of-squares test in ordinary regression? (b) How does it differ?

11. How does the quasi-likelihood version of the log-linear regression model allow for more variation than would be expected if the responses were Poisson?

12. If responses follow the Poisson log-linear regression model, the Pearson residuals should have variance approximately equal to 1. If, instead, the quasi-likelihood model with dispersion parameter ψ is appropriate, what is the approximate variance of the Pearson residuals?

13. Is it acceptable to use the quasi-likelihood model when the data actually follow the Poisson model?

14. Consider a table that categorizes 1,000 subjects into 5 rows and 10 columns. (a) If Poisson log-linear regression is used to analyze the data, what is the sample size? (That is, how many Poisson counts are there?) (b) How would one test for independence of row and column factors?

Computational Exercises

15. Elephant Mating and Age. Give an estimated model for describing the number of successful matings as a function of age, using (a) simple linear regression after transforming the number of successful matings to the square root; (b) simple linear regression after a logarithmic transformation (after adding 1); (c) log-linear regression. (d) Do the models used in parts (a) or (b) exhibit obvious inadequacies?

16. Murder–Suicides by Deliberate Plane Crash. Some sociologists suspect that highly publicized suicides may trigger additional suicides. In one investigation of this hypothesis, D. P. Phillips collected information about 17 airplane crashes that were known (because of notes left behind) to be murder–suicides. For each of these crashes, Phillips reported an index of the news coverage (circulation in nine newspapers devoting space to the crash multiplied by length of coverage) and the number of multiple-fatality plane crashes during the week following the publicized crash. The data are exhibited in Display 22.12. (Data from D. P. Phillips, "Airplane Accident Fatalities Increase Just After Newspaper Stories About Murder and Suicide," *Science* 201 (1978): 748–50.) Is there evidence that the mean number of crashes increases with increasing levels of publicity of a murder–suicide?

DISPLAY 22.12 Multiple-fatality plane crashes in the week following a murder–suicide by plane crash, and the amount of newspaper coverage given the murder–suicide

Index of coverage	Number of crashes	Index of coverage	Number of crashes
376	8	63	2
347	5	44	7
322	8	40	4
104	4	5	3
103	6	5	2
98	4	0	4
96	8	0	3
85	6	0	2
82	4		

17. Obesity and Heart Disease. Analyze the table of counts in Display 22.11 as suggested there. What is the p-value for testing independence of obesity and CVD death outcome?

18. Galapagos Islands. Reanalyze the data in Exercise 12.20 with number of native species as the response, but using log-linear regression. (a) Fit the model with log area, log elevation, log of distance from nearest island, and log area of nearest island as explanatory variables; and then check for extra-Poisson variation. (b) Use backward elimination to eliminate insignificant explanatory variables. (c) Describe the effects of the remaining explanatory variables.

19. Galapagos Islands. Repeat the previous exercise, but use the number of nonnative species as the response variable (total number of species minus the number of native species).

20. Cancer Deaths of Atomic Bomb Survivors. The data in Display 22.13 are the number of cancer deaths among survivors of the atomic bombs dropped on Japan during World War II, categorized by time (years) after the bomb that death occurred and the amount of radiation exposure that the survivors received from the blast. (Data from D. A. Pierce, personal communication.) Also listed in each cell is the *person-years at risk*, in 100's. This is the sum total of all years spent by all persons in the category. Suppose that the mean number of cancer deaths in each cell is Poisson with mean

$\mu = risk \times rate$, where *risk* is the person-years at risk and *rate* is the rate of cancer deaths per person per year. It is desired to describe this rate in terms of the amount of radiation, adjusting for the effects of time after exposure. (a) Using log(*risk*) as an offset, fit the Poisson log-linear regression model with time after blast treated as a factor (with seven levels) and with *rads* and *rads*-squared treated as covariates. Look at the deviance statistic and the deviance residuals. Does extra-Poisson variation seem to be present? Is the *rads*-squared term necessary? (b) Try the same model as in part (a); but instead of treating time after bomb as a factor with seven levels, compute the midpoint of each interval and include log(*time*) as a numerical explanatory variable. Is the deviance statistic substantially larger in this model, or does it appear that time can adequately be represented through this single term? (c) Try fitting a model that includes the interaction of log(*time*) and exposure. Is the interaction significant? (d) Based on a good-fitting model, make a statement about the effect of radiation exposure on the number of cancer deaths per person per year (and include a confidence interval if you supply an estimate of a parameter).

DISPLAY 22.13 Cancer deaths among Japanese atomic bomb survivors, categorized by estimated exposure to radiation (in rads) and years after exposure; below the number of cancer deaths are the person-years (in 100's) at risk

		Years after exposure						
exposure (rads)		0–7	8–11	12–15	16–19	20–23	24–27	28–31
0	*deaths:*	10	12	19	31	35	48	73
	risk:	262	243	240	237	233	227	220
25	*deaths:*	17	17	17	47	50	65	71
	risk:	313	290	285	280	275	269	262
75	*deaths:*	0	2	1	5	8	7	12
	risk:	38	36	35	34	34	33	32
150	*deaths:*	1	0	4	1	6	12	11
	risk:	28	26	25	25	24	24	23
250	*deaths:*	1	1	0	4	3	7	13
	risk:	13	12	12	12	11	11	10
400	*deaths:*	0	2	5	3	2	3	5
	risk:	15	14	14	14	13	13	13

21. **El Niño and Hurricanes.** Reconsider the El Niño and Hurricane data set from Exercise 10.28. Use poisson log-linear regression to describe the distribution of (a) number of storms and (b) number of hurricanes as a function of El Niño temperature and West African wetness.

Data Problems

22. **Emulating Jane Austen's Writing Style.** When she died in 1817, the English novelist Jane Austen had not yet finished the novel *Sanditon*, but she did leave notes on how she intended to conclude the book. The novel was completed by a ghost writer, who attempted to emulate Austen's style. In 1978, a researcher reported counts of some words found in chapters of books written by Austen and in chapters written by the emulator. These are reproduced in Display 22.14. (Data from A. Q. Morton, *Literary Detection: How to Prove Authorship and Fraud in Literature and Documents*, New York: Charles Scribner's Sons, 1978.) Was Jane Austen consistent in the three books in her relative uses of these words? Did the emulator do a good job in terms of matching the relative rates of occurrence of these six words? In particular, did the emulator match the relative rates that Austen used the words in the first part of *Sanditon*?

DISPLAY 22.14 Occurrences of six words in various chapters of books written by Jane Austen (*Sense and Sensibility*, *Emma*, and the first part of *Sanditon*) and in some chapters written by the writer who completed *Sanditon* (Sanditon II)

Word	Book			
	Sense and Sensibility	*Emma*	*Sanditon* I	*Sanditon* II
a	147	186	101	83
an	25	26	11	29
this	32	39	15	15
that	94	105	37	22
with	59	74	28	43
without	18	10	10	4

23. **Space Shuttle O-Ring Failures.** On January 27, 1986, the night before the space shuttle *Challenger* exploded, an engineer recommended to the National Aeronautics and Space Administration (NASA) that the shuttle not be launched in the cold weather. The forecasted temperature for the *Challenger* launch was 31°F—the coldest launch ever. After an intense 3-hour telephone conference, officials decided to proceed with the launch. Shown in Display 22.15 are the launch temperatures and the number of O-ring problems in 24 shuttle launches prior to the *Challenger* (Chapter 4). Do these data offer evidence that the number of incidents increases with decreasing temperature?

DISPLAY 22.15 Launch temperatures (°F) and numbers of O-ring incidents in 24 space shuttle flights

Launch temperature (°F)	Number of incidents	Launch temperature (°F)	Number of incidents
53	3	70	1
56	1	70	1
57	1	72	0
63	0	73	0
66	0	75	0
67	0	75	2
67	0	76	0
67	0	76	0
68	0	78	0
69	0	79	0
70	0	80	0
70	1	81	0

24. **Valve Failure in Nuclear Reactors.** Display 22.16 shows characteristics and numbers of *failures* observed in valve types from one pressurized water reactor. There are five explanatory factors: *system* (1 = containment, 2 = nuclear, 3 = power conversion, 4 = safety, 5 = process auxiliary); *operator* type (1 = air, 2 = solenoid, 3 = motor-driven, 4 = manual); *valve* type (1 = ball, 2 = butterfly, 3 = diaphragm, 4 = gate, 5 = globe, 6 = directional control); head *size* (1 = less than 2 inches, 2 = 2–10 inches, 3 = 10–30 inches); and operation *mode* (1 = normally closed, 2 = normally open). The lengths of observation periods are quite different, as indicated in the last column, *time*. Using an offset for log of observation time, identify the factors associated with large numbers of valve failures. (Data from L. M. Moore and R. J. Beckman, "Appropriate One-Sided Tolerance Bounds on the Number of Failures Using Poisson Regression," *Technometrics* 30 (1988): 283–90.)

DISPLAY 22.16 Valve characteristics and numbers of failures in a nuclear reactor; first 6 of 91 rows

System	*Operator*	*Valve*	*Size*	*Mode*	*Failures*	*Time*
1	3	4	3	1	2	4
1	3	4	3	2	2	4
1	3	5	1	1	1	2
2	1	2	2	2	0	2
2	1	3	2	1	0	2
2	1	3	2	2	0	1

25. Body Size and Reproductive Success in a Population of Male Bullfrogs. As an example of field observation in evidence of theories of sexual selection, S. J. Arnold and M. J. Wade ("On the Measurement of Natural and Sexual Selection: Applications," *Evolution* 38 (1984): 720–34) presented the following data set on size and number of mates observed in 38 male bullfrogs (Display 22.17). Is there evidence that the distribution of number of mates in this population is related to body size? If so, supply a quantitative description of that relationship, along with an appropriate measure of uncertainty. Write a brief summary of statistical findings.

DISPLAY 22.17 Body size (mm) and number of mates for 38 male bullfrogs (*Rana catesbeiana*); first 5 of 38 rows

Body size	**Mates**
144	3
150	2
144	1
154	1
132	1

26. Number of Moons. Display 22.18 shows some characteristics of terrestrial planets, gas giants, and dwarf planets in our solar system, including the known number of moons. Apparently, larger planets have more moons, but is it the volume (as indicated by diameter) or mass that are more relevant, or is it both? Answer this question and arrive at a model for describing the mean number of moons as a function of planet size. Use negative binomial regression if you have access to a computer package that does the computations. Otherwise, use quasi-likelihood analysis. (Data from Wikipedia: http://en.wikipedia.org/ wiki/Planet#Solar_System, August 10, 2011.)

Answers to Conceptual Exercises

1. No. A binomial distribution is for a count that has a precisely-defined upper limit, such as the number of heads in 10 flips of a coin (with upper limit 10) or the number of elephant legs with defects (with upper limit 4). Although there is surely some practical limit to the number of successful elephant matings, there is no precisely-defined maximum to be used as the binomial index. The Poisson distribution is more sensible.

2. In a log-linear model, the mean of Y is μ and the model is $\log(\mu) = \beta_0 + \beta_1 X_1$. Y is not transformed. If simple linear regression is used after a log transformation, the model is expressed in terms of the mean of the logarithm of Y.

3. Ratio of means. The model states that the log of the mean—not the median—has the regression form. It is not necessary to introduce the median of the Poisson distribution here.

4. No. The nonconstant variance is anticipated by the Poisson model. The maximum likelihood procedure correctly uses the information in the data to estimate the regression coefficients.

DISPLAY 22.18 Average distance from the sun, diameter, and mass (all scaled so that the values for earth are 1); and number of moons of 13 planets and dwarf planets in our solar system

Name	Distance	Diameter	Mass	Moons
Mercury	0.39	0.382	0.06	0
Venus	0.72	0.949	0.82	0
Earth	1	1	1	1
Mars	1.52	0.532	0.11	2
Ceres	2.75	0.08	0.0002	0
Jupiter	5.2	11.209	317.8	64
Saturn	9.54	9.449	95.2	62
Uranus	19.22	4.007	14.6	27
Neptune	30.06	3.883	17.2	13
Pluto	39.5	0.18	0.0022	4
Haumea	43.35	0.15	0.0007	2
Makemake	45.8	0.12	0.0007	0
Eris	67.7	0.19	0.0025	1

5. For 25-year-old elephants: Mean = 1.15; Variance = 1.15. For 45-year-old elephants: Mean = 4.53; Variance = 4.53.

6. (a) The residuals with larger means will have larger variances. Thus, if an observation has a large residual it is difficult to know whether it is an outlier or an observation from a distribution with larger variance than the others. (b) The residuals are scaled to have the same variance.

7. (a) Large Poisson means. (b) The Poisson distribution is an inadequate model, the regression model terms are inadequate, or there are a few contaminating observations. (c) Either the model is correct, or there is insufficient data to detect any inadequacies.

8. (a) Scatterplots are uninformative, the deviance goodness-of-fit test cannot be used, and the approximate normality of residuals is not guaranteed. (b) One may try to add extra terms, such as a squared term in some explanatory variable, to model a simple departure from linearity and to test its significance with a drop-in-deviance or Wald's test.

9. (a) It is based on a comparison of the magnitudes of residuals from a full to a reduced model. (b) The residuals used are deviance residuals, and a chi-squared statistic is used rather than an F-statistic.

10. (a) Same as 10(a). In this case, an F-statistic is formed just as with the usual extra-sum-of-squares test. (b) It is based on deviance residuals.

11. The variance of the responses is $\psi\mu$, where ψ is unknown. A value of ψ greater than 1 allows for more variation than anticipated by a Poisson distribution.

12. ψ.

13. Yes; the Poisson mean–variance relationship is a special case of the quasi-likelihood model (when ψ is 1). (More powerful comparisons result, however, if the Poisson model is used when it definitely applies.)

14. (a) 50. (b) Test for the significance of the 36 interaction terms in the log-linear regression with row, column, and row-by-column effects. This may be accomplished by fitting the model without interaction and comparing the Pearson statistic to a chi-squared distribution on 36 degrees of freedom.

Bibliography

Aitkin, M., Anderson, D., Francis, B., and Hinde, J. 1989. *Statistical Modelling in GLIM*. Oxford: Clarendon Press.

Agresti, A. 1990. *Categorical Data Analysis*. New York: John Wiley & Sons.

Bowerman, B. L., and O'Connell, R. T. 1993. *Forecasting and Time Series: An Applied Approach*, 3rd ed. Belmont, Calif.: Duxbury Press.

Box, G. E. P., Hunter, W. G., and Hunter, J. S. 1978. *Statistics for Experiments: An Introduction to Design, Data Analysis, and Model Building*. New York: John Wiley & Sons.

Breslow, N. E., and Day, N. E. 1980. *Statistical Methods in Cancer Research, Vol. 1: The Analysis of Case-Control Studies*. Lyon: World Health Organization.

Chatfield, C. 1984. *Analysis of Time Series: An Introduction*, 3rd ed. London: Chapman & Hall.

Cleveland, W. S. 1993. *Visualizing Data*. Murray Hill, N.J.: AT&T Bell Laboratories.

Daniel, C., and Wood, F. S. 1980. *Fitting Equations to Data: Computer Analysis of Multifactor Data*, 2nd ed. New York: John Wiley & Sons.

Devore, J. L., and Peck, R. 1993. *Statistics: The Exploration and Analysis of Data*, 2nd ed. Belmont, Calif.: Duxbury Press.

Dobson, A. J. 1983. *An Introduction to Statistical Modelling*. London: Chapman & Hall.

Draper, N. R., and Smith, H. 1981. *Applied Regression Analysis*. New York: John Wiley & Sons.

Durbin, J., and Watson, G. S. 1951. "Testing for Serial Correlation in Least Squares Regression II." *Biometrika* 38: 159–78.

Fisher, L. D., and van Belle, G. 1993. *Biostatistics: A Methodology for the Health Sciences*. New York: John Wiley & Sons.

Fisher, R. A. 1935. *The Design of Experiments*. New York: Hafner Library.

Freedman, D., Pisani, R., Purves, R., and Adhikari, A. 1991. *Statistics*. New York: W. W. Norton.

Fuller, W. A. 1987. *Measurement Error Models*. New York: John Wiley & Sons.

Good, P. 1994. *Permutation Tests: A Practical Guide to Resampling Methods for Testing Hypotheses*. New York: Springer-Verlag.

Hand, D. J., and Taylor, C. C. 1987. *Multivariate Analysis of Variance and Repeated Measures: A Practical Approach for Behavioural Scientists*. London: Chapman & Hall.

Hill, M. O. 1974. "Correspondence Analysis: A Neglected Multivariate Method." *Applied Statistics* 23: 340–54.

Appendix A

Tables

Table A.1 Probabilities of the Standard Normal Distribution

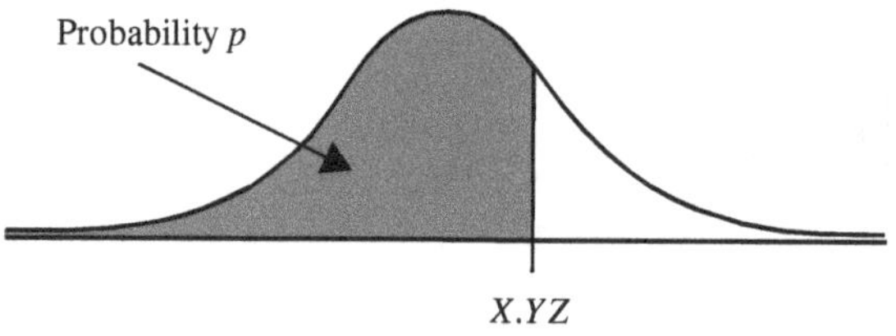

Tabled values are p, the probability that a standard normal is less than $X.YZ$

	.0Z									
X.Y	.00	.01	.02	.03	.04	.05	.06	.07	.08	.09
−3.4	.0003	.0003	.0003	.0003	.0003	.0003	.0003	.0003	.0003	.0002
−3.3	.0005	.0005	.0005	.0004	.0004	.0004	.0004	.0004	.0004	.0003
−3.2	.0007	.0007	.0006	.0006	.0006	.0006	.0006	.0005	.0005	.0005
−3.1	.0010	.0009	.0009	.0009	.0008	.0008	.0008	.0008	.0007	.0007
−3.0	.0013	.0013	.0013	.0012	.0012	.0011	.0011	.0011	.0010	.0010
−2.9	.0019	.0018	.0018	.0017	.0016	.0016	.0015	.0015	.0014	.0014
−2.8	.0026	.0025	.0024	.0023	.0023	.0022	.0021	.0021	.0020	.0019
−2.7	.0035	.0034	.0033	.0032	.0031	.0030	.0029	.0028	.0027	.0026
−2.6	.0047	.0045	.0044	.0043	.0041	.0040	.0039	.0038	.0037	.0036
−2.5	.0062	.0060	.0059	.0057	.0055	.0054	.0052	.0051	.0049	.0048
−2.4	.0082	.0080	.0078	.0075	.0073	.0071	.0069	.0068	.0066	.0064
−2.3	.0107	.0104	.0102	.0099	.0096	.0094	.0091	.0089	.0087	.0084
−2.2	.0139	.0136	.0132	.0129	.0125	.0122	.0119	.0116	.0113	.0110
−2.1	.0179	.0174	.0170	.0166	.0162	.0158	.0154	.0150	.0146	.0143
−2.0	.0228	.0222	.0217	.0212	.0207	.0202	.0197	.0192	.0188	.0183
−1.9	.0287	.0281	.0274	.0268	.0262	.0256	.0250	.0244	.0239	.0233
−1.8	.0359	.0351	.0344	.0336	.0329	.0322	.0314	.0307	.0301	.0294
−1.7	.0446	.0436	.0427	.0418	.0409	.0401	.0392	.0384	.0375	.0367
−1.6	.0548	.0537	.0526	.0516	.0505	.0495	.0485	.0475	.0465	.0455
−1.5	.0668	.0655	.0643	.0630	.0618	.0606	.0594	.0582	.0571	.0559
−1.4	.0808	.0793	.0778	.0764	.0749	.0735	.0721	.0708	.0694	.0681
−1.3	.0968	.0951	.0934	.0918	.0901	.0885	.0869	.0853	.0838	.0823
−1.2	.1151	.1131	.1112	.1093	.1075	.1056	.1038	.1020	.1003	.0985
−1.1	.1357	.1335	.1314	.1292	.1271	.1251	.1230	.1210	.1190	.1170
−1.0	.1587	.1562	.1539	.1515	.1492	.1469	.1446	.1423	.1401	.1379
−0.9	.1841	.1814	.1788	.1762	.1736	.1711	.1685	.1660	.1635	.1611
−0.8	.2119	.2090	.2061	.2033	.2005	.1977	.1949	.1922	.1894	.1867
−0.7	.2420	.2389	.2358	.2327	.2296	.2266	.2236	.2206	.2177	.2148
−0.6	.2743	.2709	.2676	.2643	.2611	.2578	.2546	.2514	.2483	.2451
−0.5	.3085	.3050	.3015	.2981	.2946	.2912	.2877	.2843	.2810	.2776
−0.4	.3446	.3409	.3372	.3336	.3300	.3264	.3228	.3192	.3156	.3121
−0.3	.3821	.3783	.3745	.3707	.3669	.3632	.3594	.3557	.3520	.3483
−0.2	.4207	.4168	.4129	.4090	.4052	.4013	.3974	.3936	.3897	.3859
−0.1	.4602	.4562	.4522	.4483	.4443	.4404	.4364	.4325	.4286	.4247
−0.0	.5000	.4960	.4920	.4880	.4840	.4801	.4761	.4721	.4681	.4641

Table A.1 continued

					.0Z					
X.Y	*.00*	*.01*	*.02*	*.03*	*.04*	*.05*	*.06*	*.07*	*.08*	*.09*
0.0	.5000	.5040	.5080	.5120	.5160	.5199	.5239	.5279	.5319	.5359
0.1	.5398	.5438	.5478	.5517	.5557	.5596	.5636	.5675	.5714	.5753
0.2	.5793	.5832	.5871	.5910	.5948	.5987	.6026	.6064	.6103	.6141
0.3	.6179	.6217	.6255	.6293	.6331	.6368	.6406	.6443	.6480	.6517
0.4	.6554	.6591	.6628	.6664	.6700	.6736	.6772	.6808	.6844	.6879
0.5	.6915	.6950	.6985	.7019	.7054	.7088	.7123	.7157	.7190	.7224
0.6	.7257	.7291	.7324	.7357	.7389	.7422	.7454	.7486	.7517	.7549
0.7	.7580	.7611	.7642	.7673	.7704	.7734	.7764	.7794	.7823	.7852
0.8	.7881	.7910	.7939	.7967	.7995	.8023	.8051	.8078	.8106	.8133
0.9	.8159	.8186	.8212	.8238	.8264	.8289	.8315	.8340	.8365	.8389
1.0	.8413	.8438	.8461	.8485	.8508	.8531	.8554	.8577	.8599	.8621
1.1	.8643	.8665	.8686	.8708	.8729	.8749	.8770	.8790	.8810	.8830
1.2	.8849	.8869	.8888	.8907	.8925	.8944	.8962	.8980	.8997	.9015
1.3	.9032	.9049	.9066	.9082	.9099	.9115	.9131	.9147	.9162	.9177
1.4	.9192	.9207	.9222	.9236	.9251	.9265	.9279	.9292	.9306	.9319
1.5	.9332	.9345	.9357	.9370	.9382	.9394	.9406	.9418	.9429	.9441
1.6	.9452	.9463	.9474	.9484	.9495	.9505	.9515	.9525	.9535	.9545
1.7	.9554	.9564	.9573	.9582	.9591	.9599	.9608	.9616	.9625	.9633
1.8	.9641	.9649	.9656	.9664	.9671	.9678	.9686	.9693	.9699	.9706
1.9	.9713	.9719	.9726	.9732	.9738	.9744	.9750	.9756	.9761	.9767
2.0	.9772	.9778	.9783	.9788	.9793	.9798	.9803	.9808	.9812	.9817
2.1	.9821	.9826	.9830	.9834	.9838	.9842	.9846	.9850	.9854	.9857
2.2	.9861	.9864	.9868	.9871	.9875	.9878	.9881	.9884	.9887	.9890
2.3	.9893	.9896	.9898	.9901	.9904	.9906	.9909	.9911	.9913	.9916
2.4	.9918	.9920	.9922	.9925	.9927	.9929	.9931	.9932	.9934	.9936
2.5	.9938	.9940	.9941	.9943	.9945	.9946	.9948	.9949	.9951	.9952
2.6	.9953	.9955	.9956	.9957	.9959	.9960	.9961	.9962	.9963	.9964
2.7	.9965	.9966	.9967	.9968	.9969	.9970	.9971	.9972	.9973	.9974
2.8	.9974	.9975	.9976	.9977	.9977	.9978	.9979	.9979	.9980	.9981
2.9	.9981	.9982	.9982	.9983	.9984	.9984	.9985	.9985	.9986	.9986
3.0	.9987	.9987	.9987	.9988	.9988	.9989	.9989	.9989	.9990	.9990
3.1	.9990	.9991	.9991	.9991	.9992	.9992	.9992	.9992	.9993	.9993
3.2	.9993	.9993	.9994	.9994	.9994	.9994	.9994	.9995	.9995	.9995
3.3	.9995	.9995	.9995	.9996	.9996	.9996	.9996	.9996	.9996	.9997
3.4	.9997	.9997	.9997	.9997	.9997	.9997	.9997	.9997	.9997	.9998

Table A.2 Selected Percentiles of t-Distributions

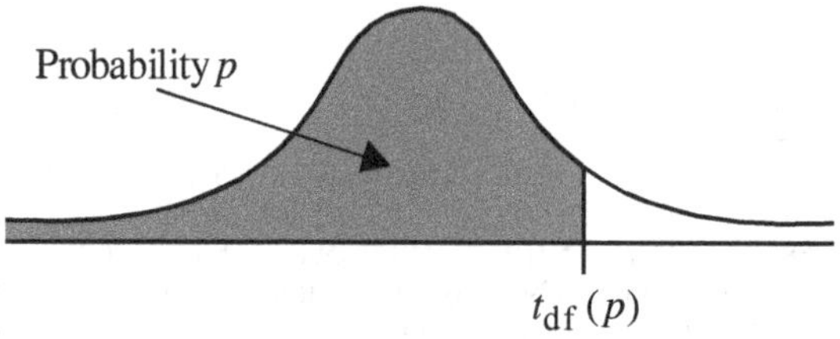

Tabled values are $t_{\text{df}}(p)$

	Probability p											
d.f.	*.75*	*.80*	*.85*	*.90*	*.95*	*.975*	*.98*	*.99*	*.995*	*.9975*	*.999*	*.9995*
1	1.000	1.376	1.963	3.078	6.314	12.71	15.89	31.82	63.67	127.3	318.3	636.6
2	0.816	1.061	1.386	1.886	2.920	4.303	4.849	6.965	9.925	14.09	22.32	31.60
3	0.765	0.978	1.250	1.638	2.353	3.182	3.482	4.541	5.841	7.453	1.215	12.92
4	0.741	0.941	1.190	1.533	2.132	2.776	2.999	3.747	4.604	5.598	7.173	8.610
5	0.727	0.920	1.156	1.476	2.015	2.571	2.757	3.365	4.032	4.773	5.893	6.869
6	0.718	0.906	1.134	1.440	1.943	2.447	2.612	3.143	3.707	4.317	5.208	5.959
7	0.711	0.896	1.119	1.415	1.895	2.365	2.517	2.998	3.499	4.029	4.785	5.408
8	0.706	0.889	1.108	1.397	1.860	2.306	2.449	2.896	3.355	3.833	4.501	5.041
9	0.703	0.883	1.100	1.383	1.833	2.262	2.398	2.821	3.250	3.690	4.297	4.781
10	0.700	0.879	1.093	1.372	1.812	2.228	2.359	2.764	3.169	3.581	4.144	4.587
11	0.697	0.876	1.088	1.363	1.796	2.201	2.328	2.718	3.106	3.497	4.025	4.437
12	0.695	0.873	1.083	1.356	1.782	2.179	2.303	2.681	3.055	3.428	3.930	4.318
13	0.694	0.870	1.079	1.350	1.771	2.160	2.282	2.650	3.012	3.372	3.852	4.221
14	0.692	0.868	1.076	1.345	1.761	2.145	2.264	2.624	2.977	3.326	3.787	4.140
15	0.691	0.866	1.074	1.341	1.753	2.131	2.249	2.602	2.947	3.286	3.733	4.073
16	0.690	0.865	1.071	1.337	1.746	2.120	2.235	2.583	2.921	3.252	3.686	4.015
17	0.689	0.863	1.069	1.333	1.740	2.110	2.224	2.567	2.898	3.222	3.646	3.965
18	0.688	0.862	1.067	1.330	1.734	2.101	2.214	2.552	2.878	3.197	3.610	3.922
19	0.688	0.861	1.066	1.328	1.729	2.093	2.205	2.539	2.861	3.174	3.579	3.883
20	0.687	0.860	1.064	1.325	1.725	2.086	2.197	2.528	2.845	3.153	3.552	3.850
21	0.686	0.859	1.063	1.323	1.721	2.080	2.189	2.518	2.831	3.135	3.527	3.819
22	0.686	0.858	1.061	1.321	1.717	2.074	2.183	2.508	2.819	3.119	3.505	3.792
23	0.685	0.858	1.060	1.319	1.714	2.069	2.177	2.500	2.807	3.104	3.485	3.768
24	0.685	0.857	1.059	1.318	1.711	2.064	2.172	2.492	2.797	3.091	3.467	3.745
25	0.684	0.856	1.058	1.316	1.708	2.060	2.167	2.485	2.787	3.078	3.450	3.725
26	0.684	0.856	1.058	1.315	1.706	2.056	2.162	2.479	2.779	3.067	3.435	3.707
27	0.684	0.855	1.057	1.314	1.703	2.052	2.158	2.473	2.771	3.057	3.421	3.690
28	0.683	0.855	1.056	1.313	1.701	2.048	2.154	2.467	2.763	3.047	3.408	3.674
29	0.683	0.854	1.055	1.311	1.699	2.045	2.150	2.462	2.756	3.038	3.396	3.659
30	0.683	0.854	1.055	1.310	1.697	2.042	2.147	2.457	2.750	3.030	3.385	3.646
40	0.681	0.851	1.050	1.303	1.684	2.021	2.123	2.423	2.704	2.971	3.307	3.551
50	0.679	0.849	1.047	1.299	1.676	2.009	2.109	2.403	2.678	2.937	3.261	3.496
60	0.679	0.848	1.045	1.296	1.671	2.000	2.099	2.390	2.660	2.915	3.232	3.460
70	0.678	0.847	1.044	1.294	1.667	1.994	2.093	2.381	2.648	2.899	3.211	3.435
80	0.678	0.846	1.043	1.292	1.664	1.990	2.088	2.374	2.639	2.887	3.195	3.416
90	0.677	0.846	1.042	1.291	1.662	1.987	2.084	2.368	2.632	2.878	3.183	3.402
100	0.677	0.845	1.042	1.290	1.660	1.984	2.081	2.364	2.626	2.871	3.174	3.390
500	0.675	0.842	1.038	1.283	1.648	1.965	2.059	2.334	2.586	2.820	3.107	3.310
1000	0.675	0.842	1.037	1.282	1.646	1.962	2.056	2.330	2.581	2.813	3.098	3.300
∞	0.674	0.842	1.036	1.282	1.645	1.960	2.054	2.326	2.576	2.807	3.090	3.291

Table A.3 Selected Percentiles of Chi-Squared Distributions

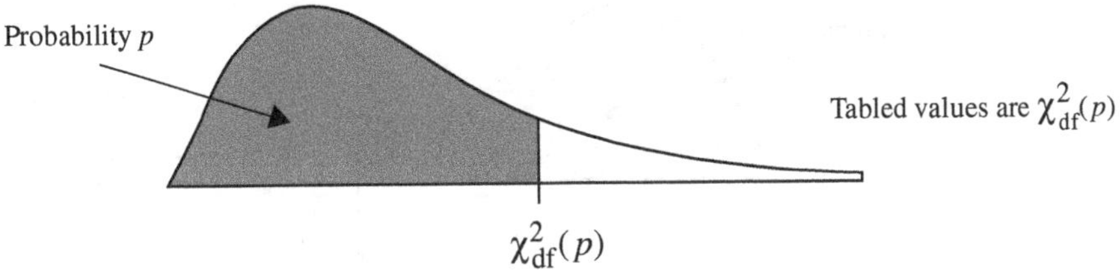

d.f.	Probability p .75	.80	.85	.90	.95	.975	.98	.99	.995	.9975	.999	.9995
1	1.32	1.64	2.07	2.71	3.84	5.02	5.41	6.63	7.88	9.14	10.83	12.12
2	2.77	3.22	3.79	4.61	5.99	7.38	7.82	9.21	10.60	11.98	13.82	15.20
3	4.11	4.64	5.32	6.25	7.81	9.35	9.84	11.34	12.84	14.32	16.27	17.73
4	5.39	5.99	6.74	7.78	9.49	11.14	11.67	13.28	14.86	16.42	18.47	20.00
5	6.63	7.29	8.12	9.24	11.07	12.83	13.39	15.09	16.75	18.39	20.52	22.11
6	7.84	8.56	9.45	10.64	12.59	14.45	15.03	16.81	18.55	20.25	22.46	24.10
7	9.04	9.80	10.75	12.02	14.07	16.01	16.62	18.48	20.28	22.04	24.32	26.02
8	10.22	11.03	12.03	13.36	15.51	17.53	18.17	20.09	21.95	23.77	26.12	27.87
9	11.39	12.24	13.29	14.68	16.92	19.02	19.68	21.67	23.59	25.46	27.88	29.67
10	12.55	13.44	14.53	15.99	18.31	20.48	21.16	23.21	25.19	27.11	29.59	31.42
11	13.70	14.63	15.77	17.28	19.68	21.92	22.62	24.72	26.76	28.73	31.26	33.14
12	14.85	15.81	16.99	18.55	21.03	23.34	24.05	26.22	28.30	30.32	32.91	34.82
13	15.98	16.98	18.20	19.81	22.36	24.74	25.47	27.69	29.82	31.88	34.53	36.48
14	17.12	18.15	19.41	21.06	23.68	26.12	26.87	29.14	31.32	33.43	36.12	38.11
15	18.25	19.31	20.60	22.31	25.00	27.49	28.26	30.58	32.80	34.95	37.70	39.72
16	19.37	20.47	21.79	23.54	26.30	28.85	29.63	32.00	34.27	36.46	39.25	41.31
17	20.49	21.61	22.98	24.77	27.59	30.19	31.00	33.41	35.72	37.95	40.79	42.88
18	21.60	22.76	24.16	25.99	28.87	31.53	32.35	34.81	37.16	39.42	42.31	44.43
19	22.72	23.90	25.33	27.20	30.14	32.85	33.69	36.19	38.58	40.88	43.82	45.97
20	23.83	25.04	26.50	28.41	31.41	34.17	35.02	37.57	40.00	42.34	45.31	47.50
21	24.93	26.17	27.66	29.62	32.67	35.48	36.34	38.93	41.40	43.78	46.80	49.01
22	26.04	27.30	28.82	30.81	33.92	36.78	37.66	40.29	42.80	45.20	48.27	50.51
23	27.14	28.43	29.98	32.01	35.17	38.08	38.97	41.64	44.18	46.62	49.73	52.00
24	28.24	29.55	31.13	33.20	36.42	39.36	40.27	42.98	45.56	48.03	51.18	53.48
25	29.34	30.68	32.28	34.38	37.65	40.65	41.57	44.31	46.93	49.44	52.62	54.95
26	30.43	31.79	33.43	35.56	38.89	41.92	42.86	45.64	48.29	50.83	54.05	56.41
27	31.53	32.91	34.57	36.74	40.11	43.19	44.14	46.96	49.64	52.22	55.48	57.86
28	32.62	34.03	35.71	37.92	41.34	44.46	45.42	48.28	50.99	53.59	56.89	59.30
29	33.71	35.14	36.85	39.09	42.56	45.72	46.69	49.59	52.34	54.97	58.30	60.73
30	34.80	36.25	37.99	40.26	43.77	46.98	47.96	50.89	53.67	56.33	59.70	62.16
40	45.62	47.27	49.24	51.81	55.76	59.34	60.44	63.69	66.77	69.70	73.40	76.09
50	56.33	58.16	60.35	63.17	67.50	71.42	72.61	76.15	79.49	82.66	86.66	89.56
60	66.98	68.97	71.34	74.40	79.08	83.30	84.58	88.38	91.95	95.34	99.61	102.70
70	77.58	79.71	82.26	85.53	90.53	95.02	96.39	100.40	104.20	107.80	112.30	115.60
80	88.13	90.41	93.11	96.58	101.90	106.60	108.10	112.30	116.30	120.10	124.80	128.30
90	98.65	101.10	103.90	107.60	113.20	118.10	119.70	124.10	128.30	132.30	137.20	140.80
100	109.10	111.70	114.70	118.50	124.30	129.60	131.10	135.80	140.20	144.30	149.50	153.20

Table A.4 Selected Percentiles of *F*-Distributions

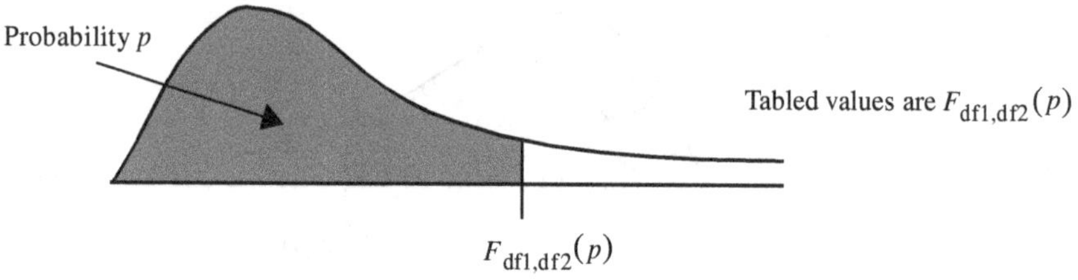

		Numerator degrees of freedom (df1)								
df2	*p*	*1*	*2*	*3*	*4*	*5*	*6*	*7*	*8*	*9*
1	.9	39.86	49.50	53.59	55.83	57.24	58.20	58.91	59.44	59.86
	.95	161.45	199.50	215.71	224.58	230.16	233.99	236.77	238.88	240.54
	.975	647.79	799.50	864.16	899.58	921.85	937.11	948.22	956.66	963.28
	.99	4052.18	4999.50	5403.35	5624.58	5763.65	5858.99	5928.36	5981.07	6022.47
	.999	405284	500000	540379	562500	576405	585937	92873	598144	602284
2	.9	8.53	9.00	9.16	9.24	9.29	9.33	9.35	9.37	9.38
	.95	18.51	19.00	19.16	19.25	19.30	19.33	19.35	19.37	19.38
	.975	38.51	39.00	39.17	39.25	39.30	39.33	39.36	39.37	39.39
	.99	98.50	99.00	99.17	99.25	99.30	99.33	99.36	99.37	99.39
	.999	998.50	999.00	999.17	999.25	999.30	999.33	999.36	999.37	999.39
3	.9	5.54	5.46	5.39	5.34	5.31	5.28	5.27	5.25	5.24
	.95	10.13	9.55	9.28	9.12	9.01	8.94	8.89	8.85	8.81
	.975	17.44	16.04	15.44	15.10	14.88	14.73	14.62	14.54	14.47
	.99	34.12	30.82	29.46	28.71	28.24	27.91	27.67	27.49	27.35
	.999	167.03	148.50	141.11	137.10	134.58	132.85	131.58	130.62	129.86
4	.9	4.54	4.32	4.19	4.11	4.05	4.01	3.98	3.95	3.94
	.95	7.71	6.94	6.59	6.39	6.26	6.16	6.09	6.04	6.00
	.975	12.22	10.65	9.98	9.60	9.36	9.20	9.07	8.98	8.90
	.99	21.20	18.00	16.69	15.98	15.52	15.21	14.98	14.80	14.66
	.999	74.14	61.25	56.18	53.44	51.71	50.53	49.66	49.00	48.47
5	.9	4.06	3.78	3.62	3.52	3.45	3.40	3.37	3.34	3.32
	.95	6.61	5.79	5.41	5.19	5.05	4.95	4.88	4.82	4.77
	.975	10.01	8.43	7.76	7.39	7.15	6.98	6.85	6.76	6.68
	.99	16.26	13.27	12.06	11.39	10.97	10.67	10.46	10.29	10.16
	.999	47.18	37.12	33.20	31.09	29.75	28.83	28.16	27.65	27.24
6	.9	3.78	3.46	3.52	3.18	3.11	3.05	3.01	2.98	2.96
	.95	5.99	5.14	4.76	4.53	4.39	4.28	4.21	4.15	4.10
	.975	8.81	7.26	6.60	6.23	5.99	5.82	5.70	5.60	5.52
	.99	13.75	10.92	9.78	9.15	8.75	8.47	8.26	8.10	7.98
	.999	35.51	27.00	23.70	21.92	20.80	20.03	19.46	19.03	18.69
7	.9	3.59	3.26	3.07	2.96	2.88	2.83	2.78	2.75	2.72
	.95	5.59	4.74	4.35	4.12	3.97	3.87	3.79	3.73	3.68
	.975	8.07	6.54	5.89	5.52	5.29	5.12	4.99	4.90	4.82
	.99	12.25	9.55	8.45	7.85	7.46	7.19	6.99	6.84	6.72
	.999	29.25	21.69	18.77	17.20	16.21	15.52	15.02	14.63	14.33
8	.9	3.46	3.11	2.92	2.81	2.73	2.67	2.62	2.59	2.56
	.95	5.32	4.46	4.07	3.84	3.69	3.58	3.5	3.44	3.39
	.975	7.57	6.06	5.42	5.05	4.82	4.65	4.53	4.43	4.36
	.99	11.26	8.65	7.59	7.01	6.63	6.37	6.18	6.03	5.91
	.999	25.41	18.49	15.83	14.39	13.48	12.86	12.4	12.05	11.77

Table A.4 continued

Numerator degrees of freedom (df1)									
10	*12*	*15*	*20*	*25*	*30*	*40*	*60*	*120*	*1000*
60.19	60.71	61.22	61.74	62.05	62.26	62.53	62.79	63.06	63.3
241.88	243.91	245.95	248.01	249.26	250.10	251.14	252.2	253.25	254.19
968.63	976.71	984.87	993.10	998.08	1001.41	1005.6	1009.8	1014.02	1017.75
6055.85	6106.32	6157.28	6208.73	6239.83	6260.65	6286.78	6313.03	6339.39	6362.68
605621	610668	615764	620908	624017	626099	628712	631337	633972	636301
9.39	9.41	9.42	9.44	9.45	9.46	9.47	9.47	9.48	9.49
19.40	19.41	19.43	19.45	19.46	19.46	19.47	19.48	19.49	19.49
39.40	39.41	39.43	39.45	39.46	39.46	39.47	39.48	39.49	39.50
99.40	99.42	99.43	99.45	99.46	99.47	99.47	99.48	99.49	99.50
999.40	999.42	999.43	999.45	999.46	999.47	999.47	999.48	999.49	999.50
5.23	5.22	5.20	5.18	5.17	5.17	5.16	5.15	5.14	5.13
8.79	8.74	8.70	8.66	8.63	8.62	8.59	8.57	8.55	8.53
14.42	14.34	14.25	14.17	14.12	14.08	14.04	13.99	13.95	13.91
27.23	27.05	26.87	26.69	26.58	26.50	26.41	26.32	26.22	26.14
129.25	128.32	127.37	126.42	125.84	125.45	124.96	124.47	123.97	123.53
3.92	3.90	3.87	3.84	3.83	3.82	3.80	3.79	3.78	3.76
5.96	5.91	5.86	5.80	5.77	5.75	5.72	5.69	5.66	5.63
8.84	8.75	8.66	8.56	8.50	8.46	8.41	8.36	8.31	8.26
14.55	14.37	14.20	14.02	13.91	13.84	13.75	13.65	13.56	13.47
48.05	47.41	46.76	46.10	45.70	45.43	45.09	44.75	44.40	44.14
3.30	3.27	3.24	3.21	3.19	3.17	3.16	3.14	3.12	3.11
4.74	4.68	4.62	4.56	4.52	4.50	4.46	4.43	4.40	4.37
6.62	6.52	6.43	6.33	6.27	6.23	6.18	6.12	6.07	6.02
10.05	9.89	9.72	9.55	9.45	9.38	9.29	9.20	9.11	9.03
26.92	26.42	25.91	25.39	25.08	24.87	24.60	24.33	24.06	23.86
2.94	2.90	2.87	2.84	2.81	2.80	2.78	2.76	2.74	2.72
4.06	4.00	3.94	3.87	3.83	3.81	3.77	3.74	3.70	3.67
5.46	5.37	5.27	5.17	5.11	5.07	5.01	4.96	4.90	4.86
7.87	7.72	7.56	7.40	7.30	7.23	7.14	7.06	6.97	6.89
18.41	17.99	17.56	17.12	16.85	16.67	16.44	16.21	15.98	15.78
2.70	2.67	2.63	2.59	2.57	2.56	2.54	2.51	2.49	2.47
3.64	3.57	3.51	3.44	3.40	3.38	3.34	3.3	3.27	3.23
4.76	4.67	4.57	4.47	4.40	4.36	4.31	4.25	4.20	4.15
6.62	6.47	6.31	6.16	6.06	5.99	5.91	5.82	5.74	5.66
14.08	13.71	13.32	12.93	12.69	12.53	12.33	12.12	11.91	11.72
2.54	2.5	2.46	2.42	2.40	2.38	2.36	2.34	2.32	2.30
3.35	3.28	3.22	3.15	3.11	3.08	3.04	3.01	2.97	2.93
4.30	4.20	4.10	4.00	3.94	3.89	3.84	3.78	3.73	3.68
5.81	5.67	5.52	5.36	5.26	5.20	5.12	5.03	4.95	4.87
11.54	11.19	10.84	10.48	10.26	10.11	9.92	9.73	9.53	9.36

Table A.4 continued

		Numerator degrees of freedom (df1)								
df2	*p*	*1*	*2*	*3*	*4*	*5*	*6*	*7*	*8*	*9*
9	.9	3.36	3.01	2.81	2.69	2.61	2.55	2.51	2.47	2.44
	.95	5.12	4.26	3.86	3.63	3.48	3.37	3.29	3.23	3.18
	.975	7.21	5.71	5.08	4.72	4.48	4.32	4.20	4.10	4.03
	.99	10.56	8.02	6.99	6.42	6.06	5.80	5.61	5.47	5.35
	.999	22.86	16.39	13.90	12.56	11.71	11.13	10.70	10.37	10.11
10	.9	3.29	2.92	2.73	2.61	2.52	2.46	2.41	2.38	2.35
	.95	4.96	4.10	3.71	3.48	3.33	3.22	3.14	3.07	3.02
	.975	6.94	5.46	4.83	4.47	4.24	4.07	3.95	3.85	3.78
	.99	10.04	7.56	6.55	5.99	5.64	5.39	5.20	5.06	4.94
	.999	21.04	14.91	12.55	11.28	10.48	9.93	9.52	9.20	8.96
11	.9	3.23	2.86	2.66	2.54	2.45	2.39	2.34	2.30	2.27
	.95	4.84	3.98	3.59	3.36	3.20	3.09	3.01	2.95	2.90
	.975	6.72	5.26	4.63	4.28	4.04	3.88	3.76	3.66	3.59
	.99	9.65	7.21	6.22	5.67	5.32	5.07	4.89	4.74	4.63
	.999	19.69	13.81	11.56	10.35	9.58	9.05	8.66	8.35	8.12
12	.9	3.18	2.81	2.61	2.48	2.39	2.33	2.28	2.24	2.21
	.95	4.75	3.89	3.49	3.26	3.11	3.00	2.91	2.85	2.80
	.975	6.55	5.10	4.47	4.12	3.89	3.73	3.61	3.51	3.44
	.99	9.33	6.93	5.95	5.41	5.06	4.82	4.64	4.50	4.39
	.999	18.64	12.97	10.80	9.63	8.89	8.38	8.00	7.71	7.48
13	.9	3.14	2.76	2.56	2.43	2.35	2.28	2.23	2.20	2.16
	.95	4.67	3.81	3.41	3.18	3.03	2.92	2.83	2.77	2.71
	.975	6.41	4.97	4.35	4.00	3.77	3.60	3.48	3.39	3.31
	.99	9.07	6.70	5.74	5.21	4.86	4.62	4.44	4.30	4.19
	.999	17.82	12.31	10.21	9.07	8.35	7.86	7.49	7.21	6.98
14	.9	3.10	2.73	2.52	2.39	2.31	2.24	2.19	2.15	2.12
	.95	4.60	3.74	3.34	3.11	2.96	2.85	2.76	2.70	2.65
	.975	6.30	4.86	4.24	3.89	3.66	3.50	3.38	3.29	3.21
	.99	8.86	6.51	5.56	5.04	4.69	4.46	4.28	4.14	4.03
	.999	17.14	11.78	9.73	8.62	7.92	7.44	7.08	6.80	6.58
15	.9	3.07	2.70	2.49	2.36	2.27	2.21	2.16	2.12	2.09
	.95	4.54	3.68	3.29	3.06	2.90	2.79	2.71	2.64	2.59
	.975	6.20	4.77	4.15	3.80	3.58	3.41	3.29	3.20	3.12
	.99	8.68	6.36	5.42	4.89	4.56	4.32	4.14	4.00	3.89
	.999	16.59	11.34	9.34	8.25	7.57	7.09	6.74	6.47	6.26
16	.9	3.05	2.67	2.46	2.33	2.24	2.18	2.13	2.09	2.06
	.95	4.49	3.63	3.24	3.01	2.85	2.74	2.66	2.59	2.54
	.975	6.12	4.69	4.08	3.73	3.50	3.34	3.22	3.12	3.05
	.99	8.53	6.23	5.29	4.77	4.44	4.20	4.03	3.89	3.78
	.999	16.12	10.97	9.01	7.94	7.27	6.80	6.46	6.19	5.98
17	.9	3.03	2.64	2.44	2.31	2.22	2.15	2.10	2.06	2.03
	.95	4.45	3.59	3.20	2.96	2.81	2.70	2.61	2.55	2.49
	.975	6.04	4.62	4.01	3.66	3.44	3.28	3.16	3.06	2.98
	.99	8.4	6.11	5.18	4.67	4.34	4.10	3.93	3.79	3.68
	.999	15.72	10.66	8.73	7.68	7.02	6.56	6.22	5.96	5.75
18	.9	3.01	2.62	2.42	2.29	2.20	2.13	2.08	2.04	2.00
	.95	4.41	3.55	3.16	2.93	2.77	2.66	2.58	2.51	2.46
	.975	5.98	4.56	3.95	3.61	3.38	3.22	3.10	3.01	2.93
	.99	8.29	6.01	5.09	4.58	4.25	4.01	3.84	3.71	3.60
	.999	15.38	10.39	8.49	7.46	6.81	6.35	6.02	5.76	5.56
19	.9	2.99	2.61	2.40	2.27	2.18	2.11	2.06	2.02	1.98
	.95	4.38	3.52	3.13	2.90	2.74	2.63	2.54	2.48	2.42
	.975	5.92	4.51	3.90	3.56	3.33	3.17	3.05	2.96	2.88
	.99	8.18	5.93	5.01	4.50	4.17	3.94	3.77	3.63	3.52
	.999	15.08	10.16	8.28	7.27	6.62	6.18	5.85	5.59	5.39

Table A.4 continued

Numerator degrees of freedom (*df1*)									
10	*12*	*15*	*20*	*25*	*30*	*40*	*60*	*120*	*1000*
2.42	2.38	2.34	2.30	2.27	2.25	2.23	2.21	2.18	2.16
3.14	3.07	3.01	2.94	2.89	2.86	2.83	2.79	2.75	2.71
3.96	3.87	3.77	3.67	3.60	3.56	3.51	3.45	3.39	3.34
5.26	5.11	4.96	4.81	4.71	4.65	4.57	4.48	4.40	4.32
9.89	9.57	9.24	8.90	8.69	8.55	8.37	8.19	8.00	7.84
2.32	2.28	2.24	2.20	2.17	2.16	2.13	2.11	2.08	2.06
2.98	2.91	2.85	2.77	2.73	2.70	2.66	2.62	2.58	2.54
3.72	3.62	3.52	3.42	3.35	3.31	3.26	3.20	3.14	3.09
4.85	4.71	4.56	4.41	4.31	4.25	4.17	4.08	4.00	3.92
8.75	8.45	8.13	7.80	7.60	7.47	7.30	7.12	6.94	6.78
2.25	2.21	2.17	2.12	2.10	2.08	2.05	2.03	2.00	1.98
2.85	2.79	2.72	2.65	2.60	2.57	2.53	2.49	2.45	2.41
3.53	3.43	3.33	3.23	3.16	3.12	3.06	3.00	2.94	2.89
4.54	4.40	4.25	4.10	4.01	3.94	3.86	3.78	3.69	3.61
7.92	7.63	7.32	7.01	6.81	6.68	6.52	6.35	6.18	6.02
2.19	2.15	2.10	2.06	2.03	2.01	1.99	1.96	1.93	1.91
2.75	2.69	2.62	2.54	2.50	2.47	2.43	2.38	2.34	2.30
3.37	3.28	3.18	3.07	3.01	2.96	2.91	2.85	2.79	2.73
4.30	4.16	4.01	3.86	3.76	3.70	3.62	3.54	3.45	3.37
7.29	7.00	6.71	6.40	6.22	6.09	5.93	5.76	5.59	5.44
2.14	2.10	2.05	2.01	1.98	1.96	1.93	1.90	1.88	1.85
2.67	2.6	2.53	2.46	2.41	2.38	2.34	2.30	2.25	2.21
3.25	3.15	3.05	2.95	2.88	2.84	2.78	2.72	2.66	2.60
4.10	3.96	3.82	3.66	3.57	3.51	3.43	3.34	3.25	3.18
6.80	6.52	6.23	5.93	5.75	5.63	5.47	5.30	5.14	4.99
2.10	2.05	2.01	1.96	1.93	1.91	1.89	1.86	1.83	1.80
2.60	2.53	2.46	2.39	2.34	2.31	2.27	2.22	2.18	2.14
3.15	3.05	2.95	2.84	2.78	2.73	2.67	2.61	2.55	2.50
3.94	3.80	3.66	3.51	3.41	3.35	3.27	3.18	3.09	3.02
6.40	6.13	5.85	5.56	5.38	5.25	5.10	4.94	4.77	4.62
2.06	2.02	1.97	1.92	1.89	1.87	1.85	1.82	1.79	1.76
2.54	2.48	2.40	2.33	2.28	2.25	2.20	2.16	2.11	2.07
3.06	2.96	2.86	2.76	2.69	2.64	2.59	2.52	2.46	2.40
3.80	3.67	3.52	3.37	3.28	3.21	3.13	3.05	2.96	2.88
6.08	5.81	5.54	5.25	5.07	4.95	4.80	4.64	4.47	4.33
2.03	1.99	1.94	1.89	1.86	1.84	1.81	1.78	1.75	1.72
2.49	2.42	2.35	2.28	2.23	2.19	2.15	2.11	2.06	2.02
2.99	2.89	2.79	2.68	2.61	2.57	2.51	2.45	2.38	2.32
3.69	3.55	3.41	3.26	3.16	3.10	3.02	2.93	2.84	2.76
5.81	5.55	5.27	4.99	4.82	4.70	4.54	4.39	4.23	4.08
2.00	1.96	1.91	1.86	1.83	1.81	1.78	1.75	1.72	1.69
2.45	2.38	2.31	2.23	2.18	2.15	2.10	2.06	2.01	1.97
2.92	2.82	2.72	2.62	2.55	2.50	2.44	2.38	2.32	2.26
3.59	3.46	3.31	3.16	3.07	3.00	2.92	2.83	2.75	2.66
5.58	5.32	5.05	4.78	4.60	4.48	4.33	4.18	4.02	3.87
1.98	1.93	1.89	1.84	1.80	1.78	1.75	1.72	1.69	1.66
2.41	2.34	2.27	2.19	2.14	2.11	2.06	2.02	1.97	1.92
2.87	2.77	2.67	2.56	2.49	2.44	2.38	2.32	2.26	2.20
3.51	3.37	3.23	3.08	2.98	2.92	2.84	2.75	2.66	2.58
5.39	5.13	4.87	4.59	4.42	4.30	4.15	4.00	3.84	3.69
1.96	1.91	1.86	1.81	1.78	1.76	1.73	1.70	1.67	1.64
2.38	2.31	2.23	2.16	2.11	2.07	2.03	1.98	1.93	1.88
2.82	2.72	2.62	2.51	2.44	2.39	2.33	2.27	2.20	2.14
3.43	3.30	3.15	3.00	2.91	2.84	2.76	2.67	2.58	2.50
5.22	4.97	4.70	4.43	4.26	4.14	3.99	3.84	3.68	3.53

Table A.4 continued

		Numerator degrees of freedom (df1)								
df2	*p*	*1*	*2*	*3*	*4*	*5*	*6*	*7*	*8*	*9*
20	.9	2.97	2.59	2.38	2.25	2.16	2.09	2.04	2.00	1.96
	.95	4.35	3.49	3.10	2.87	2.71	2.60	2.51	2.45	2.39
	.975	5.87	4.46	3.86	3.51	3.29	3.13	3.01	2.91	2.84
	.99	8.10	5.85	4.94	4.43	4.10	3.87	3.70	3.56	3.46
	.999	14.82	9.95	8.10	7.10	6.46	6.02	5.69	5.44	5.24
21	.9	2.96	2.57	2.36	2.23	2.14	2.08	2.02	1.98	1.95
	.95	4.32	3.47	3.07	2.84	2.68	2.57	2.49	2.42	2.37
	.975	5.83	4.42	3.82	3.48	3.25	3.09	2.97	2.87	2.80
	.99	8.02	5.78	4.87	4.37	4.04	3.81	3.64	3.51	3.40
	.999	14.59	9.77	7.94	6.95	6.32	5.88	5.56	5.31	5.11
22	.9	2.95	2.56	2.35	2.22	2.13	2.06	2.01	1.97	1.93
	.95	4.30	3.44	3.05	2.82	2.66	2.55	2.46	2.40	2.34
	.975	5.79	4.38	3.78	3.44	3.22	3.05	2.93	2.84	2.76
	.99	7.95	5.72	4.82	4.31	3.99	3.76	3.59	3.45	3.35
	.999	14.38	9.61	7.8	6.81	6.19	5.76	5.44	5.19	4.99
23	.9	2.94	2.55	2.34	2.21	2.11	2.05	1.99	1.95	1.92
	.95	4.28	3.42	3.03	2.80	2.64	2.53	2.44	2.37	2.32
	.975	5.75	4.35	3.75	3.41	3.18	3.02	2.90	2.81	2.73
	.99	7.88	5.66	4.76	4.26	3.94	3.71	3.54	3.41	3.30
	.999	14.20	9.47	7.67	6.70	6.08	5.65	5.33	5.09	4.89
24	.9	2.93	2.54	2.33	2.19	2.10	2.04	1.98	1.94	1.91
	.95	4.26	3.40	3.01	2.78	2.62	2.51	2.42	2.36	2.30
	.975	5.72	4.32	3.72	3.38	3.15	2.99	2.87	2.78	2.70
	.99	7.82	5.61	4.72	4.22	3.9	3.67	3.50	3.36	3.26
	.999	14.03	9.34	7.55	6.59	5.98	5.55	5.23	4.99	4.80
25	.9	2.92	2.53	2.32	2.18	2.09	2.02	1.97	1.93	1.89
	.95	4.24	3.39	2.99	2.76	2.60	2.49	2.40	2.34	2.28
	.975	5.69	4.29	3.69	3.35	3.13	2.97	2.85	2.75	2.68
	.99	7.77	5.57	4.68	4.18	3.85	3.63	3.46	3.32	3.22
	.999	13.88	9.22	7.45	6.49	5.89	5.46	5.15	4.91	4.71
26	.9	2.91	2.52	2.31	2.17	2.08	2.01	1.96	1.92	1.88
	.95	4.23	3.37	2.98	2.74	2.59	2.47	2.39	2.32	2.27
	.975	5.66	4.27	3.67	3.33	3.10	2.94	2.82	2.73	2.65
	.99	7.72	5.53	4.64	4.14	3.82	3.59	3.42	3.29	3.18
	.999	13.74	9.12	7.36	6.41	5.80	5.38	5.07	4.83	4.64
27	.9	2.90	2.51	2.30	2.17	2.07	2.00	1.95	1.91	1.87
	.95	4.21	3.35	2.96	2.73	2.57	2.46	2.37	2.31	2.25
	.975	5.63	4.24	3.65	3.31	3.08	2.92	2.80	2.71	2.63
	.99	7.68	5.49	4.60	4.11	3.78	3.56	3.39	3.26	3.15
	.999	13.61	9.02	7.27	6.33	5.73	5.31	5.00	4.76	4.57
28	.9	2.89	2.5	2.29	2.16	2.06	2.00	1.94	1.90	1.87
	.95	4.20	3.34	2.95	2.71	2.56	2.45	2.36	2.29	2.24
	.975	5.61	4.22	3.63	3.29	3.06	2.9	2.78	2.69	2.61
	.99	7.64	5.45	4.57	4.07	3.75	3.53	3.36	3.23	3.12
	.999	13.50	8.93	7.19	6.25	5.66	5.24	4.93	4.69	4.50
29	.9	2.89	2.50	2.28	2.15	2.06	1.99	1.93	1.89	1.86
	.95	4.18	3.33	2.93	2.70	2.55	2.43	2.35	2.28	2.22
	.975	5.59	4.2	3.61	3.27	3.04	2.88	2.76	2.67	2.59
	.99	7.60	5.42	4.54	4.04	3.73	3.50	3.33	3.20	3.09
	.999	13.39	8.85	7.12	6.19	5.59	5.18	4.87	4.64	4.45
30	.9	2.88	2.49	2.28	2.14	2.05	1.98	1.93	1.88	1.85
	.95	4.17	3.32	2.92	2.69	2.53	2.42	2.33	2.27	2.21
	.975	5.57	4.18	3.59	3.25	3.03	2.87	2.75	2.65	2.57
	.99	7.56	5.39	4.51	4.02	3.70	3.47	3.30	3.17	3.07
	.999	13.29	8.77	7.05	6.12	5.53	5.12	4.82	4.58	4.39

Table A.4 continued

Numerator degrees of freedom ($df1$)									
10	*12*	*15*	*20*	*25*	*30*	*40*	*60*	*120*	*1000*
1.94	1.89	1.84	1.79	1.76	1.74	1.71	1.68	1.64	1.61
2.35	2.28	2.20	2.12	2.07	2.04	1.99	1.95	1.90	1.85
2.77	2.68	2.57	2.46	2.40	2.35	2.29	2.22	2.16	2.09
3.37	3.23	3.09	2.94	2.84	2.78	2.69	2.61	2.52	2.43
5.08	4.82	4.56	4.29	4.12	4.00	3.86	3.70	3.54	3.40
1.92	1.87	1.83	1.78	1.74	1.72	1.69	1.66	1.62	1.59
2.32	2.25	2.18	2.10	2.05	2.01	1.96	1.92	1.87	1.82
2.73	2.64	2.53	2.42	2.36	2.31	2.25	2.18	2.11	2.05
3.31	3.17	3.03	2.88	2.79	2.72	2.64	2.55	2.46	2.37
4.95	4.70	4.44	4.17	4.00	3.88	3.74	3.58	3.42	3.28
1.90	1.86	1.81	1.76	1.73	1.70	1.67	1.64	1.60	1.57
2.30	2.23	2.15	2.07	2.02	1.98	1.94	1.89	1.84	1.79
2.70	2.60	2.50	2.39	2.32	2.27	2.21	2.14	2.08	2.01
3.26	3.12	2.98	2.83	2.73	2.67	2.58	2.50	2.40	2.32
4.83	4.58	4.33	4.06	3.89	3.78	3.63	3.48	3.32	3.17
1.89	1.84	1.80	1.74	1.71	1.69	1.66	1.62	1.59	1.55
2.27	2.20	2.13	2.05	2.00	1.96	1.91	1.86	1.81	1.76
2.67	2.57	2.47	2.36	2.29	2.24	2.18	2.11	2.04	1.98
3.21	3.07	2.93	2.78	2.69	2.62	2.54	2.45	2.35	2.27
4.73	4.48	4.23	3.96	3.79	3.68	3.53	3.38	3.22	3.08
1.88	1.83	1.78	1.73	1.70	1.67	1.64	1.61	1.57	1.54
2.25	2.18	2.11	2.03	1.97	1.94	1.89	1.84	1.79	1.74
2.64	2.54	2.44	2.33	2.26	2.21	2.15	2.08	2.01	1.94
3.17	3.03	2.89	2.74	2.64	2.58	2.49	2.40	2.31	2.22
4.64	4.39	4.14	3.87	3.71	3.59	3.45	3.29	3.14	2.99
1.87	1.82	1.77	1.72	1.68	1.66	1.63	1.59	1.56	1.52
2.24	2.16	2.09	2.01	1.96	1.92	1.87	1.82	1.77	1.72
2.61	2.51	2.41	2.30	2.23	2.18	2.12	2.05	1.98	1.91
3.13	2.99	2.85	2.70	2.60	2.54	2.45	2.36	2.27	2.18
4.56	4.31	4.06	3.79	3.63	3.52	3.37	3.22	3.06	2.91
1.86	1.81	1.76	1.71	1.67	1.65	1.61	1.58	1.54	1.51
2.22	2.15	2.07	1.99	1.94	1.90	1.85	1.80	1.75	1.70
2.59	2.49	2.39	2.28	2.21	2.16	2.09	2.03	1.95	1.89
3.09	2.96	2.81	2.66	2.57	2.50	2.42	2.33	2.23	2.14
4.48	4.24	3.99	3.72	3.56	3.44	3.30	3.15	2.99	2.84
1.85	1.80	1.75	1.70	1.66	1.64	1.60	1.57	1.53	1.50
2.20	2.13	2.06	1.97	1.92	1.88	1.84	1.79	1.73	1.68
2.57	2.47	2.36	2.25	2.18	2.13	2.07	2.00	1.93	1.86
3.06	2.93	2.78	2.63	2.54	2.47	2.38	2.29	2.20	2.11
4.41	4.17	3.92	3.66	3.49	3.38	3.23	3.08	2.92	2.78
1.84	1.79	1.74	1.69	1.65	1.63	1.59	1.56	1.52	1.48
2.19	2.12	2.04	1.96	1.91	1.87	1.82	1.77	1.71	1.66
2.55	2.45	2.34	2.23	2.16	2.11	2.05	1.98	1.91	1.84
3.03	2.90	2.75	2.60	2.51	2.44	2.35	2.26	2.17	2.08
4.35	4.11	3.86	3.60	3.43	3.32	3.18	3.02	2.86	2.72
1.83	1.78	1.73	1.68	1.64	1.62	1.58	1.55	1.51	1.47
2.18	2.10	2.03	1.94	1.89	1.85	1.81	1.75	1.70	1.65
2.53	2.43	2.32	2.21	2.14	2.09	2.03	1.96	1.89	1.82
3.00	2.87	2.73	2.57	2.48	2.41	2.33	2.23	2.14	2.05
4.29	4.05	3.80	3.54	3.38	3.27	3.12	2.97	2.81	2.66
1.82	1.77	1.72	1.67	1.63	1.61	1.57	1.54	1.50	1.46
2.16	2.09	2.01	1.93	1.88	1.84	1.79	1.74	1.68	1.63
2.51	2.41	2.31	2.20	2.12	2.07	2.01	1.94	1.87	1.80
2.98	2.84	2.70	2.55	2.45	2.39	2.30	2.21	2.11	2.02
4.24	4.00	3.75	3.49	3.33	3.22	3.07	2.92	2.76	2.61

Table A.4 continued

		Numerator degrees of freedom (df1)								
df2	*p*	*1*	*2*	*3*	*4*	*5*	*6*	*7*	*8*	*9*
40	.9	2.84	2.44	2.23	2.09	2.00	1.93	1.87	1.83	1.79
	.95	4.08	3.23	2.84	2.61	2.45	2.34	2.25	2.18	2.12
	.975	5.42	4.05	3.46	3.13	2.90	2.74	2.62	2.53	2.45
	.99	7.31	5.18	4.31	3.83	3.51	3.29	3.12	2.99	2.89
	.999	12.61	8.25	6.59	5.70	5.13	4.73	4.44	4.21	4.02
50	.9	2.81	2.41	2.20	2.06	1.97	1.90	1.84	1.80	1.76
	.95	4.03	3.18	2.79	2.56	2.40	2.29	2.20	2.13	2.07
	.975	5.34	3.97	3.39	3.05	2.83	2.67	2.55	2.46	2.38
	.99	7.17	5.06	4.20	3.72	3.41	3.19	3.02	2.89	2.78
	.999	12.22	7.96	6.34	5.46	4.90	4.51	4.22	4.00	3.82
60	.9	2.79	2.39	2.18	2.04	1.95	1.87	1.82	1.77	1.74
	.95	4.00	3.15	2.76	2.53	2.37	2.25	2.17	2.10	2.04
	.975	5.29	3.93	3.34	3.01	2.79	2.63	2.51	2.41	2.33
	.99	7.08	4.98	4.13	3.65	3.34	3.12	2.95	2.82	2.72
	.999	11.97	7.77	6.17	5.31	4.76	4.37	4.09	3.86	3.69
100	.9	2.76	2.36	2.14	2.00	1.91	1.83	1.78	1.73	1.69
	.95	3.94	3.09	2.70	2.46	2.31	2.19	2.10	2.03	1.97
	.975	5.18	3.83	3.25	2.92	2.70	2.54	2.42	2.32	2.24
	.99	6.90	4.82	3.98	3.51	3.21	2.99	2.82	2.69	2.59
	.999	11.50	7.41	5.86	5.02	4.48	4.11	3.83	3.61	3.44
200	.9	2.73	2.33	2.11	1.97	1.88	1.80	1.75	1.70	1.66
	.95	3.89	3.04	2.65	2.42	2.26	2.14	2.06	1.98	1.93
	.975	5.10	3.76	3.18	2.85	2.63	2.47	2.35	2.26	2.18
	.99	6.76	4.71	3.88	3.41	3.11	2.89	2.73	2.60	2.50
	.999	11.15	7.15	5.63	4.81	4.29	3.92	3.65	3.43	3.26
1000	.9	2.71	2.31	2.09	1.95	1.85	1.78	1.72	1.68	1.64
	.95	3.85	3.00	2.61	2.38	2.22	2.11	2.02	1.95	1.89
	.975	5.04	3.70	3.13	2.80	2.58	2.42	2.30	2.20	2.13
	.99	6.66	4.63	3.80	3.34	3.04	2.82	2.66	2.53	2.43
	.999	10.89	6.96	5.46	4.65	4.14	3.78	3.51	3.30	3.13

Table A.4 continued

Numerator degrees of freedom (df1)									
10	*12*	*15*	*20*	*25*	*30*	*40*	*60*	*120*	*1000*
1.76	1.71	1.66	1.61	1.57	1.54	1.51	1.47	1.42	1.38
2.08	2.00	1.92	1.84	1.78	1.74	1.69	1.64	1.58	1.52
2.39	2.29	2.18	2.07	1.99	1.94	1.88	1.80	1.72	1.65
2.80	2.66	2.52	2.37	2.27	2.20	2.11	2.02	1.92	1.82
3.87	3.64	3.40	3.14	2.98	2.87	2.73	2.57	2.41	2.25
1.73	1.68	1.63	1.57	1.53	1.50	1.46	1.42	1.38	1.33
2.03	1.95	1.87	1.78	1.73	1.69	1.63	1.58	1.51	1.45
2.32	2.22	2.11	1.99	1.92	1.87	1.80	1.72	1.64	1.56
2.70	2.56	2.42	2.27	2.17	2.10	2.01	1.91	1.80	1.70
3.67	3.44	3.20	2.95	2.79	2.68	2.53	2.38	2.21	2.05
1.71	1.66	1.60	1.54	1.50	1.48	1.44	1.40	1.35	1.30
1.99	1.92	1.84	1.75	1.69	1.65	1.59	1.53	1.47	1.40
2.27	2.17	2.06	1.94	1.87	1.82	1.74	1.67	1.58	1.49
2.63	2.50	2.35	2.20	2.10	2.03	1.94	1.84	1.73	1.62
3.54	3.32	3.08	2.83	2.67	2.55	2.41	2.25	2.08	1.92
1.66	1.61	1.56	1.49	1.45	1.42	1.38	1.34	1.28	1.22
1.93	1.85	1.77	1.68	1.62	1.57	1.52	1.45	1.38	1.30
2.18	2.08	1.97	1.85	1.77	1.71	1.64	1.56	1.46	1.36
2.50	2.37	2.22	2.07	1.97	1.89	1.80	1.69	1.57	1.45
3.30	3.07	2.84	2.59	2.43	2.32	2.17	2.01	1.83	1.64
1.63	1.58	1.52	1.46	1.41	1.38	1.34	1.29	1.23	1.16
1.88	1.80	1.72	1.62	1.56	1.52	1.46	1.39	1.30	1.21
2.11	2.01	1.90	1.78	1.70	1.64	1.56	1.47	1.37	1.25
2.41	2.27	2.13	1.97	1.87	1.79	1.69	1.58	1.45	1.30
3.12	2.90	2.67	2.42	2.26	2.15	2.00	1.83	1.64	1.43
1.61	1.55	1.49	1.43	1.38	1.35	1.30	1.25	1.18	1.08
1.84	1.76	1.68	1.58	1.52	1.47	1.41	1.33	1.24	1.11
2.06	1.96	1.85	1.72	1.64	1.58	1.50	1.41	1.29	1.13
2.34	2.20	2.06	1.90	1.79	1.72	1.61	1.50	1.35	1.16
2.99	2.77	2.54	2.30	2.14	2.02	1.87	1.69	1.49	1.22

Table A.5 Selected Percentiles of Studentized Range Distributions

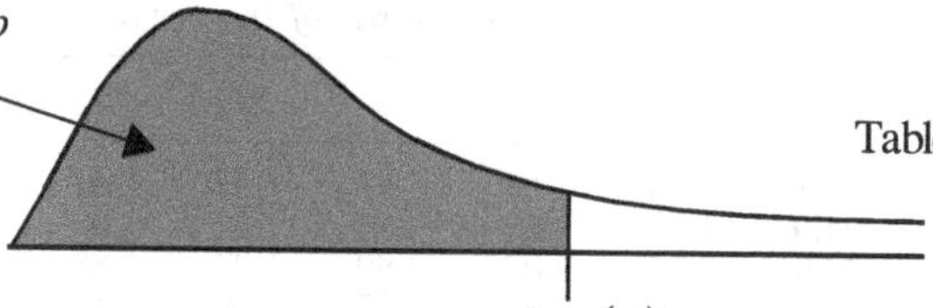

df	*p*	*Number of groups, I* 2	3	4	5	6	7	8	9	10
1	.9	8.93	13.44	16.36	18.49	20.15	21.51	22.64	23.62	24.48
	.95	17.97	26.98	32.82	37.08	40.41	43.12	45.40	47.36	49.07
	.99	90.03	135.0	164.3	185.6	202.2	215.8	227.2	237.0	245.6
2	.9	4.13	5.73	6.77	7.54	8.14	8.63	9.05	9.41	9.72
	.95	6.08	8.33	9.80	10.88	11.74	12.44	13.03	13.54	13.99
	.99	14.04	19.02	22.29	24.72	26.63	28.20	29.53	30.68	31.69
3	.9	3.33	4.47	5.20	5.74	6.16	6.51	6.81	7.06	7.29
	.95	4.50	5.91	6.82	7.50	8.04	8.48	8.85	9.18	9.46
	.99	8.26	10.62	12.17	13.33	14.24	15.00	15.64	16.20	16.69
4	.9	3.01	3.98	4.59	5.03	5.39	5.68	5.93	6.14	6.33
	.95	3.93	5.04	5.76	6.29	6.71	7.05	7.35	7.60	7.83
	.99	6.51	8.12	9.17	9.96	10.58	11.10	11.55	11.93	12.27
5	.9	2.85	3.72	4.26	4.66	4.98	5.24	5.46	5.65	5.82
	.95	3.64	4.60	5.22	5.67	6.03	6.33	6.58	6.80	6.99
	.99	5.70	6.98	7.80	8.42	8.91	9.32	9.67	9.97	10.24
6	.9	2.75	3.56	4.07	4.44	4.73	4.97	5.17	5.34	5.50
	.95	3.46	4.34	4.90	5.30	5.63	5.90	6.12	6.32	6.49
	.99	5.24	6.33	7.03	7.56	7.97	8.32	8.61	8.87	9.10
7	.9	2.68	3.45	3.93	4.28	4.55	4.78	4.97	5.14	5.28
	.95	3.34	4.16	4.68	5.06	5.36	5.61	5.82	6.00	6.16
	.99	4.95	5.92	6.54	7.01	7.37	7.68	7.94	8.17	8.37
8	.9	2.63	3.37	3.83	4.17	4.43	4.65	4.83	4.99	5.13
	.95	3.26	4.04	4.53	4.89	5.17	5.40	5.60	5.77	5.92
	.99	4.75	5.64	6.20	6.62	6.96	7.24	7.47	7.68	7.86
9	.9	2.59	3.32	3.76	4.08	4.34	4.54	4.72	4.87	5.01
	.95	3.20	3.95	4.41	4.76	5.02	5.24	5.43	5.59	5.74
	.99	4.60	5.43	5.96	6.35	6.66	6.91	7.13	7.33	7.49
10	.9	2.56	3.27	3.70	4.02	4.26	4.47	4.64	4.78	4.91
	.95	3.15	3.88	4.33	4.65	4.91	5.12	5.30	5.46	5.60
	.99	4.48	5.27	5.77	6.14	6.43	6.67	6.87	7.05	7.21
11	.9	2.54	3.23	3.66	3.96	4.20	4.40	4.57	4.71	4.84
	.95	3.11	3.82	4.26	4.57	4.82	5.03	5.20	5.35	5.49
	.99	4.39	5.15	5.62	5.97	6.25	6.48	6.67	6.84	6.99
12	.9	2.52	3.20	3.62	3.92	4.16	4.35	4.51	4.65	4.78
	.95	3.08	3.77	4.20	4.51	4.75	4.95	5.12	5.27	5.39
	.99	4.32	5.05	5.50	5.84	6.10	6.32	6.51	6.67	6.81
13	.9	2.50	3.18	3.59	3.88	4.12	4.30	4.46	4.60	4.72
	.95	3.06	3.73	4.15	4.45	4.69	4.88	5.05	5.19	5.32
	.99	4.26	4.96	5.40	5.73	5.98	6.19	6.37	6.53	6.67
14	.9	2.49	3.16	3.56	3.85	4.08	4.27	4.42	4.56	4.68
	.95	3.03	3.70	4.11	4.41	4.64	4.83	4.99	5.13	5.25
	.99	4.21	4.89	5.32	5.63	5.88	6.08	6.26	6.41	6.54
15	.9	2.48	3.14	3.54	3.83	4.05	4.23	4.39	4.52	4.64
	.95	3.01	3.67	4.08	4.37	4.59	4.78	4.94	5.08	5.20
	.99	4.17	4.84	5.25	5.56	5.80	5.99	6.16	6.31	6.44

Table A.5 continued

				Number of groups, I					
11	*12*	*13*	*14*	*15*	*16*	*17*	*18*	*19*	*20*
25.24	25.92	26.54	27.10	27.62	28.10	28.54	28.96	29.35	29.71
50.59	51.96	53.20	54.33	55.36	56.32	57.22	58.04	58.83	59.56
253.2	260.0	266.2	271.8	277.0	281.8	286.3	290.4	294.3	298.0
10.01	10.26	10.49	10.70	10.89	11.07	11.24	11.39	11.54	11.68
14.39	14.75	15.08	15.38	15.65	15.91	16.14	16.37	16.57	16.77
32.59	33.40	34.13	34.81	35.43	36.00	36.53	37.03	37.50	37.95
7.49	7.67	7.83	7.98	8.12	8.25	8.37	8.48	8.58	8.68
9.72	9.95	10.15	10.35	10.52	10.69	10.84	10.98	11.11	11.24
17.13	17.53	17.89	18.22	18.52	18.81	19.07	19.32	19.55	19.77
6.49	6.65	6.78	6.91	7.02	7.13	7.23	7.33	7.41	7.50
8.03	8.21	8.37	8.52	8.66	8.79	8.91	9.03	9.13	9.23
12.57	12.84	13.09	13.32	13.53	13.73	13.91	14.08	14.24	14.40
5.97	6.10	6.22	6.34	6.44	6.54	6.63	6.71	6.79	6.86
7.17	7.32	7.47	7.60	7.72	7.83	7.93	8.03	8.12	8.21
10.48	10.70	10.89	11.08	11.24	11.40	11.55	11.68	11.81	11.93
5.64	5.76	5.87	5.98	6.07	6.16	6.25	6.32	6.40	6.47
6.65	6.79	6.92	7.03	7.14	7.24	7.34	7.43	7.51	7.59
9.30	9.48	9.65	9.81	9.95	10.08	10.21	10.32	10.43	10.54
5.41	5.53	5.64	5.74	5.83	5.91	5.99	6.06	6.13	6.19
6.30	6.43	6.55	6.66	6.76	6.85	6.94	7.02	7.10	7.17
8.55	8.71	8.86	9.00	9.12	9.24	9.35	9.46	9.55	9.65
5.25	5.36	5.46	5.56	5.64	5.72	5.80	5.87	5.93	6.00
6.05	6.18	6.29	6.39	6.48	6.57	6.65	6.73	6.80	6.87
8.03	8.18	8.31	8.44	8.55	8.66	8.76	8.85	8.94	9.03
5.13	5.23	5.33	5.42	5.51	5.58	5.66	5.72	5.79	5.85
5.87	5.98	6.09	6.19	6.28	6.36	6.44	6.51	6.58	6.64
7.65	7.78	7.91	8.03	8.13	8.23	8.33	8.41	8.49	8.57
5.03	5.13	5.23	5.32	5.40	5.47	5.54	5.61	5.67	5.73
5.72	5.83	5.93	6.03	6.11	6.19	6.27	6.34	6.40	6.47
7.36	7.49	7.60	7.71	7.81	7.91	7.99	8.08	8.15	8.23
4.95	5.05	5.15	5.23	5.31	5.38	5.45	5.51	5.57	5.63
5.61	5.71	5.81	5.90	5.98	6.06	6.13	6.20	6.27	6.33
7.13	7.25	7.36	7.46	7.56	7.65	7.73	7.81	7.88	7.95
4.89	4.99	5.08	5.16	5.24	5.31	5.37	5.44	5.49	5.55
5.51	5.61	5.71	5.80	5.88	5.95	6.02	6.09	6.15	6.21
6.94	7.06	7.17	7.26	7.36	7.44	7.52	7.59	7.66	7.73
4.83	4.93	5.02	5.10	5.18	5.25	5.31	5.37	5.43	5.48
5.43	5.53	5.63	5.71	5.79	5.86	5.93	5.99	6.05	6.11
6.79	6.90	7.01	7.10	7.19	7.27	7.35	7.42	7.48	7.55
4.79	4.88	4.97	5.05	5.12	5.19	5.26	5.32	5.37	5.43
5.36	5.46	5.55	5.64	5.71	5.79	5.85	5.91	5.97	6.03
6.66	6.77	6.87	6.96	7.05	7.13	7.20	7.27	7.33	7.39
4.75	4.84	4.93	5.01	5.08	5.15	5.21	5.27	5.32	5.38
5.31	5.40	5.49	5.57	5.65	5.72	5.78	5.85	5.90	5.96
6.55	6.66	6.76	6.84	6.93	7.00	7.07	7.14	7.20	7.26

Table A.5 continued

		Number of groups, *I*								
df	*p*	*2*	*3*	*4*	*5*	*6*	*7*	*8*	*9*	*10*
16	.9	2.47	3.12	3.52	3.80	4.03	4.21	4.36	4.49	4.61
	.95	3.00	3.65	4.05	4.33	4.56	4.74	4.90	5.03	5.15
	.99	4.13	4.79	5.19	5.49	5.72	5.92	6.08	6.22	6.35
17	.9	2.46	3.11	3.50	3.78	4.00	4.18	4.33	4.46	4.58
	.95	2.98	3.63	4.02	4.30	4.52	4.70	4.86	4.99	5.11
	.99	4.10	4.74	5.14	5.43	5.66	5.85	6.01	6.15	6.27
18	.9	2.45	3.10	3.49	3.77	3.98	4.16	4.31	4.44	4.55
	.95	2.97	3.61	4.00	4.28	4.49	4.67	4.82	4.96	5.07
	.99	4.07	4.70	5.09	5.38	5.60	5.79	5.94	6.08	6.20
19	.9	2.45	3.09	3.47	3.75	3.97	4.14	4.29	4.42	4.53
	.95	2.96	3.59	3.98	4.25	4.47	4.65	4.79	4.92	5.04
	.99	4.05	4.67	5.05	5.33	5.55	5.73	5.89	6.02	6.14
20	.9	2.44	3.08	3.46	3.74	3.95	4.12	4.27	4.40	4.51
	.95	2.95	3.58	3.96	4.23	4.45	4.62	4.77	4.90	5.01
	.99	4.02	4.64	5.02	5.29	5.51	5.69	5.84	5.97	6.09
24	.9	2.42	3.05	3.42	3.69	3.90	4.07	4.21	4.34	4.44
	.95	2.92	3.53	3.90	4.17	4.37	4.54	4.68	4.81	4.92
	.99	3.96	4.55	4.91	5.17	5.37	5.54	5.69	5.81	5.92
30	.9	2.40	3.02	3.39	3.65	3.85	4.02	4.16	4.28	4.38
	.95	2.89	3.49	3.85	4.10	4.30	4.46	4.60	4.72	4.82
	.99	3.89	4.45	4.80	5.05	5.24	5.40	5.54	5.65	5.76
40	.9	2.38	2.99	3.35	3.60	3.80	3.96	4.10	4.21	4.32
	.95	2.86	3.44	3.79	4.04	4.23	4.39	4.52	4.63	4.73
	.99	3.82	4.37	4.70	4.93	5.11	5.26	5.39	5.50	5.60
60	.9	2.36	2.96	3.31	3.56	3.75	3.91	4.04	4.16	4.25
	.95	2.83	3.40	3.74	3.98	4.16	4.31	4.44	4.55	4.65
	.99	3.76	4.28	4.59	4.82	4.99	5.13	5.25	5.36	5.45
120	.9	2.34	2.93	3.28	3.52	3.71	3.86	3.99	4.10	4.19
	.95	2.80	3.36	3.68	3.92	4.10	4.24	4.36	4.47	4.56
	.99	3.70	4.20	4.50	4.71	4.87	5.01	5.12	5.21	5.30
∞	.9	2.33	2.90	3.24	3.48	3.66	3.81	3.93	4.04	4.13
	.95	2.77	3.31	3.63	3.86	4.03	4.17	4.29	4.39	4.47
	.99	3.64	4.12	4.40	4.60	4.76	4.88	4.99	5.08	5.16

Table A.5 continued

Number of groups, I

11	12	13	14	15	16	17	18	19	20
4.71	4.81	4.89	4.97	5.04	5.11	5.17	5.23	5.28	5.33
5.26	5.35	5.44	5.52	5.59	5.66	5.73	5.79	5.84	5.90
6.46	6.56	6.66	6.74	6.82	6.90	6.97	7.03	7.09	7.15
4.68	4.77	4.86	4.93	5.01	5.07	5.13	5.19	5.24	5.30
5.21	5.31	5.39	5.47	5.54	5.61	5.67	5.73	5.79	5.84
6.38	6.48	6.57	6.66	6.73	6.81	6.87	6.94	7.00	7.05
4.65	4.75	4.83	4.90	4.98	5.04	5.10	5.16	5.21	5.26
5.17	5.27	5.35	5.43	5.50	5.57	5.63	5.69	5.74	5.79
6.31	6.41	6.50	6.58	6.65	6.73	6.79	6.85	6.91	6.97
4.63	4.72	4.80	4.88	4.95	5.01	5.07	5.13	5.18	5.23
5.14	5.23	5.31	5.39	5.46	5.53	5.59	5.65	5.70	5.75
6.25	6.34	6.43	6.51	6.58	6.65	6.72	6.78	6.84	6.89
4.61	4.70	4.78	4.85	4.92	4.99	5.05	5.10	5.16	5.20
5.11	5.20	5.28	5.36	5.43	5.49	5.55	5.61	5.66	5.71
6.19	6.28	6.37	6.45	6.52	6.59	6.65	6.71	6.77	6.82
4.54	4.63	4.71	4.78	4.85	4.91	4.97	5.02	5.07	5.12
5.01	5.10	5.18	5.25	5.32	5.38	5.44	5.49	5.55	5.59
6.02	6.11	6.19	6.26	6.33	6.39	6.45	6.51	6.56	6.61
4.47	4.56	4.64	4.71	4.77	4.83	4.89	4.94	4.99	5.03
4.92	5.00	5.08	5.15	5.21	5.27	5.33	5.38	5.43	5.47
5.85	5.93	6.01	6.08	6.14	6.20	6.26	6.31	6.36	6.41
4.41	4.49	4.56	4.63	4.69	4.75	4.81	4.86	4.90	4.95
4.82	4.90	4.98	5.04	5.11	5.16	5.22	5.27	5.31	5.36
5.69	5.76	5.83	5.90	5.96	6.02	6.07	6.12	6.16	6.21
4.34	4.42	4.49	4.56	4.62	4.67	4.73	4.78	4.82	4.86
4.73	4.81	4.88	4.94	5.00	5.06	5.11	5.15	5.20	5.24
5.53	5.60	5.67	5.73	5.78	5.84	5.89	5.93	5.97	6.01
4.28	4.35	4.42	4.48	4.54	4.60	4.65	4.69	4.74	4.78
4.64	4.71	4.78	4.84	4.90	4.95	5.00	5.04	5.09	5.13
5.37	5.44	5.50	5.56	5.61	5.66	5.71	5.75	5.79	5.83
4.21	4.28	4.35	4.41	4.47	4.52	4.57	4.61	4.65	4.69
4.55	4.62	4.68	4.74	4.80	4.85	4.89	4.93	4.97	5.01
5.23	5.29	5.35	5.40	5.45	5.49	5.54	5.57	5.61	5.65